Contaminants of Emerging Concerns and Reigning Removal Technologies

Sustainable Water Developments – Resources, Management, Treatment, Efficiency and Reuse

Series Editor: Jochen Bundschuh

The book series addresses novel techniques and measures related to sustainable water developments with an interdisciplinary focus on different fields of water resources assessment and management, freshwater production, water and wastewater treatment and wastewater reuse as well as water-efficient technologies and water-saving measures. Particular attention is paid to the role of water issues within the water-energy-food-public health-ecosystem-climate nexus.

Water and wastewater industries are rapidly growing sectors providing significant opportunities for investments. This applies especially to those industries using sustainable water and wastewater technologies, which are increasingly favored.

The series of books constitutes an information source and facilitator for the transfer of knowledge, focusing on practical solutions and better understanding. It is aimed at academics, manufacturers and consultants, and professionals from governmental and non-governmental organizations, international funding agencies, public health, policy, regulatory and other relevant institutions, as well as the wider public, and targets communities and industries both great and small. The series includes books authored and edited by world-renowned scientists and engineers as well as by leading authorities in economics and politics. Women are particularly encouraged to contribute, either as author or editor.

Geothermal Water Management
Edited by Jochen Bundschuh and Barbara Tomaszewska

Legacy, Pathogenic and Emerging Contaminants in the Environment
Edited by Manish Kumar, Meththika Vithanage, Sanjeeb Mohapatra, and Kishor Acharya

Contaminants of Emerging Concerns and Reigning Removal Technologies
Edited by Manish Kumar, Sanjeeb Mohapatra, and Kishor Acharya

For more information about this series, please visit: www.routledge.com/Sustainable-Water-Developments—Resources-Management-Treatment-Efficiency/book-series/SWD

Contaminants of Emerging Concerns and Reigning Removal Technologies

Edited by Manish Kumar, Sanjeeb Mohapatra and Kishor Acharya

CRC Press
Taylor & Francis Group
Boca Raton London New York Leiden

CRC Press is an imprint of the
Taylor & Francis Group, an **informa** business
A BALKEMA BOOK

Cover image: ©Shutterstock

First published 2022
by CRC Press/Balkema
Schipholweg 107C, 2316 XC Leiden, The Netherlands
e-mail: enquiries@taylorandfrancis.com
www.routledge.com – www.taylorandfrancis.com

CRC Press/Balkema is an imprint of the Taylor & Francis Group, an informa business

Library of Congress Cataloging-in-Publication Data
A catalog record has been requested for this book

ISBN: 978-1-032-16279-9 (hbk)
ISBN: 978-1-032-18066-3 (pbk)
ISBN: 978-1-003-24786-9 (ebk)

DOI: 10.1201/9781003247869

Typeset in Times New Roman
by codeMantra

About the book series

Augmentation of freshwater supply and better sanitation are two of the world's most pressing challenges. However, such improvements must be done economically in an environmentally and societally sustainable way.

Increasingly, groundwater – the source that is much larger than surface water and which provides a stable supply through all the seasons – is used for freshwater supply, which is exploited from ever- deeper groundwater resources. However, the availability of groundwater in sufficient quantity and good quality is severely impacted by the increased water demand for industrial production, cooling in energy production, public water supply and in particular agricultural use, which at present consumes on a global scale about 70% of the exploited freshwater resources. In addition, climate change may have a positive or negative impact on freshwater availability, but which one is presently unknown. These developments result in a continuously increasing water stress, as has already been observed in several world regions and which has adverse implications for the security of food, water and energy supplies, the severity of which will further increase in future. This demands case-specific mitigation and adaptation pathways, which require a better assessment and understanding of surface water and groundwater systems and how they interact with a view to improve their protection and their effective and sustainable management.

With the current and anticipated increased future freshwater demand, it is increasingly difficult to sustain freshwater supply security without producing freshwater from contaminated, brackish or saline water and reusing agricultural, industrial, and municipal wastewater after adequate treatment, which extends the life cycle of water and is beneficial not only to the environment but also leads to cost reduction. Water treatment, particularly desalination, requires large amounts of energy, making energy-efficient options and use of renewable energies important. The technologies, which can either be sophisticated or simple, use physical, chemical and biological processes for water and wastewater treatment, to produce freshwater of a desired quality. Both industrial- scale approaches and smaller-scale applications are important but need a different technological approach. In particular, low-tech, cost-effective, but at the same time sustainable water and wastewater treatment systems, such as artificial wetlands or wastewater gardens, are options suitable for many small-scale applications. Technological improvements and finding new approaches to conventional technologies (e.g. those of seawater desalination) and development of innovative processes, approaches, and methods to improve water and wastewater treatment and sanitation are needed.

Improving economic, environmental and societal sustainability needs research and development to improve process design, operation, performance, automation and management of water and wastewater systems considering aims and local conditions.

In all freshwater consuming sectors, the increasing water scarcity and correspondingly increasing costs of freshwater call for a shift towards more water efficiency and water savings. In the industrial and agricultural sector, it also includes the development of technologies that reduce contamination of freshwater resources, e.g. through development of a chemical-free agriculture. In the domestic sector, there are plenty of options for freshwater saving and improving efficiency such as water-efficient toilets, water-free toilets, or on-site recycling for uses such as toilet flushing, which alone could provide an estimated 30% reduction in water use for the average household. As already mentioned, in all water-consuming sectors, the recycling and reuse of the respective wastewater can provide an important freshwater source. However, the rate at which these water-efficient technologies and water-saving applications are developed and adopted depends on the behavior of individual consumers and requires favorable political, policy and financial conditions.

Due to the interdependency of water and energy (water-energy nexus), i.e. water production needs energy (e.g. for groundwater pumping) and energy generation needs water (e.g. for cooling), the management of both commodities should be more coordinated. This requires integrated energy and water planning, i.e. management of both commodities in a well-coordinated form rather than managing water and energy separately as is routine at present. Only such integrated management allows reducing trade-offs between water and energy use.

However, water is not just linked to energy but must be considered within the whole of the water–energy–food–ecosystem–climate nexus. This requires consideration of what a planned water development requires from the other sectors or how it affects – positively or negatively – the other sectors. Such integrated management of water and the other interlinked resources can implement synergies, reduce trade-offs and optimize resources use and management efficiency, all in all improving security of water, energy, and food security and contributing to protection of ecosystems and climate. Corresponding actions, policies and regulations that support such integral approaches, as well as corresponding research, training and teaching are necessary for their implementation.

The fact that in many developing and transition countries women are disproportionately disadvantaged by water and sanitation limitation requires special attention to this aspect in these countries. Women (including schoolgirls) often spend several hours a day fetching water. This time could be much better used for attending school or working to improve knowledge and skills as well as to generate income and so to reduce gender inequality and poverty. The absence of in-door sanitary facilities exposes women to potential harassment. Moreover, missing single-sex sanitation facilities in schools and the absence of clean water contribute to diseases. This is why solving water and sanitation problems is a critical factor in women empowerment in these countries and necessitates that men and women work side by side to address the water- and wastewater-related operations for improvement of economic, social and sustainable freshwater provision and sanitation.

Individual volumes published in the series span the wide spectrum between research, development and practice in the topic of freshwater and related areas such as gender and social aspects as well as policy, regulatory, legal and economic aspects of water. The series covers all fields and facets in optimal approaches to the:

- Assessment, protection, development and sustainable management of groundwater and surface water resources thereby optimizing their use.
- Improvement of human access to water resources in adequate quantity and good quality.
- Meeting of the increasing demand for drinking water, and irrigation water needed for food and energy security, protection of ecosystems and climate and contribution to a socially and economically sound human development.

Treatment of water and wastewater also including its reuse.
Implementation of water-efficient technologies and water-saving measures.

A key goal of the series is to include all countries of the globe in jointly addressing the challenges of water security and sanitation. Therefore, we aim for a balanced selection of authors and editors originating from developing and developed countries as well as for gender equality. This will help society to provide access to freshwater resources in adequate quantity and of good quality, meeting the increasing demand for drinking water, domestic water and irrigation water needed for food security while contributing to socially and economically sound development.

This book series aims to become a state-of-the-art resource for a broad group of readers, including professionals, academics and students dealing with ground and surface water resources, their assessment, exploitation and management as well as the water and wastewater industry. This comprises especially hydrogeologists, hydrologists, water resources engineers, wastewater engineers, chemical engineers and environmental engineers and scientists.

The book series provides a source of valuable information on surface water but especially on aquifers and groundwater resources in all their facets. As such, it covers not only the scientific and technical aspects but also environmental, legal, policy, economic, social and gender aspects of groundwater resources management. Without departing from the larger framework of integrated groundwater resources management, the topics are centered on water, solute and heat transport in aquifers, hydrogeochemical processes in aquifers, contamination, protection, resources assessment and use.

The book series constitutes an information source and facilitator for the transfer of knowledge, both for small communities with decentralized water supply and sanitation as well as large industries that employ hundreds or thousands of professionals in countries worldwide, working in the different fields of freshwater production, wastewater treatment and water reuse as well as those concerned with water-efficient technologies and water-saving measures. In contrast to many other industries, suffering from the global economic downturn, water and wastewater industries are rapidly growing sectors providing significant opportunities for investments. This applies especially to those using sustainable water and wastewater technologies, which are increasingly favored. The series is also aimed at communities, manufacturers and consultants as well as a range of stakeholders and professionals from governmental and non-governmental organizations, international funding agencies, public health, policy, regulating and other relevant institutions, and the broader public. It is designed to increase awareness of water resources protection and understanding of sustainable water and wastewater solutions including the promotion of water and wastewater reuse and water savings.

By consolidating international research and technical results, the objective of this book series is to focus on practical solutions in better understanding groundwater and surface water systems, the implementation of sustainable water and wastewater treatment and water reuse and the implementation of water-efficient technologies and water-saving measures. Failing to improve and move forward would have serious social, environmental and economic impacts on a global scale.

The book series includes books authored and edited by world-renowned scientists and engineers and by leading authorities in economics and politics. Women are particularly encouraged to contribute, either as author or editor.

Jochen Bundschuh
(Series Editor)

Editorial board

Contents

About the editors

Manish Kumar, a fellow of the Royal Society of Chemistry (FRSC), has earned a PhD in Environmental Engineering from the University of Tokyo, Japan, and is currently engaged as faculty at IIT Gandhinagar, Gujarat, India. He supervised six PhD theses and more than 25 master's dissertations. He contributed more than 100 Science Citation Indexed journal articles, more than 50 book chapters and edited/co-edited six books published by Springer, INTECH and others. He has 20 years' experience in research and teaching with H-index = 24, i10-index = 44 with total citation (~2500)-Google Scholar. Researchgate rates his profile better than 95% of their members of all streams and age groups (as of April, 2020). He is a member of global collaboration of 51 institutes on Wastewater-Based Epidemiology of COVID-19. He is an executive committee member of IWA-India Chapter and represents South-Asia in IWA specialist group for METRAL and related substances. He was selected as judge for IWA young scientist award 2020. He won several accolades, viz. Best Research Award (2013) at 4th Asia Pacific Water Young Professional Conference, two Best Poster Awards (2013 & 2016), DST young scientist grant (2012–2015), JSPS Research Fellowship (2010–2012), COE Young Researcher Fund (2007) and Linnaeus-Palme Grant from SIDA, Sweden (2005). He is associate editor for Elsevier *Journal of Groundwater for Sustainable Development* and Japan Society for Water's journal *Hydrological Research Letter*. He is the lead editor of four virtual special issues for international journals including *Journal of Hazardous Material* (Elsevier).

Sanjeeb Mohapatra, after finishing his PhD at Environmental Science and Engineering Department, IIT Bombay, India, joined the National University of Singapore (NUS), Singapore, which is one of the top ten universities in the world. His research interest broadly covers the monitoring of emerging contaminants (ECs), photo-degradation and enzymatic degradation of ECs and the role of dissolved organic matter in deciding the fate of such contaminants. He is a recipient of the Water Advanced Research Innovation (WARI) Fellowship awarded by the Department of Science and Technology (DST), India, University of Nebraska Lincoln, USA, Daugherty Water for Food Global Institute (DWFI), USA, and Indo-U.S. Science and Technology Forum (IUSSTF). He is a recipient of Newton-Bhabha Fellowship jointly awarded by DST, India, and British Council, UK. He is also a recipient of DST-INSPIRE fellowship offered by DST, India. Dr. Mohapatra also won several international and local awards for his research work for the last couple of years and has many publications in reputed journals, book chapters and refereed conferences to his credit.

Kishor Acharya is a postdoctoral researcher at the School of Engineering, Newcastle University, UK, specialized in the field of public health engineering. His research interest varies from developing and applying portable molecular tools for on-site microbial hazard assessment to design novel techniques/methods for bio-remediation for removal of organic pollutants. Dr. Acharya is a recipient of prestigious Rapid Response Grant award by the Global Challenges Academy to help the global dissemination of research skills and tools and capacity building among researchers of Nepal working in WASH sector. Dr Acharya was recently endorsed as exceptional talent by Royal Society. He has published his research work in leading environmental engineering journals such as *Water Research* and *Environmental Science and Technology.*

Contributors

SR. Abhishek, Department of Information Science and Engineering, NMAM Institute of Technology, Nitte, Udupi -574110, Karnataka, India

Kishor Acharya, School of Engineering, Newcastle University, Newcastle upon Tyne, NE1 7RU, United Kingdom

Shaikh Ziauddin Ahammad, Department of Biochemical Engineering and Biotechnology, Indian Institute of Technology Delhi, Hauz Khas, New Delhi, India

Basanta Kumar Biswal, Energy Research Institute @ NTU, Nanyang Technological University, 50 Nanyang Drive, 637553, Singapore & School of Civil and Environmental Engineering, Nanyang Technological University, 50 Nanyang Avenue, 639798, Singapore

Jonathan M. Bloor, Wolfson Nanomaterials & Devices Laboratory, School of Computing, Electronics and Mathematics, University of Plymouth, Plymouth PL4 8A, United Kingdom

Sasmita Chand, Centre of Sustainable Built Environment, MSAP, Manipal Academy of Higher Education, Manipal-576104, Karnataka, India

Sujata Chand, Kalinga Institute of Industrial Technology, School of Fashion Technology, Bhubaneswar, Odisha 751024, India

Johnson Christian, Department of Environmental Science & Engineering, MEFGI, Rajkot

Rupesh M. Das, Environmental Sciences and Biomedical Metrology Division, CSIR-National Physical Laboratory, Dr. K. S. Krishnan Marg, New Delhi 110012, India

Russell J. Davenport, School of Engineering, Newcastle University, Newcastle upon Tyne, NE1 7RU, United Kingdom

Aswathy E V, Department of Civil Engineering, National Institute of Technology Calicut, Calicut 673601, India

Sai S. Gorthi, Optics and Microfluidics Instrumentation Lab, Department of Instrumentation and Applied Physics, Indian Institute of Science, Bangalore, 560012, India

Richard D. Handy, School of Biological and Marine Sciences, University of Plymouth, Plymouth PL4 8A, United Kingdom

Umesh U. Jadhav, Energy Research Institute @ NTU, Nanyang Technological University, 50 Nanyang Drive, 637553, Singapore, School of Civil and Environmental Engineering, Nanyang Technological University, 50 Nanyang Avenue, 639798, Singapore & Department of Microbiology, Savitribai Phule Pune University, Pune, 411007, India

Tushali Jagwani, M.tech. Environment Department, Malviya National Institute of Technology, Jaipur, Rajasthan

David F. L. Jenkins, Wolfson Nanomaterials & Devices Laboratory, School of Computing, Electronics and Mathematics, University of Plymouth, Plymouth PL4 8A, United Kingdom

David Jenkins, School of Engineering, Computing and Mathematics (Faculty of Science and Engineering), University of Plymouth, United Kingdom

Awadhesh N. Jha, School of Biological and Marine Sciences, University of Plymouth, Plymouth PL4 8A, United Kingdom

Deepchandra Joshi, Department of Biochemical Engineering and Biotechnology, Indian Institute of Technology Delhi, Hauz Khas, New Delhi, India

P.B.H. Karthik, Department of Information Science and Engineering, NMAM Institute of Technology, Nitte, Udupi -574110, Karnataka, India

Nathan Keighley, FERA science ltd., Newcastle University, United Kingdom

Mahesh Kumar, Advanced Materials and Device Metrology Division, CSIR-National Physical Laboratory, Dr. K. S. Krishnan Marg, New Delhi 110012, India

Bing Li, Department of Brain Sciences, Imperial College London, W12 0NN, UK & Care Research & Technology Centre, UK Dementia Research Institute, W12 0NN, United Kingdom

Tapas K. Mallick, Environment and Sustainability Institute, University of Exeter, Penryn Campus, Cornwall, TR10 9FE, United Kingdom

Alok Kumar Meher, Instrumentation Laboratory, Central Pollution Control Board (CPCB), Ministry of Environment, Forest & Climate Change (MoEF & CC, Parivesh Bhawan, East Arjun Nagar, Shahdara, Delhi-110032, India

Devendra Mohan, Department of Civil Engineering, Indian Institute of Technology, Banaras Hindu University, Varanasi-221005, India

Wojciech Mrozik, School of Engineering, Newcastle University, Newcastle upon Tyne, NE1 7RU, United Kingdom

Navneet Munoth, Department of Architecture and Planning, Maulana Azad National Institute of Technology Bhopal, Madhya Pradesh, India

Sushanta Kumar Naik, ICAR Research Complex for Eastern Region, Farming System Research Centre for Hill and Plateau region, Ministry of Agriculture, Ranchi-834010, Jharkhand, India

Nayanathara O S, Department of Civil Engineering, National Institute of Technology Calicut, Calicut 673601, India

Sukdeb Pal, Wastewater Technology Division, CSIR-National Environmental Engineering Research Institute, Nagpur 440020, India & Academy of Scientific and Innovative Research (AcSIR), Ghaziabad- 201002, India

Deepak Panchal, Wastewater Technology Division, CSIR-National Environmental Engineering Research Institute, Nagpur 440020, India & Academy of Scientific and Innovative Research (AcSIR), Ghaziabad- 201002, India

Deeksha Patil, Department of Microbiology, Savitribai Phule Pune University, Pune, 411007, India

Jidapa Plaimart, School of Engineering, Newcastle University, Newcastle upon Tyne, NE1 7RU, United Kingdom

Chandra Prabha, Department of Architecture and Planning, Maulana Azad National Institute of Technology Bhopal, Madhya Pradesh, India

Bhubaneswar Pradhan, ICAR-Indian Institute of Agricultural Biotechnology, Ranchi-834010, Jharkhand

Om Prakash, Wastewater Technology Division, CSIR-National Environmental Engineering Research Institute, Nagpur 440020, India

Shivaraju Harikaranahalli Puttaiah, Department of Water & Health, JSS AHER, Mysuru 570 015, India

Vikram S. Raghaven, Optics and Microfluidics Instrumentation Lab, Department of Instrumentation and Applied Physics, Indian Institute of Science, Bangalore, 560012, India

Anurag Roy, Environment and Sustainability Institute, University of Exeter, Penryn Campus, Cornwall, TR10 9FE, United Kingdom

Koushik Singha Roy, Assistant Professor, Department of Environmental Science, SRM University, Delhi-NCR campus

Shiva Shankar Y, Department of Civil Engineering, Ujjain Engineering College, Ujjain, Madhya Pradesh, India

Abhishek Sharma, Wastewater Technology Division, CSIR-National Environmental Engineering Research Institute, Nagpur 440020, India & Academy of Scientific and Innovative Research (AcSIR), Ghaziabad- 201002, India

Shruti Sharma, Centre for Water Engineering & Management, Central University of Jharkhand, Ranchi, India

Harikaranahalli Puttaiah Shivaraju, Department of Environmental Sciences, Faculty of Natural Sciences, JSS Academy of Higher Education & Research, Mysuru-570015, India & Center for Water, Food and Energy, GREENS trust, Harikaranahalli Village, Dombaranahalli Post, Turuvekere Taluka, Tumkur-572215, India

Ajai Singh, Centre for Water Engineering & Management, Central University of Jharkhand, Ranchi, India

Neeraj Kumar Singh, Deputy Manager (Environment), Central Mine Planning and design Institute Limited, Bhubaneswar, India

Nitin Kumar Singh, Assistant Professor, Department of Environmental Science and Engineering, Marwadi University, Rajkot, Gujarat, India – 360003

N. Sneha, Department of Biotechnology Engineering, NMAM Institute of Technology, Nitte, Udupi -574110, Karnataka, India

T.R. Sreekrishnan, Department of Biochemical Engineering and Biotechnology, Indian Institute of Technology Delhi, Hauz Khas, New Delhi, India

Senthilarasu Sundaram, Electrical and Electronics Engineering, School of Engineering and the Built Environment, Edinburgh Napier University, Merchiston Campus, Edinburgh EH10 5DT, United Kingdom

Sasireka Velusamy, Environment and Sustainability Institute, University of Exeter, Penryn Campus, Cornwall, TR10 9FE, United Kingdom

Dharmik Vyas, Department of Environmental Science & Engineering, MEFGI, Rajkot

David Werner, School of Engineering, Newcastle University, Newcastle upon Tyne, NE1 7RU, United Kingdom

Manish Yadav, Deputy Manager (Environment), Central Mine Planning and Design Institute Limited, Bhubaneswar, India

Suman Yadav, MSc, APS University, Rewa, Madhya Pradesh, India

En-Hua Yang, School of Civil and Environmental Engineering, Nanyang Technological University, 50 Nanyang Avenue, 639798, Singapore

Shivamurthy Yashas, Department of Environmental Sciences, Faculty of Natural Sciences, JSS Academy of Higher Education & Research, Mysuru-570015, India

Preface

It has been an exciting endeavour to be the editors for the book *Contaminants of Emerging Concerns and Reigning Removal Technologies* published by CRC Press, Taylor & Francis Group. At the same time, it was a daunting task for us editors to thoroughly and select only the ones we thought were unique and matched the scope of this book. With an increased demand for wastewater reuse, groundwater recharge with treated wastewater has been practised worldwide.

Emerging contaminants are inadequately removed at conventional wastewater treatment plants and have become a source of surface water and groundwater pollution. Even at a low level (ng/L), they may pose both chronic and acute harmful effects on human health and ecosystem. Specifically, due to their recalcitrant nature and reduced biodegradability, emerging contaminants induce chronic effects ranging from antibiotic resistance, endocrine disruption and even known to increase chances of Alzheimer's disease and cancer. Thus, the first section on 'Mitigation Strategies for Emerging Contaminants' deals with an array of emerging contaminants focusing on the occurrence of pharmaceutical and personal care products, food additives, pesticides, biocides and chemicals used for manufacturing in groundwater. Several treatment options have also been presented, but not limited to, the applications of traditional treatment technologies, membrane filtration, adsorption, electrochemical oxidation, ozonation, Fenton and photo-Fenton processes, catalytic oxidation, photocatalysis and heterogeneous photocatalysis. Additionally, this section highlights the process limitations of conventional treatments technologies and emphasizes the factors affecting advanced oxidation processes and their sustainable operation. Nature-based subsurface flow constructed wetlands (SFCWs) as a polishing step with an insight to process parameters is also discussed in depth. Finally, the formation of several transformed products due to said treatment technologies, their transformation pathway, toxicity, and fate in the aquatic environment are elucidated.

Besides emerging contaminants, an elevated concentration of the geogenic contaminants in the aquatic matrices is an emerging concern, owing to their adverse impact on the ecosystem and public health. Groundwater pollution caused by anthropogenic activities includes organic and inorganic pollutants, which have been detected at unsafe levels in groundwater, making the groundwater practically unusable and poses a significant threat to humanity. Thus, remediation is necessary to regulate or eliminate these contaminants that exceed the permission levels. Several technologies, namely coagulation, ion exchange, adsorption and membrane separations, can effectively remove geogenic contaminants from the aqueous matrices. Thus, the second section on the 'Removal of Geo- and Anthropogenic contaminants' majorly targets source, occurrence, fate and removal of legacy and geogenic contaminants with a particular emphasis on the solid waste and landfill leachate as a transient source of contamination, heavy metal-ion and their strategic detection and removal options, physicochemical and biological treatment methods, novel nanoparticles such as hydroxyapatite nanoparticles and nanostructured adsorbents for the removal of uranium and fluoride contamination.

Finally, the last section on 'Environmental Assessment Pathways and Socio-Ecosystem Framework' depicts the potential concern of emerging contaminants to understand surface water pollution from metaldehyde-based molluscicide, as well as to elucidate the vulnerability of groundwater quality to agriculture and surface runoff. This section also presents a comprehensive review under the backdrop of future scenario of climate change and enumerates the root causes of groundwater contamination for better resource management and policy framework. Legal aspects of pesticides and fertilizers' uses and related problems have also been dealt with the perspective of groundwater and health effects. Finally, emerging treatment technologies are explored and presented from the exploration and optimization point of view, wherein general nanomaterials and allied devices seem to be emerging as better techniques. With the advent of Deep Machine Learning and Computer Vision, a revolution seems around the corner.

This book explains the occurrence, fate and removal of emerging and geogenic contaminants in the groundwater. Thereby, it covers the scientific and technical aspects and environmental, legal and policy aspects for groundwater resources management. Without departing from the broader framework of integrated groundwater resources management, the topics are centred on sources of contaminants, partitioning, formation of transformed products, several mitigation strategies and, finally, environmental assessment pathways, socio-ecosystem framework and policy issues.

Acknowledgements

This book project would not have been possible without the financial contribution from UNICEF-India, Asia Pacific Network (APN)-Grant Number CRRP2016-06MY-Kumar under the aegis of which all contributed chapters have been compiled as a comprehensive reference book on emerging and geogenic contaminant status and management in the environment of post-COVID-19 Anthropocene. We acknowledge the constant support and co-operation of the Indian Institute of Technology (IIT) Gandhinagar, India, National University of Singapore, Singapore, and Newcastle University, UK. We would like to express our sincere thanks to the **CRC Press/Balkema book series** editor on sustainable water development, Prof. Jochen Bundschuh of the University of Southern Queensland, Australia; and gratitude to all the CRC employee, Mr. Janjaap Blom, Kaustav Ghosh and others for their persistent and hard efforts in the successful coordination of the current book project.

The editors would like to express their sincere gratitude to the authors to submit their work on time and revise it appropriately at short notice. We would like to express our sincere thanks and gratitude to all the domain experts: Dr. Gayathri Menon, Dr. MS Priyanka, Dr. Santanu Mukherjee, Dr. Richa Singh, Dr. Vaibhav Srivastava, Dr. Ravi Sankannavar, Dr. Ansaf V Karim, and Dr. Neha Garg, who reviewed various chapters and provided their valuable suggestions to improve the book. We also acknowledge every small help received from our colleagues, students, peers and notable academicians without whom we would have not completed our editorial responsibilities on time. Finally, we acknowledge the constant care and support of our families, parents and well-wishers with which this book project came into reality.

Dr. Manish Kumar
Dr. Sanjeeb Mohapatra
Dr. Kishor Acharya

Section I

Mitigation strategies for emerging contaminants

Chapter 1

Occurrence, fate, and plasma treatment of emerging contaminants in groundwater

Deepchandra Joshi, Shaikh Ziauddin Ahammad, and T.R. Sreekrishnan
Indian Institute of Technology Delhi

CONTENTS

1.1 INTRODUCTION

Water pollution is a serious concern for a long time, and this problem has now further elevated with the finding of the latest class of contaminants known as emerging contaminants (ECs) (Gavrilescu et al., 2015). The main kinds of ECs, found in wastewater, are pharmaceutical and personal care products (PPCPs), endocrine-disrupting compounds (EDCs), pesticides, antibiotic-resistant bacteria (ARB), and antibiotic-resistant gene (ARG) (ME Stuart et al., 2011; Chen et al., 2017, Lopez et al., 2015). These ECs are highly toxic, variable, persistent, and bioaccumulating in nature (B. M. Sharma et al., 2019). After contaminating almost all of our surface water resources, we have now started finding their fingerprints in the groundwater (Postigo & Barceló, 2015). Rapid urbanization, industrialization, and all other irresponsible anthropogenic activities are responsible for groundwater contamination (Naidu et al., 2016). Out of the total freshwater present on earth, the groundwater contains 68.70% of it as shown in Figure 1.1, and its central region is in surface aquifers and below bedrocks (Sorensen et al., 2015). This hidden water has an essential visible role in maintaining a healthy water chain, maintaining subsurface aquatic ecosystems, and maintaining the water level in rivers and other water bodies (Taylor et al., 2013; Pinasseau et al., 2019).

DOI: 10.1201/9781003247869-2

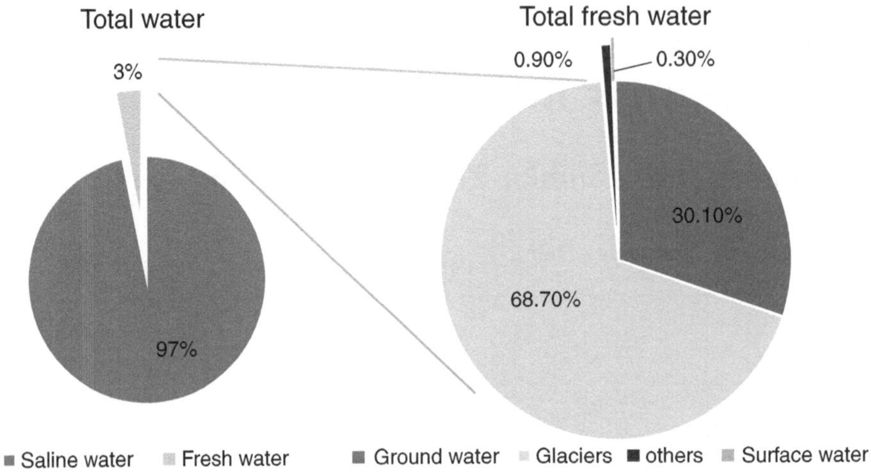

Figure 1.1 Distribution of water resources on earth (e-education.psu.edu, n.d.).

Groundwater is generally considered pure and free from any contamination, and its contamination is a severe issue affecting the natural ecosystem and humans. The continuously depleting surface water resources and increasing population will increase our dependency on groundwater in the near future (Díaz-Cruz & Barceló, 2008). Therefore, it is essential to emphasize groundwater contamination, and proper study needs to be done to know the actual groundwater condition (Banzhaf et al., 2017). Only a limited groundwater study has been done and with the specific pollution source in a limited geographical region. The data from this preliminary study have shown that ECs' concentration is relatively lower in groundwater than surface water (Sui et al., 2015). With this much low concentration of the order below 1 ng/L, there is no acute health risk, but the continuous exposure of a mixture of ECs to humans can cause some chronic health effects (Lapworth et al., 2012). These ECs reach groundwater either from a point source or a diffused source (Tran et al., 2019).

The point source is responsible for the entry of some specific contaminants such as pesticides from agricultural land, antibiotics, and pharmaceuticals from hospitals. In contrast, the diffused sources spread over a wider area and contribute various ECs to groundwater. The ECs must bypass multiple natural attenuations and filtration processes before reaching groundwater. Only the most mobile, persistent, and degradation by-products can reach groundwater (Ramakrishnan et al., 2015). The environmental condition in groundwater is different from the surface water and is not favorable for ECs' degradation, further increasing ECs' persistence. In low degrading conditions, the ECs can reside in groundwater for multiple decades and act as a legacy source of pollution (Sardiña et al., 2019). The excessive use of chemicals, improper disposal techniques, and inefficient ECs degradation result in the continuous increase in the concentration of legacy pollutants in groundwater. These groundwater contaminants, then for an extended period by the regular natural process of water exchange, continuously pollute the freshwater bodies coming in contact (Sophocleous, 2002). The water exchange happening at aquifers is the primary transport mechanism of legacy ECs from groundwater to other freshwater bodies in contact. The shallow water table regions are at greater risk than the deep-water table regions due to the less natural barriers resulting in easy reach. With the geography, the climate condition also affects the groundwater pollution, and in the rainy season, the increased water helps increase the mobility of ECs (Charbeneau, 2006; Jyrkama & Sykes, 2007). Proper and reliable studies are not present, and therefore regular studies including all the aspects of seasonal variation, accidental spillage, pandemic outbreak, etc. are needed. In high groundwater use and high contamination load regions, regular monitoring must be ensured as they are at a higher risk of groundwater contamination. ECs' transformational products and the microbial contaminants (ARB and ARG) should also be given special attention along with the chemical contaminants. Currently, the fecal bacterial test is the main microbial test, whose primary purpose is to know the microbial load in order to prevent the spread of any infectious disease (Pantha et al., 2021; Szekeres et al., 2018), while the risk by ARB and ARG is more as they can transfer antibiotic resistance in the environment (Bouki et al., 2013). Such a groundwater study will help determine actual contamination conditions, ECs' behavior, transport mechanism, and ultimate fate. The results of this study will also be used to prioritize the most toxic ECs based on their occurrence, physicochemical properties, and toxicological effects.

All this information will be beneficial and is very much needed by the researchers and regulating bodies. Using these data, proper rules and regulations, standards, and thresholds will be set, helping preserve groundwater. In the absence of any such rules and regulations, the ECs continuously get added to the groundwater. This is the right time to protect groundwater quality and quantity, as once it gets contaminated, there is no rectification method available for its treatment. The present wastewater treatment system is ineffective against the ECs' complete removal (Bolong et al., 2009). Thus, researchers are working on new and advanced wastewater treatment methods such as plasma techniques for ECs' complete removal.

1.2 GROUNDWATER CONTAMINATION AND ITS EFFECT

The continuous addition of ECs to groundwater and their long-term persistence will convert this hidden water source to a hidden pollution source. The ECs persist in the groundwater due to unfavorable groundwater conditions for their degradation, due to which the groundwater pollutant becomes a legacy pollutant source. Between the groundwater and surface water, natural exchange of water occurs, and during this process, the ECs also get exchanged. The ECs found in groundwater are of two types: recent and old legacy contaminants (Díaz-Cruz & Barceló, 2008). The primary mechanism of ECs' degradation in groundwater is biodegradation, transformation, and sorption. A significant portion of recent ECs gets attenuated by these methods, and the remaining nondegraded or partially degraded compounds get added up to the class of legacy contaminants. The ECs' concentration in groundwater is low, but with the continuous addition of contaminants, the ECs in groundwater continuously increase and deteriorate groundwater health. The continuously increasing groundwater pollution has reached the stage where ECs are found to be present in almost all the worldwide aquifers. This needs to be stopped as the aquifers have limited capability to take care of ECs passing through them (Dodgen et al., 2017).

The majority of ECs are unregulated, and only a few are regulated, but in the future, many more ECs will get added to the list of regulated compounds (Richardson & Kimura, 2017). The pesticides and many industrial compounds are currently regulated for the groundwater under the Groundwater Regulations and the Drinking Water Directive (Li, 2018). The ECs' regulation needs to be done correctly as ECs in groundwater are variable and depend on their geographical region, local use, its frequency, and discharge. Groundwater has a lower concentration of ECs than surface water (Meffe & de Bustamante, 2014) as surface water is a direct receiver and has low ECs residence time (Ritter et al., 2002). The ECs, due to improper disposal techniques, reach different water bodies after bypassing the wastewater treatment plant due to limited efficiency against their removal (Petrović et al., 2003). These ECs, after following multiple routes of direct and indirect contacts, finally came back to the humans only. The direct route example includes contact with treated municipal water and indirect route via the food chain. The ECs show bioaccumulation and biomagnification after entering the human body (Bolong et al., 2009). The acute and chronic health effect of ECs-contaminated groundwater is not well known to us. Researchers have reported undesirable consequences due to continuous exposure of the mixture of ECs on the non-targeted sources, mainly the humans and the ecosystem, due to the continuous exposure of a mixture of ECs (Sui et al., 2015; Pal et al., 2014). Most of such studies are done taking a relatively higher ECs concentration than the one found in groundwater. More of the actual study is needed to take the same ECs concentration as found in groundwater. In the toxicity assessment, the selected test species and their sensitivity against the pollutant also matter. Generally, in toxicity assessment, aquatic animals are selected, which are more prone to any contamination and environmental effect than humans (Bossart, 2011). Even in low concentration, the antibiotic is highly dangerous as the microbes in a low sublethal concentration of antibiotics develop antibiotic resistance (Andrade et al., 2020; Poole, 2017). The antibiotic resistance is so severe that WHO has declared it the biggest challenge of the 21st century (WHO, 2020). There are many types of ECs, and Table 1.1 contains a major class of ECs, and their example, which are generally found in wastewater and groundwater. The ECs' list is continuously increasing with

Table 1.1 Main ECs found in the wastewater and groundwater

Compound group	*Compound class*
Pharmaceuticals	Veterinary and human antibiotics; analgesics, anti-inflammatory and anti-histamine drugs; psychiatric drugs; lipid regulators; β-blockers and antihypertensives; X-ray contrasts; steroids and hormones; blood-viscosity affecting agents; antidiabetics; antidepressants; abuse drugs; stimulants
Personal care products	Fragrances; sun-screen agents; insect repellents; antiseptics, biocides; moth repellents; surfactants
Pesticides	Insecticides; fungicides; herbicides; nematocides; biocides
Food additives	Antioxidants; sweeteners
Manufacturing additives	Corrosion inhibitors; flame retardants; gas propellants, plasticizers; plastic additives; stain repellents; surfactants; antioxidants; solvents; paraffin
Biocides	Biocides

Source: Adopted from Lamastra et al. (2016).

the addition of many new compounds and their transformational by-products. Risk assessment of each compound is not possible, so attention should be given to the priority pollutant about whose harmful effect is known and which are known to be found frequently in higher concentration (Stefanakis & Becker, 2016). To date, limited studies taking a defined point source have been performed on groundwater, giving only sparse knowledge. No actual information on ECs' contamination level, behavior, diversity, and effects is obtained from these specific studies. Therefore, the contaminants' concentration from these studies should not be taken as the absolute concentration, and to get much more accurate data, a general study is needed. In the study, the effects of sudden events such as explosions, spillage, and disease outbreaks must be included; these events also transfer a significant amount of chemicals and contaminants to the water bodies. Therefore, a proper and more generalized study over a large geographical area will give much exciting information about ECs in groundwater and help us understand the actual groundwater contamination condition. The data obtained from these studies will help regulatory bodies make proper groundwater regulations and set up proper threshold values and groundwater contaminants standards. The information about ECs' behavior known to us has been obtained only from the surface water studies where the conditions are different from groundwater. Various standards and threshold values are needed for surface water and groundwater.

1.3 SOURCES OF GROUNDWATER POLLUTION

The contamination source is the main factor deciding the types of contaminants, their physicochemical properties, and their exposure level. The groundwater contamination is mainly caused by two primary sources: point and diffused sources (P. Sharma et al., 2020).

1.3.1 Point source

It is a well-defined contamination source with a high pollution load and a constrained pollution area. It mainly includes pollution from factories, food processing plants, hospitals, municipal sewage plants, stormwater overflow, mining sites, landfill sites, industrial impoundments, and septic tanks. The pollutants from these sources seep out and reach groundwater (Lapworth et al., 2012; Stuart et al., 2012).

1.3.2 Diffused source

It is a poorly defined contamination source, with a presence over a large geographical area, and has a comparatively low pollution load. In this contamination source, there is no direct link between the pollution and the polluter. It mainly includes the contamination from waste runoff from agriculture, farm sites, stormwater, biosolid and manure source, leakage in sewage network, and household septic tanks (Lapworth et al., 2012). Most of the groundwater study available has been done by taking the point source of pollution. This is due to an easy study with a defined pollution source whose contamination load is high, and its effect is visible. These studies will also help to develop an engineered solution for the source of high environmental risk.

In the urban area, solid waste disposal is a main challenge, and generally, dumping of the generated waste is done for which dumping and landfill sites are constructed in or around the city. Due to improper construction and inefficient disposal mechanisms, landfill sites are becoming a pollution source in urban areas. Waste coming from different places containing almost all kinds of ECs is dumped at these sites, and by the leaching process, contaminants from these sites can reach the soil and nearby water bodies. Therefore, careful selection and construction of the landfill sites are very important. The site must be selected away from the human population, not nearby to any surface and groundwater source, and must not be a shallow water table region. Other than location, the age and working mode of landfill site also affects ECs' leaching. To avoid leaching, the proper water discharge piping system should be present below the dumping site. If necessary, proper covering must be done because during the rainy season, the chances of contaminants leaching with rainwater increase (Han et al., 2016). Like this, a great degree of pollution exists below the industrial area due to industrial wastes' improper disposal techniques. The ECs from these sites can easily reach groundwater as the subsurface presence bypass the natural attenuation (Mohapatra et al., 2021).

Households are also an important source of PPCPs due to the improper disposal technique used in the household of unused medicines and personal care products. The medicines are generally flushed down to toilets, and personal care products are added to regular solid waste. Septic tank, especially the shallow tanks, is one of the small and diffused points contamination sources. Due to many septic tanks, their overall load is very high, and various pollutants, antibiotics, caffeine, etc. enter the groundwater (Sorensen et al., 2015). The number of septic tanks is very large and is spread over a broad region; therefore, regulating them is complicated. Wastewater and wastewater treatment plants are also important point sources of groundwater pollution. By infiltration, the contaminated water from this source reaches groundwater and transfers the maximum variety of ECs (Lapworth et al., 2012).

Another important urban point source of contamination is recharging stormwater, which was built as the flood management system, now; due to the studies showing their role in ECs' transfer to water, they are becoming a problem instead of a resource (Pinasseau et al., 2019; Tran et al., 2019). The contaminants reach the karst zone from the various contaminated surface waste sources, characterized as the underground drainage system. Due to their rapid flow system, there is low retention of water resulting in less ECs' attenuation. The unattenuated ECs reach the aquifers and other connected water bodies (Lapworth et al., 2012; Burri et al., 2019). Agriculture farms are a well-known source of pesticides and other related contaminants to groundwater. The sludge from the wastewater treatment plant used as the manure also contains the partially degraded ECs, and, on its application, to fields, the ECs' transport to water bodies occurs (Díaz-Cruz & Barceló, 2008). By the application of antibiotics containing manure, the antibiotics are making their way to reach groundwater. The antibiotic concentration is lower than the river water and is of the same order as found in drinking water (Hu et al., 2010).

Knowing these two important EC sources to groundwater, these practices cannot be stopped, making them a continuous pollution source, as suddenly restricting farmers from using pesticides is not possible; similarly, manure made from sludge is common agriculture practice, which cannot be stopped. Compared to contamination from agriculture source, the groundwater pollution by municipal wastewater have more varieties of ECs due to the presence of almost all kinds of ECs (PPCPs, endocrine disrupting compounds, ARB, and ARG) in municipal water (Tran, Reinhard et al., 2018). The transfer of PPCPs by these indirect and attenuated routes is less than the direct route of contaminated surface water and groundwater exchange by either the natural or artificial process (Lapworth et al., 2012). The mining and patrolling industries are two important and less studied groundwater contamination sources. In the deep oil and gas wells, hydraulic fracturing or fracking directly becomes a source of groundwater pollution due to subsurface spillage of fracking fluids. These fracking fluids are extremely harmful, with no exact knowledge of their composition as their composition is kept secret. However, they contain harmful chemicals, metals, and other contaminants in high concentrations in the range of mg/L (Postigo & Barceló, 2015).

With proper practices, standards, rules, and regulations, groundwater pollution can be controlled or minimized from point sources. There is more incredible difficulty controlling groundwater pollution from diffused sources, as the large geographical area they cover makes the contaminant's assessment and monitoring difficult. The broad geography has a significant advantage in that it also helps in the natural attenuation of ECs.

1.4 FATE AND PATHWAYS OF GROUNDWATER POLLUTION

A simple source pathway receptor model can describe contaminants' movement to groundwater (Lapworth et al., 2012). Before reaching groundwater, the ECs have to bypass many natural attenuation and physical barriers. During this movement, lots of ECs get attenuated or partially metabolized, and only the most mobile and persistent can make their way to groundwater. The ECs' pathway from various surface sources to groundwater is represented in Figure 1.2. The natural attenuation affects most ECs but is not complete, and many persistent ECs bypass it. Once these ECs reaches groundwater, they remain there for a more extended period due to the subsurface geochemical conditions of low microbial diversity, low temperature, variable redox condition, limited oxygen, and nutrient conditions all these factors favor the high HRT of ECs (Díaz-Cruz & Barceló, 2008; Postigo & Barceló, 2015). With the environmental condition, the physicochemical characteristic of ECs, mainly their solubility, polarity, volatility, etc., plays a vital role in their persistence and mobility (Stefanakis & Becker, 2016). In groundwater conditions, the primary attenuation mechanism is transformation and sorption. In the absence of microbial diversity and low chemical transformation conditions in the long term, sorption becomes the primary attenuation mechanism. With time, a portion of the ECs in groundwater gets completely attenuated, and the remaining portion joins the pool of legacy pollutant sources. The concentration and variety of this legacy pollutant keep increasing with the continuous addition of ECs due to increasing water pollution and the absence of any effective ECs removal system. The water recharging process is the primary mechanism through which the exchange of ECs from contaminated surface water to groundwater takes place. With the ECs physicochemical property, the geographical conditions and the groundwater system also affect the fate of ECs in groundwater. The climatic conditions also affect the ECs' mobility to the ground; with more water during the rainy season, the ECs' mobility increases, whereas the excess water also helps dilute groundwater (Jyrkama & Sykes,

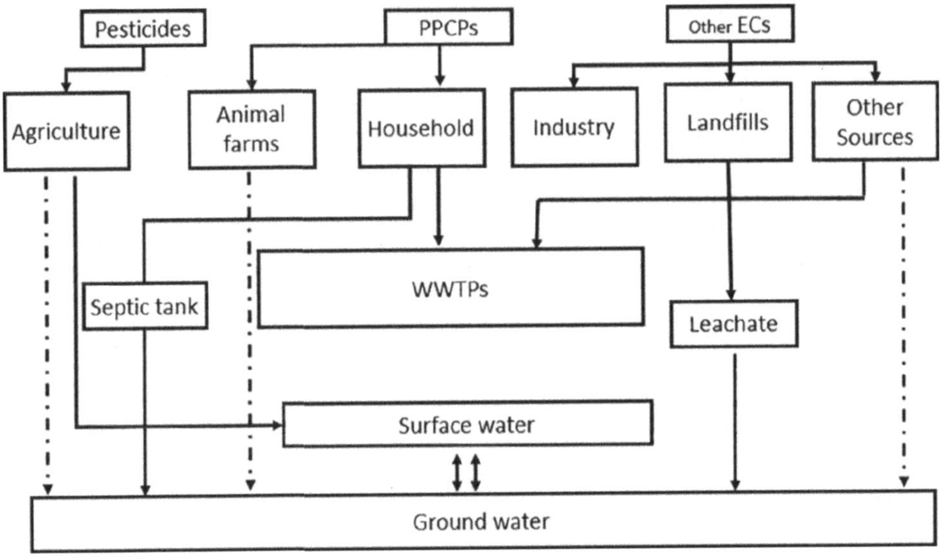

Figure 1.2 Potential pathway of ECs entry into groundwater. Adopted from (Lapworth et al., 2012.)

2007). The regions of shallow groundwater, where the sandy soil and fractured limestones are present, are at a greater risk of ECs' contamination due to the easy reach of unattenuated ECs, whereas in the regions of deep groundwater or where thick clay soil is present, ECs get filtered out and attenuated by natural processes before reaching the groundwater (Díaz-Cruz & Barceló, 2008).

The water recharge mainly occurring in the hyporheic zone has a significant role in contaminating groundwater, but it also dilutes the contaminant concentration. The conditions of heavy rain and floods also dilute the groundwater, but many times with them, they also bring the ECs, and the excess water increases the ECs' movement (Corada-Fernández et al., 2017). Therefore, the seasonal variation has a significant effect on the variation of groundwater composition.

There are multiple pathways through which the ECs enter groundwater, and the main pathways are the natural water recharging process, subsurface leakage, and direct contaminants dumping into the soil. The groundwater system is generally connected to all the major water bodies, so pollution at one particular site can pollute all other connected water resources. Often, geographical conditions help prevent the connected water bodies' pollution by restricting the ECs' movement to a particular localized region only (Sorensen et al., 2015). The other factors influencing the ECs' movement are hydrology, groundwater depth, flow rate and pattern, and sediment porosity. These factors are limited to a particular region only, making different pathway study requirements in different geographical regions (Díaz-Cruz & Barceló, 2008).

The large area, geography, and seasonal variation in water flow make the ECs' pathway study of groundwater very difficult. For the pathway study of water contamination conventionally, the nutrients, microbial indicators, and isotopes of boron, oxygen, nitrogen, and other stable isotopes are used. The groundwater contamination from multiple sources, generally containing similar contaminant types, makes these tracers less effective. The identification of finding actual contamination source becomes difficult. More reliable tracers like radioactive, fluorinated compounds and ECs are being used to overcome this difficulty. Whichever tracer is selected, it must have sufficient concentration in both non-treated and treated water, should be persistent in groundwater, and its analysis should be easy (McCance et al., 2018). These new tracers are highly effective in the complex groundwater system in qualitative and quantitative contamination analysis (White et al., 2016).

The significant ECs found to be present in groundwater can also help decide the pollution source, such as pesticides from agriculture farms, pharmaceuticals from the hospitals, and the pharma industry. It has also been found that pesticides and other agriculture and farm-related ECs have higher wastewater concentrations in developing countries. In contrast, in developing nations, the PPCPs are found to be in higher concentration (Tran et al., 2018; Terzić et al., 2008; Sorensen et al., 2015; Fekadu et al., 2019). These ECs, due to improper disposal again through different routes, finally came into human contact either directly or indirectly from the food chain. The extent of interaction between groundwater, humans, and ecosystems increases, and many other lesser-known modes such as vapor intrusion, will be involved (Marianne Stuart et al., 2012).

1.5 ECs TREATMENT

For wastewater treatment, there are many treatment technologies available, especially for the conventional pollutant (Khuntia et al., 2021; 2022). The conventional treatment system, including physical, chemical, and biological processes, is highly effective against conventional pollutants and shows a limited efficiency against the ECs (Priyanka & Mohapatra, 2020; Priyanka et al., 2021; Eregowda & Mohapatra, 2020; Mackay & Cherry, 1989). The concentration of ECs in groundwater at present is not at a significant level. However, their concentration is regularly increasing (Francisco et al., 2019; B. M. Sharma et al., 2019). In a study on the Ganga river and its surrounding groundwater by Sharma et al., it has been found that the concentration of PPCPs was in the range of 34–293 ng/L for groundwater and 54.7–826 ng/L for river water (Sharma et al. 2019). The low concentration reports do not mean the groundwater is fit for human consumption; the effluent analysis only shows the absence of parent compounds, but the transformational products are still present. Significantly less attention has been given to groundwater with no specific treatment technology for its rectification. Currently, the groundwater is treated on a large scale only to remove harmful conventional pollutants, metals, and hydrocarbons (Cecconet et al., 2020). The treatment method used for the removal of these contaminants prefers to use the ex situ method over in situ. The water is first pumped out and then treated using the biological, chemical process, air sparging, and permeable reactive barriers (Careghini et al., 2013) treatment systems.

In a large-scale groundwater treatment study (18,000 L per treatment session) studied by Poi et al., (2018), the bioaugmentation process is highly effective in reducing petroleum-contaminated groundwater. The petroleum residues also have many recalcitrant organic pollutants, and the bioaugmentation process was effective in reducing the petroleum concentration from 1,564 to 89 mg/L in 32 days (Poi et al., 2018). Such large-scale groundwater treatment methods have the limitation of the long duration of treatment and the high cost involved, and this is the main reason they are not used at the community level (Thakur et al., 2020). Compared to these pollutants, the ECs' treatment is very challenging in both the surface and groundwater. Researchers have worked with many advanced treatment methods such as adsorption, membrane filtration, and the advanced oxidation process (AOPs). All these treatment systems have been studied only with the contaminants in surface water, but the results obtained from these studies will be equally crucial for tackling the problem of ECs in groundwater. The groundwater has a lower ECs concentration than surface water but has the most recalcitrant type of ECs. Therefore, the groundwater's selected treatment system has to treat a low concentration of highly recalcitrant compounds. The treatment method should be highly effective. Additionally, such a treatment system with high removal must have high mineralization efficiency. Out of all the advanced treatment systems, the AOPs are most effective; they are based on OH radical's oxidation potential (2.8 V). In Figure 1.3, the main types of AOPs with other ECs treatment methods are shown. The different AOPs, despite having effectivity against the ECs, have some limitations; for example, the Fenton reaction needs a highly acidic pH of >3 which is not safe to use on a large scale. The photocatalytic process uses the costly catalyst whose recovery needs to be ensured. In the UV/H_2O_2/O_3-based approach, the generation of UV and O_3 and storage of H_2O_2 is problematic; these processes do not have a high mineralization efficiency. A novel method of using plasma energy produces OH radical in high concentration, unlike other AOPs.

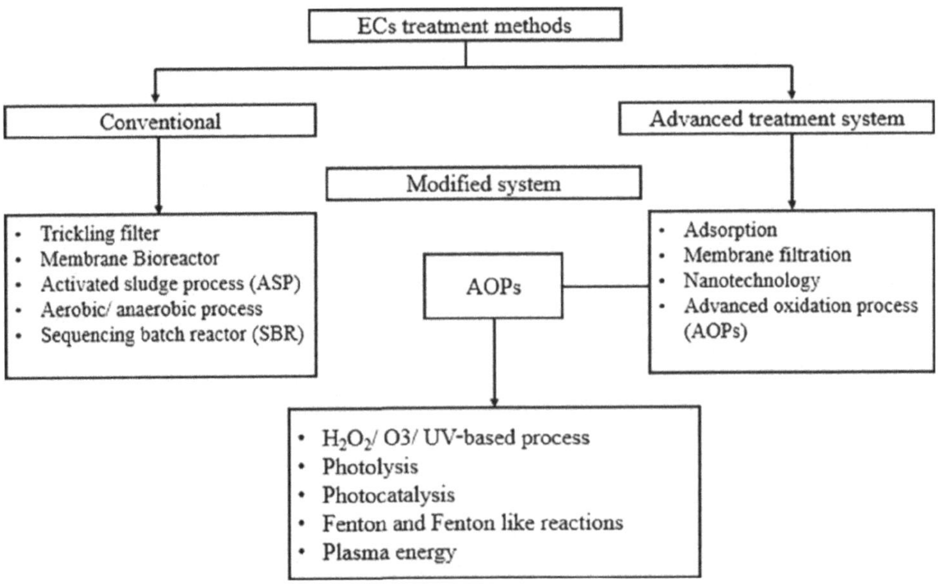

Figure 1.3 The different methods available for the ECs' removal.

1.5.1 ECs' treatment using plasma energy

Plasma is the fourth state of matter which comprises highly energetic and charged reactive species. The collision of these reactive high-energy electrons and radicals produces various reactive species such as OH radicals, $O_2^{\bullet-}$, O=O, O_3, and other reactive oxygen and nitrogen species (RONS). With these chemical reactive species, other plasma effects such as UV, O_3, heat, plasma energy, and sonication, all these multiple effecting acting simultaneously ensures a high degradation and mineralization of the contaminants (Magureanu et al., 2015). Plasma energy can be classified mainly into two types: the high-temperature (10^6–10^8 K) plasma and the low-temperature plasma. The low-temperature plasma is further subdivided into thermal (4,000–20,000 K) and non-thermal plasma (range of 300–1,000 K) (Lee, et al. 2017). The thermal plasma is more of use in solid decomposition and sludge treatment (Gomez et al., 2009); for wastewater treatment, the non-thermal or cold plasma is mainly used (Locke et al., 2006). The low-temperature range of non-thermal plasma also makes this process safe when used on a large scale. The different types of thermal and non-thermal plasma reactors used for wastewater treatment are shown in Figure 1.4. Table 1.2 shows the viability of plasma against the different PPCPs and pesticides. The PPCPs by the different plasma types can be removed entirely in less than an hour of treatment time with a few tens kV of the input voltage. The GPSD and DBD plasma types are energy efficient and can remove the high concentration of PPCPs in significantly less treatment time. While the pulsed plasma can achieve a high PPCP removal, its energy requirement and treatment time are more than the GPSD and DBD. With high removal efficiency against chemical contaminants, the plasma also provides equal effectivity against the microbial contaminants (Bourke et al., 2017) and ARG (Yang et al., 2020; Liao et al., 2018; Liao et al., 2021; Gaunt et al., 2006). From the various studies reported in Table 1.2, it can be seen that more than 3 logs of microbial reduction are achieved in all the cases. Also, in the study by (Yang et al., 2020), where a high 4.7 log reduction of multiple antibiotic-resistant *E. coli* was achieved using the 50W GDP system, the same system was

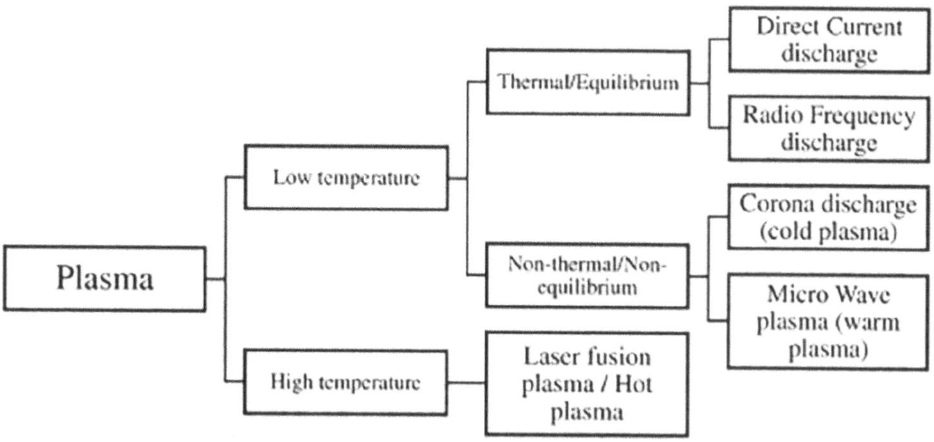

Figure 1.4 The different types of plasma in use for wastewater treatment. Adopted from (Ruj & Ghosh, 2014.)

Table 1.2 The efficiency of the plasma system against the ECs

Compound	*Plasma Type*	*Power(W)/ Voltge (kV)*	*Initial Concentration (ppm)*	*Time (min)*	*Removal (%)*	*Reference*
Diclofenac	Pulsed corona	20 kV	50	15	100	Dobrin et al. 2013
Diclofenac Carbamazepine Ciprofloxacin	Corona discharge	101.5 W	1	4 to 8	100	Singh et al. 2017
Paracetamol	DBD	20 kV	25	15	100	Iervolino et al. 2019
Alachlor	DBD	40 W/8kV	1	30	80	Wardenier et al. 2019
Diclofenac Ibuprofen	DBD	150 W	50	30 20	100	Aziz et al. 2017
Norfloxacin	DBD	10 kV	200	2	Oxygen and air > 90Nitrogen 50	Zhang et al. 2018

DBD, Dielectric barrier discharge.

Table 1.3 The efficiency of the plasma system against microbial contaminants

Microbe	*Plasma type*	*Power (W)/voltage (kV)*	*Initial concentration*	*Time (min)*	*Removal*	*Reference*
B. atrophaeus spore	HVACP	70 kV	6.36 $\log_{10}$ CFU/mL	1	6 log10	Patil et al. (2014)
E. coli ATCC 25922	Discharge plasma	9 kV	7.78 $\log_{10}$ CFU/mL	20	2.14 log 10	Gurol et al. (2012)
Staphylococcus aureus	NPP	6 kV	10^7 CFU/mL	5	3.5 log 10	(Hoon et al. 2015)
	Ar DBD	0.6 kV			log 3	
Multidrug-resistant bacteria	NPP	6 kV			2.34 log 10	
	Ar DBD	0.6 kV			Log 3	
E. coli	GDP	50W/500 V	7.6 logs	15	4.7 logs	Yang et al. (2020)

DBD, dielectric barrier discharge; GDP, glow discharge plasma (GDP); HVACP, high-voltage atmospheric cold plasma; NPP, nano-second pulsed plasma.

effective in 5 log reduction of tetA, tetR, aphA, and tnpA genes in 30 min. This plasma property will help prevent the proliferation and development of ARBs in wastewater treatment plants (Rizzo et al., 2013). The existing wastewater treatment plants are mainly based on a biological process, and suddenly all the treatment methods cannot be replaced with the advanced treatment system. The plasma system provides the flexibility that it can be used with the existing treatment methods. The modification of the conventional treatment process will increase their effectiveness against the ECs (Jiang et al., 2018). In the modification of conventional treatment plants, a plasma system can be used in both pre- and post-biological processes. When used in the pre-treatment, it will increase the biodegradability of the recalcitrant compounds, and when used as the post-treatment, it will eliminate the requirement of the disinfectant unit due to its effectivity against microbes.

1.5.2 Large-scale groundwater treatment

The plasma technique is still in the early stage and has been mainly studied under lab scale only (Singh et al., 2019; Ajo et al., 2018; Aziz et al., 2017); further, it needs technical advancement and support of pilot-scale studies for its full-scale implementation. When the plasma is used on a large scale with the effectiveness of the economy and environment, friendliness will be another critical factor. In terms of environmental friendliness, plasma is a clean technology which does not require any toxic chemicals, and there is no secondary waste formation. The only concern with its large-scale use will be the continuous supply of electricity and the initial investment for technology and infrastructure development. There is no effective treatment system for ECs, so the main focus should be on technology development rather than its cost. Once the technology will be well developed and established, its optimization in terms of energy requirement and cost can be done. The commercialization of the plasma treatment system and its public acceptability will take time, and by preventing the ECs' entry into groundwater, one can prevent the groundwater from further contamination.

1.6 GROUNDWATER PRESERVATION

At present, with the effective treatment method, we also do not have the proper knowledge of groundwater contamination sources and the actual groundwater contamination level. Therefore, firstly we need a dedicated study focused on understanding groundwater contamination and its source. These studies will help to know the actual groundwater condition and the sources, which require special attention. The data from such studies will help in finding the ECs' concentration and frequency, and toxicity will help prioritize and set the threshold value of ECs. By regular monitoring of priority ECs, we will know the behavior or any trend shown by the ECs. The increased knowledge will also help the agencies in establishing the standards for use and discharge. Countries like the USA and Europe have already started working toward groundwater prevention by making priority lists and regulating many compounds. The US EPA regularly releases the contaminant candidate list (CCL) in which various harmful products are added, and in their recent list, they have included many of the ECs that are being studied and monitored on a priority basis (Lesser et al., 2018).

The European Directive (2006/118/EC), which regulates the chemicals and nitrated pesticides, and their transformational products, ensure the proper regulation of enlisted compounds. The European Union is also giving their special attention to groundwater and made dedicated legislation for it, Groundwater Directive (2006/118/EC) under their EU Water Framework Directive (WFD 2000/60/EC), which ensures the protection of groundwater and the ecosystem dependent on it (Lapworth et al., 2012). This dedicated groundwater directive for groundwater protection develops the necessary methods for preventing ECs and other pollutant entry. It also ensures the continuous monitoring and characteristic of local water bodies, forming quality criteria, and implements the needed improvement (Lamastra et al., 2016). In the USA, along with US EPA, various other agencies, including National Oceanic and Atmospheric Administration, US geographical survey , and the Department of Defense, are involved in water protection (Lapworth et al., 2012).

The US geographical survey studied the subsurface water, the environment present there, and its contaminants for over a century and found that the natural conditions are ineffective in rectifying groundwater, and preventing pollution is the only way to protect this source of water. Their study also found that dilution has a significant role in controlling groundwater pollution (Díaz-Cruz & Barceló, 2008). By developing certain habits and following safe practices, further entry of ECs into groundwater can be avoided. The use of chemical contaminants can be minimized, and wherever they are necessarily required, their efficient use must be done. This can be done at the product designing stage, where we develop such products with the same effectiveness. However, the required dosage will be less; products will be made available in various sizes; the packaging will have complete information about their expiry date and proper disposal. The drugs and antibiotics will be built more target-specific so that their complete metabolism will take place inside the body. Proper risk assessment of all the chemical contaminants will be done. In agriculture, the use of biopesticides will be promoted in animal farms, and in poultry, less medicine will be used by improving the hygiene conditions to prevent animal diseases. An effective water treatment system effective in ECs' removal will be developed, and regular monitoring of all water resources and wastewater treatment plants (WWTPs) must be done. In this, the biosensors with rapid testing will be used. The sewage pipelines, septic tanks, landfill sites, and other such primary groundwater sources will be appropriately designed and built to avoid leakage, and the facility must be developed to know the leakages and the methods to take care of them (Marianne Stuart et al., 2012). The industrial injection wells for wastewater disposal must be stopped entirely or only practiced after the proper treatment with efficient ECs removal. Good-quality pesticides or biopesticides must be promoted with their proper application technique. Experimental and modeling studies to develop a proper contamination map will help understand contamination degrees and use proper preservation and protection plans. This map and continuous monitoring will also be used to classify the contamination according to various degrees of low, medium, and high pollution sources (Barzegar et al., 2018). We also need improvement in our groundwater sampling methods as the majorly used spot process, including the grab method, gives only a limited knowledge of pollution at a specific time, and the actual pollution level is not known to us, so better sampling techniques for regular analysis is required (Pinasseau et al., 2019).

1.7 CONCLUSION AND FUTURE PERSPECTIVE

Groundwater contamination by ECs is a severe concern as it's an essential source of fresh water to a significant global population. With the limiting surface water resources due to water pollution and continuously increasing population, our dependency on this hidden water resource will further increase in the future. Therefore, its preservation is important as once the groundwater gets contaminated, its rectification will be very difficult. The ECs have to cross multiple barriers, and only the most persistent one can reach the groundwater. Once they reach, they remain there for multiple decades and become a legacy source of pollution. During the movement, many ECs are transformed, and these transformed products can have the same or higher toxicity than the parent compound. These transformed products must also be analyzed with the parent compounds in regular water monitoring. The prioritization of ECs based on their concentration, frequency, persistence, and ecotoxicity will help in making the ECs study easy. Further, the risk assessment study of contaminated water will help us in knowing the exact threat due to ECs. In risk assessment studies, the concentration of different ECs must be selected according to the concentration which is present in the groundwater. Selecting both the point and diffused groundwater pollution sources is important for any study as it will give much more realistic data. In tracing the pathway followed by ECs from different sources, the tracer study is very much useful. With the conventional tracer study, the use of some new and more reliable tracers will make the study more effective. The ECs themselves can also be used as the tracers. The use of computational modeling with the tracer study will further increase our understanding of ECs' behavior, transport, and fate. All these studies will ultimately help in developing an effective wastewater treatment system against ECs' treatment. The plasma treatment system has shown high effectiveness against the ECs, but it's still on a lab scale, and scaling up the technology will take time.

Till this time, there are no effective treatment systems and preventing techniques for the ECs' entry into groundwater. Another problem with ECs is the discrepancy in their reporting by various researchers. This problem is present due to the absence of any standardized worldwide ECs analysis method and reporting results. A standardized method will create uniform data helping in better comparison of various studies. A dedicated online database for ECs having all the data from various sources will be beneficial. The governments will use such a database to develop the regulations and increasing awareness against the ECs.

REFERENCES

Ajo, P., Preis, S., Vornamo, T., Mänttäri, M., Kallioinen, M., & Louhi-Kultanen, M. J. J. (2018). Hospital wastewater treatment with pilot-scale pulsed corona discharge for removal of pharmaceutical residues. *Journal of Environmental Chemical Engineering, 6*(2), 1569–1577.

Andrade, L., Kelly, M., Hynds, P., Weatherill, J., Majury, A., & O'Dwyer, J. (2020). Groundwater resources as a global reservoir for antimicrobial-resistant bacteria. *Water Research, 170*, 115360.

Aziz, K. H. H., Miessner, H., Mueller, S., Kalass, D., Moeller, D., Khorshid, I., & Rashid, M. A. M. (2017). Degradation of pharmaceutical diclofenac and ibuprofen in aqueous solution, a direct comparison of ozonation, photocatalysis, and non-thermal plasma. *Chemical Engineering Journal, 313*, 1033–1041.

Banzhaf, S., Filipovic, M., Lewis, J., Sparrenbom, C. J., & Barthel, R. J. A. (2017). A review of contamination of surface-, ground-, and drinking water in Sweden by perfluoroalkyl and polyfluoroalkyl substances (PFASs). *Ambio, 46*(3), 335–346.

Barzegar, R., Moghaddam, A. A., Deo, R., Fijani, E., & Tziritis, E. (2018). Mapping groundwater contamination risk of multiple aquifers using multi-model ensemble of machine learning algorithms. *Science of the Total Environment, 621*, 697–712.

Bolong, N., Ismail, A. F., Salim, M. R., & Matsuura, T. (2009). A review of the effects of emerging contaminants in wastewater and options for their removal. *Desalination, 239*(1–3), 229–246.

Bossart, G. D. (2011). Marine mammals as sentinel species for oceans and human health. *Veterinary Pathology, 48*(3), 676–690.

Bouki, C., Venieri, D., Diamadopoulos, E. (2013). Detection and fate of antibiotic resistant bacteria in wastewater treatment plants: A review. *Ecotoxicology and Environmental Safety, 91*, 1–9.

Bourke, P., Ziuzina, D., Han, L., Cullen, P. J., & Gilmore, B. F. (2017). Microbiological interactions with cold plasma. *Journal of Applied Microbiology, 123*(2), 308–324.

Burri, N. M., Weatherl, R., Moeck, C., & Schirmer, M. (2019). A review of threats to groundwater quality in the anthropocene. *Science of the Total Environment, 684*, 136–154.

Careghini, A., Saponaro, S., & Sezenna, E. (2013). Biobarriers for groundwater treatment: A review. *Water Science and Technology, 67*(3), 453–468.

Cecconet, D., Sabba, F., Devecseri, M., Callegari, A., & Capodaglio, A. G. (2020). In situ groundwater remediation with bioelectrochemical systems: A critical review and future perspectives. *Environment International, 137*, 105550.

Charbeneau, R. J. (2006). *Groundwater hydraulics and pollutant transport*. Waveland Press. https://www.google.co.in/books/edition/Groundwater_Hydraulics_and_Pollutant_Tra/scQSAAAAQBAJ?hl=en&gbpv=0

Chen, Q.-L., Li, H., Zhou, X.-Y., Zhao, Y., Su, J.-Q., Zhang, X., & Huang, F.-Y. (2017). An underappreciated hotspot of antibiotic resistance: The groundwater near the municipal solid waste landfill. *Science of the Total Environment, 609*, 966–973.

Corada-Fernández, C., Candela, L., Torres-Fuentes, N., Pintado-Herrera, M. G., Paniw, M., & González-Mazo, E. (2017). Effects of extreme rainfall events on the distribution of selected emerging contaminants in surface and groundwater: The Guadalete River basin (SW, Spain). *Science of the Total Environment, 605*, 770–783.

Díaz-Cruz, M. S., & Barceló, D. (2008). Trace organic chemicals contamination in ground water recharge. *Chemosphere, 72*(3), 333–342.

Dobrin, D., Bradu, C., Magureanu, M., Mandache, N. B., & Parvulescu, V. I. (2013). Degradation of diclofenac in water using a pulsed corona discharge. *Chemical Engineering Journal, 234*, 389–396.

Dodgen, L., Kelly, W., Panno, S., Taylor, S., Armstrong, D., Wiles, K., ... & Zheng, W. (2017). Characterizing pharmaceutical, personal care product, and hormone contamination in a karst aquifer of southwestern Illinois, USA, using water quality and stream flow parameters. *Science of the Total Environment, 578*, 281–289.

e-education.psu.edu (2019.). *Earth in the future*. Retrieved from https://www.e-education.psu.edu/earth103/

Eregowda, T., Mohapatra, S. (2020). Fate of micropollutants in engineered and natural environment. In Kumar M., Munoz-Arriola F., Furumai H., Chaminda T. (eds) *Resilience, Response, and Risk in Water Systems* (pp. 283–301). Singapore: Springer.

Fekadu, S., Alemayehu, E., Dewil, R., & Van der Bruggen, B. (2019). Pharmaceuticals in freshwater aquatic environments: A comparison of the African and European challenge. *Science of the Total Environment, 654*, 324–337.

Francisco, L. F. V., do Amaral Crispim, B., Spósito, J. C. V., Solórzano, J. C. J., Maran, N. H., Kummrow, F., ... & Barufatti, A. (2019). Metals and emerging contaminants in groundwater and human health risk assessment. *Environmental Science and Pollution Research, 26*(24), 24581–24594.

Gaunt, L. F., Beggs, C. B., & Georghiou, G. E. (2006). Bactericidal action of the reactive species produced by gas-discharge non-thermal plasma at atmospheric pressure: A review. *IEEE Transactions on Plasma Science, 34*(4), 1257–1269.

Gavrilescu, M., Demnerová, K., Aamand, J., Agathos, S., & Fava, F. (2015). Emerging pollutants in the environment: Present and future challenges in biomonitoring, ecological risks and bioremediation. *New Biotechnology, 32*(1), 147–156.

Gomez, E., Rani, D. A., Cheeseman, C. R., Deegan, D., Wise, M., & Boccaccini, A. R. (2009). Thermal plasma technology for the treatment of wastes: A critical review. *Journal of Hazardous Materials, 161*(2–3): 614–626.

Gurol, C., Ekinci, F. Y., Aslan, N., & Korachi, M. (2012). Low temperature plasma for decontamination of E. coli in milk. *International Journal of Food Microbiology, 157*(1), 1–5.

Han, Z., Ma, H., Shi, G., He, L., Wei, L., & Shi, Q. (2016). A review of groundwater contamination near municipal solid waste landfill sites in China. *Science of the Total Environment, 569*, 1255–1264.

Hoon Park, J., Kumar, N., Hoon Park, D., Yusupov, M., Neyts, E. C., Verlackt, C. C., ... & Attri, P. (2015). A comparative study for the inactivation of multidrug resistance bacteria using dielectric barrier discharge and nano-second pulsed plasma. *Scientific Reports, 5*(1), 1–14.

Hu, X., Zhou, Q., & Luo, Y. (2010). Occurrence and source analysis of typical veterinary antibiotics in manure, soil, vegetables and groundwater from organic vegetable bases, northern China. *Environmental Pollution, 158*(9), 2992–2998. doi:10.1016/j.envpol.2010.05.023

Iervolino, G., Vaiano, V., & Palma, V. (2019). Enhanced removal of water pollutants by dielectric barrier discharge non-thermal plasma reactor. *Separation and Purification Technology*, 215, 155–162.

Jiang, L., Li, S., Cheng, Z., Chen, J., & Nie, G. (2018). Treatment of 1, 2-dichloroethane and n-hexane in a combined system of non-thermal plasma catalysis reactor coupled with a biotrickling filter. *Journal of Chemical Technology & Biotechnology, 93*(1), 127–137.

Jyrkama, M. I., & Sykes, J. F. (2007). The impact of climate change on spatially varying groundwater recharge in the grand river watershed (Ontario). *Journal of Hydrology, 338*(3–4), 237–250.

Khuntia, H. K., Mohapatra, S., Sukhani, S., Chanakya, H. N., & Gin, K. Y. H. (2021). Editor: Maulin P Shah. Anaerobic Membrane Bioreactors for Industrial Wastewater Treatment. In *Biological Treatment of Industrial Wastewater* (pp. 294–327). UK: RSC.

Khuntia, H. K., Mohapatra, S., Sukhani, S., & Chanakya, H. N. (2022). Role of biofiltration in the treatment of sewage. In *An Innovative Role of Biofiltration in Wastewater Treatment Plants (WWTPs)* (pp. 411–438). Netherlands: Elsevier.

Lamastra, L., Balderacchi, M., & Trevisan, M. (2016). Inclusion of emerging organic contaminants in groundwater monitoring plans. *MethodsX, 3*, 459–476.

Lapworth, D., Baran, N., Stuart, M. E., & Ward, R.S. (2012). Emerging organic contaminants in groundwater: A review of sources, fate and occurrence. *Environmental Pollution, 163*, 287–303.

Lee, J., Lee, C. W., Yong, H. I., Lee, H. J., Jo, C., & Jung, S. (2017). Use of atmospheric pressure cold plasma for meat industry. *Korean Journal for Food Science of Animal Resources*, 37(4): 477.

Lesser, L. E., Mora, A., Moreau, C., Mahlknecht, J., Hernández-Antonio, A., Ramírez, A. I., & Barrios-Piña, H. (2018). Survey of 218 organic contaminants in groundwater derived from the world's largest untreated wastewater irrigation system: Mezquital Valley, Mexico. *Chemosphere, 198*, 510–521.

Li, Z. (2018). A health-based regulatory chain framework to evaluate international pesticide groundwater regulations integrating soil and drinking water standards. *Environment International, 121*, 1253–1278.

Liao, X., Cullen, P., Liu, D., Muhammad, A. I., Chen, S., Ye, X., ... & Ding, T. (2018). Combating *Staphylococcus aureus* and its methicillin resistance gene (*mecA*) with cold plasma. *Science of the Total Environment, 645*, 1287–1295.

Liao, X., Liu, D., Chen, S., Ye, X., & Ding, T. (2021). Degradation of antibiotic resistance contaminants in wastewater by atmospheric cold plasma: Kinetics and mechanisms. *Environmental Technology, 42*(1), 58–71.

Locke, B. R., Sato, M., Sunka, P., Hoffmann, M. R., & Chang, J. S. (2006). Electrohydraulic discharge and non-thermal plasma for water treatment. *Industrial & Engineering Chemistry Research*, 45(3): 882–905.

Lopez, B., Ollivier, P., Togola, A., Baran, N., & Ghestem, J.-P. (2015). Screening of French groundwater for regulated and emerging contaminants. *Science of the Total Environment, 518*, 562–573.

Mackay, D. M., & Cherry, J. A. (1989). Groundwater contamination: Pump-and-treat remediation. *Environmental Science & Technology, 23*(6), 630–636.

Magureanu, M., Mandache, N. B., & Parvulescu, V. I. (2015). Degradation of pharmaceutical compounds in water by non-thermal plasma treatment. *Water Research, 81*, 124–136.

McCance, W., Jones, O., Edwards, M., Surapaneni, A., Chadalavada, S., & Currell, M. (2018). Contaminants of emerging concern as novel groundwater tracers for delineating wastewater impacts in urban and peri-urban areas. *Water Research, 146*, 118–133.

Meffe, R., & de Bustamante, I. (2014). Emerging organic contaminants in surface water and groundwater: A first overview of the situation in Italy. *Science of the Total Environment, 481*, 280–295.

Mohapatra S., Menon N., Padhye L.P., Tatiparti S.S.V., &Mukherji S. (2021) Natural attenuation of pharmaceuticals in the aquatic environment and role of phototransformation. In: Kumar M., Snow D., Honda R., & Mukherjee S. (eds) *Contaminants in Drinking and Wastewater Sources. Springer Transactions in Civil and Environmental Engineering*. Singapore: Springer. https://doi.org/10.1007/978-981-15-4599-3_3

Naidu, R., Espana, V. A. A., Liu, Y., & Jit, J. (2016). Emerging contaminants in the environment: Risk-based analysis for better management. *Chemosphere, 154*, 350–357.

Pal, A., He, Y., Jekel, M., Reinhard, M., & Gin, K. Y.-H. (2014). Emerging contaminants of public health significance as water quality indicator compounds in the urban water cycle. *Environment International, 71*, 46–62.

Pantha, K., Acharya, K., Mohapatra, S., Khanal, S., Amatya, N., Ospina-Betancourth, C., ... & Werner, D. (2021). Faecal pollution source tracking in the holy Bagmati River by portable 16S rRNA gene sequencing. *npj Clean Water*, 4(1): 1–10.

Patil, S., Moiseev, T., Misra, N. N., Cullen, P. J., Mosnier, J. P., Keener, K. M., & Bourke, P. (2014). Influence of high voltage atmospheric cold plasma process parameters and role of relative humidity on inactivation of Bacillus atrophaeus spores inside a sealed package. *Journal of Hospital Infection, 88*(3), 162–169.

Petrović, M., Gonzalez, S., & Barceló, D. (2003). Analysis and removal of emerging contaminants in wastewater and drinking water. *TrAC Trends in Analytical Chemistry, 22*(10), 685–696.

Pinasseau, L., Wiest, L., Fildier, A., Volatier, L., Fones, G. R., Mills, G. A., ... & Vulliet, E. (2019). Use of passive sampling and high resolution mass spectrometry using a suspect screening approach to characterise emerging pollutants in contaminated groundwater and runoff. *Science of the Total Environment, 672*, 253–263.

Poi, G., Shahsavari, E., Aburto-Medina, A., Mok, P. C., & Ball, A. S. (2018). Large scale treatment of total petroleum-hydrocarbon contaminated groundwater using bioaugmentation. *Journal of Environmental Management, 214*, 157–163.

Poole, K. (2017). At the nexus of antibiotics and metals: The impact of Cu and Zn on antibiotic activity and resistance. *Trends in Microbiology 25*(10), 820–832.

Postigo, C., & Barceló, D. (2015). Synthetic organic compounds and their transformation products in groundwater: Occurrence, fate and mitigation. *Science of the Total Environment, 503*, 32–47.

Priyanka, M. & Mohapatra, S. (2020). Shifts and trends in analysis of contaminants of emerging concern: Sulfonamides. In Kumar M., Munoz-Arriola F., Furumai H., & Chaminda T. (eds) *Resilience, Response, and Risk in Water Systems* (pp. 261–281), Singapore: Springer.

Priyanka, M. S., Mohapatra, S., & Menon, N. G. (2021). Editor: Maulin P. Shah and Susana Rodriguez-Couto. Hybrid membrane technology: an alternative to industrial wastewater treatment. In *Membrane-Based Hybrid Processes for Wastewater Treatment* (pp. 481–501), Netherlands: Elsevier.

Ramakrishnan, A., Blaney, L., Kao, J., Tyagi, R. D., Zhang, T. C., & Surampalli, R. Y. (2015). Emerging contaminants in landfill leachate and their sustainable management. *Environmental Earth Sciences 73*(3), 1357–1368.

Richardson, S. D., & Kimura, S. Y. (2017). Emerging environmental contaminants: Challenges facing our next generation and potential engineering solutions. *Environmental Technology & Innovation, 8*, 40–56.

Ritter, K. S., Solomon, K., Sibley, P., Hall, K., Keen, P., Mattu, G., & Linton, B. (2002). Sources, pathways, and relative risks of contaminants in surface water and groundwater: A perspective prepared for the Walkerton inquiry. *Journal of Toxicology and Environmental Health Part A. 65*(1), 1–142.

Rizzo, L., Manaia, C., Merlin, C., Schwartz, T., Dagot, C., Ploy, M., ... & Fatta-Kassinos, D. (2013). Urban wastewater treatment plants as hotspots for antibiotic resistant bacteria and genes spread into the environment: A review. *Science of the Total Environment, 447*, 345–360.

Ruj, B., & Ghosh, S. (2014). Technological aspects for thermal plasma treatment of municipal solid waste—A review. *Fuel Processing Technology, 126*, 298–308.

Sardiña, P., Leahy, P., Metzeling, L., Stevenson, G., & Hinwood, A. (2019). Emerging and legacy contaminants across land-use gradients and the risk to aquatic ecosystems. *Science of the Total Environment, 695*, 133842.

Sharma, B. M., Bečanová, J., Scheringer, M., Sharma, A., Bharat, G. K., Whitehead, P. G., ... & Nizzetto, L. (2019). Health and ecological risk assessment of emerging contaminants (pharmaceuticals, personal care products, and artificial sweeteners) in surface and groundwater (drinking water) in the Ganges River Basin, India. *Science of the Total Environment, 646*, 1459–1467.

Sharma, P. K., Mayank, M., Ojha, C. S. P., & Shukla, S. K. (2020). A review on groundwater contaminant transport and remediation. *ISH Journal of Hydraulic Engineering, 26*(1), 112–121.

Singh, R. K., Philip, L., & Ramanujam, S. (2019). Continuous flow pulse corona discharge reactor for the tertiary treatment of drinking water: Insights on disinfection and emerging contaminants removal. *Chemical Engineering Journal, 355*, 269–278.

Singh, R. K., Philip, L., & Ramanujam, S. (2017). Rapid degradation, mineralization and detoxification of pharmaceutically active compounds in aqueous solution during pulsed corona discharge treatment. *Water research, 121*, 20–36.

Sophocleous, M. (2002). Interactions between groundwater and surface water: The state of the science. *Hydrogeology Journal, 10*(1), 52–67.

Sorensen, J., Lapworth, D., Nkhuwa, D., Stuart, M., Gooddy, D., Bell, R., ... & Pedley, S. (2015). Emerging contaminants in urban groundwater sources in Africa. *Water Research, 72*, 51–63.

Stefanakis, A. I., & Becker, J. A. (2016). A review of emerging contaminants in water: Classification, sources, and potential risks. In: Damia B., & Jim B (eds) *Impact of water pollution on human health and environmental sustainability* (pp. 55–80). IGI Global. https://www.sciencedirect.com/science/article/pii/S0048969711013908

Stuart, M., Lapworth, D., Crane, E., & Hart, A. (2012). Review of risk from potential emerging contaminants in UK groundwater. *Science of the Total Environment, 416*, 1–21.

Stuart, M., Manamsa, K., Talbot, J., & Crane, E. (2011). Emerging contaminants in groundwater.

Sui, Q., Cao, X., Lu, S., Zhao, W., Qiu, Z., & Yu, G. (2015). Occurrence, sources and fate of pharmaceuticals and personal care products in the groundwater: A review. *Emerging Contaminants, 1*(1), 14–24.

Szekeres, E., Chiriac, C. M., Baricz, A., Szőke-Nagy, T., Lung, I., Soran, M.-L., ... & Coman, C. J. E. P. (2018). Investigating antibiotics, antibiotic resistance genes, and microbial contaminants in groundwater in relation to the proximity of urban areas. *Environmental Pollution, 236*, 734–744.

Taylor, R. G., Scanlon, B., Döll, P., Rodell, M., Van Beek, R., Wada, Y., ... & Treidel, H. (2013). Ground water and climate change. *Nature Climate Change, 3*(4), 322–329.

Terzić, S., Senta, I., Ahel, M., Gros, M., Petrović, M., Barcelo, D., ... & Jabučar, D. (2008). Occurrence and fate of emerging wastewater contaminants in Western Balkan Region. *Science of the Total Environment, 399*(1–3), 66–77.

Thakur, A. K., Vithanage, M., Das, D., & Kumar, M. (2020). A review on design, material selection, mechanism, and modeling of permeable reactive materials for community-scale groundwater treatment. *Environmental Technology & Innovation, 19*, 100917.

Tran, N. H., Reinhard, M., & Gin, K. Y.-H. (2018). occurrence and fate of emerging contaminants in municipal wastewater treatment plants from different geographical regions-a review. *Water Research, 133*, 182–207.

Tran, N. H., Reinhard, M., Khan, E., Chen, H., Nguyen, V. T., Li, Y., ... & Gin, K. Y.-H. (2019). Emerging contaminants in wastewater, stormwater runoff, and surface water: Application as chemical markers for diffuse sources. *Science of the Total Environment, 676*, 252–267.

Wardenier, N., Gorbanev, Y., Van Moer, I., Nikiforov, A., Van Hulle, S. W., Surmont, P., ... & Vanraes, P. (2019). Removal of alachlor in water by non-thermal plasma: reactive species and pathways in batch and continuous process. *Water Research, 161*, 549–559.

White, D., Lapworth, D., Stuart, M., & Williams, P. J. S. (2016). Hydrochemical profiles in urban groundwater systems: New insights into contaminant sources and pathways in the subsurface from legacy and emerging contaminants. *Science of the Total Environment, 562*, 962–973.

WHO. (2020). *Antibiotic resistance*. Retrieved from https://www.who.int/news-room/fact-sheets/detail/antimicrobial-resistance

Yang, Y., Wan, K., Yang, Z., Li, D., Li, G., Zhang, S., ... & Yu, X. (2020). Inactivation of antibiotic resistant *Escherichia coli* and degradation of its resistance genes by glow discharge plasma in an aqueous solution. *Chemosphere, 252*, 126476.

Zhang, Q., Zhang, H., Zhang, Q., & Huang, Q. (2018). Degradation of norfloxacin in aqueous solution by atmospheric-pressure non-thermal plasma: Mechanism and degradation pathways. *Chemosphere, 210*, 433–439.

Chapter 2

Subsurface flow constructed wetlands as a post-treatment unit for emerging contaminants in municipal wastewater

Jidapa Plaimart, Kishor Acharya, Wojciech Mrozik, Russell J. Davenport, and David Werner
Newcastle University

CONTENTS

2.1 INTRODUCTION

Centralised wastewater treatment plants and decentralised (on-site) systems are the two main treatment systems used to treat municipal wastewater (Opher and Friedler, 2016). Centralised treatment plants normally utilise an activated sludge process, while on-site domestic wastewater treatment systems use cesspits and septic tanks. Septic tanks are commonly used as a primary treatment step in several developing countries and in some areas in developed countries. However, the effluent from the activated sludge process and the on-site treatment systems are inadequately treated, particularly for emerging contaminants (ECs), thus being a source of surface water and groundwater contamination (Dubey et al., 2021; Eregowda and Mohapatra, 2020; Schaider et al.,

DOI: 10.1201/9781003247869-3

2017). A post-treatment process is therefore recommended. A wide range of emerging contaminants such as pharmaceuticals, hormones and personal care products (PCPs) may remain in treated effluent raising several concerns, for example, about the presence of micropollutants such as antibiotics (Mailler et al., 2016; Wang et al., 2018). Antibiotics may give rise to the development of antibiotic-resistant microorganisms, which reduce the therapeutic potential of antibiotics against animal and human pathogens (Tran et al., 2016; Tran et al., 2018). Various strategies of micropollutant reduction have been investigated to tackle this problem. Constructed wetlands, a nature-based water treatment system, implemented as a post-treatment unit to remove pharmaceutical residuals from wastewater discharges have intrigued many researchers. Wetlands can be with surface or subsurface flow, but subsurface flow constructed wetlands (SFCWs) have been reported to show higher treatment efficiencies for organic matter, nutrients, pharmaceutical contaminants and pathogens (Dotro et al., 2017). SFCWs could therefore be a promising effluent polishing option for centralised or on-site wastewater treatment systems. However, there is lack of information on the implement of such option as well as the underlying pollutant removal mechanisms. This chapter therefore aims to review the different factors that affect the performance of SFCWs with a focus on emerging contaminant removal from municipal wastewater.

2.2 EMERGING CONTAMINANTS IN MUNICIPAL WASTEWATER

Emerging contaminants, also known as contaminants of emerging concerns (CECs), refer to a broad range of chemicals such as pharmaceuticals, hormones, PCPs and pesticides (Mailler et al., 2016; McLain and Gachomo, 2019), which have a potential of endangering ecosystems and human health (Geissen et al., 2015). ECs can be classified into more than 20 classes based on their origin, use, potential effects or environmental fate (Nassar and Younis, 2019). Several classifications such as pharmaceuticals and personal care products (PPCPs); persistent, bio-accumulative, toxic; and endocrine disrupting chemicals are in use, but no standard system exists to classify ECs; thus some categories can occasionally overlap (Nassar and Younis, 2019; Priyanka and Mohapatra, 2020).

Most reported EC groups include: (1) pharmaceuticals, e.g., human and veterinary antibiotics, analgesics, anti-inflammatory drugs, and β-Blockers, (2) PCPs, e.g., fragrances and insect repellents, (3) hormones and steroids, (4) disinfectants, (5) flame retardants, (6) herbicides and pesticides, (7) industrial additives and agents and (8) gasoline additives (Stefanakis and Becker, 2015; Wang and Wang, 2016; Tran et al., 2018; Nassar and Younis, 2019). ECs are not commonly included in routine water monitoring programmes, and consequently their fate and behaviour are often not well-understood (Geissen et al., 2015). They can be released to the environment through several routes including wastewater treatment plants (WWTPs), landfill leachate, agricultural activities, improper industrial disposal and direct discharge of household wastewater in some countries (Tewari et al., 2013; Geissen et al., 2015; Yang et al., 2017b; Tran et al., 2018). Their release into the environment has likely occurred for many years but possibly has not been noticed until proper detection methods were developed (Geissen et al., 2015).

Among the aforementioned routes, ECs released from WWTPs are of particular interest due to their continuous discharge to waterbodies. ECs can be detected in both influent and treated effluent of WWTPs at trace levels (ng/L–μg/L) (Tran et al., 2018). Many of them, e.g., organic micropollutants, are known to be persistent in the environment and not easily degradable (Stefanakis and Becker, 2015). Moreover, although some ECs are not lasting in the environment and can be transformed or eliminated through natural processes, their continuous release from WWTPs maintains their existence in natural waterbodies and thus adversely affects aquatic ecosystems (Stefanakis and Becker, 2015).

Tran et al. (2018) compared the occurrence of ECs in municipal wastewater from different geographical locations and found that similar pharmaceuticals were present in both European and Asian municipal wastewater, only differing in the detected concentration levels. A wide range of antibiotics was detected in WWTP effluents at higher concentrations in Asian countries than those found in European countries, which is presumably because in several Asian countries such as Thailand, India and China antibiotics can be purchased without a doctor's prescription giving easier accessibility and subsequently higher consumption rates.

Apart from antibiotics, anti-inflammatory drugs particularly nonsteroidal types (non-steroidal anti-inflammatory drugs (NSAIDs)), e.g., ibuprofen, acetaminophen, diclofenac and salicylic acid, are often found in typical domestic wastewater as they are taken to relieve pain (Tran et al., 2018). For instance, diclofenac was detected at a considerable level (49–121 ng/L) in the canals receiving effluent from WWTPs in Bangkok with an analysed hazard quotient of nearly 1 indicating potential ecological risk. In addition, this compound is not easily biodegraded, thus being a significant threat to aquatic ecosystems (Hijosa-Valsero et al., 2010; Tewari et al., 2013).

2.3 WASTEWATER TREATMENT TECHNOLOGIES

Wastewater treatment schemes consist of a range of treatment levels. They can be normally categorised into four steps involving preliminary, primary, secondary and tertiary treatment (Crini and Lichtfouse, 2019; Khuntia et al., 2021; 2022; Mogili et al., 2021; Tabraiz et al., 2021). The requirement for each step relies on the source, characteristics and intended use of wastewater (Mohapatra et al., 2021). Each treatment step can treat different pollutants (Ullah et al., 2020). Preliminary treatment is used to remove coarse suspended solids (Ullah et al., 2020). Primary treatment is mainly for separation of solid organic matter via a sedimentation or flotation unit (Hreiz et al., 2015; Rout et al., 2020). Wastewater then flows to a secondary treatment governed by biological degradation which is used to remove the remaining solids and soluble organic matter escaping from a primary treatment (Rout et al., 2020). In this process, a microbial biomass suspension (the activated sludge) is responsible for the removal of pollutants under aerobic conditions (Hreiz et al., 2015). Secondary effluent is then occasionally treated further in a tertiary treatment step. This may be known as a polishing unit which is mainly employed to remove nutrients, suspended solids and pathogens (Rout et al., 2020). These numerous treatment methods have been further categorised as physical, chemical and biological techniques (Crini and Lichtfouse, 2019). Examples of the most widely applied technologies based on these three categories are listed in Table 2.1.

Table 2.1 Classification and examples of wastewater treatment technologies

Classification	*Example technologies*
Physical	Adsorption: activated carbon/zeolites Membrane separation: reverse osmosis, nanofiltration
Chemical	Chemical oxidation: ozonation Photo-oxidation Coagulation: aluminium sulphate Ion exchange
Biological	Electrochemical reactors: microbial fuel cell Biodegradation: activated sludge, membrane bioreactor Constructed wetlands

Source: Details were derived from Crini and Lichtfouse (2019) and El-Gendy et al. (2020).

2.3.1 Activated sludge process

The activated sludge process (ASP) is normally applied as a secondary treatment step in WWTPs and it is mainly governed by microbiological activities to degrade organic matter and nutrients in wastewater (Islam et al., 2013; Meerburg et al., 2015). Chemicals are not required for this process or maybe required in insignificant amounts (Islam et al., 2013). The basic ASP consists of two main units including (1) a single aerated-bioreactor operated continuously where suspended microbes consume the dissolved and colloidal organic matter (Hreiz et al., 2015), and (2) a settling tank (also known as secondary clarifier) where gravitational separation of activated sludge and treated wastewater is employed. The effluent overflows from the settling tank into the receiving watercourses or perhaps undergoes tertiary treatment such as disinfection or media filtration in some WWTPs before being discharged (Hreiz et al., 2015). There is also a sludge recycle line and a sludge waste line attached at the bottom of the settling tank. The former is to return the major proportion of the settled sludge to the bioreactor for maintaining a high microbial concentration in the bioreactor, while the latter is to dispose a small proportion of the sludge. Thus, the biomass concentration in the bioreactor is stabilised and an adequate solid retention time (SRT) is achieved (Hreiz et al., 2015). Due to its efficiency and simplicity, ASP is by far the most well-known secondary wastewater treatment system (Hreiz et al., 2015; Meerburg et al., 2015; Guven et al., 2019). In terms of pollutant removal, ASP is not only capable of removing macro-organic matter but also organic micropollutants (Alvarino et al., 2018). ECs can be removed either through biomineralisation by microbes or sorption onto particulate matter (sludges) (Alvarino et al., 2018). The removal performance of ECs varies by different factors and ASP operating conditions, such as SRT, food to mass ratio, temperature and redox conditions or even the recalcitrant behaviour of ECs themselves (Collado et al., 2014). For example, Alvarino et al. (2018) reported the removal efficiency of different pharmaceutical by an ASP. It was found that ibuprofen and naproxen were removed more than 80%, and antibiotics like sulfamethoxazole were only removed by 20%–50%, while a recalcitrant compound like diclofenac was removed less than 20%. Similarly, Collado et al. (2014) reported a high removal efficiency for ibuprofen and naproxen and moderate removal for sulfamethoxazole. They also reported that the removal performance varies by ambient temperature, with higher removal performance being observed in the summer than in colder periods. Furthermore, other biological processes can have an impact on the EC removal. There were reports on the enhanced elimination of ECs by nitrification (Fernandez-Fontaina et al., 2012; Collado et al., 2014; Alvarino et al., 2016).

2.3.2 On-site wastewater treatment systems

The on-site (household) systems are nowadays also gaining attention as a result of a continually growing urban population and increasing water scarcity worldwide (Opher and Friedler, 2016). The centralised WWTPs normally consist of a biological treatment like the ASP, while the on-site municipal wastewater treatment systems generally use cesspits and septic tanks (Yates et al., 2019). On-site domestic wastewater treatment systems such as cesspits and septic tanks are a kind of anaerobic reactor. The main aim of septic tank systems is to provide primary treatment for domestic sewage by intercepting and separating solid faecal matter from the liquid and the system needs to be de-sludged every 4 years to prevent sewage overflow to watercourses (Ting et al., 2013; Singh et al., 2019). Septic tank systems consist of two main components involving a septic tank and a drainfield, may also be known as soil absorption system, in which the septic tank effluent is distributed via gravity or pressure from pipes or pits into well-draining unsaturated zone soils (Schaider et al., 2017). The systems are normally suitable for a village population of up to 500 (Withers et al., 2011). In developed countries, e.g., the UK and the USA, such systems are still being used in some rural areas (Withers et al., 2011; Schaider et al., 2017). In developing countries, the septic tank system is a common treatment technique as a result of an unproperly managed sewerage systems and lack of centralised treatment facilities in the rural area (Singh et al., 2019). For instance, in a warm climate country like Thailand, septic tank systems are generally used to treat domestic wastewater (85%) even in the areas served by centralised WWTPs (Tsuzuki et al., 2010; WHO, 2018).

Conventional septic tanks are capable of removing settleable solids, oils, greases and floating debris from the raw sewage at a range of 60%–80% and the removed solids are stored in sludge and scum layers (Nasr and Mikhaeil, 2015). Nevertheless, faecal coliforms and some organic compounds are poorly removed (Nasr and Mikhaeil, 2015; Schaider et al., 2017). The effluent from septic tank systems known as settled sewage is normally allowed to seep into soil or directly discharged into nearby watercourses, thereby becoming a source of groundwater and surface water contamination (Surinkul and Koottatep, 2009; Withers et al., 2011; Yang et al., 2017a). For example, in Thailand high levels of several pharmaceuticals were detected in the canal system around Bangkok, as a result of inadequately treated effluent from septic tanks and possibly also the direct discharge from households, suggesting potential ecological risks (Tewari et al., 2013). In the USA, endocrine disrupting compounds were detected in the septic tank effluent at trace concentrations ranging from a few ng/L to several µg/L (Yang et al., 2017a). More common ECs like caffeine and ibuprofen were detected at levels ranging from 0.12 to 12.04 µg/L in the groundwater downgradient from septic tank systems in North Carolina (Del Rosario et al., 2014). Schaider et al. (2017) reported that although some organic micropollutants may be removed within the septic tank via physical separation of solid particles and oil-associated organic micropollutants as well as anaerobic degradation, the greatest micropollutant removal seems to occur in drainfields via sorption, volatilisation and aerobic degradation. However, another study found that some pharmaceutical compounds increased or remained relative constant 30m downgradient of a drainfield (Phillips et al., 2015). It is therefore desirable that a post-treatment unit is installed after septic tank systems to further reduce the remaining contaminants of concerns to acceptable levels before discharging treated effluent into natural waterbodies.

2.3.3 Nature-based wastewater treatment systems

Nature-based wastewater treatment systems rely mostly on natural processes such as gravity forces and natural components such as plant, sunlight and biological microorganisms for wastewater treatment in a relatively passive manner (Crites et al., 2014). The systems sometimes also apply pumps and pipe works for wastewater conveyance and distribution, however require little to no external energy (Crites et al., 2014). Consequently, they can reduce the need for energy from fossil fuels as well as create wildlife habitats. Natural treatment systems such as ponds, lagoons and wetlands are some of the oldest and have been the most globally used technologies in the past decades (Verbyla, 2017). There are several types of ponds that are used to treat wastewater at different stages in the treatment process. The most common pond types include anaerobic ponds, facultative ponds and maturation ponds (Verbyla, 2017). There are also combined natural and engineered systems (Zawadzka et al., 2019). Such systems involve constructed wetlands, riverbank filtration and managed aquifer recharge (Zawadzka et al., 2019). The systems can be used for primary, secondary and/or tertiary wastewater treatment depending on individual purposes (Verbyla, 2017). Constructed wetlands are very effective as a tertiary treatment system after activated sludge plants (Hoffmann et al., 2011). Systems such as sand filtration and constructed wetlands have gained recognition for wastewater treatment in small communities as compared to conventional WWTPs due to their efficiency, low construction and operating/maintenance costs (Adrados et al., 2014). Nevertheless, in order to effectively treat wastewater, numerous factors have to be taken into account, including the system's capacity, plants used, microbial characteristics and the interactions of microbial-mediated and wastewater contaminants (Adrados et al., 2014). Constructed wetlands (CWs) can perform a vital role in ecosystem service supply if they are implemented in suitable environmental settings, and if land cover management for ecological functioning is in place (Zawadzka et al., 2019). Moreover, people's perception in terms of aesthetic value for CWs is highly positive as compared to their engineered equivalents (Zawadzka et al., 2019). De Feo and Ferrara (2017) conducted a life cycle assessment on two on-site systems namely CWs and an activated sludge compact system. It was found that CWs were claimed to be a better option than an activated sludge compact system in terms of environmental impacts due to their lower electricity consumption.

Several forms and modifications of CW technology have been invented (Nuamah et al., 2020). Generally, CW configurations are based on wetland hydrology and flow pattern which include free-water surface flow subsurface flow and a hybrid system (Stefanakis, 2016; Verbyla, 2017) (Figure 2.1). Free-water surface flow CWs are mostly similar to wastewater treatment ponds with only the difference that they contain submerged/emergent vegetation and floating macrophytes (Verbyla, 2017). SFCWs can be classified into two types relative to flow pattern which are horizontal and vertical flow

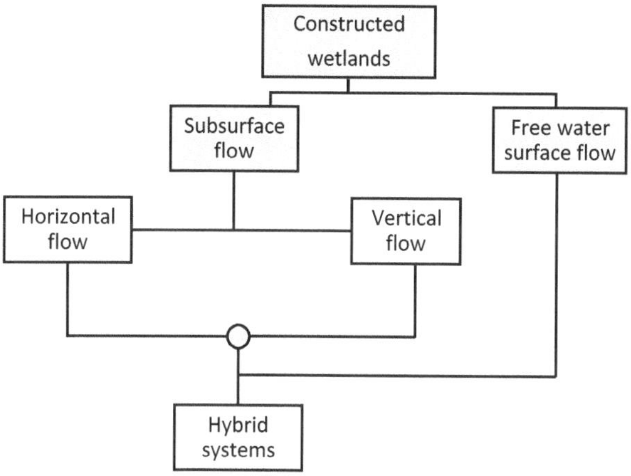

Figure 2.1 Classification of constructed wetlands. (Adapted from Stefanakis 2016.)

(Wang et al., 2017). Hybrid CWs consist of those aforementioned types of CWs connected in sequences and they have now been implemented in many countries across Europe and Asia (Vymazal, 2013). More details on SFCWs and hybrid systems are provided in Section 2.4.

In a technology comparison, among SFCWs, free surface flow CWs and ponds, there is no universal conclusion that one type of these systems is better than the others; hence, the advantages and disadvantages of these three systems are still a subject of debate. Sperling and Chernicharo (2005) stated that SFCWs can be susceptible to clogging if improperly managed. Mara (2004) found that a horizontal SFCW required 22% larger area than a secondary facultative pond for the same biological oxygen demand (BOD) removal at 25°C ambient temperature in a tropical climate country like Brazil. Contrarily, Mburu et al. (2013) carried out a pilot-scale study in Kenya, which also has a tropical climate, and it was found that a facultative pond required a three times larger area than a horizontal SFCW when water was treated to the same standard. Moreover, Kadlec (2009) revealed that a horizontal SFCW performed well in cold climates as it is less cold sensitive and easier to insulate for winter operation than a free-water surface treatment system. SFCWs also provide some other advantages over surface flow wetlands and waste stabilisation ponds such as lower risk of mosquito breeding and unpleasant odours (Mara, 2004; Austin and Yu, 2016). The main characteristics of CWs and conventional treatment systems are compared in Table 2.2.

Nature-based wastewater treatment systems, particularly CWs, would therefore be a low-cost alternative solution for wastewater quality improvement while providing ecological, technical and societal benefits.

2.4 SUBSURFACE FLOW CONSTRUCTED WETLANDS

SFCWs are low-cost and efficient alternative water treatment systems. They have been earlier mentioned to some extent in Section 2.3. More detailed information can be found in this section.

Table 2.2 Comparisons of the main features of constructed wetlands (CWs) and conventional wastewater treatment systems (Stefanakis, 2016)

	Constructed wetlands	*Conventional wastewater treatment systems*
Infrastructure	No mechanical components (or only pumps)	Many and complex mechanical parts
Investment/Operational costs	Low particularly when there is available land	High
Demand of land area	High	Low
Raw materials	Nearly exclusive use of renewable resources (wind, solar, etc.)	Use of non-renewable materials for construction (steel, concrete) and operation (electricity, chemicals)
Energy consumption	Low	High
Staff during operation	No specialised personnel needed	Demand for specialised personnel
Performance	Similar to conventional systems with small fluctuations depending on temperature variations	Continuous effluent of high quality
By-products	No by-products	Massive daily volumes of sludge production that require daily management

2.4.1 The design

SFCWs mainly contain shallow basins with a seepage barrier and water inlet/outlet (Crites et al., 2014). SFCWs are normally designed for horizontal or vertical flow. The former requires a larger area than the latter, but is easier to design and suitable for locations without energy supply (Vymazal, 1998; Mara, 2004). Nevertheless, if evapotranspiration is an issue, a vertical flow bed (VFB) is preferable to a horizontal flow bed (HFB), because it has an unsaturated upper layer in the bed, which is more suitable for warm climate countries. Additionally, it also has been reported to show higher treatment efficiency (Vymazal, 1998; Hoffmann et al., 2011). SFCWs traditionally consist of a sand or gravel bed, which acts as a filter medium where the water level is kept below its surface (Austin and Yu, 2016). The gravel in the bed does not perform filtering function, but is mainly used to cover the influent distribution and drainage pipes, and avoid puddles on the surface (Hoffmann et al., 2011). SFCWs require a pre-treatment prior to wetland treatment, which can be for example a pond or septic tanks (Crites et al., 2014). The main aim for this is to decrease the concentrations of suspended solids and organic matters that otherwise would accumulate in the inlet zone and clog SFCW systems (Hoffmann et al., 2011; Crites et al., 2014).

2.4.1.1 Horizontal flow beds

HFBs are the most common type of SFCWs. They are usually planted with common reeds (*Phragmites australis*) in Europe and the USA, while *Cyperus*, *Typha* and

Heliconia are commonly used in tropical climate countries (Vymazal and Kropfelova, 2008; Dotro et al., 2017). In HFBs the wastewater flows horizontally through the filter bed medium under the surface of the bed until it approaches the outlet area, and the water level is kept 5–10 cm below the gravel layer surface (Hoffmann et al., 2011) (Figure 2.2). The water level at the outlet is controlled with an adjustable standpipe (Hoffmann et al., 2011; Dotro et al., 2017). The systems consist of an integration of a plastic liner and a geotextile membrane to isolate the whole bed from surrounding land (Dotro et al., 2017). Generally, a maximum width of 25–30 m is in use to enable even flow distribution into a single wetland cell. The depth of HFBs is normally designed at 60 cm with an additional 15 cm freeboard for water accumulation, which is based on the assumed maximum root depth penetration (Dotro et al., 2017). However, the implementation of a maximum cross-sectional area loading (the load applied at the inlet per width and depth) is instead considered as a design criterion nowadays, providing an opportunity to alter bed depth and length (Dotro et al., 2017). The main recommended design parameters are listed in Table 2.3.

2.4.1.2 Vertical flow beds

VFBs are usually implemented where there is a space constraint as they only require about half of the area of HFBs (Austin and Yu, 2016). The most common design for VFBs consists of a basin where the wetland influent is intermittently applied onto the surface using pumps and then percolates vertically down through the filter medium (sand/gravel) to the drainage area at the bottom (Austin and Yu, 2016; Dotro et al., 2017) (Figure 2.3). The influent needs to be pre-treated to avoid wetland systems failure from filter bed clogging. The wetlands are typically receiving wastewater at 4–12 doses daily and go through long resting periods when the influent percolates down to the bottom and the surface dries out (Hoffmann et al., 2011). The intermittent loading increases the oxygen transfer and aerobic degradation processes in the filter bed (Dotro et al., 2017). VFBs have high capability of removing organic carbon because of the highly oxidising conditions in their filter bed. They are also appropriate when

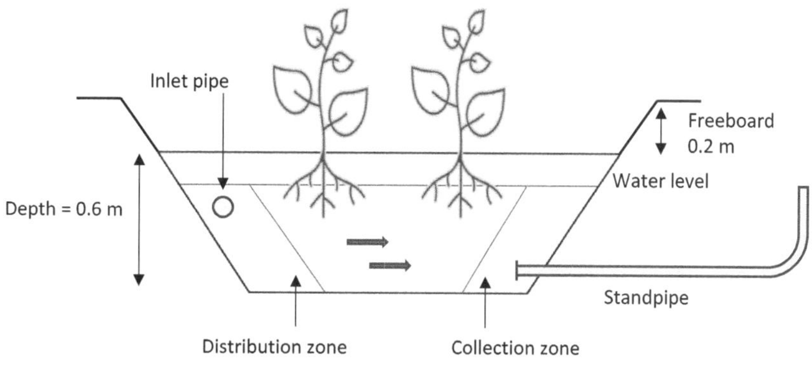

Figure 2.2 Horizontal subsurface flow constructed wetlands design. (Adapted from Dotro et al. 2017.)

Table 2.3 Main design components of horizontal subsurface flow constructed wetlands as a secondary or tertiary treatment step (Dotro et al., 2017)

	Treatment purpose	
	Secondary	*Tertiary*
Pre-treatment	Septic tank	Primary settling + biological treatment
Specific surface area requirement (m^2/population equivalent)	5–10	0.7
Maximum areal organic loading rate (g BOD_5/m^2.d)	4–8	2–13
Hydraulic loading rate (mm/d)	20–40	200
Gravel size (mm)	>4	10–12
General filter bed depth (m)	0.5–0.7	0.6

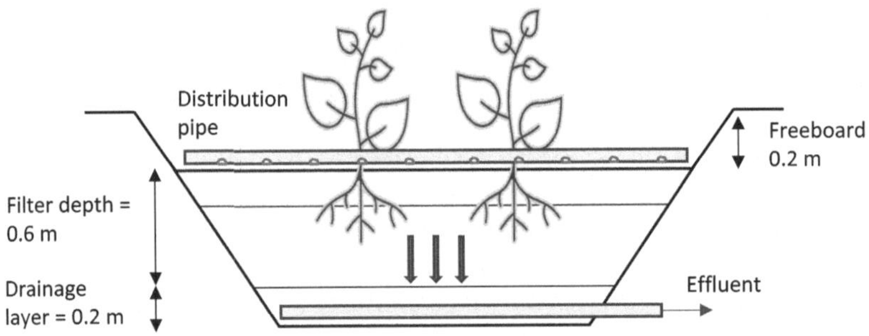

Figure 2.3 Vertical subsurface flow constructed wetlands design. (Adapted from Dotro et al. 2017.)

a stringently aerobic activity such as nitrification is required (Dotro et al., 2017; Wang et al., 2017). VFBs are usually planted with *Phragmites australis* (common reed), the same as the HFB type. The roles of vegetation with regard to pollutant removal in VFBs are mostly related to physical mechanisms such as protecting surface from erosion, maintaining hydraulic properties of the filter and providing surface area for biofilms attached to plant roots (Dotro et al., 2017).

Basic design recommendations for VFBs treating domestic wastewater are as follows:

1. The distribution pipes should provide even distribution of wastewater over the VFB. The drainage pipes are covered with gravel to allow good drainage.
2. The depth of the sand filter bed should be at least 50 cm with an additional 10 cm of gravel at the top to avoid free water accumulation on the surface, and 10 cm gravel at the bottom (Hoffmann et al., 2011). However, the top layer also has disadvantages because it can reduce oxygen supply to the main layer leading to less

Table 2.4 Main design components of vertical subsurface flow constructed wetlands depending on cold or warm climate condition (Hoffmann et al., 2011; Dotro et al., 2017)

Design parameter	*Cold climate*	*Warm climate*
Specific surface area requirement (m^2/population equivalent)	3–4	1–2
Maximum organic loading rate (g COD/m^2.d)	20	60–70
Hydraulic loading rate (mm/d)	100–120	200

organic matter degradation at the main layer hence higher risk of clogging (Dotro et al., 2017). More specific design components depending on climate conditions are provided in Table 2.4.

2.4.1.3 Hybrid systems

A sequence of free-water surface flow CWs and HFB/VFB SFCWs as a hybrid system is gaining interest, and has been utilised in Europe and Asia to achieve the highest water quality for high-strength wastewater (Nuamah et al., 2020). Integrating different types of CWs capitalises on the capacity of each to efficiently remove specific compounds, which can often be undertaken at a much lower capital and operating cost than for conventional wastewater treatment systems (Austin and Yu, 2016). Hybrid systems are initially designed to enhance nitrogen removal as the different types of wetland environments allow various redox conditions which are suitable for nitrification (oxidation of ammonia to nitrate) and denitrification (reduction of nitrate to N_2O and N_2) (Vymazal, 2013; Wang et al., 2017). In general, HFBs can provide good anaerobic conditions for denitrification; however, the capability for nitrification is very limited. In contrast, VFBs can remove ammonia due to their better aerobic condition; however, denitrification hardly occurs in these systems. A combination of both types therefore strengthens the performance of the individual systems (Wang et al., 2017). A VFB-HFB arrangement was reported to be the most widely used type for both industrial and municipal wastewater treatment among other combinations with high efficiency in ammonium removal (Vymazal, 2013; Nuamah et al., 2020). Melián et al. (2010) conducted a pilot-scale VFB-HFB SFCWs in the Canary Islands, Spain. It was found that such system achieved very high treatment efficiency for BOD and ammonium. In addition, high (>99.5%) removal of faecal coliforms was also reported. The same combination has also been implemented in a warm tropical climate country like Thailand, and it performed very well for BOD and nutrient removal, but unfortunately the lack of staff taking responsibility for the system management caused problems later (Brix et al., 2011).

2.4.2 Filter media

Filter media, substrates or support materials are intentionally used for water filtration, and they are also one of the key components that define a successful operation of SFCWs (Austin and Yu, 2016; Yang et al., 2018; Khalifa et al., 2020). They play an important role as follows: (1) serving as a substrate for the growth of bacteria that perform the biochemical treatment of the wastewater that pass through the systems (Austin and Yu, 2016); (2) providing physical support for the materials to directly interact with the contaminants via sorption mechanisms (Gupta et al., 2016); and (3) supporting plant growth (Stefanakis, 2016). A suitable permeable filter media should be selected in accordance with the hydraulic and organic loading to the systems as most of the treatment issues occur when the permeability is inadequately chosen for the applied load (Austin and Yu, 2016). Additionally, Khalifa et al. (2020) suggested that the media should offer both aerobic and anaerobic pores for microbial processes as well as an internal carbon source.

Most studies have classified CW substrates based on their origin including natural substrates (soil, sand and gravel), industrial substrates (slag and coal fly ash) and man-made substrates (activated carbon and ceramsite) (Cheng et al., 2018; Yang et al., 2018). Natural substrates such as sand and gravel have been mostly used among other types due to their abundance and efficiency for pollutant retention with low purchasing cost (Yang et al., 2018). Gravel did not perform well in nutrient removal but was found to have efficiency in heavy metal removal (Buddhawong et al., 2005; Wang et al., 2020). Coarse sand was recommended to be the most suitable filter material for SFCWs for wastewater treatment in developing countries (Hoffmann et al., 2011). The sand should have a hydraulic capacity of about 10^{-4} to 10^{-3} m/s with grain size of 0.6 mm (Hoffmann et al., 2011; Austin and Yu, 2016). Sand-like material revealed higher removal efficiency for ammonium and total phosphorus relative to gravel in SFCWs (Abdelhakeem et al., 2016). A wide range of alternative substrates has also been considered based on local availability together with achieving enhanced universal or specific pollutants treatment performance with cost effectiveness (Yang et al., 2018). An emerging substrate, zeolite, has gained wide recognition as it contains high micropore and macropore volumes which increase its ability of nitrogen adsorption (Wang et al., 2020). It has been recommended as an ideal substrate for nitrogen removal in CWs (Zou et al., 2012). Another example is activated carbon. It is an amorphous carbon obtained by processing that provides a highly adsorption capacity for organic or inorganic matters in the wastewater due to its large specific surface area (Wang et al., 2020). It has gained increasing recognition in the field of environmental pollution control (Wang et al., 2020). Concurrently, in recent years a more sustainable sorbent material known as biochar sourced from agricultural wastes has also gained interest in water remediation in addition to activated carbon, as it is a cost-effective product with a considerably potential in various pollutant removal applications (Mohan et al., 2014). Activated carbon and biochar amendments are also implemented in wetlands to reduce the mobility and availability of persistent chemicals for uptake into the aquatic or terrestrial food chains (Ghosh et al., 2011). Such emerging materials could be integrated with natural media to optimise SFCW performance. Ulrich et al. (2015) showed that only a small amount (0.2%–2% weight) of biochar or activated carbon added to sand significantly enhances trace organic micropollutant retention. On the other hand, de Castro et al. (2018) compared polishing units for AS effluent with the filter media sand, or 9:1 v:v mixtures of sand with vermiculite, charcoal or granular activated carbon, and found similar EC removal efficiencies by all four filter media, ranging from >90% for oestrogens to 10%–30% for ibuprofen, diclofenac and paracetamol. Selection of filter media is one of the major concerns in CWs wastewater treatment. Suitable media selection leads to a better treatment efficiency and a lower risk of system failure (Wang et al., 2020). When selecting media, key points to consider include their source and cost, pollutant removal performance, hydraulic and engineering feasibility, support for microbial and plant growth, substrate clogging and recovery/disposal of exhausted substrate (Wang et al., 2020). Substrate clogging is one of the main concerns when designing SFCWs. The main contributors to clogging are the small particle sizes, low porosity and poor hydraulic conductivities which can be prevented by choosing big particle media and trying to avoid using substrates that can react with the pollutants to form precipitates, thus occupying the pore space (Yang et al., 2018).

2.4.3 Plants

The plants or macrophytes play an important role in SFCWs by maintaining the hydraulic conductivity of the substrate, reducing excessive nutrients, removing contaminants and also facilitating growth of bacterial communities and other microorganisms which form a biofilm attached to the surface of roots and substrate particles (Shelef et al., 2013; Verlicchi and Zambello, 2014). Additionally, the plants play an indirect role in transferring oxygen from leaves to the rhizosphere via the roots, releasing chelating agents and antibiotics that increase precipitation of metals present in wastewater (Khalifa et al., 2020). Other functions of plants include improving aesthetic appearance, and the elimination of pathogens, insects and offensive odours (Shelef et al., 2013). Nevertheless, the presence of plants can also provide some adverse effects such as increased mosquito reproduction, enhanced methane emissions and nitrous oxide emissions produced by anaerobic denitrification via aerenchyma tissue (Shelef et al., 2013). The selection of CW plants is dependent on several factors such as climate conditions, types and characteristics of influent water, and types of SFCW (Austin and Yu, 2016; Nuamah et al., 2020). In any cases, native plant species should be specified before selection (Austin and Yu, 2016). *Phragmites australis* is the most used plant for both subsurface flow and free-water surface CWs. It can be widely found in Europe and Asia, and tolerates municipal wastewater that has been pre-treated in a septic tank (Austin and Yu, 2016; Dotro et al., 2017). *Typha latifolia* or broadleaf cattail is a widely distributed North American native. It is capable of treating sewage and other effluents high in organic nitrogen (Austin and Yu, 2016). The removal effect of plants on wastewater pollutants varies greatly by environmental conditions, number and type of plants, and the nature and chemical structure of such pollutants (Wang et al., 2014). Several studies found higher removal efficiency of conventional water quality parameters such as BOD and nutrients in planted SFCWs than unplanted beds (Shelef et al., 2013; Abdelhakeem et al., 2016; Marín-Muñiz et al., 2017; Marín-Muñiz et al., 2020). Contrarily, some studies also revealed that SFCWs without plants showed higher or equal removal of total organic carbon (TOC), dissolved oxygen (DO), chemical oxygen demand (COD), BOD as well as some pharmaceuticals like carbamazepine and diclofenac compared to those with plants (Tanner, 2001; Baptista et al., 2003; Zhang et al., 2011).

2.4.4 Microbial communities in SFCWs

Although filtration is considered as a key process for pollutant elimination in SFCWs, extra interactions occur among water, plants and media through microbial processes (Adrados et al., 2014). The most stable microbiota in SFCWs are found in the biofilm attached to plant's roots and/or to the surface of filter media (Adrados et al., 2014). Sánchez (2017) explained that the wetland rhizosphere tends to perform a key responsibility in the formation of oxic-anoxic interfaces where aerobic and anaerobic microbes can perform in proximity, increasing elemental cycling thus enhancing microbial activity. Microbial density and activity were enhanced in planted CWs (Sánchez, 2017). Sidrach-Cardona et al. (2015) found that plants in CWs such as common reeds influenced the microbial community of the rhizoplane, interstitial water and gravel-related microbes beyond the plant's roots. However, many studies also reported a lack of significant effect of plants on microbial community structure (Baptista et al., 2003; Iasur-Kruh et al., 2010; Sánchez, 2017). Sidrach-Cardona et al. (2015) suggested that such insignificant effects of plants on microbial communities could have resulted from inadequate redox conditions from root development. In any case, the underlying mechanisms of plants on microbial activities in different components of CW systems are still unclear.

Microbial communities can additionally be differentiated by other factors such as temperature, dissolved oxygen, hydraulic design, availability of organic matter and filter material type (Sánchez, 2017). Microbial activities involved in wastewater treatment are strongly influenced by climate and seasonal changes (Sperling and Chernicharo, 2005; Sánchez, 2017). The high-temperature condition provides a conductive environment for higher treatment efficiencies particularly of such nature-based systems compared to those in cold temperature (Sperling and Chernicharo, 2005; Bateganya et al., 2016). Wang et al. (2016) revealed that the treatment performance and microbial structure of SFCWs varied greatly with seasonal changes for both planted and unplanted beds. Planted beds showed more dominant phyla (e.g., *Bacteroidetes*, *Proteobacteria* and *Actinobacteria*) in the summer than winter while unplanted beds revealed higher *Cyanobacteria* and photosynthetic bacteria in the summer than winter. Additionally, the abundance of microbes is generally decreased during wintertime.

Substrate is also another factor that shapes microbial population in CWs. It provides optimal conditions for microbial growth, thus a variety of microbial species are present and contribute to a better treatment system performance (Meng et al., 2014; Rajan et al., 2019). Guan et al. (2015) conducted a study on microbial communities in different wetland substrate types (sand, zeolite, gravel). It was found that structure and community of bacteria showed prominent spatial variations in zeolite and sand wetlands, but only slightly changed in gravel wetlands. Several phyla were present in CWs, including *Bacteroidetes*, *Proteobacteria*, *Nitrospirae*, *Cyanobacteria*, etc. *Proteobacteria* dominated among other phyla, and this group of microbes drives more effective organic pollutant reduction.

Moreover, enhanced nitrogen removal efficiency is one of the important targets for further development of CWs. Microbes involved in the N-cycle, particularly nitrification and denitrification, play a crucial role in nitrogen transformation in CWs (Adrados et al., 2014). Rajan et al. (2019) reported that *Nitrosomonas* species of nitrifying bacteria were found in the vicinity of plant roots which therefore had better nitrogen removal than unplanted beds. Paranychianakis et al. (2016) reported on the seasonal shift in the composition of denitrifying microbes towards to a community with a lower genetic potential for N_2O emission in planted SFCWs.

2.4.5 Treatment efficiency for conventional water quality parameters

SFCWs showed high treatment efficiencies for organic matter, nutrients and pathogens (Hoffmann et al., 2011; Wu et al., 2015). In general, removal of BOD and total suspended solid (TSS) were found to be highly efficient and consistent across all types of SFCWs (Zhang et al., 2015). In tropical environmental conditions, HFBs could efficiently reduce on average more than 80% of BOD and COD, 70% of total phosphorus (TP) and 50% of total nitrogen (TN) (Varma et al., 2021). For VFBs, the average BOD and COD removal was reported to be around 87% which is slightly higher than HFBs (Varma et al., 2021). Similar removal (80%) in the two system types was found for TSS (Zhang et al., 2015). Although low denitrification occurred in VFBs as a result of prevailing aerobic conditions, TN removal was still observed at 68% which is higher than HFBs (Varma et al., 2021). In terms of the hybrid systems, higher efficiency was found for removal of BOD, COD, TN, ammonia, nitrate, nitrite, etc., as compared to other systems. The average BOD, COD and TN reduction was approximately 90%, 80% and 70%, respectively (Vymazal, 2013; Zhang et al., 2015; Varma et al., 2021). The high removal rate of BOD and COD could be attributed to filtration or sedimentation of suspended solids as well as biodegradation. VFBs performed better than HFBs because the former is intermittently loaded resulting in a higher oxygen transfer to the filter medium as compared to HFBs (Zhang et al., 2015). Besides, the removal efficiency is highly dependent on ambient temperature. Poorer treatment efficiency can be found at low temperature as a consequence of multiple variables such as the microbial metabolism rate, the plant's oxygen transfer capability, the sedimentation and adsorption velocity of the filter material, etc. (Ji et al., 2020).

For pathogen removal, SFCWs and hybrid systems were reported to eliminate a considerably higher number of bacterial pathogens than surface flow wetlands (Shingare et al., 2019). Three-log removal of faecal/total coliforms was typically found in SFCWs (Rajan et al., 2019; Shingare et al., 2019; Khalifa et al., 2020). SFCWs also showed highest removal performance for helminths eggs. This is because parasitic eggs are typically bigger than bacterial pathogens hence a satisfying filtration with attachment areas for such eggs was found in CWs with filter media like SFCWs (Shingare et al., 2019).

2.5 SFCWs AS A POST-TREATMENT UNIT FOR ECs

CWs have been widely investigated for EC removal from wastewater over the past decades (Li et al., 2014; Dotro et al., 2017; Ilyas and van Hullebusch, 2020). SFCWs implemented as a post-treatment system to remove pharmaceutical residuals from wastewater are receiving increased attention in many countries across the world (Li et al., 2014). SFCWs can be potentially used as both secondary and tertiary treatment unit based on treatment requirements and suitability in each location. Reported removal efficiencies of the most commonly monitored pharmaceuticals found in secondary/tertiary SFCW unit are summarised in Table 2.5.

Even though SFCWs are not specifically designed for EC treatment, the systems were still capable of removing such micropollutants. As shown in Table 2.5, in general, SFCWs performed well for most of the pharmaceuticals reported. Between HFBs and VFBs, VFBs showed higher removal performance for most reported compounds especially in secondary treatment. Better removal efficiencies were observed for most reported compounds in secondary than tertiary treatment for both types of wetlands.

Removal efficiencies can be affected by various conditions (temperature, plants, substrate) in each SFCW system. Removal of ECs happens in a combination of complex physical, chemical and biological interactions by several wetland components. Such mechanisms include biodegradation, sorption and plant interactions (Verlicchi and Zambello, 2014). Nevertheless, the primary removal mechanism is still not easily ascertained (Zhang et al., 2011). For example Zhang et al. (2011) found that planted SFCWs, through rhizosphere effects, revealed significantly increased removal efficiencies of ibuprofen and naproxen compared to unplanted beds. However, planted beds showed no significant difference for removal of recalcitrant compounds like diclofenac and carbamazepine, if compared with unplanted beds. The removal mechanism for these compounds likely came from sorption to organic surfaces instead of plant effects. For the reported pharmaceuticals (Table 2.5), three mechanisms took place for removal, but the dominant mechanism was thought to be biodegradation (Ilyas and van Hullebusch, 2020). SFCWs were also found to perform well in elimination of antibiotic resistance genes (ARGs) whereby biodegradation and adsorption onto filter medium were observed to be the major removal mechanisms; however, this is still being researched (Pei et al., 2019). Hybrid wetlands are also increasingly investigated for ECs treatment in recent years. A full-scale study of a hybrid wetland system in southern Spain by Ávila et al. (2015) showed that such a system was capable of removing 89%–99% of pharmaceuticals such as ibuprofen and diclofenac. Ilyas and van Hullebusch (2020) who conducted a critical comparison of different types of SFCWs revealed that hybrid systems performed comparatively better for EC removal as compared to HFBs or VFBs alone. This is because of the synchronisation of aerobic/anaerobic conditions

Table 2.5 Removal efficiencies (%) of pharmaceuticals in SFCWs applied as an alternative secondary or tertiary wastewater treatment system

Pharmaceuticals	*Secondary treatment*		*Tertiary treatment*	
	HFBs	*VFBs*	*HFBs*	*VFBs*
Diclofenac	0%–55%	63%–73%	5%–7.5%	78%–79%
Ibuprofen	50%–70%	85%–99%	28%–96%	66%–72%
Naproxen	50%–90%	84%–92%	14%–36%	39%–45%
Acetaminophen	47%–99%	94%–97%	45%	96%–98%
Salicylic acid	50%–98%	60%–90%	0%–25%	0%–25%
Carbamazepine	20%–30%	20%–26%	60%–88%	0%–26%
Caffeine	60%–90%	82%–99%	0%–25%	0%–25%
Sulfamethoxazole	73%–87%	0%–25%	0%–25%	0%–25%

Data were derived from Verlicchi and Zambello (2014), Li et al. (2014), Ávila et al. (2015) and Ilyas and van Hullebusch (2020).

and longer hydraulic retention time from multiple compartments of the hybrid systems. Moreover, in comparison with conventional treatment systems, Dotro et al. (2017) reported that SFCWs could remove a range of micropollutants better than an activated sludge plant, which still can be attributed to the coexisting aerobic-anoxic-anaerobic microenvironments of biofilms and a long sludge retention time in the CW systems. Furthermore, researchers also investigated the impact of nitrification on ECs removal. Co-metabolic oxidation by the ammonia oxidation enzyme (ammonium monooxygenase (AMO)), taking part in the first step of nitrification processes, is thought to be initiating the biotransformation of several ECs (Fernandez-Fontaina et al., 2012; Alvarino et al., 2016). However, it should be noted that nitrification is not the only metabolic pathway that controls EC biodegradation (Rattier et al., 2014).

2.6 FURTHER RESEARCH OPPORTUNITIES FOR SFCW DESIGN OPTIMISATION

The optimal treatment performance of SFCWs relies on SFCW design. Some of further research needs and opportunities that could be considered for SFCW design optimisation are summarised below.

- Use of molecular microbiology methods to enhance understanding of microbial ecology in SFCWs.
- Integration of additives (biochar/activated carbon) in natural filter media to improve SFCW performance.
- Filter media optimisation to prevent rapid EC breakthrough during periods of biological instability.
- Hybrid SFCW designs to enhance nitrogen and micropollutant removal.
- Additional supply of artificial aeration to enhance pollutant removal efficiency of SFCWs.

2.7 CONCLUSIONS

Occurrence of ECs in treated wastewater discharged into natural waterbodies is increasingly recognised as an environmental issue. Several wastewater treatment technologies have been investigated to tackle this problem including nature-based strategies. Among them, SFCWs have intrigued many researchers, particularly for EC removal from wastewater. Following the reviews of several publications, SFCWs showed a remarkable potential not only because of their high EC removal efficiency, but also in terms of economic value and environmental sustainability. Biodegradation has been reported to be the most crucial mechanism for EC removal in SFCWs. VFBs showed better performance than HFBs, and the combination of both types (hybrid systems) showed the highest effectiveness. Final treatment by means of SFCWs would therefore be a promising solution for removing EC residual from wastewater. However, further research is still required to better understand the underlying mechanisms involved in the EC removal processes. Such knowledge would enable optimised SFCW treatment design for effective removal for the most critical EC residuals that pose significant threats to aquatic ecosystems.

REFERENCES

Abdelhakeem, S.G., Aboulroos, S.A. and Kamel, M.M. (2016) Performance of a vertical subsurface flow constructed wetland under different operational conditions. *Journal of Advanced Research*, 7(5), pp. 803–814.

Adrados, B., Sánchez, O., Arias, C.A., Becares, E., Garrido, L., Mas, J., Brix, H. and Morató, J. (2014) Microbial communities from different types of natural wastewater treatment systems: Vertical and horizontal flow constructed wetlands and biofilters. *Water Research*, 55, pp. 304–312.

Alvarino, T., Komesli, O., Suarez, S., Lema, J.M. and Omil, F. (2016) The potential of the innovative SeMPAC process for enhancing the removal of recalcitrant organic micropollutants. *Journal of Hazardous Materials*, 308, pp. 29–36.

Alvarino, T., Suarez, S., Lema, J. and Omil, F. (2018) Understanding the sorption and biotransformation of organic micropollutants in innovative biological wastewater treatment technologies. *Science of the Total Environment*, 615, pp. 297–306.

Austin, G. and Yu, K. (2016) *Constructed wetlands and sustainable development*. London; New York: Routledge, Taylor & Francis Group.

Ávila, C., Bayona, J.M., Martín, I., Salas, J.J. and García, J. (2015) Emerging organic contaminant removal in a full-scale hybrid constructed wetland system for wastewater treatment and reuse. *Ecological Engineering*, 80, pp. 108–116.

Baptista, J., Donnelly, T., Rayne, D. and Davenport, R. (2003) Microbial mechanisms of carbon removal in subsurface flow wetlands. *Water Science and Technology*, 48, pp. 127–134.

Bateganya, N.L., Kazibwe, A., Langergraber, G., Okot-Okumu, J. and Hein, T. (2016) Performance of subsurface flow constructed wetland mesocosms in enhancing nutrient removal from municipal wastewater in warm tropical environments. *Environmental Technology*, 37(8), pp. 960–974.

Brix, H., Koottatep, T., Fryd, O. and Laugesen, C.H. (2011) The flower and the butterfly constructed wetland system at Koh Phi Phi—System design and lessons learned during implementation and operation. *Ecological Engineering*, 37(5), pp. 729–735.

Buddhawong, S., Kuschk, P., Mattusch, J., Wiessner, A. and Stottmeister, U. (2005) Removal of Arsenic and Zinc Using Different Laboratory Model Wetland Systems. *Engineering in Life Sciences*, 5(3), pp. 247–252.

Cheng, G., Li, Q., Su, Z., Sheng, S. and Fu, J. (2018) Preparation, optimization, and application of sustainable ceramsite substrate from coal fly ash/waterworks sludge/oyster shell for phosphorus immobilization in constructed wetlands. *Journal of Cleaner Production*, 175, pp. 572–581.

Collado, N., Rodriguez-Mozaz, S., Gros, M., Rubirola, A., Barceló, D., Comas, J., Rodriguez-Roda, I. and Buttiglieri, G. (2014) Pharmaceuticals occurrence in a WWTP with significant industrial contribution and its input into the river system. *Environmental Pollution*, 185, pp. 202–212.

Crini, G. and Lichtfouse, E. (2019) Advantages and disadvantages of techniques used for wastewater treatment. *Environmental Chemistry Letters*, 17(1), pp. 145–155.

Crites, R.W., Middlebrooks, E.J., Bastian, R.K. and Reed, S.C. (2014) *Natural wastewater treatment systems*. Boca Raton, FL: CRC Press.

de Castro, L.V., Brandt, E.M.F., Campos, A.C.V., de Aquino, S.F., Werner, D., Afonso, R. J.d.C.F. and Mota Filho, C.R. (2018) Behavior of micropollutants in polishing units that combine sorption and biodegradation mechanisms to improve the quality of activated sludge effluent. *Water, Air, & Soil Pollution*, 229(6), p. 189.

De Feo, G. and Ferrara, C. (2017) A procedure for evaluating the most environmentally sound alternative between two on-site small-scale wastewater treatment systems. *Journal of Cleaner Production*, 164, pp. 124–136.

Del Rosario, K.L., Mitra, S., Humphrey, C.P. and O'Driscoll, M.A. (2014) Detection of pharmaceuticals and other personal care products in groundwater beneath and adjacent to onsite wastewater treatment systems in a coastal plain shallow aquifer. *Science of the Total Environment*, 487, pp. 216–223.

Dotro, G., Langergraber, G., Molle, P., Nivala, J., Puigagut, J., Stein, O. and von Sperling, M. (2017) *Treatment wetlands*. IWA Publishing.

Dubey, M., Mohapatra, S., Tyagi, V.K., Suthar, S. and Kazmi, A.A., 2021. Occurrence, fate, and persistence of emerging micropollutants in sewage sludge treatment. *Environmental Pollution*, 273, p.116515.

El-Gendy, A., El Shafie, A., Issa, A., Al-Meer, S., Al-Saad, K. and El-Azazy, M. (2020) 'Carbon-based materials (CBMs) for determination and remediation of antimicrobials in different substrates: Wastewater and infant foods as examples', in Bartoli, M., Frediani, M., Rosi, L. (eds.). *Carbon-based material for environmental protection and remediation*, London: IntechOpen, pp. 103–122.

Eregowda T., Mohapatra S. (2020) 'Fate of micropollutants in engineered and natural environment', in Kumar M., Munoz-Arriola F., Furumai H., Chaminda T. (eds) *Resilience, response, and risk in water systems. Springer transactions in civil and environmental engineering.* Springer, Singapore. https://doi.org/10.1007/978-981-15-4668-6_15

Fernandez-Fontaina, E., Omil, F., Lema, J.M. and Carballa, M. (2012) Influence of nitrifying conditions on the biodegradation and sorption of emerging micropollutants. *Water Research*, 46(16), pp. 5434–5444.

Geissen, V., Mol, H., Klumpp, E., Umlauf, G., Nadal, M., van der Ploeg, M., van de Zee, S.E.A.T.M. and Ritsema, C.J. (2015) Emerging pollutants in the environment: A challenge for water resource management. *International Soil and Water Conservation Research*, 3(1), pp. 57–65.

Ghosh, U., Luthy, R.G., Cornelissen, G., Werner, D. and Menzie, C.A. (2011) In-situ sorbent amendments: A new direction in contaminated sediment management. *Environmental Science & Technology*, 45(4), pp. 1163–1168.

Guan, W., Yin, M., He, T. and Xie, S. (2015) Influence of substrate type on microbial community structure in vertical-flow constructed wetland treating polluted river water. *Environmental Science and Pollution Research*, 22(20), pp. 16202–16209.

Gupta, P., Ann, T.-w. and Lee, S.-M. (2016) Use of biochar to enhance constructed wetland performance in wastewater reclamation. *Environmental Engineering Research*, 21(1), pp. 36–44.

Guven, H., Ozgun, H., Ersahin, M.E., Dereli, R.K., Sinop, I. and Ozturk, I. (2019) High-rate activated sludge processes for municipal wastewater treatment: the effect of food waste addition and hydraulic limits of the system. *Environmental Science and Pollution Research*, 26(2), pp. 1770–1780.

Hijosa-Valsero, M., Matamoros, V., Martín-Villacorta, J., Bécares, E. and Bayona, J.M. (2010) Assessment of full-scale natural systems for the removal of PPCPs from wastewater in small communities. *Water Research*, 44(5), pp. 1429–1439.

Hoffmann, H., Platzer, C., Winker, M. and von Muench, E. (2011) *Technology review of constructed wetlands: Subsurface flow constructed wetlands for greywater and domestic wastewater treatment*. Deutsche Gesellschaft für Internationale Zusammenarbeit (GIZ) GmbH, Eschborn, Germany, pp. 1–35.

Hreiz, R., Latifi, M.A. and Roche, N. (2015) Optimal design and operation of activated sludge processes: State-of-the-art. *Chemical Engineering Journal*, 281, pp. 900–920.

Iasur-Kruh, L., Hadar, Y., Milstein, D., Gasith, A. and Minz, D. (2010) Microbial population and activity in wetland microcosms constructed for improving treated municipal wastewater. *Microbial Ecology*, 59(4), pp. 700–709.

Ilyas, H. and van Hullebusch, E.D. (2020) Performance comparison of different types of constructed wetlands for the removal of pharmaceuticals and their transformation products: A review. *Environmental Science and Pollution Research*, 27(13), pp. 14342–14364.

Islam, M.A., Amin, M.S.A. and Hoinkis, J. (2013) Optimal design of an activated sludge plant: theoretical analysis. *Applied Water Science*, 3(2), pp. 375–386.

Ji, M., Hu, Z., Hou, C., Liu, H., Ngo, H.H., Guo, W., Lu, S. and Zhang, J. (2020) New insights for enhancing the performance of constructed wetlands at low temperatures. *Bioresource Technology*, 301, p. 122722.

Kadlec, R.H. (2009) Comparison of free water and horizontal subsurface treatment wetlands. *Ecological Engineering*, 35(2), pp. 159–174.

Khalifa, M.E., El-Reash, Y.G.A., Ahmed, M.I. and Rizk, F.W. (2020) Effect of media variation on the removal efficiency of pollutants from domestic wastewater in constructed wetland systems. *Ecological Engineering*, 143, p. 105668.

Khuntia, H. K., Mohapatra, S., Sukhani, S., Chanakya, H. N., & Gin, K. Y. H. (2021). Editor: Maulin P Shah. Anaerobic Membrane Bioreactors for Industrial Wastewater Treatment. In *Biological Treatment of Industrial Wastewater*, UK: RSC, pp. 294–327.

Khuntia, H. K., Mohapatra, S., Sukhani, S., & Chanakya, H. N. (2022). Editors: Maulin Shah, Susana Rodriguez-Couto, Jayanta Biswas. Role of biofiltration in the treatment of sewage. In *An Innovative Role of Biofiltration in Wastewater Treatment Plants (WWTPs)*, Netherlands: Elsevier, pp. 411–438.

Li, Y., Zhu, G., Ng, W.J. and Tan, S.K. (2014) A review on removing pharmaceutical contaminants from wastewater by constructed wetlands: Design, performance and mechanism. *Science of the Total Environment*, 468–469, pp. 908–932.

Mailler, R., Gasperi, J., Coquet, Y., Derome, C., Buleté, A., Vulliet, E., Bressy, A., Varrault, G., Chebbo, G. and Rocher, V. (2016) Removal of emerging micropollutants from wastewater by activated carbon adsorption: Experimental study of different activated carbons and factors influencing the adsorption of micropollutants in wastewater. *Journal of Environmental Chemical Engineering*, 4(1), pp. 1102–1109.

Mara, D.D. (2004) *Domestic wastewater treatment in developing countries*. London; Sterling, VA: Earthscan Publications.

Marín-Muñiz, J.L., García-González, M.C., Ruelas-Monjardín, L.C. and Moreno-Casasola, P. (2017) Influence of different porous media and ornamental vegetation on wastewater pollutant removal in vertical subsurface flow wetland microcosms. *Environmental Engineering Science*, 35(2), pp. 88–94.

Marín-Muñiz, J.L., Hernández, M.E., Gallegos-Pérez, M.P. and Amaya-Tejeda, S.I. (2020) Plant growth and pollutant removal from wastewater in domiciliary constructed wetland microcosms with monoculture and polyculture of tropical ornamental plants. *Ecological Engineering*, 147, p. 105658.

Mburu, N., Tebitendwa, S.M., van Bruggen, J.J.A., Rousseau, D.P.L. and Lens, P.N.L. (2013) Performance comparison and economics analysis of waste stabilization ponds and horizontal subsurface flow constructed wetlands treating domestic wastewater: A case study of the Juja sewage treatment works. *Journal of Environmental Management*, 128, pp. 220–225.

McLain, N.K. and Gachomo, E.W. (2019) Chemicals of emerging concern in treated wastewater impact microbial growth. *Frontiers in Environmental Science*, 7, p. 156.

Meerburg, F.A., Boon, N., Van Winckel, T., Vercamer, J.A.R., Nopens, I. and Vlaeminck, S.E. (2015) Toward energy-neutral wastewater treatment: A high-rate contact stabilization process to maximally recover sewage organics. *Bioresource Technology*, 179, pp. 373–381.

Melián, J.A.H., Rodríguez, A.J.M., Araña, J., Díaz, O.G. and Henríquez, J.J.G. (2010) Hybrid constructed wetlands for wastewater treatment and reuse in the Canary Islands. *Ecological Engineering*, 36(7), pp. 891–899.

Meng, P., Pei, H., Hu, W., Shao, Y. and Li, Z. (2014) How to increase microbial degradation in constructed wetlands: Influencing factors and improvement measures. *Bioresource Technology*, 157, pp. 316–326.

Mogili, N.V. and Mohapatra, S. (2021) Microbial community analysis of wastewater during blackwater treatment. In *Wastewater Treatment Reactors*, Netherlands: Elsevier, pp. 483–507.

Mohan, D., Sarswat, A., Ok, Y.S. and Pittman, C.U. (2014) Organic and inorganic contaminants removal from water with biochar, a renewable, low cost and sustainable adsorbent – A critical review. *Bioresource Technology*, 160, pp. 191–202.

Mohapatra, S., Sharma, N., Mohapatra, G., Padhye, L.P. and Mukherji, S., 2021. Seasonal variation in fluorescence characteristics of dissolved organic matter in wastewater and identification of proteins through HRLC-MS/MS. *Journal of Hazardous Materials*, 413, p. 125453.

Nasr, F.A. and Mikhaeil, B. (2015) Treatment of domestic wastewater using modified septic tank. *Desalination and Water Treatment*, 56(8), pp. 2073–2081.

Nassar, H. and Younis, S. (2019) 'From priority contaminants to emerged threat: Risk and occurrence-based analysis for better water management strategies in present and future', in: Fosso-Kankeu, E. (ed.) *Nano and bio-based technologies for wastewater treatment: Prediction and control tools for the dispersion of pollutants in the environment*. Hoboken, NJ: John Wiley & Sons Inc, pp. 41–103.

Nuamah, L.A., Li, Y., Pu, Y., Nwankwegu, A.S., Haikuo, Z., Norgbey, E., Banahene, P. and Bofah-Buoh, R. (2020) Constructed wetlands, status, progress, and challenges. The need for critical operational reassessment for a cleaner productive ecosystem. *Journal of Cleaner Production*, 269, p. 122340.

Opher, T. and Friedler, E. (2016) Comparative LCA of decentralized wastewater treatment alternatives for non-potable urban reuse. *Journal of Environmental Management*, 182, pp. 464–476.

Paranychianakis, N.V., Tsiknia, M. and Kalogerakis, N. (2016) Pathways regulating the removal of nitrogen in planted and unplanted subsurface flow constructed wetlands. *Water Research*, 102, pp. 321–329.

Pei, M., Zhang, B., He, Y., Su, J., Gin, K., Lev, O., Shen, G. and Hu, S. (2019) State of the art of tertiary treatment technologies for controlling antibiotic resistance in wastewater treatment plants. *Environment International*, 131, p. 105026.

Phillips, P.J., Schubert, C., Argue, D., Fisher, I., Furlong, E.T., Foreman, W., Gray, J. and Chalmers, A. (2015) Concentrations of hormones, pharmaceuticals and other micropollutants in groundwater affected by septic systems in New England and New York. *Science of the Total Environment*, 512–513, pp. 43–54.

Priyanka M.S., Mohapatra S. (2020) 'Shifts and Trends in Analysis of Contaminants of Emerging Concern: Sulfonamides', in Kumar, M., Munoz-Arriola, F., Furumai, H., Chaminda, T. (eds) *Resilience, Response, and Risk in Water Systems*. Singapore: Springer Transactions in Civil and Environmental Engineering: Springer.

Rajan, R.J., Sudarsan, J.S. and Nithiyanantham, S. (2019) Microbial population dynamics in constructed wetlands: Review of recent advancements for wastewater treatment. *Environmental Engineering Research*, 24(2), pp. 181–190.

Rattier, M., Reungoat, J., Keller, J. and Gernjak, W. (2014) Removal of micropollutants during tertiary wastewater treatment by biofiltration: Role of nitrifiers and removal mechanisms. *Water Research*, 54, pp. 89–99.

Rout, P.R., Zhang, T.C., Bhunia, P. and Surampalli, R.Y. (2020) Treatment technologies for emerging contaminants in wastewater treatment plants: A review. *Science of the Total Environment*, 753, p. 141990.

Sánchez, O. (2017) Constructed wetlands revisited: Microbial diversity in the –omics era. *Microbial Ecology*, 73(3), pp. 722–733.

Schaider, L.A., Rodgers, K.M. and Rudel, R.A. (2017) Review of organic wastewater compound concentrations and removal in onsite wastewater treatment systems. *Environmental Science & Technology*, 51(13), pp. 7304–7317.

Shelef, O., Gross, A. and Rachmilevitch, S. (2013) Role of plants in a constructed wetland: current and new perspectives. *Water*, 5(2), pp. 405–419.

Shingare, R.P., Thawale, P.R., Raghunathan, K., Mishra, A. and Kumar, S. (2019) Constructed wetland for wastewater reuse: Role and efficiency in removing enteric pathogens. *Journal of Environmental Management*, 246, pp. 444–461.

Sidrach-Cardona, R., Sánchez, O., Garrido, L., Mas, J. and Becares, E. (2015) 'Molecular characterization of microbial communities in constructed wetlands: The effect of plant species, organic matter and hydraulic design,' in Barrett, LM. (ed.) *Wastewater treatment: Processes, management strategies and environmental/health impacts*. Hauppauge, NY: Nova Science Publishers, Inc., pp. 45–67.

Singh, R.P., Kun, W. and Fu, D. (2019) Designing process and operational effect of modified septic tank for the pre-treatment of rural domestic sewage. *Journal of Environmental Management*, 251, p. 109552.

Sperling, M.V. and de Lemos Chernicharo, C.A. (2005) *Biological wastewater treatment in warm climate regions. Volume one*. London: IWA Publishing.

Stefanakis, A. (2016) 'Constructed wetlands: Description and benefits of an eco-tech water treatment system', in McKeown, A.E. and George, B. (eds.) *Impact of water pollution on human health and environmental sustainability*. Hershey, PA: IGI Global, pp. 281–303.

Stefanakis, A. and Becker, J.A. (2015) 'A review of emerging contaminants in water: Classification, sources and potential risks', in McKeown, E.A., and Bugyi, G. (eds.) *Impact of water pollution on human health and environmental sustainability*, Hershey, PA: IGI Global, pp. 57–82.

Surinkul, N. and Koottatep, T. (2009) Advanced sanitation planning tool with health risk assessment: Case study of a peri-urban community in Thailand. *Human and Ecological Risk Assessment: An International Journal*, 15(5), pp. 1064–1077.

Tanner, C.C. (2001) Plants as ecosystem engineers in subsurface-flow treatment wetlands. *Water Science and Technology*, 44(11–12), pp. 9–17.

Tabraiz, S., Petropoulos, E., Shamurad, B., Quintela-Baluja, M., Mohapatra, S., Acharya, K., ... & Sallis, P. J. (2021). Temperature and immigration effects on quorum sensing in the biofilms of anaerobic membrane bioreactors. *Journal of Environmental Management*, 293, pp. 112947.

Tewari, S., Jindal, R., Kho, Y.L., Eo, S. and Choi, K. (2013) Major pharmaceutical residues in wastewater treatment plants and receiving waters in Bangkok, Thailand, and associated ecological risks. *Chemosphere*, 91(5), pp. 697–704.

Ting, S.C., Ismail, A.R. and Malek, M.A. (2013) Development of effluent removal prediction model efficiency in septic sludge treatment plant through clonal selection algorithm. *Journal of Environmental Management*, 129, pp. 260–265.

Tran, N.H., Chen, H., Reinhard, M., Mao, F. and Gin, K.Y.-H. (2016) Occurrence and removal of multiple classes of antibiotics and antimicrobial agents in biological wastewater treatment processes. *Water Research*, 104, pp. 461–472.

Tran, N.H., Reinhard, M. and Gin, K.Y.-H. (2018) Occurrence and fate of emerging contaminants in municipal wastewater treatment plants from different geographical regions-a review. *Water Research*, 133, pp. 182–207.

Tsuzuki, Y., Koottatep, T., Jiawkok, S. and Saengpeng, S. (2010) Municipal wastewater characteristics in Thailand and effects of 'soft intervention' measures in households on pollutant discharge reduction. *Water Science and Technology*, 62, pp. 231–244.

Ullah, A., Hussain, S., Wasim, A. and Jahanzaib, M. (2020) Development of a decision support system for the selection of wastewater treatment technologies. *Science of the Total Environment*, 731, p. 139158.

Ulrich, B.A., Im, E.A., Werner, D. and Higgins, C.P. (2015) Biochar and activated carbon for enhanced trace organic contaminant retention in stormwater infiltration systems. *Environmental Science & Technology*, 49(10), pp. 6222–6230.

Varma, M., Gupta, A.K., Ghosal, P.S. and Majumder, A. (2021) A review on performance of constructed wetlands in tropical and cold climate: Insights of mechanism, role of influencing factors, and system modification in low temperature. *Science of the Total Environment*, 755, p. 142540.

Verbyla, M.E. (2017) *Ponds, lagoons, and wetlands for wastewater management*. New York, NY: Momentum Press.

Verlicchi, P. and Zambello, E. (2014) How efficient are constructed wetlands in removing pharmaceuticals from untreated and treated urban wastewaters? A review. *Science of the Total Environment*, 470–471, pp. 1281–1306.

Vymazal, J. (1998) *Constructed wetlands for wastewater treatment in Europe*. Leiden: Backhuys.

Vymazal, J. (2013) The use of hybrid constructed wetlands for wastewater treatment with special attention to nitrogen removal: A review of a recent development. *Water Research*, 47(14), pp. 4795–4811.

Vymazal, J. and Kropfelova, L. (2008) *Wastewater treatment in constructed wetlands with horizontal sub-surface flow*. Dordrecht: Springer.

Wang, C., Zheng, S.-S., Wang, P.-F. and Qian, J. (2014) Effects of vegetations on the removal of contaminants in aquatic environments: A review. *Journal of Hydrodynamics*, 26(4), pp. 497–511.

Wang, J., Tian, Z., Huo, Y., Yang, M., Zheng, X. and Zhang, Y. (2018) Monitoring of 943 organic micropollutants in wastewater from municipal wastewater treatment plants with secondary and advanced treatment processes. *Journal of Environmental Sciences*, 67, pp. 309–317.

Wang, J. and Wang, S. (2016) Removal of pharmaceuticals and personal care products (PPCPs) from wastewater: A review. *Journal of Environmental Management*, 182, pp. 620–640.

Wang, M., Zhang, D.Q., Dong, J.W. and Tan, S.K. (2017) Constructed wetlands for wastewater treatment in cold climate—A review. *Journal of Environmental Sciences*, 57, pp. 293–311.

Wang, Q., Xie, H., Ngo, H.H., Guo, W., Zhang, J., Liu, C., Liang, S., Hu, Z., Yang, Z. and Zhao, C. (2016) Microbial abundance and community in subsurface flow constructed wetland microcosms: Role of plant presence. *Environmental Science and Pollution Research*, 23(5), pp. 4036–4045.

Wang, Y., Cai, Z., Sheng, S., Pan, F., Chen, F. and Fu, J. (2020) Comprehensive evaluation of substrate materials for contaminants removal in constructed wetlands. *Science of the Total Environment*, 701, p. 134736.

WHO (2018) *A global overview of national regulations and standards for drinking-water quality*. Available at: https://apps.who.int/iris/handle/10665/272345 (Accessed: 3 March 2021).

Withers, P.J.A., Jarvie, H.P. and Stoate, C. (2011) Quantifying the impact of septic tank systems on eutrophication risk in rural headwaters. *Environment International*, 37(3), pp. 644–653.

Wu, H., Zhang, J., Ngo, H.H., Guo, W., Hu, Z., Liang, S., Fan, J. and Liu, H. (2015) A review on the sustainability of constructed wetlands for wastewater treatment: Design and operation. *Bioresource Technology*, 175, pp. 594–601.

Yang, Y., Ok, Y.S., Kim, K.-H., Kwon, E.E. and Tsang, Y.F. (2017b) Occurrences and removal of pharmaceuticals and personal care products (PPCPs) in drinking water and water/sewage treatment plants: A review. *Science of the Total Environment*, 596–597, pp. 303–320.

Yang, Y., Zhao, Y., Liu, R. and Morgan, D. (2018) Global development of various emerged substrates utilized in constructed wetlands. *Bioresource Technology*, 261, pp. 441–452.

Yang, Y.-Y., Toor, G.S., Wilson, P.C. and Williams, C.F. (2017a) Micropollutants in groundwater from septic systems: Transformations, transport mechanisms, and human health risk assessment. *Water Research*, 123, pp. 258–267.

Yates, C.A., Johnes, P.J. and Spencer, R.G.M. (2019) Characterisation of treated effluent from four commonly employed wastewater treatment facilities: A UK case study. *Journal of Environmental Management*, 232, pp. 919–927.

Zawadzka, J., Gallagher, E., Smith, H. and Corstanje, R. (2019) Ecosystem services from combined natural and engineered water and wastewater treatment systems: Going beyond water quality enhancement. *Ecological Engineering*, 2, p. 100006.

Zhang, D.-Q., Jinadasa, K.B.S.N., Gersberg, R.M., Liu, Y., Tan, S.K. and Ng, W.J. (2015) Application of constructed wetlands for wastewater treatment in tropical and subtropical regions (2000–2013). *Journal of Environmental Sciences*, 30, pp. 30–46.
Zhang, D.Q., Tan, S.K., Gersberg, R.M., Sadreddini, S., Zhu, J. and Tuan, N.A. (2011) Removal of pharmaceutical compounds in tropical constructed wetlands. *Ecological Engineering*, 37(3), pp. 460–464.
Zou, J., Guo, X., Han, Y., Liu, J. and Liang, H. (2012) Study of a novel vertical flow constructed wetland system with drop aeration for rural wastewater treatment. *Water, Air, & Soil Pollution*, 223(2), pp. 889–900.

Chapter 3

Occurrence and fate of CECs transformation products

Nayanathara O S and Aswathy E V
National Institute of Technology Calicut

CONTENTS

3.1 OVERVIEW OF CHEMICALS OF EMERGING CONCERN

The wastewater originating from various domestic, industrial, and agricultural sources contains chemical compounds that are not commonly monitored in the environment (Ramírez-Malule et al., 2020). These chemical compounds, which include pharmaceutically active compounds, personal care products (PPCs), pesticides, and many other synthetic organic compounds, have gained increased interest among the scientific community in recent times. These compounds are collectively termed as "emerging contaminants (ECs)" or "emerging pollutants (EPs)" (Wilkinson et al., 2017). These ECs can interfere with the endocrine systems and are also known as endocrine disruptors (Bolong et al., 2009). ECs can be broadly classified into pharmaceutical and personal care products (PPCPs), agrochemicals (pesticides, growth agent and fertilizer), and industrial compounds (Volatile Organic Compounds (VOCs), corrosion inhibitors, flame retarders, plasticizers, antioxidants) (Menon et al., 2020; Mohapatra et al., 2018, 2021b; Priyanka and Mohapatra, 2020).

In India, pharmaceutical products are given more importance among different ECs as India ranks top among countries for pharmaceutical production. They are used for the treatment of both humans and animals. These pharmaceutical compounds reach soil and water bodies through different pathways. Gani et al. (2017) studied

DOI: 10.1201/9781003247869-4

the contamination of ECs in Indian aquatic sources. By reviewing 41 publications, they found that pesticides constitute the largest EC in India (around 57%), followed by pharmaceuticals (17%). Only a few studies were conducted regarding the presence of pharmaceuticals in Indian aquatic sources. Certain compounds like cetirizine, ciproflaxin, ampicillin, etc. were detected at a higher concentration in Indian aquatic sources. Balakrishna et al. 2017 studied the presence of PPCPs in Indian water bodies. In this paper, around 19 reports on the occurrence of PPCPs in India were reviewed. They found that some compounds are present at a very high concentration when compared to other countries in Europe, Asia, Australia, and North America. The concentration of ciprofloxacin and sulfamethoxazole present in the hospital effluent from India is around ten times higher than that present in Portugal and Italy. They also found that the concentration of certain pharmaceutical compounds is more in India when compared to China. There are several other studies conducted in India on the presence of pharmaceutical compounds in surface water bodies.

Wastewater treatment plants, sewage treatment plants, and landfill leaching form the major source of PPCPs to the environment (Ebele et al., 2017; Eregowda and Mohapatra, 2020). Due to incomplete removal during the treatment process, their concentration varied in the range of several μg/L in reclaimed surface water (Chen et al., 2013). PPCPs administered by humans comes out of their body via excretion and are discharged into sewage treatment plants or septic tanks. The treated water from STPs is used for irrigation, and the sludge generated is used as fertilizers (Dubey et al., 2021; Yang and Toor, 2015). They ultimately reach groundwater or surface water resulting in its contamination. Veterinary drugs, which are used for the prevention of diseases in animals, find their way to the environment through manure or as their excreta. Manure from livestock that contains unmetabolized pharmaceutical compounds is used for agricultural purposes. They, in turn, reach groundwater and surface water via leaching or runoff (La Farré et al., 2008). PPCs include sunscreen, soap, lotions, toothpaste, etc. (Peck, 2006). They reach the water treatment plant through bathing, shower waste, washing sinks, and swimming, and ultimately, they are discharged into the environment (Ebele et al., 2017). Agrochemicals are mainly used for increasing crop yield and reducing pests. They include pesticides, fertilizers, insecticides, and growing agents. These agrochemicals remain in the soil for a longer time and get added up during each time application (Rajput et al., 2020). They find their way to the water body through different pathways. Leaching, runoff, and infiltration are the different ways by which agrochemicals reach the water body. Surface runoff of salts results in salinization of water bodies. Runoff of pesticides and fertilizers reaches surface water, and it bioaccumulates in aquatic organisms affecting aquatic life severely. Other agricultural activities like silviculture and aquaculture also affect the quality of ground and surface water (Thatai et al., 2018, Mazumder et al., 2020). In general, the major problems associated with the increase in the use of agrochemicals are water and soil pollution, ecological instability, development of strains resistant to pests and weeds, and adverse health impact on human and aquatic life (Majeed, 2018).

Plasticizers, flame retardants, polyfluorinated compounds, VOCs, antifouling compounds, corrosion inhibitors, antioxidants, etc. are some of the relevant industrial chemicals that are considered as ECs (Calvo-flores, 2017, Gahlot et al., 2020, Kumar et al., 2020). The effluent containing industrial compounds reaches the treatment plant. However, complete removal does not take place, and ultimately, they reach soil and groundwater. The major source by which industrial compounds reach the environment is from the discharge of wastewater from the production unit of industrial compounds. This wastewater passes through the water treatment plant. However, due to incomplete removal of these contaminants using existing treatment facilities, they ultimately reach surface water resulting in its contamination. The sludge from the treatment plant which also carries ECs are used as manure for agricultural purpose. They also find their way to the food chain or surface water or groundwater via consumption of agricultural products or runoff or leaching, respectively.

Exposure to ECs can lead to chronic and acute illness, which can even lead to death. It affects both aquatic and human health (Rajput et al., 2020). ECs do not have any regulatory status, and their impact on the environment and human health is to be explored (Deblonde et al., 2011). Table 3.1 summarizes the chronic health effect of agrochemicals, industrial compounds, and PPCPs.

3.1.1 Fate of emerging contaminants and formation of transformation products (TPs)

The conventional secondary treatment process used widely in wastewater treatment systems is the trickling filter and activated sludge process. However, these are not designed for the removal of ECs. Hence, they ultimately reach the environment (Petrie et al., 2015). With recent scientific advancement in the analysis and detection of trace compounds, the presence of ECs has been identified and quantified in diverse environments like sewage effluent, surface, and groundwater, drinking water, river sediments, etc. Most of the ECs and their metabolites are found to be non-biodegradable and hence bioaccumulates in the environment. This ultimately enters into the food web (Rodriguez-Narvaez et al., 2017). So, it is essential to identify the fate of these ECs reaching the environment. Generally, the different methods of transformation of ECs reaching the aquatic environment are sorption to microplastic, sorption to sediments, sorption to suspended particulate matter, bioaccumulation, and formation of transformation products (TPs). The future study mainly aims at the formation of TPs from ECs as a large number of unknown TPs are formed even at the time of wastewater treatment, biological degradation, and natural photochemical transformation (Wilkinson et al., 2017). In some cases, the TPs formed can be more toxic and persistent than the parent compound (La Farré et al., 2008). While the knowledge on the occurrence and fate of contaminants of emerging concern (CECs) is increasing, the potential risk caused by their TPs is rarely investigated. Hence, the objective of this chapter is to have a better understanding on the occurrence and fate of CEC TPs and to assess their impacts on human health and the environment.

3.2 PATHWAYS FOR THE FORMATION OF TPs

In order to understand the adverse effects caused by the TPs on human health and the environment, the first step is to identify the different mechanisms from which they are formed. A large number of TPs are formed as a result of biotic and abiotic processes. The biotic process includes degradation using microorganisms, which can be both aerobic and anaerobic. The abiotic transformation includes processes like hydrolysis, photolysis, sorption, chlorine disinfection, advanced oxidation process, etc. (Yin et al., 2017). Figure 3.1 shows the different mechanisms which form TPs from their parent compound.

Table 3.1 Health impacts of different ECs

Emerging contaminants	*Classification*	*Compounds*	*Health impact*	*References*
PPCPs	Steroidal	Estrogen (Estrone, 17β estradiol), progesterone, tamoxifen, testosterone, phytosterols	Infertility, sexual underdevelopment, altered or reduced sexual behavior, attention deficit or hyperactivity, altered thyroid or adrenal cortical function, increased incidents of certain cancers, birth defects	Caliman and Gavrilescu (2009)
	Personal care products	Disinfectants, fragrances (musk xylol, tonalide, galaxolide), UV screens (homosalate, benzophenone-3)		
	Non-steroidal	Antibiotics (penicillin, amoxycillin, tetracycline), analgesics (paracetamol), anti-inflammatories (ibuprofen, naproxen, diclofenac)		
Agrochemicals	Fertilizer	Zuari fertilizer, Bayer, Chambal	Irritation of skin and eyes, nervous system depression, dry and cracked hands, ulceration of the skin	Rajput et al. (2020)
	Insecticide	Orthene, dursban, thiodan, cythion	Seizures, pulmonary edema, muscle twitching, nausea	
	Pesticide	Orthocide, daconil, terraclor, cumin, vancide		
Industrial compounds	Antioxidants	2,6-Di-*tert*-butylphenol	Tumors, neurodevelopmental toxicity, endocrine disruption, thyroid hormone imbalance, reproductive problems. In animal toxicity, it causes hepatotoxicity, embryotoxicity, and thyroid effects as well as maternal toxicity during gestation	Calvo-flores (2017)
	Plasticizers	Bisphenol A, phthalates		
	Flame retardants	PBDE, tris (2-chloroethyl) phosphate		
	Corrosion inhibitors	1,2,3-Benzotriazole, benzothiazole 2-sulfonoc acid		
	VOCs	Benzene, CCl4, styrene, xylenes		

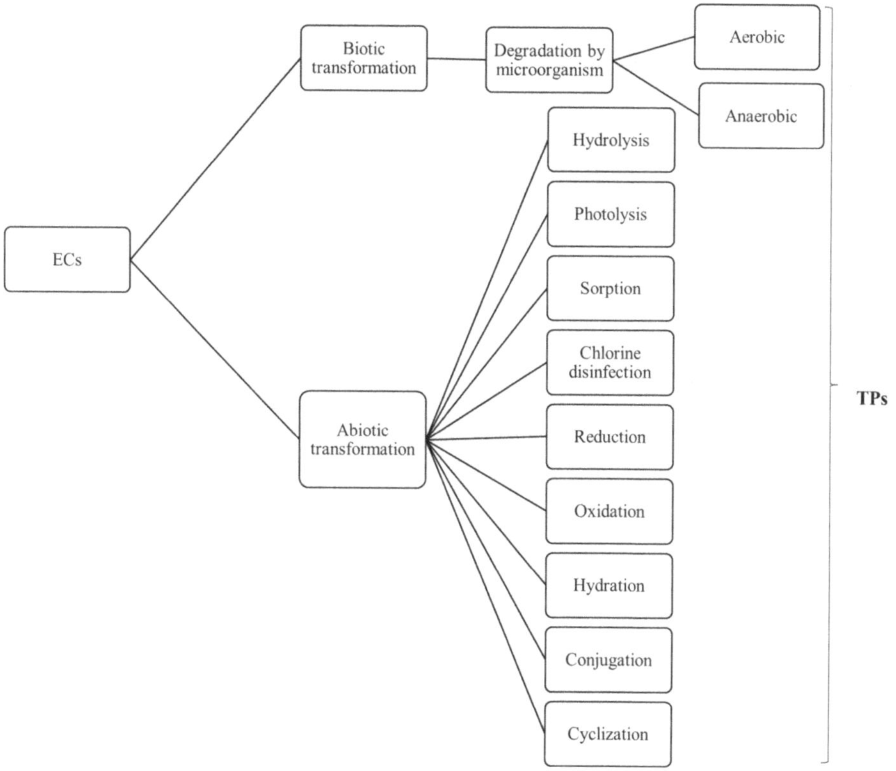

Figure 3.1 Different mechanisms for the formation of TPs from the parent compound. (From Yin et al., 2017.)

3.2.1 Biotic transformation

The biotic transformation uses microbial populations like bacteria and fungi. Estrone, 17-β estradiol-17-glucuronide, and 17-β estradiol-13-glucuronide are the biotransformation products of 17-β estradiol (Ternes et al., 1999). Certain enzymes also catalyze the transformation of ECs. During water treatment, horseradish peroxidase, a commercially available enzyme, was found to transform tetracycline (TC), an antibiotic widely used in humans and animals. The biological toxicity of TC was found to reduce during this transformation (Leng et al., 2020). Triphenyl phosphate (TPHP) is one of the widely used organophosphate fire retardants. Hence, their concentration in the aquatic environment is increasing gradually. It undergoes biotransformation in *Daphnia magna*, a standard species used for studying toxicity. A total of nine biotransformation products are formed from TPHP. Five of them exhibited high toxicity in aquatic organisms (Choi et al., 2020). Dydrogesterone is a synthetic progestogenic hormone. Its biotransformation using *Macrophomina phaseolina*, a plant fungus yielded 3β,11α-dihydroxy-5β,9β,10α- pregna-7-ene-6,20-dione, 8α-hydroxy-9β,10α-pregna4,6-diene-3,20-dione and 15β-hydroxy-9β,10α-pregna-4,6-diene-3,20-dione (Wajid et al., 2019). However, the formation of TPs through biotic transformation is limited because of the bio persistence of ECs, which are organic in nature.

3.2.2 Abiotic processes

Abiotic processes make use of various physical and chemical factors. This includes processes like hydrolysis, photolysis, sorption, oxidation, reduction, cyclization, conjugation, hydration, etc. Abiotic processes are responsible for the formation of TPs from PPCPs, pesticides, and industrial compounds.

Degradation of parent compounds using sunlight is known as photodegradation or photolysis. It can be direct or indirect. In direct photodegradation, the compound absorbs sunlight directly, resulting in the breakup of a molecule. Whereas in indirect photolysis, the active compounds present in water (nitrate ion) absorbs the energy and forms certain reactive species like hydroxyl radical, singlet oxygen, etc. These active species help in the formation of TPs (Fatta-Kassinos et al., 2011). Difloxacin, a veterinary drug under photolysis, produces seven TPs (Kusari et al., 2009). Figure 3.2 shows some of the photolytic TPs formed from difloxacin.

Quinmerac, an herbicide, forms 7-chloro-3-methyl quinoline as TP under photolysis (Agüera López et al., 2014). Hydrolysis is an important pathway for the degradation of veterinary drugs. In hydrolysis, the formation of TP takes place as a result of the reaction of the parent compound with water. Monensin is one of the commonly used veterinary drugs. Its transformation, via electrochemical oxidation, hydrolysis, and metabolism with microsomes yielded twelve different TPs (Kotthoff et al., 2019). Hydrolysis is also one of the primary paths for the degradation of the pesticide molecule (Coats, 1991). Diaminozide is a ripening/plant growth regulator that is widely used. It undergoes transformation to 1,1-dimethylhydrazine via hydrolysis. This TP formed is carcinogenic in nature (Brown and Casida, 1988).

Another pathway for the formation of TPs is the advanced oxidation process. In this method, the transformation of the parent compound occurs with the help of hydroxyl radical. It comprises different techniques like Fenton, TiO_2 photocatalysis, photo-Fenton, etc. Paracetamol, naproxen, diclofenac (DCF), carbamazepine (CBZ), sulfathiazole are some of the pharmaceutical compounds that are degraded by advanced oxidation process (Kosjek and Heath, 2008). TPs are also formed as a result of the sorption of soil particles. Soils that are having high organic matter exhibits sorption that is nonlinear and stronger. Ibuprofen (IBP) and naproxen (NAP) are two anti-inflammatory drugs that undergo sorption onto soil rich in organic matter. This resulted in the formation of two soil TPs: IBP-T1 and NAP-T2 (Vulava et al., 2016).

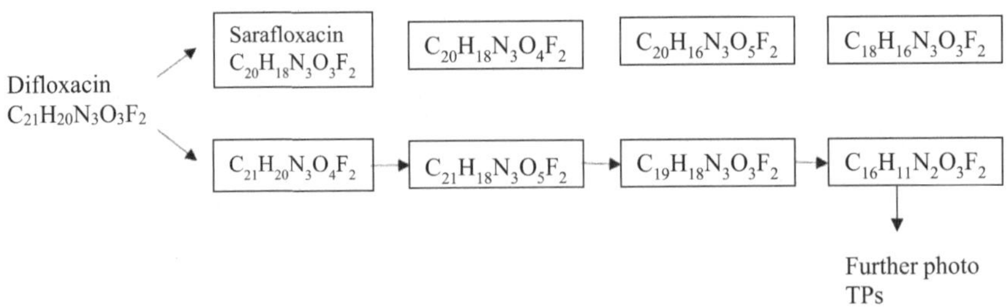

Figure 3.2 Photolytic TPs formed from difloxacin. (From Kusari et al., 2009.)

Chlorination is one of the most commonly adopted disinfection methods in water treatment plants. Ciprofloxacin, ofloxacin, and levofloxacin are three antibiotic compounds coming under the group fluoroquinolone. During chlorination, these three antibiotics form TPs, which are anti-bacterial in nature. They need to be carefully examined as these TPs in the environment are sources of antibiotic resistance (Kennedy Neth et al., 2019). Hydroxylation, dehalogenation, isomerization, dealkylation, dimerization, cyclization, ring-opening, and decarboxylation are the most commonly observed pathway for the formation of TPs of pharmaceuticals (Michael et al., 2014), whereas hydration, conjugation, oxidation, reduction, hydrolysis, cyclization, and isomerization are the most commonly observed mechanisms for pesticides (Coats, 1991).

3.3 METHODS TO IDENTIFY TPs

It is difficult to identify the TPs formed from the parent compound since they are present in minor fractions. However, some methods are developed to identify unknown compounds in complex matrices at low concentrations. One method is by using liquid chromatography (LC)-high-resolution mass spectroscopy (HRMS). It helps in determining the molecular formula of TPs formed from parent compounds (Krauss et al., 2010). This method comes under non-target screening, where no prior information on the TPs formed is available. Screening based on suspect chemical lists is another method for identifying TP. This comes under suspect screening. High-performance liquid chromatography-quadrupole time-of-flight mass spectrometry (HPLC-QToF-MS) is another advanced method used for non-target and suspect screening (Wang et al., 2020). In the majority of cases, the absolute configuration of TPs can be determined by using a combination of ^{1}H-NMR (proton nuclear magnetic resonance), LC-MS (liquid chromatography coupled to mass spectrometry), and ^{13}C- NMR (carbon-13 nuclear magnetic resonance). For the identification of polar TPs, HPLC in combination with electron spray ionization MS is used (Skotnicka-Pitak et al., 2008). A study on different methods to identify the TPs formed from 17β-estradiol and 17α-ethinyl estradiol was performed (Skotnicka-Pitak et al., 2008). Figure 3.3 shows some of the proposed methods to determine the TPs formed from parent compound estradiol via different methods of transformation. Wang et al. (2020) identified 14 pharmaceuticals and 46 pesticides from three wastewater treatment plants by suspect screening. In the treatment unit, these parent compound undergoes transformation to form TPs. Around 57 TPs were identified from 60 parent compounds by non-target and suspect screening. There are different methods for identifying TPs. Table 3.2 shows some compounds and identification of their TPs using different methods.

3.4 TPs FORMED FROM DIFFERENT ECs

TPs formed via different pathways may be more dangerous than their parental compound under three conditions (Escher and Fenner, 2011):

1. The yield of TPs formed is high when compared to the parental compound.
2. The persistence and mobility of TPs formed are more than the parental compound.
3. The TPs formed are more toxic than the parental compound.

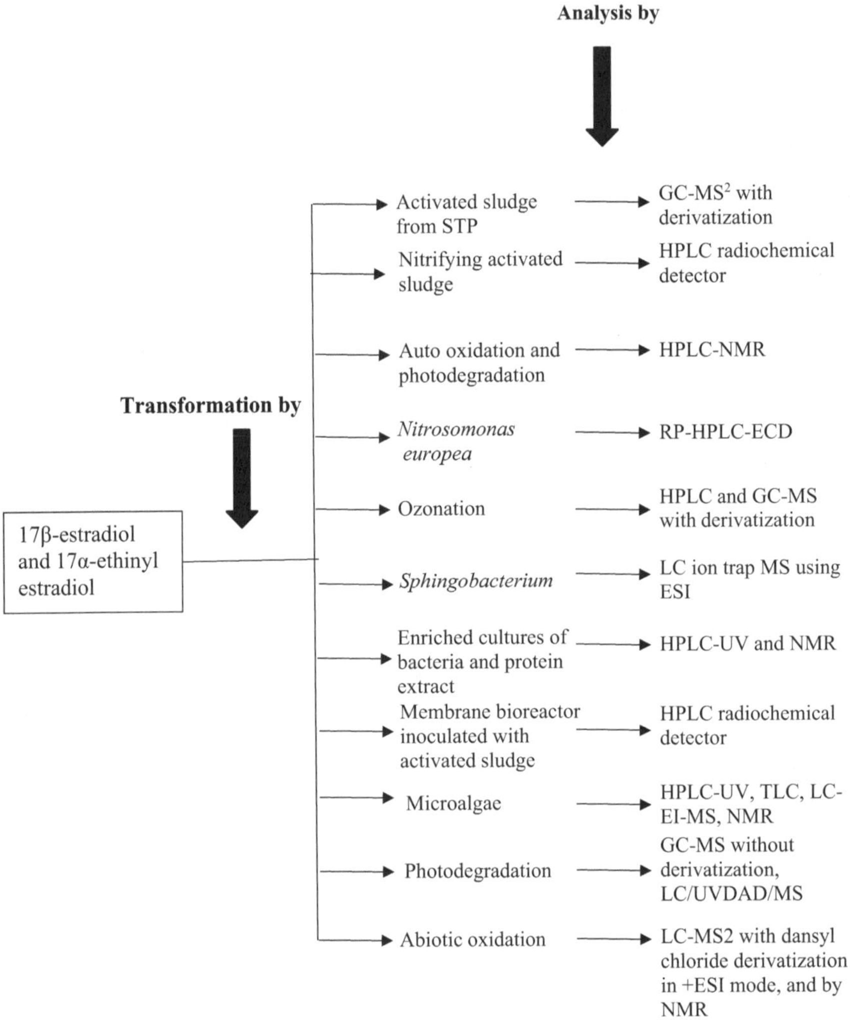

Figure 3.3 Different methods used for the analysis of TPs formed from 17β-estradiol and 17α-ethinyl estradiol via different mechanisms. (From Skotnicka-Pitak et al., 2008.)

3.4.1 TPs formed from PPCPs

Different mechanisms from which TPs are formed from pharmaceutical products are photo transformation, transformation with human liver microsomes, microbial degradation, etc. (Lege et al., 2020). Table 3.3 summarizes some of the TPs formed from ECs. Figure 3.4 shows the TP formed from a pharmaceutical compound, torasemide. In a study conducted by Lege et al. (2020), around 16 TPs were formed from torasemide through different mechanisms. However, out of 16, only 3 were found to have a high yield. TP 362, one of the TPs formed from torasemide, was found to be toxicologically relevant as this TP is capable of interacting with proteins and DNA. The persistence of

Table 3.2 Methods to identify TPs from the parent compounds

Parent compound	*Degradation mechanism*	*TPs formed*	*Methods to detect TPs*	*References*
Acyclovir	Biotransformation	Carboxy acyclovir	Hybrid Linear Ion Trap- FT Mass Spectrometry (LTQ-Orbitrap Velos) and 1D (^{1}H NMR, ^{13}C NMR) and 2D (^{1}H, ^{1}H-COSY, ^{1}H-^{13}C- HSQC) NMR Spectroscopy	Prasse et al. (2011)
Metoprolol (MTP)	Biological transformation	MTP acid, x- hydroxy MPT, O-desmethyl MTP, and some unknown TPs	LC coupled to HRMS	Rubirola et al. (2014)
Methotrixate	Biotransformation in aerobic activated sludge	2,4-Diamino-N10- methyl-pteroic acid and 8 other TPs	Ultra-HPLC– quadrupole – Orbitrap-MS	Kosjek et al., (2015)
Triclosan	Aerobic degradation	Triclosan-O- sulfate and other TPs	LTQ-Orbitrap- MS	Chen et al. (2015)
Tetracycline and erythromycin	Enzymatic degradation	Five TPs for erythromycin and 3 TPs for tetracycline	LC coupled to HRMS LTQ-Orbitrap	Llorca et al. (2015)
Nine pharmaceutical compounds	Microbial transformation in river sediments	16 TPs	UHPLC/QToF- MS	Li et al. (2014)
Macrolide, fluoroquinolone	Phosphorylation and acylation	Ciprofloxacin and norfloxacin (TP of fluoroquinolone) azithromycin, roxithromycin, and erythromycin (TP of macrolide)	UHPLC/QToF- MS	Terzic et al. (2011)
Sildenafil	Photo transformation	9TPs	UPLC-Q Exactive-MS	Aceña et al. (2014)
Carbamazepine (CBZ)	Photocatalysis and sonophotocatalysis	Several TPs of keto and hydroxy derivatives of CBZ	UPLC–(+ESI)- QToF-MS	Jelic et al. (2013)
Iopromide	Electrochemical treatment	Eight TPs	HRMS and MS^{n}-fragmentations	Lütke Eversloh et al. (2014)

Table 3.3 TPs formed from pharmaceuticals

Sl. No.	*Pharmaceutical*	*Mechanism*	*TP.*	*Reference*
1	Ibuprofen	Treatment in wastewater treatment plants	IbSW2, IbB4, and Ib1	Boix et al. (2016)
	Gemfibrozil		GSWB1	
2	Diclofenac	Biotic and abiotic transformation	4-OH-diclofenac, diclofenac acyl-β-D- glucuronide	López-Serna et al. (2013)
	Atorvastatin		2OH atorvastatin Acyl-β-D-glucuronide	
	Ranitidine		Ranitidine N oxide	
	Erythromycin		Anhydroerythromycin	
	Glyburide		Cis-3OH-glyburide, trans-4OH-glyburide	
3	Acetaminophen	Biodegradation	3-Hydroxyacetaminophen, hydroquinone, 1,4-benzoquinone, N-acetyl-p-benzoquinone imine, p-acetanisidide, 4-methoxyphenol, 2-hexenoic acid, and 1, 4-dimethoxybenzene	J. Li et al. (2014)
4	Carbamazepine	Treatment in wastewater treatment plants	10,11-Epoxy-carbamazepine, 2-OH-carbamazepine	García-Galán et al. (2016)

TP 378a is more than the parental compound. These compounds are microbially less degraded and hence are present more frequently in treated wastewater. From different pharmaceutical compounds, different TPs are also formed. N-Acetyl-p-benzoquinone-imine is the TP formed from acetaminophen (paracetamol), which is one of the most commonly used pharmaceutical compounds without any prescription (Nematollahi et al., 2009). Two to three percent of CBZ is excreted via urine as its original form. More than 50% of the parent compound is getting converted into carbamazepine 10,11 epoxide. Hence, the concentration of this TP in water was found to be more than the precursor (Gómez et al., 2007).

3.4.2 TPs formed from pesticide

The transformation of pesticides is highly complex, and it depends on various environmental factors. Reduction, oxidation, hydrolysis, and biological and photochemical processes are mainly responsible for the degradation of pesticides in the environment (Agüera López et al., 2014). Around 450 pesticides and their TPs were investigated in the wastewater and surface water (Rousis et al., 2017). In a study conducted on the occurrence of commonly used herbicides and their TPs in streams of Iowa, Illinois, and Minnesota, it was seen that the total concentration of TPs in these streams was approximately 17 times greater than the concentration of parent compounds (Kalkhoff et al., 1997). Table 3.4 summarizes the TPs formed from different pesticides.

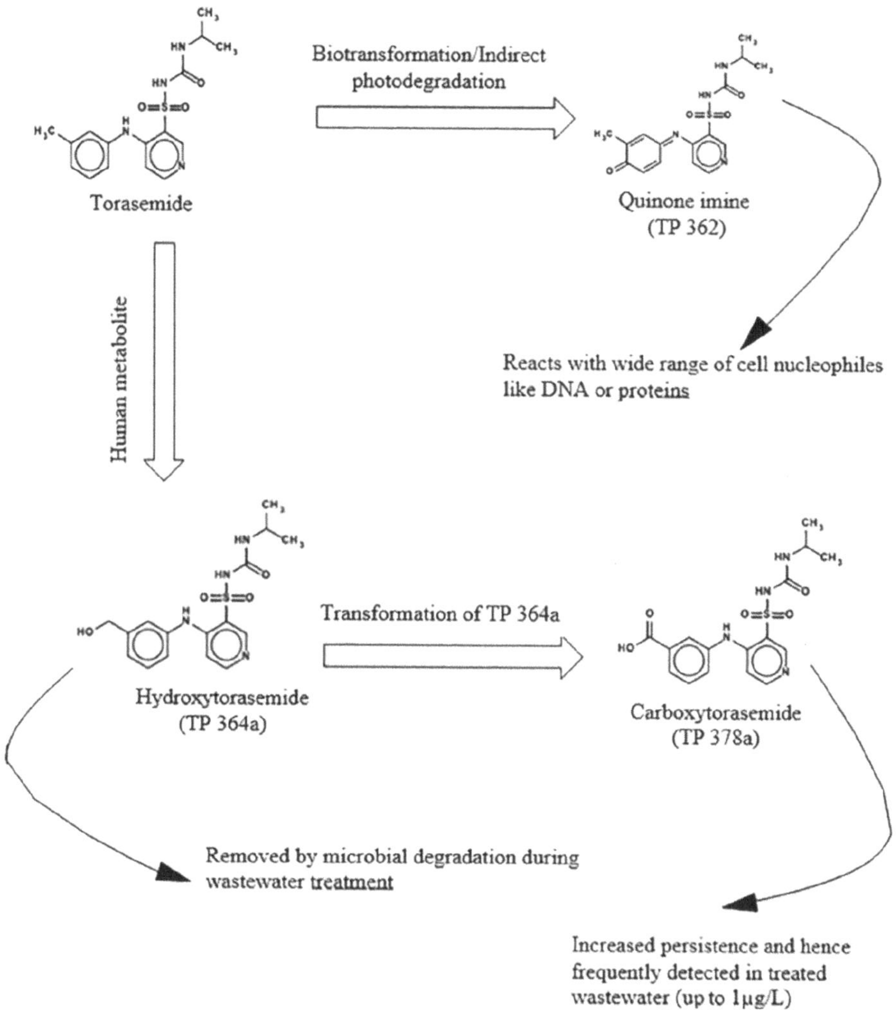

Figure 3.4 Example for TP formed from pharmaceutical compound torasemide and its relevance. (From Lege et al., 2020.)

3.4.3 TPs formed from industrial chemicals

For the protection of public from accidental fires, flame retardants are widely used in different products and commodities such as furniture, plastics, mattresses electronic, and electrical equipment. More than 175 types of flame retardants are available in market. Hence, a large amount of that can be emitted to the environment during their entire life time (Zhang et al., 2016). They may get transformed into different products that are more toxic and persistent than the parent compound. Biotic and abiotic transformation of various industrial chemicals like VOC, corrosion inhibitors, plasticizers, etc. also results in the formation of various TPs. Table 3.5 summarizes some of the TPs formed from industrial chemicals.

Table 3.4 TPs formed from pesticide

Pesticide	*Mechanism*	*TPs*	*References*
Chlorpyripos	Biodegradation	O,O-diethyl thiophosphate	Avila et al. (2020)
Imazalil (IMA), thiabendazole (TBZ), propiconazole (PYR)	Treatment in wastewater treatment plants	IMA-257, IMA-331, IMA-188, IMA-273, IMA-274, IMA-329, IMA-313, TBZ-218, TBZ-147, PYR-136, PYR-230, PYR-216a, PYR-216b, PYR-234	Campos-Mañas et al. (2019)
Dichlorvos	Ozonation	Desmethyl dichlorvos (d-DDVP), dichloroacetic acid (DCA) and dimethyl phosphate (DMP)	Cruz-Alcalde et al. (2018)
Fenitrothion, ethyl-parathion, methylparathion, and pentachlorophenol	Photolytic	(4-Nitrophenol, 3-methyl-4-nitrophenol, paraoxon ethyl, paroxon methyl, fenitrooxon and S-methylisomer of fenitrothion)	Castillo et al. (1997)
Atrazine	Sorption and degradation	Desethylatrazine	Hillebrand et al. (2014)
Glyphosate	Microbial degradation	Aminomethylphosphonic acid	Battaglin et al. (2005)
Thiamethoxam Fluroxypyr	Treatment in wastewater treatment plants	1-(2-Chlorothiazol-5-yl)-N-methylmethanamine 4-Amino-3,5-dichloro-6-fluoro-2-pyridone	Wang et al. (2020)

Table 3.5 TPs formed from industrial chemicals

Industrial chemical	*Mechanism*	*TPs*	*Reference*
1H-benzotriazole 5-methyl-1Hbenzotriazole	Biotransformation	4-Hydroxy-1H-benzotriazole, 5-hydroxy-1H-benzotriazole 1H-benzotriazole-5-carboxylic acid	Huntscha et al. (2014)
Brominated flame retardants	Fires and combustion	Polybrominated dibenzo-p-dioxins and dibenzofurans, mixed polybromochloro-dibenzo-p-dioxins, dibenzofurans	Zhang et al. (2016)
Alkylphenol Ethoxylates	Wastewater treatment	Alkylphenoxyacetic acids	Chiu et al. (2010)
Nonylphenol ethoxylate	Degradation	Nonylphenol (NP), nonylphenolmono-t-triethoxylates, and nonylphenol mono- and diethoxycarboxylates	Loos et al. (2010)
Bisphenol A	O- methylation	Bisphenol A monomethyl ether and bisphenol A dimethyl ether	Ashfaq et al. (2018)

3.5 THE FATE OF TPs FORMED FROM ECs

TPs are formed from ECs via different mechanisms as discussed in Section 6.2. The fate of these TPs formed can be of two ways: (1) They are distributed to various environmental compartments like air, water, and soil sediments. (2) They are degraded and further transformed. Figure 3.5 shows the formation and fate of TPs formed from ECs. It can be seen that ECs formed from various sources reaches water or soil or water treatment plant through different pathways. These ECs undergo various transformations in the presence of sunlight or interaction with water or different treatment methods resulting in their conversion to TPs. TPs formed from ECs are sometimes more stable or more persistent than the parent compound. Sometimes the mobility of TPs formed from ECs is more in a water–soil environment. Hence, they reach groundwater at a faster rate. Some TPs can also accumulate in plants which in turn reaches humans if the plant is edible. Sometimes the TPs formed are having more affinity toward the soil. In that cases, the horizontal mobility of soil will be less when compared to the parent compound (Boti et al., 2014).

Studies revealed that except some TPs (hydroxy metronidazole, carbamazepine epoxide, and 4-hydroxy diclofenac) formed from four pharmaceutical compounds like naproxen, CBZ, tramadol, and ibuprofen, all others are hydrolytically stable (Toński et al., 2019). Atenolol, used for the treatment of cardiovascular diseases, is transformed to form atenolol acid. This undergoes complete degradation in the soil after 61 days. However, an anticonvulsant drug CBZ and its TP remain persistent and ultimately reach groundwater (Koba et al., 2016). Acetaminophen is one of the most commonly used analgesics. Its concentration was found to be highest in influent samples and sewers. However, a membrane bioreactor system was found to be 100% effective in removing acetaminophen from wastewater samples. Some other pharmaceutical compounds like sulfapyridine (SPY), sulfamethoxazole, sulfamethazine, venlafaxine, sulfadiazine, diazepam , DCF (sodium salt), fluoxetine, CBZ, ranitidine (RTD), metoprolol (MTP) that are partially removed using membrane bioreactor system were found to be removed completely using reverse osmosis (RO) membrane system. Nanofiltration membranes are also effective in the complete removal of pharmaceuticals and their metabolites. But their removal efficiency is less than that of RO membranes (Mamo et al., 2017). Chiu et al. (2010) suggest that the removal of TPs can be enhanced by improving the secondary treatment processes. It includes optimization of operating conditions in a biological treatment system. Table 3.6 summarizes some of the studies performed on the formation and fate of TPs from different ECs.

3.6 TOXICITY OF TPs

TPs formed from ECs can be either more toxic or less toxic than their parent compound. Cosmetic products/PPCs contain around 10% of Ultraviolet (UV) filters for protection against UV light. Phytotoxicological study was conducted by performing a cell reproduction inhibition study on *Scenedesmus vacuolatus*. Under UV radiation, the compounds present in UV filters like ethylhexyl-methoxycinnamate, amylmethoxy-cinnamate, and 2-ethylhexyl-4-(dimethylamino) benzoate are degraded

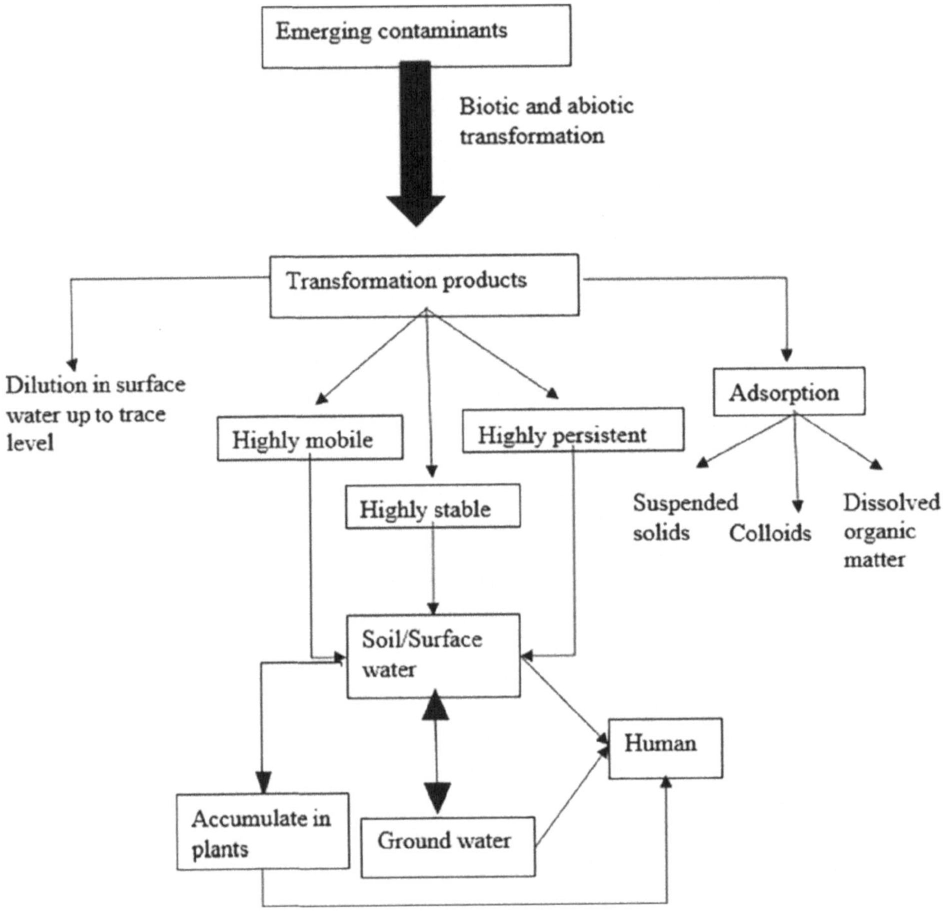

Figure 3.5 Occurrence and fate of TPs.

at a faster rate, and the phytotoxicity was found to reduce after the degradation (Rodil et al., 2009). The reaction of components present in sunscreen agents like octyl-p-methoxycinnamate and octyl dimethyl-p-aminobenzoate with hypochlorite during chlorination was conducted by examining the mutagenicity of *Salmonella typhimurium*. In the first stage of reaction, the TPs formed were found to exhibit higher mutagenicity. However, at the end of 6h, the mutagenicity was found to decrease. This shows that the mutagenic TPs formed during the initial part of the reaction were unstable and were converted to less mutagenic TPs at the end of the reaction (Nakajima et al., 2009).

The toxicity of the TPs formed from alprazolam and diazepam was obtained by studying the percentage inhibition of bacteria *Vibrio fischeri* with respect to photocatalytic degradation. In the first few minutes of photocatalytic degradation, the toxicity was found to decrease. However, after 10min of irradiation, an increase in toxicity was observed. This indicates the formation of TPs that are more toxic than the parent compounds (Mitsika et al., n.d.). Anthelmintics is one of the most commonly used classes

Table 3.6 Formation and fate of TPs formed from different ECs

Emerging contaminants	*Mechanism*	*T.P.s*	*Fate of TPs*	*Remarks*	*Reference*
Triclosan (TCS) and Triclocarban (TCC)	Nitrification	Methyl triclosan, 2,4-dichlorophenol a photolysis product of TCS. TCC did not degrade	No TP is formed from TCC. The parent compound TCC remains as such	Nitrification is not an effective method for the reduction of these compounds	Armstrong et al. (2019)
Triclocarban (TCC)	Reductive dechlorination	Dichlorocarbanilide (DCC), Tetra-chlorinated TCC (4-Cl- TCC) Monochlorocarbanilide (MCC), Nonchlorinated carbanilide (NCC), monochloroanilines, 4-chlorocatechol	Conversion of the parent compound to TPs are as follows TCC → DCC → MCC → NCC	DCC and MCC were unquantified in the biosolid sample. NCC was first quantified in a biosolid sample	Kor-Bicakci et al. (2020)
Bisphenol A	O-methylation	Bisphenol A monomethyl ether (BPA-MME) and Bisphenol A dimethyl ether (BPA- DME)	BPA-DME was widely detected in sludge, wastewater, and suspended solids. BPA-MME was detected in wastewater, but its concentration was less in sludge and suspended solids	In river surface water, the concentration of both TPs was almost the same. They pose a potential risk to the fishes in the surface water of the Jiulong River and its estuary	Ashfaq et al. (2018)
Metformin	Photodegradation	Guanylurea	Complete mineralization of guanylurea in the wastewater treatment plant	Neither metformin nor guanylurea is accumulated in fish	Straub et al. (2019)
Dextromethorphan	Photolysis	19TPs	The concentration of N-demethyldextrorphan was found to be higher than the parent compound in the river water and WWTP effluent	This study gives information on the fate of the opioid group of pharmaceutical compounds in environmental waters	Campos-Mañas et al. (2020)
Isoxaflutole		Diketonitrile	The TP formed is more water-soluble than the parent compound. Hence its transport to the environment is a greater concern	The half-life of isoxaflutole is very less. Hence it is readily degraded to diketonitrile	Lin et al. (2002)

of veterinary drugs. A large number of anthelmintics are commercially available. Some of them are fenbendazole, flubendazole, mebendazole, doramectin, eprinomectin, etc. Hydroxy-mebendazole, one of the benzimidazole metabolites, is found to be more toxic than mebendazole, the parent drug.

The two dominant metabolites of doramectin are still toxic; however, they are less toxic when compared to the parent compound (Horvat et al., 2012). Arvaniti et al. (2020) conducted a study on the sonochemical degradation of trimethoprim, a widely used antibiotic. Degradation resulted in the formation of several TPs, of which nine were identified successfully using LC-QToF-MS analysis. *In silico* toxicity test indicated that some TPs are more toxic to aquatic organisms than the parent compound (Arvaniti et al., 2020).

Han et al. (2020) studied the degradation of TC and oxytetracycline (OTC) using the photo-Fenton process. Degradation of TC and OTC yielded four and nine TPs, respectively. *V. fischeri* toxicity test showed that TC degradation produces less toxic TPs, while at the beginning, OTC degradation leads to higher toxicity; further, it yielded nontoxic TPs (Han et al., 2020). Chatzimpaloglou et al. (2021) studied the photocatalytic and photolytic degradation of antineoplastic drug irinotecan. A total of 19 TPs and 32 TPs were identified during the photolytic and photocatalytic processes, respectively. The toxicity of the parent compound was found to reduce drastically during the degradation process due to the formation of TPs that are not easily degraded (Chatzimpaloglou et al., 2021). Han et al. (2019) studied the toxicity of TPs formed due to the chloramination of CBZ, a widely used antiepileptic drug. From previous studies, around 12 TPs are formed during chloramination of CBZ. Three of the TPs exhibited cytotoxicity, chromosome damage, and greater DNA damage. TPs like 9 (10H)-acridone and acridine showed higher genotoxic or cytotoxic effects than the parent compound. These TPs are found to exhibit a greater toxic risk (Han et al., 2019).

Malathion, pyriproxyfen, fenoxaprop-ethyl, and benalaxyl are four widely used pesticides. Around 21 TPs are formed from all these parent compounds. It was seen that more than 50% of these TPs exhibited endocrine-disrupting effects, which result in alteration of gene expression and increase in hormone secretion (Ji et al., 2020). Polychlorinated biphenyls and their TPs are found to exhibit high toxicity. These toxic effects include immunotoxicity, dermal damage, carcinogenicity, and adverse effects on endocrine functions, reproduction, and development (Van Den Berg et al., 1998). Tak et al. (2020) conducted a study on the identification of various ECs and their TPs formed in an MBBR (moving bed bioreactor) in a drinking water-based treatment plant around the river Yamuna. They observed that using an MBBR system, although some of the ECs were degraded, a wide range of TPs formed were found to be more toxic than the parent compound, and they were present even at the end of treatment (Tak et al., 2020). González-Mariño et al. studied the photodegradation of nitenpyram, a neonicotinoid insecticide. Eighteen TPs were identified during the degradation. An *in silico* toxicity test revealed that the majority of TPs exhibited higher toxicity than the parent compound (González-Mariño et al., 2018).

In general, the degradation of CECs yields TPs that are sometimes more toxic than their parent compound. The persistence of these TPs in water is a serious issue that is to be resolved.

3.7 CONCLUSION AND FUTURE PERSPECTIVE

ECs that are released into the environment are transformed into various products known as TPs via different mechanisms. These TPs formed maybe sometimes more persistent and toxic than the parent compound. Hence, they ultimately reach water or soil or accumulate in plants. As a result, there is a chance that these TPs can reach back to the human being or aquatic organisms. TPs being toxic can impart adverse health impacts to humans, and it also affects the aquatic ecosystem. Studies on TPs formed are very limited. Hence, scientific advancement in the analysis and detection of trace compounds has to be improved to identify all TPs that are originating from CECs in diverse environments like sewage effluent, surface and groundwater, drinking water, etc. Conventional treatment systems should be modified with advanced methods that can remove the TPs from wastewater. More research is to be conducted in this area as the formation and fate of TPs still remain as a less explored field.

REFERENCES

Aceña, J., Pérez, S., Gardinali, P., Abad, J.L., Eichhorn, P., Heuett, N., Barceló, D., 2014. Structure elucidation of phototransformation products of unapproved analogs of the erectile dysfunction drug sildenafil in artificial freshwater with UPLC-Q Exactive-MS. *J. Mass Spectrom.* 49, 1279–1289. https://doi.org/10.1002/jms.3461

Agüera López, A., del Mar Gómez Ramos, M., & Fernández-Alba, A. R., 2014. Transformation products of pesticides in the environment: Analysis and occurrence. In: Dimitra A Lambropoulou, Nollet LML (eds) *Transformation Products of Emerging Contaminants in the Environment: Analysis, Processes, Occurrence, Effects and Risks.* John Wiley and Sons. pp. 387–412. ttps://doi.org/10.11647/obp.0218.15

Arstrong, D.L., Lozano, N., Rice, C.P., Ramirez, M., Torrents, A., 2019. Fate of triclosan, triclocarban, and their transformation products in wastewater under nitrifying conditions. *J. Water Process Eng.* 28, 144–151. https://doi.org/10.1016/j.jwpe.2019.01.015

Arvaniti, O.S., Frontistis, Z., Nika, M.C., Aalizadeh, R., Thomaidis, N.S., Mantzavinos, D., 2020. Sonochemical degradation of trimethoprim in water matrices: Effect of operating conditions, identification of transformation products and toxicity assessment. *Ultrason. Sonochem.* 67, 105139. https://doi.org/10.1016/j.ultsonch.2020.105139

Ashfaq, M., Sun, Q., Zhang, H., Li, Y., Wang, Y., Li, M., Lv, M., Liao, X., Yu, C.P., 2018. Occurrence and fate of bisphenol A transformation products, bisphenol A monomethyl ether and bisphenol A dimethyl ether, in wastewater treatment plants and surface water. *J. Hazard. Mater.* 357, 401–407. https://doi.org/10.1016/j.jhazmat.2018.06.022

Avila, R., Peris, A., Eljarrat, E., Vicent, T., Blánquez, P., 2020. Biodegradation of hydrophobic pesticides by microalgae: Transformation products and impact on algae biochemical methane potential. *Sci. Total Environ.* 754, 142114. https://doi.org/10.1016/j.scitotenv.2020.142114

Balakrishna, K., Rath, A., Praveenkumarreddy, Y., Siri, K., 2017. A review of the occurrence of pharmaceuticals and personal care products in Indian water bodies. *Ecotoxicol. Environ. Saf.* 137, 113–120. https://doi.org/10.1016/j.ecoenv.2016.11.014

Battaglin, W.A., Kolpin, D.W., Scribner, E.A., Kuivila, K.M., Sandstrom, M.W., 2005. Glyphosate, other herbicides and transformation products in midwestern streams, 2002. *J. Am. Water Resour. Assoc.* 41, 323–332.

Boix, C., Ibáñez, M., Sancho, J. V., Parsons, J.R., Voogt, P. de, Hernández, F., 2016. Biotransformation of pharmaceuticals in surface water and during waste water treatment: Identification and occurrence of transformation products. *J. Hazard. Mater.* 302, 175–187. https://doi.org/10.1016/j.jhazmat.2015.09.053

Bolong, N., Ismail, A.F., Salim, M.R., Matsuura, T., 2009. A review of the effects of emerging contaminants in wastewater and options for their removal. *Desalination* 239, 229–246. https://doi.org/10.1016/j.desal.2008.03.020

Boti, V., Sakkas, V., Albanis, T., 2014. Assessment of the occurrence and fate of transformation products of endocrine disrupting compounds EDCs in the environment. In: Dimitra A Lambropoulou, Nollet LML (eds) *Transformation Products of Emerging Contaminants in the Environment: Analysis, Processes, Occurrence, Effects and Risks*, John Wiley & Sons, pp. 648–673. https://doi.org/10.1002/9781118339558

Brown, M.A., Casida, J.E., 1988. Oxidative conversion of daminozide to methylating and acylating agents. *J. Agric. Food Chem.* 36, 819–822.

Caliman, F.A., Gavrilescu, M., 2009. Pharmaceuticals, personal care products and endocrine disrupting agents in the environment - A review. *CLEAN- Soil Air Water* 37, 277–303. https://doi.org/10.1002/clen.200900038

Calvo-flores, F.G., 2017. Industrial chemicals as emerging pollutant. In: Francisco G. Calvo-Flores Isac-García, Joaquín, Dobado, José A. (eds) *Emerging Pollutants: Origin, Structure and Properties*, Wiley-VCH Verlag GmbH & Co. KGaA, Weinheim, Germany, pp. 265–340. https://doi.org/10.1002/9783527691203.ch9

Campos-Mañas, M.C., Cuevas, S.M., Ferrer, I., Thurman, E.M., Sánchez-Pérez, J.A., Agüera, A., 2020. Determination of dextromethorphan and dextrorphan solar photo-transformation products by LC/Q-TOF-MS: Laboratory scale experiments and real water samples analysis. *Environ. Pollut.* 265, 114722. https://doi.org/10.1016/j.envpol.2020.114722

Campos-Mañas, M.C., Plaza-Bolaños, P., Martínez-Piernas, A.B., Sánchez-Pérez, J.A., Agüera, A., 2019. Determination of pesticide levels in wastewater from an agro-food industry: Target, suspect and transformation product analysis. *Chemosphere* 232, 152–163. https://doi.org/10.1016/j.chemosphere.2019.05.147

Castillo, M., Domingues, R., Alpendurada, M.F., Barceló, D., 1997. Persistence of selected pesticides and their phenolic transformation products in natural waters using off-line liquid solid extraction followed by liquid chromatographic techniques. *Anal. Chim. Acta* 353, 133–142. https://doi.org/10.1016/S0003–2670(97)00353–X

Chatzimpaloglou, Christophoridis, C., Fountoulakis, I., Antonopoulou, M., Vlastos, D., Bais, A., Fytianos, K., 2021. Photolytic and photocatalytic degradation of antineoplastic drug irinotecan. Kinetic study, identification of transformation products and toxicity evaluation. *Chem. Eng. J.* 405. https://doi.org/10.1016/j.cej.2020.126866

Chen, W., Xu, J., Lu, S., Jiao, W., Wu, L., Chang, A.C., 2013. Fates and transport of PPCPs in soil receiving reclaimed water irrigation. *Chemosphere* 93, 2621–2630. https://doi.org/10.1016/j.chemosphere.2013.09.088

Chen, X., Casas, M.E., Nielsen, J.L., Wimmer, R., Bester, K., 2015. Identification of Triclosan-O-Sulfate and other transformation products of Triclosan formed by activated sludge. *Sci. Total Environ.* 505, 39–46. https://doi.org/10.1016/j.scitotenv.2014.09.077

Chiu, T.Y., Paterakis, N., Cartmell, E., Scrimshaw, M.D., Lester, J.N., 2010. A critical review of the formation of mono-and dicarboxylated metabolic intermediates of alkylphenol polyethoxylates during wastewater treatment and their environmental significance. *Crit. Rev. Environ. Sci. Technol.* 40, 199–238. https://doi.org/10.1080/10643380802219517

Choi, Y., Jeon, J., Choi, Y., Kim, S.D., 2020. Characterizing biotransformation products and pathways of the flame retardant triphenyl phosphate in Daphnia magna using non-target screening. *Sci. Total Environ.* 708, 135106. https://doi.org/10.1016/j.scitotenv.2019.135106

Coats, J.R., 1991. Pesticide degradation mechanisms and environmental activation. In: Somasundaram and Coats (eds) *Pesticide Transformation Products.* Chapter 2, ACS Symposium series,

Vol 459, American Chemical Society, Washington DC pp. 10–30. https://doi.org/10.1021/bk-1991-0459.ch002

Cruz-Alcalde, A., Sans, C., Esplugas, S., 2018. Priority pesticide dichlorvos removal from water by ozonation process: Reactivity, transformation products and associated toxicity. *Sep. Purif. Technol.* 192, 123–129. https://doi.org/10.1016/j.seppur.2017.09.069

Deblonde, T., Cossu-Leguille, C., Hartemann, P., 2011. Emerging pollutants in wastewater: A review of the literature. *Int. J. Hyg. Environ. Health* 214, 442–448. https://doi.org/10.1016/j.ijheh.2011.08.002

Dubey, M., Mohapatra, S., Tyagi, V.K., Suthar, S. and Kazmi, A.A., 2021. Occurrence, fate, and persistence of emerging micropollutants in sewage sludge treatment. *Environ. Pollut.* 273, p. 116515.

Ebele, A.J., Abou-Elwafa Abdallah, M., Harrad, S., 2017. Pharmaceuticals and personal care products (PPCPs) in the freshwater aquatic environment. *Emerg. Contam.* 3, 1–16. https://doi.org/10.1016/j.emcon.2016.12.004

Eregowda T., Mohapatra S. (2020) Fate of micropollutants in engineered and natural environment. In: Kumar M., Munoz-Arriola F., Furumai H., Chaminda T. (eds) *Resilience, Response, and Risk in Water Systems. Springer Transactions in Civil and Environmental Engineering.* Springer, Singapore, pp. 283–301. https://doi.org/10.1007/978-981-15-4668-6_15

Escher, B.I., Fenner, K., 2011. Recent advances in environmental risk assessment of transformation products. *Environ. Sci. Technol.* 45, 3835–3847. https://doi.org/10.1021/es1030799

Fatta-Kassinos, D., Vasquez, M.I., Kümmerer, K., 2011. Transformation products of pharmaceuticals in surface waters and wastewater formed during photolysis and advanced oxidation processes: Degradation, elucidation of byproducts and assessment of their biological potency. *Chemosphere* 85, 693–709. https://doi.org/10.1016/j.chemosphere.2011.06.082

García-Galán, M.J., Petrovic, M., Rodríguez-Mozaz, S., Barceló, D., 2016. Multiresidue trace analysis of pharmaceuticals, their human metabolites and transformation products by fully automated on-line solid-phase extraction-liquid chromatography-tandem mass spectrometry. *Talanta* 158, 330–341. https://doi.org/10.1016/j.talanta.2016.05.061

Gahlot, R., Taki, K., Kumar, M. 2020. Efficacy of nanoclays as the potential adsorbent for dyes and metal removal from the wastewater: A review. *Environ. Nanotechnol. Monit. Manag.* 14, 100339.

Gómez, M.J., Martínez Bueno, M.J., Lacorte, S., Fernández-Alba, A.R., Agüera, A., 2007. Pilot survey monitoring pharmaceuticals and related compounds in a sewage treatment plant located on the Mediterranean coast. *Chemosphere* 66, 993–1002. https://doi.org/10.1016/j.chemosphere.2006.07.051

González-Mariño, I., Rodríguez, I., Rojo, L., Cela, R., 2018. Photodegradation of nitenpyram under UV and solar radiation: Kinetics, transformation products identification and toxicity prediction. *Sci. Total Environ.* 644, 995–1005. https://doi.org/10.1016/j.scitotenv.2018.06.318

Han, C.H., Park, H.D., Kim, S.B., Yargeau, V., Choi, J.W., Lee, S.H., Park, J.A., 2020. Oxidation of tetracycline and oxytetracycline for the photo-Fenton process: Their transformation products and toxicity assessment. *Water Res.* 172. https://doi.org/10.1016/j.watres.2020.115514

Han, Y., Ma, M., Oda, Y., Rao, K., Wang, Z., Yang, R., Liu, Y., 2019. Insight into the generation of toxic products during chloramination of carbamazepine : Kinetics, transformation pathway and toxicity. *Sci. Total Environ.* 679, 221–228. https://doi.org/10.1016/j.scitotenv.2019.04.423

Hillebrand, O., Nödler, K., Geyer, T., Licha, T., 2014. Investigating the dynamics of two herbicides at a karst spring in Germany: Consequences for sustainable raw water management. *Sci. Total Environ.* 482–483, 193–200. https://doi.org/10.1016/j.scitotenv.2014.02.117

Horvat, A.J.M., Petrović, M., Babic, S., Pavlović, D.M., Ašperger, D., Pelko, S., Mance, A., Kaštelan-Macan, M., 2012. Analysis, occurrence and fate of anthelmintics and their transformation products in the environment. *TrAC – Trends Anal. Chem.* 31, 61–84. https://doi.org/10.1016/j.trac.2011.06.023

Huntscha, S., Hofstetter, T.B., Schymanski, E.L., Spahr, S., Hollender, J., 2014. Biotransformation of benzotriazoles: Insights from transformation product identification and compound-specific isotope analysis. *Environ. Sci. Technol.* 48, 4435–4443. https://doi.org/10.1021/es405694z

Jelic, A., Michael, I., Achilleos, A., Hapeshi, E., Lambropoulou, D., Perez, S., Petrovic, M., Fatta-Kassinos, D., Barcelo, D., 2013. Transformation products and reaction pathways of carbamazepine during photocatalytic and sonophotocatalytic treatment. *J. Hazard. Mater.* 263, 177–186. https://doi.org/10.1016/j.jhazmat.2013.07.068

Ji, C., Song, Q., Chen, Y., Zhou, Z., Wang, P., Liu, J., Sun, Z., Zhao, M., 2020. The potential endocrine disruption of pesticide transformation products (TPs): The blind spot of pesticide risk assessment. *Environ. Int.* 137. https://doi.org/10.1016/j.envint.2020.105490

Kalkhoff, S.J., Lee, K.E., Porter, S.D., Terrio, P.J., Thurman, E.M., 1997. Herbicides and herbicide degradation products in upper midwest agricultural streams during august base- flow conditions. *J. Environ. Qual.* 32, 1025–1035.

Kennedy Neth, N.L., Carlin, C.M., Keen, O.S., 2019. Emerging investigator series: Transformation of common antibiotics during water disinfection with chlorine and formation of antibacterially active products. *Environ. Sci. Water Res. Technol.* 5, 1222–1233. https://doi.org/10.1039/c9ew00182d

Koba, O., Golovko, O., Kodešová, R., Klement, A., Grabic, R., 2016. Transformation of atenolol, metoprolol, and carbamazepine in soils: The identification, quantification, and stability of the transformation products and further implications for the environment. *Environ. Pollut.* 218, 574–585. https://doi.org/10.1016/j.envpol.2016.07.041

Kor-Bicakci, G., Abbott, T., Ubay-Cokgor, E., Eskicioglu, C., 2020. Occurrence and fate of antimicrobial triclocarban and its transformation products in municipal sludge during advanced anaerobic digestion using microwave pretreatment. *Sci. Total Environ.* 705, 135862. https://doi.org/10.1016/j.scitotenv.2019.135862

Kosjek, T., Heath, E., 2008. Applications of mass spectrometry to identifying pharmaceutical transformation products in water treatment. *TrAC – Trends Anal. Chem.* 27, 807–820. https://doi.org/10.1016/j.trac.2008.08.014

Kosjek, T., Negreira, N., de Alda, M.L., Barceló, D., 2015. Aerobic activated sludge transformation of methotrexate: Identification of biotransformation products. *Chemosphere* 119, S42–S50. https://doi.org/10.1016/j.chemosphere.2014.04.081

Kotthoff, L., Lisec, J., Schwerdtle, T., Koch, M., 2019. Prediction of transformation products of monensin by electrochemistry compared to microsomal assay and hydrolysis. *Molecules* 48, 24. https://doi.org/10.3390/molecules24152732

Krauss, M., Singer, H., Hollender, J., 2010. LC-high resolution MS in environmental analysis: From target screening to the identification of unknowns. *Anal. Bioanal. Chem.* 397, 943–951. https://doi.org/10.1007/s00216-010-3608–9

Kumar, M., Chaminda, G. T., Honda, R., 2020. Seasonality impels the antibiotic resistance in Kelani River of the emerging economy of Sri Lanka. *NPJ Clean Water* 3(1), 1–8.

Kusari, S., Prabhakaran, D., Lamshöft, M., Spiteller, M., 2009. In vitro residual anti-bacterial activity of difloxacin, sarafloxacin and their photoproducts after photolysis in water. *Environ. Pollut.* 157, 2722–2730. https://doi.org/10.1016/j.envpol.2009.04.033

La Farré, M., Pérez, S., Kantiani, L., Barceló, D., 2008. Fate and toxicity of emerging pollutants, their metabolites and transformation products in the aquatic environment. *TrAC – Trends Anal. Chem.* 27, 991–1007. https://doi.org/10.1016/j.trac.2008.09.010

Lege, S., Sorwat, J., Yanez Heras, J.E., Zwiener, C., 2020. Abiotic and biotic transformation of torasemide - Occurrence of degradation products in the aquatic environment. *Water Res.* 177, 115753. https://doi.org/10.1016/j.watres.2020.115753

Leng, Y., Bao, J., Xiao, H., Song, D., Du, J., Mohapatra, S., Werner, D., Wang, J., 2020. Transformation mechanisms of tetracycline by horseradish peroxidase with/without redox mediator ABTS for variable water chemistry. *Chemosphere* 258, 127306. https://doi.org/10.1016/j.chemosphere.2020.127306

Li, J., Ye, Q., Gan, J., 2014. Degradation and transformation products of acetaminophen in soil. *Water Res.* 49, 44–52. https://doi.org/10.1016/j.watres.2013.11.008

Li, Z., Maier, M.P., Radke, M., 2014. Screening for pharmaceutical transformation products formed in river sediment by combining ultrahigh performance liquid chromatography/high resolution mass spectrometry with a rapid data-processing method. *Anal. Chim. Acta* 810, 61–70. https://doi.org/10.1016/j.aca.2013.12.012

Lin, C.H., Lerch, R.N., Thurman, E.M., Garrett, H.E., George, M.F., 2002. Determination of isoxaflutole (balance) and its metabolites in water using solid phase extraction followed by high-performance liquid chromatography with ultraviolet or mass spectrometry. *J. Agric. Food Chem.* 50, 5816–5824. https://doi.org/10.1021/jf025622d

Llorca, M., Rodríguez-Mozaz, S., Couillerot, O., Panigoni, K., de Gunzburg, J., Bayer, S., Czaja, R., Barceló, D., 2015. Identification of new transformation products during enzymatic treatment of tetracycline and erythromycin antibiotics at laboratory scale by an on-line turbulent flow liquid-chromatography coupled to a high resolution mass spectrometer LTQ-Orbitrap. *Chemosphere* 119, 90–98. https://doi.org/10.1016/j.chemosphere.2014.05.072

Loos, R., Locoro, G., Comero, S., Contini, S., Schwesig, D., Werres, F., Balsaa, P., Gans, O., Weiss, S., Blaha, L., Bolchi, M., Gawlik, B.M., 2010. Pan-European survey on the occurrence of selected polar organic persistent pollutants in ground water. *Water Res.* 44, 4115–4126. https://doi.org/10.1016/j.watres.2010.05.032

López-Serna, R., Jurado, A., Vázquez-Suñé, E., Carrera, J., Petrović, M., Barceló, D., 2013. Occurrence of 95 pharmaceuticals and transformation products in urban groundwaters underlying the metropolis of Barcelona, Spain. *Environ. Pollut.* 174, 305–315. https://doi.org/10.1016/j.envpol.2012.11.022

Lütke Eversloh, C., Henning, N., Schulz, M., Ternes, T.A., 2014. Electrochemical treatment of iopromide under conditions of reverse osmosis concentrates - Elucidation of the degradation pathway. *Water Res.* 48, 237–246. https://doi.org/10.1016/j.watres.2013.09.035

Majeed, A., 2018. Application of agrochemicals in agriculture : Benefits, risks and responsibility of stakeholders. *J. Food Sci. Toxicol.* 2, 3.

Mamo, J., García-Galán, M.J., Stefani, M., Rodríguez-Mozaz, S., Barceló, D., Monclús, H., Rodriguez-Roda, I., Comas, J., 2017. Fate of pharmaceuticals and their transformation products in integrated membrane systems for wastewater reclamation. *Chem. Eng. J.* 331, 450–461. https://doi.org/10.1016/j.cej.2017.08.050

Mazumder, P., Sharma, S. K., Taki, K., Kalamdhad, A. S., & Kumar, M., 2020. Microbes involved in arsenic mobilization and respiration: A review on isolation, identification, isolates and implications. *Environ. Geochem. Health* 2, 3443–3469.

Menon N.G., Mohapatra S., Padhye L.P., Tatiparti S.S.V., Mukherji S., 2020. Review on occurrence and toxicity of pharmaceutical contamination in Southeast Asia. In: Kumar M., Snow D., Honda R. (eds) *Emerging Issues in the Water Environment during Anthropocene. Springer Transactions in Civil and Environmental Engineering.* Springer, Singapore. https://doi.org/10.1007/978-981-32-9771-5_4

Michael, I., Vasquez, M.I., Hapeshi, E., Haddad, T., Baginska, E., Kümmerer, K., Fatta-Kassinos, D., 2014. Metabolites and transformation products of pharmaceuticals in the aquatic environment as contaminants of emerging concern. In: Dimitra A Lambropoulou, Nollet LML (eds) *Transformation Products of Emerging Contaminants in the Environment: Analysis, Processes, Occurrence, Effects and Risks.* John Wiley & Sons The Atrium, Southern Gate, Chichester, West Sussex, United Kingdom, pp. 425–470. https://doi.org/10.1002/9781118339558.ch14

Mitsika, E.E., Christophoridis, C., Kouinoglou, N., Lazaridis, N., Zacharis, C.K., Fytianos, K., n.d. Optimized photo-fenton degradation of psychoactive pharmaceuticals alprazolam and diazepam using a chemometric approach structure and toxicity of transformation products. *J. Hazard. Mater.* 403, 123819. https://doi.org/10.1016/j.jhazmat.2020.123819

Mohapatra S., Padhye L.P., Mukherji S., 2018. Challenges in detection of antibiotics in wastewater matrix. In: Gupta T., Agarwal A., Agarwal R., Labhsetwar N. (eds) *Environmental Contaminants. Energy, Environment, and Sustainability.* Springer, Singapore. https://doi.org/10.1007/978-981-10-7332-8_1

Mohapatra S., Menon N., Padhye L.P., Tatiparti S.S.V., Mukherji S. 2021b. Natural attenuation of pharmaceuticals in the aquatic environment and role of phototransformation. In: Kumar M., Snow D., Honda R., Mukherjee S. (eds) *Contaminants in Drinking and Wastewater Sources. Springer Transactions in Civil and Environmental Engineering.* Springer, Singapore. https://doi.org/10.1007/978-981-15-4599-3_3

Mohapatra, S., Sharma, N., Mohapatra, G., Padhye, L.P. and Mukherji, S., 2021a. Seasonal variation in fluorescence characteristics of dissolved organic matter in wastewater and identification of proteins through HRLC-MS/MS. *J. Hazard. Mater.* 413, 125453.

Nakajima, M., Kawakami, T., Niino, T., Takahashi, Y., Onodera, S., 2009. Aquatic fate of sunscreen agents octyl-4-methoxycinnamate and octyl-4-dimethylaminobenzoate in model swimming pools and the mutagenic assays of their chlorination byproducts. *J. Heal. Sci.* 55, 363–372. https://doi.org/10.1248/jhs.55.363

Nematollahi, D., Shayani-Jam, H., Alimoradi, M., Niroomand, S., 2009. Electrochemical oxidation of acetaminophen in aqueous solutions: Kinetic evaluation of hydrolysis, hydroxylation and dimerization processes. *Electrochim. Acta* 54, 7407–7415. https://doi.org/10.1016/j.electacta.2009.07.077

Peck, A.M., 2006. Analytical methods for the determination of persistent ingredients of personal care products in environmental matrices. *Anal. Bioanal. Chem.* 386, 907–939. https://doi.org/10.1007/s00216-006-0728–3

Petrie, B., Barden, R., Kasprzyk-Hordern, B., 2015. A review on emerging contaminants in wastewaters and the environment: Current knowledge, understudied areas and recommendations for future monitoring. *Water Res.* 72, 3–27. https://doi.org/10.1016/j.watres.2014.08.053

Prasse, C., Wagner, M., Schulz, R., Ternes, T.A., 2011. Biotransformation of the antiviral drugs acyclovir and penciclovir in activated sludge treatment. *Environ. Sci. Technol.* 45, 2761– 2769. https://doi.org/10.1021/es103732y

Priyanka M.S., Mohapatra S., 2020. Shifts and trends in analysis of contaminants of emerging concern: Sulfonamides. In: Kumar M., Munoz-Arriola F., Furumai H., Chaminda T. (eds) *Resilience, Response, and Risk in Water Systems. Springer Transactions in Civil and Environmental Engineering.* Springer, Singapore, pp 261–281. https://doi.org/10.1007/978-981-15-4668-6_14

Rajput, P., Thakur, A., Devi, P., 2020. *Emerging Agrochemicals Contaminants: Current Status, Challenges, and Technological Solutions*In: Prasad M.N.V. (ed) *Agrochemicals Detection, Treatment and Remediation.* Elsevier, pp. 117–142 https://doi.org/10.1016/b978-0-08-103017-2.00005–2

Ramírez-Malule, H., Quiñones-Murillo, D.H., Manotas-Duque, D., 2020. Emerging contaminants as global environmental hazards. A bibliometric analysis. *Emerg. Contam.* 6, 179–193. https://doi.org/10.1016/j.emcon.2020.05.001

Rodil, R., Moeder, M., Altenburger, R., Schmitt-Jansen, M., 2009. Photostability and phytotoxicity of selected sunscreen agents and their degradation mixtures in water. *Anal. Bioanal. Chem.* 395, 1513–1524. https://doi.org/10.1007/s00216-009-3113–1

Rodriguez-Narvaez, O.M., Peralta-Hernandez, J.M., Goonetilleke, A., Bandala, E.R., 2017. Treatment technologies for emerging contaminants in water: A review. *Chem. Eng. J.* 323, 361–380. https://doi.org/10.1016/j.cej.2017.04.106

Rousis, N.I., Bade, R., Bijlsma, L., Zuccato, E., Sancho, J. V., Hernandez, F., Castiglioni, S., 2017. Monitoring a large number of pesticides and transformation products in water samples from Spain and Italy. *Environ. Res.* 156, 31–38. https://doi.org/10.1016/j.envres.2017.03.013

Rubirola, A., Llorca, M., Rodriguez-Mozaz, S., Casas, N., Rodriguez-Roda, I., Barceló, D., Buttiglieri, G., 2014. Characterization of metoprolol biodegradation and its transformation products generated in activated sludge batch experiments and in full scale WWTPs. *Water Res.* 63, 21–32. https://doi.org/10.1016/j.watres.2014.05.031

Skotnicka-Pitak, J., Garcia, E.M., Pitak, M., Aga, D.S., 2008. Identification of the transformation products of 17α-ethinylestradiol and 17β-estradiol by mass spectrometry and other instrumental techniques. *TrAC – Trends Anal. Chem.* 27, 1036–1052. https://doi.org/10.1016/j.trac.2008.10.003

Straub, J.O., Caldwell, D.J., Davidson, T., D'Aco, V., Kappler, K., Robinson, P.F., Simon-Hettich, B., Tell, J., 2019. Environmental risk assessment of metformin and its transformation product guanylurea. I. Environmental fate. *Chemosphere* 216, 844–854. https://doi.org/10.1016/j.chemosphere.2018.10.036

Tak, S., Tiwari, A., Vellanki, B.P., 2020. Identification of emerging contaminants and their transformation products in a moving bed biofilm reactor (MBBR)–based drinking water treatment plant around River Yamuna in India. *Environ. Monit. Assess.* 192, 1–23. https://doi.org/10.1007/s10661-020-08303–4

Ternes, T.A., Kreckel, P., Mueller, J., 1999. Behaviour and occurrence of estrogens in municipal sewage treatment plants - II. Aerobic batch experiments with activated sludge. *Sci. Total Environ.* 225, 91–99. https://doi.org/10.1016/S0048–9697(98)00335–0

Terzic, S., Senta, I., Matosic, M., Ahel, M., 2011. Identification of biotransformation products of macrolide and fluoroquinolone antimicrobials in membrane bioreactor treatment by ultrahigh-performance liquid chromatography/quadrupole time-of-flight mass spectrometry. *Anal. Bioanal. Chem.* 401, 353–363. https://doi.org/10.1007/s00216-011-5060–x

Thatai, S., Verma, R., Khurana, P., Goel, P., Kumar, D., 2018. *Water Quality Standards, Its Pollution and Treatment Methods,* In: Mu Naushad (eds) *A New Generation Material Graphene: Applications in Water Technology.* Springer International Publishing, Gewerbestrasse 11, 6330 Cham, Switzerland pp. 21–42 https://doi.org/10.1007/978–3–319–75484–0

Toński, M., Dołżonek, J., Stepnowski, P., Białk-Bielińska, A., 2019. Hydrolytic stability of selected pharmaceuticals and their transformation products. *Chemosphere* 236. https://doi.org/10.1016/j.chemosphere.2019.06.206

Van Den Berg, M., Birnbaum, L., Bosveld, A.T.C., Brunström, B., Cook, P., Feeley, M., Giesy, J.P., Hanberg, A., Hasegawa, R., Kennedy, S.W., Kubiak, T., Larsen, J.C., Van Leeuwen, F.X.R., Liem, A.K.D., Nolt, C., Peterson, R.E., Poellinger, L., Safe, S., Schrenk, D., Tillitt, D., Tysklind, M., Younes, M., Wærn, F., Zacharewski, T., 1998. Toxic equivalency factors (TEFs) for PCBs, PCDDs, PCDFs for humans and wildlife. *Environ. Health Perspect.* 106, 775–792. https://doi.org/10.1289/ehp.98106775

Vulava, V.M., Cory, W.C., Murphey, V.L., Ulmer, C.Z., 2016. Sorption, photodegradation, and chemical transformation of naproxen and ibuprofen in soils and water. *Sci. Total Environ.* 565, 1063–1070. https://doi.org/10.1016/j.scitotenv.2016.05.132

Wajid, A., Ahmad, M.S., Yousuf, S., Jabeen, A., Choudhary, M.I., 2019. Biotransformation of progestonic hormone dydrogesterone with *Macrophomina phaseolina*, and study of the effect of biotransformed products on phagocytes oxidative burst. *Steroids* 143, 67–72. https://doi.org/10.1016/j.steroids.2018.12.009

Wang, X., Yu, N., Yang, J., Jin, L., Guo, H., Shi, W., Zhang, X., Yang, L., Yu, H., Wei, S., 2020. Suspect and non-target screening of pesticides and pharmaceuticals transformation products in wastewater using QTOF-MS. *Environ. Int.* 137. https://doi.org/10.1016/j.envint.2020.105599

Wilkinson, J., Hooda, P.S., Barker, J., Barton, S., Swinden, J., 2017. Occurrence, fate and transformation of emerging contaminants in water: An overarching review of the field. *Environ. Pollut.* 231, 954–970. https://doi.org/10.1016/j.envpol.2017.08.032

Yang, Y., Toor, G.S., 2015. Contaminants in the Urban Environment : Pharmaceuticals and Personal Care Products (PPCPs) – Part 2. SL420. U S Department of Agriculture and Soil.

Yin, L., Wang, B., Yuan, H., Deng, S., Huang, J., Wang, Y., Yu, G., 2017. Pay special attention to the transformation products of PPCPs in environment. *Emerg. Contam.* 3, 69–75. https://doi.org/10.1016/j.emcon.2017.04.001

Zhang, M., Buekens, A., Li, X., 2016. Brominated flame retardants and the formation of dioxins and furans in fires and combustion. *J. Hazard. Mater.* 304, 26–39. https://doi.org/10.1016/j.jhazmat.2015.10.014

Chapter 4

Advanced treatment technologies for removal of contaminants of emerging concern

Om Prakash
Wastewater Technology Division, CSIR-National Environmental Engineering Research Institute

Deepak Panchal, Abhishek Sharma, and Sukdeb Pal
Wastewater Technology Division, CSIR-National Environmental Engineering Research Institute
Academy of Scientific and Innovative Research (AcSIR)

CONTENTS

DOI: 10.1201/9781003247869-5

4.1 OCCURRENCE OF CONTAMINANTS OF EMERGING CONCERNS (CECs) IN ENVIRONMENT

The new class of highly persistent pollutants commonly referred to as 'Contaminants of Emerging Concern' (CECs) have become a growing environmental concern. CECs are categorized as compounds that were previously not considered or known to be significantly distributed in the environment and are now being detected as compounds that may be a risk to human health and ecosystems. Several pollutants come under CECs, such as pharmaceutical and personal care products (PPCPs), hormones, perfluorinated compounds, pesticides, flame retardants, and musk fragrances (Dubey et al., 2021). In particular, PPCPs such as analgesics, antibiotics, anticonvulsants, hormones, fragrances, β-blockers, antidepressants, and antiepileptics are major contaminants (Cincinelli et al., 2015). Their intermediates and metabolites are continuously discharged into the environmental systems through various point- and nonpoint sources. Uncontrolled discharge of domestic wastewater (sewage) and agricultural runoffs are the nonpoint sources from where CECs enter into the aquatic systems and even in aerial environment, whereas point sources are industrial wastewater discharged from sectors like mining, cement plants, refineries, smelters, pharmaceutical industries, landfills, combined sewer outfalls, incinerators, power plants, leaking underground storage facility, and spilling and contaminated sites (Ambade et al., 2021). CECs can also enter the environment in the form of secondary pollutants, and their characteristics of sorbing onto the solids (soils and sediments) and biotransformation further contribute to their load in environmental matrices. CECs undergo a transformation in two steps: In the first step, hydrolysis and redox (oxidation and reduction) reactions occur, resulting in the formation of products with enhanced polarity and solubility in water. The second step involves the modification of the contaminant or the products formed in the first step by conjugation reactions. This in turn facilitates the degradation of products or intermediates with or without any use of catalyst/biocatalyst. Some conjugation reactions occur before the first step, and this results in the formation of metabolized/unmetabolized compounds that could directly be excreted through urine and feces. Human excreta contains both metabolized and nonmetabolized medicinal drugs. A high percentage of ingested drugs/pharmaceuticals reach unmetabolized into the sewage *via* human excreta (Carlsson et al., 2006). Therefore, CECs are indirectly introduced in the aquatic environment by human urine and feces, direct disposal of discarded medicinal products and other PPCPs, and runoff through agricultural field or aquaculture. It was revealed that ~66% of the population keep pharma products as household waste, 12% of the population directly flush them, and only 22% return them for their proper disposal. In this scenario, inadequately managed landfills produce highly toxic leachates that further contaminate the groundwater resources (Ariffin and Zakili, 2019).

Pharmaceutical processing industries contaminate the areas nearby the production units and therefore are designated as point source for CECs. Processing generates exceptionally high-loaded effluent with several pharmaceuticals and hence is of huge concern. The situation is worse in developing countries that lack adequate wastewater treatment facilities. Conventional treatment facilities can reduce the concentration of these CECs up to a certain level. However, complete removal has not been achieved till now. Therefore, the use of treated wastewaters in agricultural soils can result in bioaccumulation of CECs in crops (Wu et al., 2015). Additionally, nonpolar and less-polar CECs are adsorbed onto the biosolids and can be removed with biological sludge during the secondary (biological) treatment of wastewater. Eventually, this sludge is used as fertilizers in agricultural land that can certainly act as a source of CECs. This can further reach the surface water, resulting in contamination of surface and groundwater (Jones et al., 2005). Apart from industrial wastewater, hospital effluents are also highly concentrated mixture of many CECs and, therefore, considered as the second largest CECs emission source. Animal waste also contains several CECs and their metabolites that make animal husbandry as the third largest source of CECs (Aus der Beek et al., 2016).

Many pharmaceutical products like anesthetics, antibiotics, anti-inflammatory, anti-fungal, anti-parasitic, and sedatives are used as veterinary medicines. Thus, these drugs can easily reach up to environmental settings through urine and feces of livestock. Various veterinary pharmaceuticals such as antibiotics and hormones have already been detected in animal wastes, and in agricultural fields where these wastes have been applied as fertilizer. This again contributes to build-up of CECs in nearby terrestrial (soil) and aquatic environments (ground and surface water) (Bártíková et al., 2016). A significant number of CECs are used in aquaculture with fish foods to produce beneficial effects (Mo et al., 2017) that indirectly contributes to the accumulation of CECs in an aquatic environment, soil, and sediments. Therefore, rearing center, animal husbandry, and aquacultures are prominent sectors that use and discharge various CECs in huge quantities, leading to increased chances of occurrence of CECs in the environment (Mo et al., 2017). In the aqueous system, transport of CECs is controlled by the source–pathway–receptor model. Significant CEC transport procedures are leaching, sorption, desorption, and degradation. However, in case of soil–water matrices, texture, size, solubility, hydrophobicity, sorption-desorption, dissociation constants and division coefficients at various pH values contribute are the major factors for transport and mobility of CECs (Zhang et al., 2016).

As of now, there is no drinking or clean water guidelines for CECs except in some of the developed countries. These guidelines are formed on the basis of existing knowledge of toxicity and acute health effects caused by CECs. Even at nanogram level, these contaminants can pose acute as well as long-term deleterious/adverse effects on living beings. These effects can be hormonal changes or disruption of metabolic pathways including inhibition of nucleic acid formation and protein synthesis. CECs' Log K_{ow} values are greater than 3 that can result into bioaccumulation of CECs in the organisms' tissues (Rizzo et al., 2009). The presence of endocrine disruptors in the aquatic environment and potential adverse effects on humans and biota require a constant monitoring of their fates in wastewater. Certainly, the existing knowledge is limited, and there is an urgent need for extensive investigation before setting up guidelines for CECs in drinking water. On the other hand, elimination of these CECs is of utmost importance and necessary in order to reduce the toxicity toward public health and environment.

Removal of CECs from wastewater matrix has been extensively investigated through several physical, chemical, and biological treatment technologies. However, these available treatment options have not been able to completely degrade or remove CECs from the environmental matrix. The subsequent section of this chapter will provide insight into various treatment technologies for the removal of CECs from environmental settings. Further, various detection methodologies and recent advanced treatment technologies such as combined treatment and heterogenous photocatalysis that have proven to be a better choice for complete removal of CECs from aqueous medium have been reviewed. The future research directions for the implementation of combined treatment technologies and heterogenous (hybrid) nanocomposites for removal of CECs have been discussed.

4.2 DETECTION OF CECs IN ENVIRONMENTAL AQUEOUS MEDIA

CECs have been present in the water matrix for years, but their quantification has recently gained attention because they are extremely hazardous to human as well as the environment. Several CECs are hydrophilic in nature and, therefore, they can easily dissolve in water. Such chemical characteristic allows easy transportation of CECs through the water system. This can cause irreversible changes toward aquatic as well as human lives. The diverse, complex chemical nature and the trace level (usually ng/L) existence of CECs in the environmental matrices are the major challenges for their precise identification and quantification. Consequently, due to lack of adequate analytical techniques, these contaminants remained undetected. Therefore, the indiscriminate use of CECs and their undetectable (unnoticeable) factor have led to enormous number of CECs and their transformation products in the wastewater, surface water, groundwater, and drinking water. Thus, there is an urgent need for an appropriate analytical technique that can facilitate rapid, wide-ranging monitoring and easy screening of CECs in aqueous as well as in other environmental samples. The recent development in the analytical tools, especially in chromatography and mass spectrophotometry (MS), has made detection easier even at very low concentrations (μg/L–ng/L). Different chromatographic methods used for detection of CECs in different water matrixes and detected concentrations are summarized in Table 4.1.

The detection of CECs and metabolites is generally performed using gas chromatography/MS (GC/MS). Techniques like liquid chromatography (LC)/GC-MS are used in combination with C_{18}-based solid-phase extraction (SPE). Some polar compounds have low volatilization, so derivatization is required for identification using GC/MS. This paved the path for advancement of the methods, and therefore, LC/MS and advanced LC/MS/MS are widely accepted techniques for CECs detection in terms of sample preparation that does not demand any further derivatization.

Chromatography is a separation method, and MS is used for identification of unknown compound and quantification of known elements based on their mass to charge ratio. This facilitates enhanced selectivity, sensitivity, and specificity of a detection system. Advancement in MS was required for quantification at ng/L level and therefore technologies like ion trap (IT) and triple quadrupole were developed. Some more recent advancements of triple quadrupole, linear ITs, quadrupole-linear IT, and quadrupole time of flight have been used for structural interpretation of many transformation products. Other techniques like microbiological assays or immunoanalytical techniques and capillary electrophoresis have also been tested for determination of CECs. Available methods for identification and quantification of CECs have been optimized, validated, and used to analyse almost 3,000 chemical compounds in various environmental samples (Patel et al., 2019). Sample preparation is another important

Table 4.1 Different methods of detection of CECs in different matrices along with the detectable concentration

Method of detection	*Matrix*	*Contaminants*	*Concentration (ng/g)*	*References*
LC/MS	Groundwater	Sulfamethoxazole	<0.1–34	Lapworth et al. (2018)
		Perfluoroalkyl	<0.1–33	
		Phenoxyacetic acid	<20–210	
HPLC-MS/MS	Spring water	Gemfibrozil	5	Doummar and Aoun (2018)
		Iohexol	10	
UPLC	River	Caffeine	743	Sharma et al. (2019)
		Ketoprofen	107	
		Saccharine	2.4–85	
		Sucralose	0.5–24	
HPLC and ESI-MS/MS	River	Caffeine	1870	Dai et al. (2016)
		Carbamazepine	24.7	
		Chloramphenicol	17	
		Ketoprofen	77.6	
		Metoprolol	115	
HPLC and ESI-MS/MS	River	Caffeine	2130	Dai et al. (2015)
		Carbamazepine	56.4	
		Chloramphenicol	8.2	
		Ketoprofen	67.6	
		Metoprolol	91.4	
HPLC-MS/MS	Drinking water treatment plant	Paracetamol	0.07–0.12	Cai et al. (2015)
		Bezafibrate	0.20–0.77	
		Antipyrine	1.80–2.71	
		Isopropylantipyrine	0.84–5.55	
		Sulfamethoxazole	1.39–2.48	
	Tap water	Carbamazepine	38.24	
		Isopropylantipyrine	0.72	
		Antipyrine	0.28	
UPLC	Reclaimed water	Carbamazepine	103.2	Li et al. (2015)
		Caffeine	109.5	
		Ketoprofen	11.9	
		Nalidixic acid	5.4	
HPLC-tandem-MS	Groundwater (agricultural area)	Ciprofloxacin	BDL–0.519	Lee et al. (2019)
		Difloxacin	BDL–0.464	
		Sulfadimethoxine	BDL–0.449	
		Spiramycin	BDL–0.687	
		Oxfendazole	BDL–22.8	
		Acetaminophen	BDL–0.647	
		Crotamiton	BDL–3.86	

ESI-MS, electrospray ionization mass spectroscopy; HPLC, high-pressure liquid chromatography; LC/MS, liquid chromatography mass spectroscopy, ND, not detectable; UPLC, ultra-performance liquid chromatography.

step for accurate measurement of aqueous samples. Some of the widely used sample preparation methods are liquid-liquid extraction, SPE, on-line SPE, magnetic SPE, molecular imprinted polymers, and solid-phase micro-extraction. In various environmental samples, signal intensity for compound of interest (analyte) can be considerably reduced due to complex matrix resulting into reduced limit of detection. In such cases, sample preparation can be done using chelating agents that can further remove unwanted matrix chemicals, providing more accurate results.

4.3 REMOVAL OF CECs: CONVENTIONAL WASTEWATER TREATMENT PROCESSES

Wastewater treatment plants are generally considered to be standard treatment option for the removal of various pollutants including colloidal particles, organics, heavy metals, nutrients, and pathogens from effluent. However, these treatment plants are not adequately designed for the removal of CECs. Operating conditions of treatment facility, physicochemical characteristics, and persistence of CECs play a major role in the removal of CECs. Treatment facilities generally consist of primary, secondary, and an optional tertiary treatment stage, where primary treatment steps facilitated maximum removal of colloidal and suspended particles/solids. CECs are also expected to be removed to some extent through the process of adsorption onto sludge in the primary stage. It is noteworthy that adsorption can only facilitate removal of a few pharmaceutical compounds, whereas CECs like naproxen, sulfamethoxazole (SMX), iopromide, and ibuprofen (IBU) remained dissolved in the wastewater. The secondary treatment stage involves biological transformation and mineralization of CECs through the action of various processes such as activated sludge process (ASP), trickling filters, microalgae-/fungi-driven treatments, moving bed biofilm reactors, and aerobic/anaerobic/facultative microbiological treatments. The microbial/fungal based treatments have been found to be efficient for PPCPs removal (95%–100%) but are generally ineffective for pesticides removal. The coupled treatment systems such as biological active processes facilitate the biodegradation and adsorption simultaneously that effectively remove the betablockers and pesticides, except a few CECs (Jin et al.,

Table 4.2 Removal efficiencies of conventional treatment processes toward various CECs

Treatment system	*Contaminants*	*Removal efficiency*	*References*
Conventional wastewater treatment plants	Bisphenol A Ibuprofen Bezafibrate	>90%	Clara et al. (2005)
Biological removal	Ibuprofen Naproxen Diclofenac	>90% 50%–80% 20%–40%	Joss et al. (2005)
Primary treatment	Diclofenac Estriol	~28%	Behera et al. (2011)
Secondary treatment	Carbamazepine Naproxen	33.8%–89.9%	Stamatis and Konstantinou (2013)
Biological treatment	Caffeine N, N-diethyl-meta-toluamide	<40%	Wang et al. (2014)
Secondary treatment	Sulfamerazine Oxytetracycline Lincomycin Sulfadiazine	>99% 64% 2% 22%	Gao et al. (2012)
Secondary treatment	Naproxen Ketoprofen Crotamiton	46% 45% 20%	Nakada et al. (2006)
Coagulation and flocculation	Triclosan Paracetamol Phenylbutazone	89% 81.3 79.3	Azzouz and Ballesteros (2013)

2013). Sorption of CECs onto the suspended particles allowed the easy removal of sorbed CECs along with sludge removal during the biological processes. Some of the CECs (betablockers) showed poor removal during the ASP process as the dominant removal mechanism was not biodegradation and only adsorption onto suspended particles playing the major role (Ahmed et al., 2017). Trickling filter was observed to be less effective and showed <70% removal efficiency when compared with ASP (>85%). Given the physicochemical stability of these compounds, microorganisms alone cannot metabolize them as a source of energy. This may eventually inhibit the activity of microorganisms that can lead to bioaccumulation of CECs in the food chain.

In MBR system, CECs can be removed by simultaneous action of microbial community for biodegradation of the higher molecular weight molecules at the membrane surface (Luo et al., 2014). However, clogging and fouling of membranes, operational failures, and high maintenance cost are some of the major issues that are generally associated with MBR. Chlorination at the tertiary stage rapidly reacts with pharmaceutical compounds containing amines, resulting in the formation of chlorinated by-products and other numerous identified toxic sub-products. Chlorine dioxide can only react with certain compounds and found to be ineffective for oxidation/oxidative degradation of various pharmaceutical compounds like bezafibrate, carbamazepine (CBZ), diazepam, or IBU. Breakdown of CECs does not necessarily mean the eradication of toxicity as the by-products or intermediate metabolites can have unidentified and increased toxicity with higher persistence level. The reported removal efficiencies of CECs achieved at primary, secondary, and tertiary treatment stages are listed in Table 4.2.

Therefore, it can be said that CECs remain in effluents even after the action of treatment stages in wastewater treatment plants and thus can further contaminate the surface and groundwater.

4.4 ADVANCED TREATMENT PROCESSES

4.4.1 Membrane filtration

Membrane is a selective barrier of semipermeable material that is circumscribed by the driving forces across the filtration system. It can be categorized as microfiltration (MF), ultrafiltration (UF), nanofiltration (NF) and reverse osmosis (RO) on the basis of pore size of the membrane. Properties of membranes (surface charges, hydrophobicity, pore size, and permeability) and operating conditions (flux rate, transmembrane pressure, feed water quality, and the rejection/recovery potential of the membrane) play a pivotal role in treatment mechanisms of CECs. On the contrary, physiochemical characteristics of CECs such as hydrophobicity, polarity, molecular size, charge, diffusivity, and solubility are equally important for their removal.

NF and RO have the smallest membrane pore size followed by UF and MF. MF and UF are very effective process for water treatment. But, surprisingly, both MF and UF are inconvenient for the elimination of pharmaceuticals from water because most of the CECs have smaller molecular weight than the molecular weight cut-off (MWCO) of MR and UF. Usually, the pore size of MF and UF is in the range of 0.1 and 0.01 to 0.1 μm, respectively. Further, NF and RO have a small pore diameter which facilitates efficient removal of CECs in water. MWCO of NF and RO comes in the range of 200–1,000 and 100–200 Da, respectively (Ganiyu et al., 2015).

4.4.1.1 Nanofiltration (NF)

High-pressure NF is an efficient method for the removal of CECs. There are a lot of factors that influence the removal mechanism and efficiency of NF. Hence, removal efficiency of CECs depends upon the removal mechanisms. NF could effectively remove more than 90% of bisphenol, phthalate acid ester including di-n-octyl phthalate, diethyl phthalate, dibutyl phthalate, dimethyl phthalate, and diethylhexyl phthalate (Wei et al., 2016). Removal of estradiol (85%) and estrone (65%) was investigated using NF270, and it was concluded that negatively charged CECs show high removal efficiency than that of neutral CECs (Braeken and Van der Bruggen, 2009). Such as, SMX and carbadox were rejected during nanofiltration (90% to 95%) effectively due to the negative surface charge of the membrane that developed at $pH > 5.6$ (Shah et al., 2012).

4.4.1.2 Reverse osmosis (RO)

RO is the most efficient filtration process capable of extracting dissolved solids through semipermeable membrane under high pressure (exceeding the osmotic pressure). The efficiency of RO depends on the physical property, type, mechanical resistance, porosity, and chemical nature of the membrane. It has gained much attention for CECs control due to its ability of high percentage removal. However, RO has some limitations as the membranes are costly and they could easily get clogged when wastewater containing high suspended solids (SS) is passed through it. RO is not appropriate for wide-scale wastewater treatment due to the vulnerability of membranes toward oxidizing and fouling agents. Moreover, RO reject contains very high total dissolved solids and toxic and hazardous materials, thereby making its disposal a very solicitude material (Homem and Santos. 2011).

RO has effectively eliminated >90%–95% of CECs (>90%) and various other pharmaceuticals from potable water (Heberer. 2002). There are some studies that reported low removal of ketoprofen (64.3%) and diclofenac (DIC) (55.2%–60%) using RO (Verlicchi, 2010). It was further investigated that high removal efficiency was observed when large pore-sized membrane was used. Therefore, removal efficiency of RO is proportionate to the pore diameter of the RO membrane (Heberer, 2002). Polyamide- and cellulose acetate -based membranes achieved salt removal efficiency of 90% and 94%, respectively (Kimura et al., 2004). Cyclophosphamide (a drug used in chemotherapy) was effectively removed by RO (>90%) while NF could only achieve a limited removal efficiency (20%–40%) (Wang et al., 2009).

4.4.2 Adsorption

Adsorption is one of the widely used and accepted industrial separation technologies for the purification of water because of the ease of operation and efficiency of the treatment process. With due design consideration, the adsorption process can adequately treat effluent better than other chemical methods such as coagulation. Adsorption is a surface phenomenon identified by the mass transfer of a matter at the interface of two specific phases such as liquid–solid. In water treatment, it involves the interphase accumulation of dissolved adsorbates (pollutants) onto the solid adsorbent surface from an aqueous solution. On the basis of nature and interaction of interfaces, adsorption is differentiated in two types: physisorption and chemisorption. Both chemisorption and physisorption facilitate the removal of CECs. If the interaction occurs between adsorbate and adsorbent by physical nature, it is termed as physisorption. It is primarily accomplished through van der Waals' forces, hydrogen bonds, dipole–dipole interactions, polarity, and steric and π–π interactions. On the other hand, chemisorption occurs under the effect of chemical bond as a force of attraction between adsorbate and adsorbent. Chemisorption occurs due to the exchange of electrons and bonds between various functional moieties of adsorbate (CECs) and the adsorbent.

Over the last few decades, an enormous number of studies have been done for the removal of emerging pollutants (CECs) by the use of various adsorbent materials. It includes activated carbon (AC), biochar, carbon nanotubes, graphene, and clay. Among them, AC is one of the prominent adsorbents for the removal of CECs from aquatic environment. Generally, ACs are synthesized by pyrolysis of biomass under inert atmosphere. The activation of carbon can occur by chemical or physical mechanisms. In chemical activation, the impregnated organic and inorganic precursors (KOH, $ZnCl_2$, $FeCl_3$, etc.) are pyrolyzed in furnace. Further, the inorganic matter is extracted from the pyrolyzed product to produce AC. In physical activation, the initial composites are firstly pyrolyzed at 500°C–650°C under nitrogen flow and then the temperature is further elevated to 700°C–1,000°C under carbon dioxide (Prauchner et al., 2016). The above manufacturing process provides high porosity, high surface area, and adequate adsorption capacity that make them appropriate as an adsorbent for the removal of CECs.

The adsorption efficacy of NH_4Cl-AC (NAC) was studied by removing diazinon (organophosphate insecticide) from an aqueous solution. NAC (0.3 g/L) showed optimum adsorption rate (97.5%) of diazinon (20 mg/L) in shorter contact time (30 min) at pH 7. The adsorption of diazinon using NAC was best fitted to the Langmuir model giving an adsorption capacity of 250 mg/g (Moussavi et al., 2013). Similarly, Yaghmaeian et al. (2014) investigated the removal efficiency and regeneration of NAC in an immobile bed column after the removal of amoxicillin (AMX). NAC immobile bed column has performed better than the standard AC column. The chemical-activated (H_3PO_4) peach stones-based ACs was successfully used for the removal of CBZ (psychiatric drug), caffeine (a stimulant), and DIC (an anti-inflammatory drug) from aqueous solution. P CBZ was adsorbed with the highest adsorption capacity of 335 mg/g due to its low water solubility and hydrophilic properties. Olive waste cake-based AC was synthesized for the elimination of four pharmaceutical compounds such as ketoprofen, naproxen, DIC, and IBU. The adsorbent efficacy was best fitted in the Langmuir isotherm model, resulting in high adsorption capacities for ketoprofen (24.7 mg/g), naproxen (39.5 mg/g), DIC (56.2 mg/g), and IBU (12.6 mg/g) (Baccar et al., 2012). A waste rubber tire was turned into ACs and used for the adsorption of pesticides such as methyl parathion, methoxychlor, and atrazine from wastewater (Gupta et al., 2011). The adsorption obeyed the Langmuir isotherm model with maximum sorption capacity for methyl parathion (104.9 mg/g), methoxychlor (112 mg/g), and atrazine (88.9 mg/g) for the reaction time of 1 h at pH 2. The high adsorption capacities were achieved due to the high porosity of ACs. A mixture of plant material (coffee grounds, eucalyptus, and soyabean oil), sawdust, and $Ca(OH)_2$-based ACs was prepared at 800°C (Rovani et al., 2014). The adsorbent was used for the removal of 17α-ethinylestradiol (EE2) and

17β-estradiol (E2) from the aqueous system. The maximum sorption capacities were 7.584 and 7.883 mg/g for E2 and EE2, respectively, that fitted best on the Sips isotherm model at 298 K.

4.4.3 Advanced oxidation processes (AOPs)

Advanced oxidation processes (AOPs) occur due to the production of extremely reactive free radicals like HO•, $O_2{•}^-$, and HO_2• in the aqueous media. These reactive species are produced using H_2O_2, cavitation, ultraviolet (UV), and O_3 either individually or in combination. High reactivity and non-selectivity of these oxidative species promote complete degradation of organic pollutants, making AOPs versatile and efficient for CECs removal. AOPs used in conjunction such as $UV + H_2O_2$, $O_3 + H_2O_2$, $Fe^{2+} + H_2O_2$, electro-Fenton, and photo-Fenton facilitate an improved degradation of CECs. AOPs can successfully attain a very high removal efficiency (~99%) for a variety of CECs in aqueous media. AOPs have been successfully employed for the elimination of IBU, DIC, ciprofloxacin (CIP), tetracycline (TC), CBZ, AMX, and sulfamethoxazole7-α-ethinylestradiol (Jiang and Zhou, 2013).

Compared to the individual application of ozone-assisted AOPs (O_3/H_2O_2, O_3/UV) and photon-conjugated AOPs, their combination is more effective for cleaving carbon–halogen bonds. Occasionally, the oxidation can also result in the formation of intermediates and by-products that are more toxic than their parent compounds that may persist in treated water and causes more problems/toxicity. For example, the oxidation of triclocarban and triclosan (TCS) generates highly persistent and toxic chloroanilines and chlorodioxins. Similarly, oxidation of CBZ generates products like epoxy CBZ and many other intermediates that are mutagenic, carcinogenic, or more toxic acridine.

4.4.3.1 Electrochemical oxidation (EO)

Electrochemical oxidation (EO) is one of the commonly used and widely accepted technologies for the purification of water. It works on *in situ* production of robust oxidants such as hydroxyl, sulfate, and chloride radicals in the aqueous phase. An electrochemical process usually includes the anode (oxidation reaction) and the cathode (reducing reaction) with a power source. Fundamentally, EO processes facilitate redox reactions that occur at both anode (pollutant oxidation) and the cathode (pollutants reduction), resulting in complete removal of various contaminants. Anodes or anodic materials mineralize or degrade organic pollutants by anodic oxidation reactions. Hence, anode and anodic materials can play significant roles in the oxidative elimination of CECs. Pollutants can be electrochemically oxidized using two separate oxidation mechanisms, viz. indirect and direct anodic oxidation. The direct EO process can be explained as the electrolysis (anodic oxidation) performed over the anode, involving transmission of charges between the anode and organic pollutants. The mass transfer of pollutants and electron movement at anode at specific potential (higher than oxygen evolution reaction OER) potential) is the only mechanism to oxidize the organic pollutants. If the applied potential is lower than the potential of water oxidation reaction, EO process cannot be performed. In indirect oxidation, oxidants are generated at anode interfaces and efficiently dissociate the CECs in the solution. Chlorine is the most common oxidant generated by chloride oxidation and persulfate at anode interfaces.

Niobium coated with boron-doped diamond (Nb/BDD) mesh anode was used to mineralize the SMX, propranolol, and CBZ in water. Oxygen injection along with sodium sulfate in the reaction mixture for 1.5 h and at 2.5 A allowed adequate mineralized SMX, CBZ, and propranolol up to 86%, 82%, and 85%, respectively (García-Espinoza and Nacheva, 2019). Transformation via desulfuration, hydroxylation, deamination, and bond rupture were the chief mechanisms of mineralization. Acetaminophen (ACT), DIC, and SMX were removed using graphite electrodes in both direct and indirect electro-oxidation processes. In direct EO system, removal efficiencies of 81.6%, 90%, and 74.3% were achieved within 1 h for SMX, DIC, and ACT, respectively, while, in indirect EO system, SMX, DIC, and ACT were mineralized with the efficacy of 99.6%, 92.6%, and 89.6% within 30 min of reaction time, respectively (Liu et al., 2019). A different set of experiments were carried out using BDD anode for the elimination of CIP, SMX, and salbutamol (SM) with two salts of sulfates and chlorides (Lan et al., 2017). Sulfate enhanced the elimination of SMX and CIP, but no effect on the removal of SM was observed, while in the presence of chloride, only CIP was adequately removed in comparison to SMX and SM. Further, removal of AMX was investigated using various electrode materials such as carbon-felt, carbon fiber, carbon-graphite, platinum, lead dioxide, dimensionally stable anode, ($Ti/RuO_2–IrO_2$), and BDD on the anode electrode to know their oxidation capacities in aqueous solution. Among all, BDD electrode demonstrated to be the best anode material at high current densities due to the formation of large quantities of hydroxyl radicals and other oxidants, such as hydrogen peroxide, ozone, and persulfates, which help increase in oxidation and mineralization of AMX by means of controlled oxidation.

4.4.3.2 Ozonation

Ozonation is the process in which ozone is directly used for water treatment without the addition of any other surfactants or catalysts. O_3 is electrophilic and can directly react to electron-rich species. It can directly be used against carbon compounds having π-bonds. Ozone on reaction with hydroxide anions forms hydroxyl radical (•OH) in water. The reactions taking place in an aqueous medium can be summarized in the steps below (Equations 4.1–4.7):

$$O_3 + OH^- \rightarrow O_2 + HO_2^- \quad (4.1)$$

$$O_3 + HO_2^- \rightarrow HO_2^{\circ} + O_3^{\circ-} \quad (4.2)$$

$$HO_2^{\cdot} \rightarrow H^+ + O_3^{\circ-} \quad (4.3)$$

$$O_2^{\cdot-} + O_3 \rightarrow O_2 + O_3^{\circ-} \quad (4.4)$$

$$O_3^{\cdot-} + H^+ \rightarrow HO_3^{\circ} \quad (4.5)$$

$$HO_3^{\circ} \rightarrow HO^{\circ} + O_2 \quad (4.6)$$

$$HO^{\circ} + O_3 \rightarrow HO_2^{\circ} + O_2 \quad (4.7)$$

pH of the solution plays an important role throughout the whole process, while hydroxide anions initiate these chain reactions. After evaluating several studies, it can be inferred that ozonation is a feasible option for eliminating IBU and naproxen. Ozonation alone can degrade carbon–carbon bonded compounds (carbamazepine), a secondary amine group (diclofenac), and amine (sulfamethoxazole) with a rapid removal rate (10^5 to 10^6 L/mol·s) (Huber et al., 2003). While other CECs like IBU, iopromide, diazepam (without susceptible reactive sites) were found to have low degradation rates during ozonation. In order to destroy a wide variety of CECs, ozone has proven to be more successful than chlorine dioxide.

CECs can be categorized into three categories based on reaction rate with ozone and chlorine. The three groups are as follows:

1. **Category I:** It comprises aromatic rings along with hydroxyl and amine functions and low pK_a values. For both ozone and chlorine, they undergo high oxidative removal (>95%).
2. **Category II:** It comprises compounds that undergo negligible oxidation (<20%) with both ozone and chlorine.
3. **Category III:** It comprises compounds that respond preferably to ozone but not to chlorine like CBZ, IBU, and meprobamate. They have nucleophilic groups that can react with electrophilic ozone.

Several variables, including oxidant dosage, CECs concentration, mode of activity, and physicochemical parameters of water such as SS, total dissolved solids, and chemical oxygen demand influence the oxidation rate. The concentration of SS in the secondary clarifier's outlet merely affects ozonation efficiency. Whereas dissolved solids show 100 times faster oxidation than suspended particles due to diffusion limitations. Dosage of ozone depends on the chemical oxygen demand concentration in wastewater, whereas the contact time is generally maintained from 15 to 30 min. CBZ, clarithromycin, DIC, erythromycin, indomethacin, roxithromycin, SMX, and trimethoprim are readily reactive toward ozone, while others such as venlafaxine, diazepam, and IBU showed a limited response. Hospital wastewater contaminated with cyclophosphamide requires 256 mg/L of ozone for >97% removal (Ferre-Aracil et al., 2016). A dose of 0.2 mg/L of ozone was needed for the oxidation of many CECs in pure water, while bezafibrate drug can be oxidized at a concentration almost 100 times lower than other CECs. The degradation rate for antibiotics, spectinomycin, and lincomycin during ozonation increases with the rise in pH of the reaction solution. This could be due to the rise in •OH concentration formed from enhanced hydroxide concentration at alkaline pH (as shown in Equations (4.1–4.7)). Similar behavior was observed during the ozonation of clofibric acid. During ozonation of wastewater, a high concentration of suspended particles reduced the removal efficiency for iopromide, iohexol, and iopamide, and degradation followed the order as iopamide > iohexol > iopromide (Zucker et al., 2015).

Ozonation can remove >90% of CECs and therefore a complete removal is not achieved. A residual concentration of CECs was also measured in drinking water at ng/L level after the process of ozonation. Thus, ozonation is not a completely effective treatment option for purification of drinking water. Further, DIC and CBZ release toxic metabolites (e.g., CIP) whose degradation was inhibited due to low reactivity of O_3 molecule toward their quinolone moiety. Implementation and maintenance of ozonation is expensive for fluctuating effluent flows. The major drawback of ozonation is its constant energy requirements for ozone production. Mass transport of ozone in the form of gas again raises its operational cost.

4.4.3.3 Oxidation using ozone/UV/hydrogen peroxide

To increase the efficiency of ozonation, addition of a catalyst, exposure to light, and hydrogen peroxide treatment are being used in combination. Exposure to UV light can photo-catalytically dissociate certain compounds, but alone it is not sufficient for complete degradation of CECs. Hydrogen peroxide combined with UV light generates nonselective hydroxyl radicals at a high rate that quickly degrade CECs. This combination effectively eliminates several CECs, including recalcitrants, pesticides, PPCPs, and plasticizers. During the processes of UV irradiation, degradation majorly depends on the wavelength (λ), quantum yield ($\varnothing$), and pollutant molar adsorption capability (ε) as given in Equation (4.8).

$$\varnothing\lambda = \frac{k\lambda}{2.303E\lambda\varepsilon\lambda} \tag{4.8}$$

- Here, $\varnothing\lambda$ is quantum yield at λ (mol/Einstein),
- $k\lambda$ is the kinetic constant of degradation at the given λ (s^{-1}),
- $\varepsilon\lambda$ is the adsorption coefficient of the compound at λ (m^2/mol), and
- $E\lambda$ is the energy emitted by the light source (Einstein/s/m^2).

With this combination of reagents, different reaction mechanisms occur in AOPs:

I. Photo-driven release of OH• and homolytic H_2O_2 cleavage (Equation 4.9).

$$H_2O_2 + hv \rightarrow 2HO^{\cdot} \tag{4.9}$$

II. In combination with O_3, and H_2O_2, a rapid formation of OH is initiated in solution (Equations 4.10–4.13):

$$O_3 + H_2O + hv \rightarrow O_2 + H_2O_2 \tag{4.10}$$

$$H_2O_2 \rightarrow HO_2^- + H^+ \tag{4.11}$$

$$O_3 + HO_2^- + hv \rightarrow O_3^- + HO_2^{\cdot} \tag{4.12}$$

$$O_3^- + H^+ + hv \rightarrow O_2 + HO^{\cdot} \tag{4.13}$$

III. H_2O_2 improves ozonation through similar mechanisms of O_3/UV. Two pathways of hydroxyl radical development during ozonation are possible:

Pathway 1 (Equations 4.14–4.16):

$$H_2O_2 + H_2O \rightarrow H_3O^+ + HO_2^- \tag{4.14}$$

$$O_3 + HO^- \rightarrow HO_2^- + O_2 \tag{4.15}$$

$$O_3 + HO_2^- \rightarrow HO^{\cdot}O_2^- + O_2 \tag{4.16}$$

Pathway 2 (Equations 4.17 and 4.18):

$$O_3 + O_2^{\cdot -} \rightarrow O_3^{\cdot -} + O_2 \quad (4.17)$$

$$O_3^{\cdot -} + H_2O \rightarrow HO^{\cdot} + HO^{-} + O_2 \quad (4.18)$$

Most processes do not entirely degrade naproxen, IBU, DIC, and clofibric acid. However, the combined approach of O_3/H_2O_2 facilitates the effective elimination of these CECs with 98% mineralization (Zwiener and Frimmel, 2000).

4.4.3.4 Fenton and photo-Fenton processes

The process involves an oxidizing agent composed of iron ions and H_2O_2 in solution, performing in both homogeneous and heterogeneous conditions. This technology favors heterogenous reactions in association with immobilized catalyst. The production of hydroxyl radicals' powers oxidation during the Fenton process which is summarized in the reactions (Equations 4.19–4.23):

$$Fe^{2+} + H_2O_2 \rightarrow Fe^{3+} + HO^{-} + HO^{\cdot} \quad (4.19)$$

$$Fe^{3+} + H_2O_2 \rightarrow H^{+} + Fe(HO_2)^{2+} \quad (4.20)$$

$$Fe(HO_2)^{2+} \rightarrow Fe^{2+} + HO_2^{\cdot} \quad (4.21)$$

$$FeOH^{2+} + H_2O_2 \rightarrow Fe(OH)(HO_2)^{+} + H^{+} \quad (4.22)$$

$$Fe(OH)(HO_2)^{+} \rightarrow Fe^{2+} + HO_2^{\cdot} + HO^{-} \quad (4.23)$$

Once the hydroxyl radicals are formed, they immediately degrade the CECs into non-toxic products (Equation 4.24):

$$HO^{\cdot} + \text{organic substances} \rightarrow H_2O + CO_2 \text{ (degradation products)} \quad (4.24)$$

Solar radiation can also be used as a light source in photo-Fenton process. The solar irradiation during the photo-Fenton process completely oxidized DIC within 60 min. However, the degradation of residual organics was achieved within 100 min resulting in maximum disappearance of dissolved organic carbon. DIC degradation occurred in the homogenous medium controlled by precipitation-dissolution equilibrium. HPLC and GC/MS coupled with TOF-MS were used to recognize the intermediates formed through hypothesized degradation pathways. Photo-Fenton oxidation process resulted in complete degradation of IBU at only 40% of TOC removal. Complexation of the IBU carboxylic acid moiety to Fe^{2+} entails the degradation process (Ravina et al., 2002).

Fenton and photo-Fenton reactions are not suitable for wastewater containing high concentrations of organic, Cl^{-}, HCO_3^{-}, NO_3^{2-}, and CO_3^{2-}, as they all can react with hydroxyl radicals.

4.4.3.5 Photocatalysis

Photolysis can be defined as the oxidation of any compound (compatible for oxidation) in the presence of light irradiation. It can be divided into two photolytic transformation groups such as direct and indirect photolysis. Direct photolysis encompasses the decomposition of organic compounds by direct absorption of UV light, whereas, in indirect decomposition, UV light facilitated the formation of reactive species (radicals) involving catalysts, which in turn oxidize the organic by chemical reactions. There are a number of factors affecting the photolysis such as light intensity, absorption spectra of the compound, quantum yields, formation or availability of O_3, H_2O_2, nascent oxygen [O], and the solution matrix. Photolysis using polychromatic UV light for the degradation of CIP, oxytetracycline (OTC), and SMX was found to be pH dependent. An increase in pH from 5 to 7 increases the degradation of OTC and CIP while the degradation of SMX decreases (Avisar et al., 2010). Photoreduction of isopropanol (1%) initiated the formation of many radicles, improving the degradation of IBU, naproxen, clofibric acid, and DIC (Packer et al., 2003). There was a significant increment (58%–67%) in the photo-degradation of metronidazole when UV/H_2O_2 was employed, while direct exposure to UV only resulted in 6%–12% degradation (Shemer et al., 2006). UV irradiation initiates the formation of the hydroxyl radicals from H_2O_2, which then attacked metronidazole.

Total degradation and mineralization were achieved during the combined process of photocatalysis (via TiO_2) and ozonation (Mendez-Arriaga et al., 2011). Photocatalysts such as TiO_2 have improved the photo-degradation of CECs through charge carriers' formation (electron-hole pair) (Equation 4.25).

$$TiO_2 + hv \rightarrow TiO_2\left(e^- + h^+\right) \quad (4.25)$$

The holes have an extremely high potential for the degradation of CECs. Upon reacting with H_2O molecules, these holes (h^+) produce hydroxyl radical, facilitating organic compounds' degradation. TiO_2-led photo-oxidations facilitated 90% removal of OTC under the sunlight and simulated light (Zhao et al., 2013). pH of the reaction mixture is also significant for the degradation of CBZ photocatalysis using TiO_2, N-doped TiO_2, and ZnO_2 (Avisar et al., 2013). Ions of phosphate hindered the degradation process of OTC by TiO_2 catalysis; however, no changes were observed due to ammonium, nitrate, sulfate, chloride, and bicarbonate ions (Pereira et al., 2013). Under simulated solar irradiation, photocatalyst $BiOCl_{0.875}Br_{0.125}$ removed CBZ, propranolol, IBU, and bezafibrate following the order propranolol > bezafibrate > IBU > IBU > CBZ for the kinetics. Removal of propranolol and CBZ was primarily carried out by photocatalysis, but IBU and bezafibrate were depleted by adsorption phenomenon (Lester et al., 2014).

Moxifloxacin removal using TiO_2/UV photocatalyst occurred through oxidation of carbonyl group leading to hydroxylation and ring-opening at diazobicyclo-substituent (Van Doorslaer et al., 2015). Humic acid significantly affected photocatalytic action of TiO_2 for the removal of IBU as the removal efficiency reduced from 80% to 30% (Zakersalehi et al., 2013). N-doping doubled the photocatalytic activity of the TiO_2 membrane during CBZ degradation (Horovitz et al., 2016). Upon increasing the reaction solution's pH, the photocatalytic action was reduced and only 40% degradation was achieved for CBZ due to the increased alkalinity (100 mg/L as $CaCO_3$) (Avisar et al., 2013). Further, Table 4.3 shows some recently developed technologies for the degradation of CECs.

Table 4.3 Removal of CECs achieved using different treatment technologies

Contaminants	*Treatment technologies*	*Removal percentage (%) and capacities*	*References*
Carbamazepine	Hydrodynamic cavitation (HC) H_2O_2+O_3	100%	Thanekar et al. (2018)
	HC+O_3	91.4%	
	HC+H_2O_2	58.3%	
	HC+ UV	52.9%	
Amoxicillin Cephalexin Ofloxacin Sulfadiazine Sulfamethazine Tetracycline	Graphene	100%	Peng et al. (2016)
Naproxen Ibuprofen Diclofenac	3D reduced graphene oxide hydrogel	70%–80%	Umbreen et al. (2018)
Diclofenac	Mixed-linker Zr-MOFs	99% 2.04 mmol/g	Gao et al. (2019)
sulfamethoxazole	$Cu_2O/Cu_2V_2O_7$ nanocomposite	90.1%	Kumar et al. (2021)
Ciprofloxacin	ZnO nanoparticles	96%	Malakootian et al. (2019)
Ofloxacin	Ag modified ZnO	98%	Kaur et al. (2018)
Ciprofloxacin	Ag/Fe_2O_3/ZnO	76.4%	Kaur et al. (2019)
Antipyrine	ZnO/sepiolite	70%	Akkari et al. (2018)

4.5 RECENT DEVELOPMENTS IN TREATMENT TECHNOLOGIES

4.5.1 Combined processes

Evidently, there is no particular treatment technology that is able to remove CECs from wastewater completely; however, a combination of treatment technologies can undoubtedly act as a prominent solution to this problem. A careful selection of treatment technologies also plays an important role for the effective removal of CECs. During the wastewater treatment stages, toxic effluent can destroy the microbial community (used for biodegradation) if the effluent is directly subjected to the biological treatment system. However, the prior application of AOPS to such effluents can secure the valuable microbial culture from dying, facilitating an effective biodegradation of residual CECs. This indicates that combining technologies can be promising for effective removal of CECs/pollutants, but it does require further research and validation. Organic material can clog the membranes of RO; therefore, applying a carbon-based filter as a pretreatment process can reduce the organic load and increase the life of the RO system. RO and Fenton process were combined together to check the removal of AMX and it was observed that almost 99% removal was achieved (Zhang et al., 2006). Solar-driven photo-Fenton and biological treatment together eliminated 95% of the quinolone antibiotics; however, Fenton alone can only remove 33% of antibiotics (Sirtori et al., 2009). Removal of imidazole was achieved up to 90%–100% using the combined ozonation and adsorption approach simultaneously (Sánchez-Polo et al., 2008). Catalytic ozonation and adsorption together have the potential for effective removal of pharmaceuticals. Here, degradation of CECs was facilitated by ozonation and residual CECs and its intermediates were adsorbed by the adsorbent. Fe_2O_3/CeO_2-loaded AC successfully removed the SMX using both catalytic ozonation and adsorption process (Akhtar et al., 2011). TC was efficiently adsorbed and photo-degraded catalytically using mesoporous bio-microspheres (Hao et al., 2012).

4.5.2 Integration of membrane technology

Conventional high-pressure polymer membranes, NF and RO, have shown effective rejection of various classes of CECs; however, these membranes, in particular NF, with an MWCO of 200 Da or higher offer poor performance toward polar organic molecules with low molecular weight. Membrane-driven removal of CECs follows complex removal mechanisms governed by the targeted compounds' physicochemical properties and their specific and synergistic interactions with the water components. Studies on the use of membranes decorated with nanocomposite to remove newly emerging organic pollutants are very few. NF membrane coated with chitosan and silver showed an increase in antibacterial activity and up to 84% repulsion of selected CECs. The nanocomposite-decorated membranes were positively charged. The CECs removal followed a sequence as atenolol > CBZ > IBU, which was positively, neutrally, and negatively charged at an experimental pH of 7 (Huang et al., 2016). A single-walled carbon nanotube (SWNT) reactor coupled with a commercial UF membrane unit (bench-scale dead-end stirred cell) was investigated for the removal of bisphenol A (BPA) and 17βestradiol (E2) from synthetic laboratory water. It was observed that the retention of endocrine-disrupting compounds increased by 77% due to the presence of SWNTs (Heo et al., 2012). Photo-regenerable multi-walled carbon nanotubes-embedded titanium dioxide (MWNT-TiO_2) nanocomposite membranes (having porous, spherical morphology) was used to remove pharmaceuticals such as CBZ, IBU, and ACT from water. Initially, the MWNT-TiO_2 membrane tremendously removed CBZ, IBU, and ACT with 80%, 45%, and 24% of removal efficiency, respectively. However, after regeneration of membranes, the removal potential was decreased to 55%, 32%, and 19% for CBZ, IBU, and ACT, respectively, because of exhausting active adsorptive sites of the membrane (Zaib et al., 2013). Likewise, combined SWNT and MWNT membrane was used in the filtration of TCS, ACT, and IBU. The removal varied from ~10% to 95% and increased with an increasing number of aromatic rings (AAP ≈ IBU < TCS). The removal of PPCPs was also influenced by varying pH from 4 to 10, leading to up to 70% removal at higher pH. It was also observed that the neutral PPCP molecules exhibited more excellent removal with ions because of reduced electrostatic repulsion (TCS and IBU) or formation of H_2 bond (AAP) (Wang et al., 2015). A spherical nanoscale polymer adsorbent was synthesized in polyethersulfone matrices and charged with trimethylolpropane trimethacrylate (hydrophobic agent) and 2-hydroxy-3-methacryloxypropyltrimethylammonium chloride (lipophilic agent). The modified membrane has a high adsorption capacity for BPA (98.6%) because of hydrophobic nature of membrane. Membrane also exhibited good regeneration properties and demonstrated absorption of 97.3% and 97.8% of BPA in the second and thirds cycles, respectively (Niedergall et al., 2014). CEC rejection by nanocomposite membranes seems to be compound-specific and dependent on the properties of the incorporated nanomaterials and the matrix of the substrate membranes. Research on CECs removal using membranes (both conventional polymer and decorated with nanocomposites) was carried out mainly in synthetic waters enriched with compounds with specific properties. For pilot-scale implementation of membranes decorated with nanocomposites for CECs removal, it is necessary to investigate the effect of the components/ionic species present in the aqueous matrix.

4.5.3 Heterogenous photocatalysis

The major disadvantage of using the TiO_2 as a photocatalyst is its wide bandgap (3.2 eV), limiting its usage to the UV region only. However, the photocatalytic activity of such photocatalysts could be enhanced by incorporating secondary metal/metal oxide, extending the light absorption to the visible region. The coupling of metal/metal oxide with TiO_2 can significantly enhance the photogeneration of charge carriers (electron-hole pair) and interfacial electron transfer rates and reduce the recombination rate of charge carriers. The photoactivity of platinum (Pt)-loaded titania (Pt/TiO_2) was compared to gold (Au)- and silver (Ag)-loaded TiO_2 under solar irradiations. It was observed that Pt/TiO_2 exhibited the highest photo-degradation capacity for ACT under sunlight attributing to its enhanced electron trapping capability (Nasr et al., 2019). Further, it was also revealed that the amount of components in photocatalyst could also have an impact on removal efficiency. For instance, the highest CBZ removal was achieved using graphene-TiO_2-P25 nanocomposite when 1.5 g of graphene was used as weight percentage (Appavoo et al., 2014). A highly efficient visible light active heterostructure nitrogen (N), sulfur (S) co-doped TiO_2@MoS_2 nanocomposite was fabricated for the photo-degradation of DIC. MoS_2 was coupled with N, S-doped TiO_2 (in the lattice), forming a heterojunction with huge photoactivity in visible light and showed removal five times and four times higher than that of MoS_2 and N, S co-doped TiO_2, respectively (Irandost et al., 2019). The nanocomposite graphene oxide/titania catalyst (GO/TiO_2) was found as a potential visible light-driven photocatalyst for removing the mixture of multiple emerging contaminants. A heterogenous photoexcited sodium decatungstate ($Na_4W_{10}O_{32}$) was employed for the photo-degradation of atenolol, levofloxacin, and trimethoprimin in aqueous matrix operating under an ambient environment (Molinari et al., 2017). Some of the more advancements have also been made such as development of magnetic nanocomposites with simultaneous photocatalytic action. Such materials can allow their easy regeneration from the water matrix upon application of an external magnetic field and reduce the unintentional residual concentration in the environment (Fanourakis et al., 2020). Recently, magnetic photocatalysts $FeNi_3/SiO_2/CuS$ showed removal of TC, while materials like TiO_2-GO-Fe_3O_4 and mesoporous graphitic carbon nitride showed enhanced reduction for AMX (Nasseh et al., 2018).

4.6 COMPARISON OF DIFFERENT ADVANCED TREATMENT TECHNOLOGIES

To meet the current demands of potable water, industrial and domestic requirements, and to reduce the water-associated health impacts, we urgently need some substantial treatment options (advanced water treatments) to remove these pollutants. Various traditional methods are available to eliminate these contaminants, but they are still ineffective in removing these pollutants. Simultaneously, advanced treatment technologies such as adsorption, membrane filtration, EO, ozonation, sonication, Fenton reaction, and photocatalysis have gained much attention for CECs degradation. The reduction percentage of CECs using O_3 was assigned to various categories such as A, B, and C (group A: azithromycin, BPA, CBZ, CIP, clarithromycin, DIC, erythromycin, metoprolol, SMX, and the hormones 17-alpha-ethinylestradiol and 17-beta-estradiol; group B: benzotriazole, bezafibrate, mecoprop, and methylene benzotriazole; and group C: acesulfame, iopromide, and primidone) with specific ozone doses in the range 0.1–1.6 g O_3/g of dissolved organic carbon that was enough to abate >80%, 50%–80%, and <50%, respectively (Lee et al., 2014). Adsorption of 22 pharmaceuticals was achieved using powdered AC in pilot plant scale setup. Powdered AC tremendously attained 95% removal of 22 pharmaceuticals (Kårelid et al., 2017). Engineered biochar with high specific surface area and active binding sites exhibited very tremendous adsorption capacity against DIC, TCS, and naproxen for 92.7, 113, and 127 mg/g, respectively (Czech et al., 2021). The ultrasonication with different frequencies (28, 580, and 1,000 kHz) was used to degrade various endocrine disruptors such as BPA, 17β-estradiol (E2), and 17α-ethinyl estradiol (EE2). It was observed that the different frequencies showed different degradation percentage: 580 kHz (92%–97%) > 1,000 kHz (90%–94%) > 28 kHz (62%–67%) at an initial concentration of 1 μM of each compound with a contact time of 30 min (Park et al., 2011). A series of ultrathin h-BN/Bi_2MoO_6 heterojunction with excellent photocatalytic activity exhibited effective photo-degradation of TC, OTC, and doxycycline up to 99.19%, 95.28%, and 91.04% under visible light irradiation, respectively, due to the synergistic effect among heterojunction and pharmaceuticals (Du et al., 2021). CIP and fluoxetine were removed from hospital effluent by using Photo-Fenton process. At pH 4.5, the fluoxetine was significantly removed up to 80% with iron citrate after 20 min, whereas only 36% degradation was attained with iron nitrate under similar experimental conditions. However, CIP was degraded up to 86% with iron citrate, while with iron nitrate, the CIP degradation achieved only up to 75% after 20 min (de Lima et al., 2017). A comparative analysis is given in Table 4.4, highlighting the pros and cons of CECs removal technologies.

4.7 CONCLUSIONS AND FUTURE PERSPECTIVES

The advanced treatment systems have become a mounting and vigorous research area as they can effectively treat a wide range of CECs; however, the complete removal of many CECs has not been achieved. Various treatment (physicochemical and biological) technologies have been investigated for the effective removal of CECs. Physical processes that transfer the solute (pollutant) from aqueous to the surface of an adsorbent, such as AC by adsorption phenomenon and physical separation driven by RO/NF membranes, were useful for many CECs removal but only up to a certain extent. Biological treatment technologies include traditional ASP, MBR, and microalgae-based treatment; they also exhibit potential for the biodegradation of various pharmaceuticals. These removal techniques have many disadvantages, such as high retention time for the treatment (ASP) and sludge disposal containing CECs. AOPs (chemical-based), including Fenton, photo-Fenton, ozonation, and photocatalysis, are

Table 4.4 Advantages and disadvantages of technologies (Crini and Lichtfouse, 2019)

Technology	*Mechanisms*	*Advantages*	*Disadvantages*
Adsorption	Surface phenomena	• Wide range of target pollutants • High proficient (adsorption) method with fast kinetics • High removal efficiency	• Costly method • Nonselective processes • Adsorbent dependent • Costly regeneration
Membrane Filtration	Separation through semipermeable membrane	• Fast and efficient process • Viable to work at high concentration • Zero chemical requirement • Low waste generation	• Confined flow rate • High power required • High service and function cost • Rapid membrane clogging
Electrochemical Oxidation	Electrolysis (occurrence of redox reaction)	• Efficient technology for the recovery/ recycling • Low power required • Adaptation to various pollutants load and flow rates	• High-cost maintenance (sacrificial anodes) • Chemicals required • Anode passivation • Sludge deposition
Ozonation	Reactive oxygen species generation	• Fast and efficient process • On-site ozone generation • Integrated physiochemical process • Enhance biodegradability of product	• Chemical intensive • Dependent on production, transportation, and management of the ROS • Efficiency influenced by oxidant • Costly method
Photocatalysis	Reactive oxygen species generation	• Easy operation • Highly efficient • Destruct organic pollutants	• Costly method • Depends on range of light • Costly regeneration
Fenton Reaction	Haber–Weiss reaction	• Easy operation • Biodegradability improvement • Established technology	• Sludge deposition • Requires neutralization of pH • Operation in acidic condition

among the most efficient technologies for removing various classes of the CECs. However, large-scale implementation has many limitations and challenges due to their high operational and maintenance costs.

To overcome these limitations, hybrid systems are being used in recent time. AOPs generate many by-products or metabolites, which exhibit toxicity, and the adsorption process can effectively remove these metabolites resulting in achieving 100% removal of CECs. Therefore, advanced tertiary treatment processes can be used in combinations with adsorption. The hybrid systems consisting of AOPs, such as MBR + UV,

UF + ozonation, and AC, have proven to be effective CECs removal procedures, providing near-complete removal of many CECs. Some advanced treatment options such as combinations of UV/H_2O_2/ozone, photo-Fenton, biological treatment, and heterogenous photocatalysts (nanocomposites) have resulted in the efficient removal of CECs. Therefore, it can be concluded that several treatment technologies can be implemented in combination to achieve an effective removal of CECs.

However, knowingly and unknowingly, new CECs are continuously being detected in environmental settings. Due to the insufficient knowledge of their environmental fate, toxicity, persistence, health impacts, and remediation, their eradication is of enormous concern. At this moment, there is a dearth of drinking water guidelines for the regulation of CECs. Therefore, we need to develop regulative policies and stringently implement them to recycle the water for meeting increasing current clean water demand worldwide. Some of the recommendations for CECs regulation and removal are listed below:

- Special attention is required in developing countries for monitoring of CECs.
- The analytical methods for the detection of CECs should be developed and applied.
- Continued research is required to understand the designs, kinetics, effect of the operational factors, and the interaction mechanisms related to the degradation of CECs.
- Strict regulations for point-source effluents must be implemented.
- The acute and chronic effect of CECs exposure toward humans and their environment should be explored thoroughly.
- Integration of new hybrid treatment technologies (integration of Fenton, ozonation, UV, and nanocomposites) can be taken into consideration.
- CECs guidelines for effluent discharge from industries must be formulated and implemented.
- Treatment facilities should be implemented with more advanced technologies, such as hybrid systems.

ACKNOWLEDGMENTS

DP and AS acknowledge the University Grant Commission, New Delhi, India for providing Senior & Junior Research Fellowship, respectively. The Director of CSIR-NEERI is thankfully acknowledged for giving the opportunity to pursue the work in CSIR-NEERI, Nagpur, India. The chapter is checked for plagiarism using the iThenticate software and recorded in the Knowledge Resource Center, CSIR-NEERI, Nagpur for anti-plagiarism (KRC No.: CSIR-NEERI/KRC/2020/NOV/WWT/1).

NOTE

Sukdeb Pal is the corresponding author.

REFERENCES

Ahmed, M.B., Zhou, J.L., Ngo, H.H., Guo, W., Thomaidis, N.S. and Xu, J., 2017. Progress in the biological and chemical treatment technologies for emerging contaminant removal from wastewater: A critical review. *Journal of Hazardous Materials*, 323, pp. 274–298. https://doi.org/10.1016/j.jhazmat.2016.04.045

Akhtar, J., Amin, N.S. and Aris, A., 2011. Combined adsorption and catalytic ozonation for removal of sulfamethoxazole using Fe_2O_3/CeO_2 loaded activated carbon. *Chemical Engineering Journal*, 170(1), pp. 136–144. https://doi.org/10.1016/j.cej.2011.03.043

Akkari, M., Aranda, P., Belver, C., Bedia, J., Amara, A.B.H. and Ruiz-Hitzky, E., 2018. ZnO/sepiolite heterostructured materials for solar photocatalytic degradation of pharmaceuticals in wastewater. *Applied Clay Science*, 156, pp. 104–109. https://doi.org/10.1016/j.clay.2018.01.021

Ambade, B., Sethi, S.S., Kurwadkar, S., Kumar, A. and Sankar, T.K., 2021. Toxicity and health risk assessment of polycyclic aromatic hydrocarbons in surface water, sediments and groundwater vulnerability in Damodar River Basin. *Groundwater for Sustainable Development*, 13, p. 100553. https://doi.org/10.1016/j.gsd.2021.100553

Appavoo, I.A., Hu, J., Huang, Y., Li, S.F.Y. and Ong, S.L., 2014. Response surface modeling of Carbamazepine (CBZ) removal by Graphene-P25 nanocomposites/UVA process using central composite design. *Water Research*, 57, pp. 270–279. https://doi.org/10.1016/j.watres.2014.03.007

Ariffin, M. and Zakili, T.S.T., 2019. Household pharmaceutical waste disposal in Selangor, Malaysia—policy, public perception, and current practices. *Environmental Management*, 64(4), pp. 509–519. https://doi.org/10.1007/s00267-019-01199–y

Aus der Beek, T., Weber, F.A., Bergmann, A., Hickmann, S., Ebert, I., Hein, A. and Küster, A., 2016. Pharmaceuticals in the environment—Global occurrences and perspectives. *Environmental Toxicology and Chemistry*, 35(4), pp. 823–835. https://doi.org/10.1002/etc.3339

Avisar, D., Horovitz, I., Lozzi, L., Ruggieri, F., Baker, M., Abel, M.L. and Mamane, H., 2013. Impact of water quality on removal of carbamazepine in natural waters by N-doped TiO_2 photo-catalytic thin film surfaces. *Journal of Hazardous Materials*, 244, pp. 463–471. https://doi.org/10.1016/j.jhazmat.2012.09.058

Avisar, D., Lester, Y. and Mamane, H., 2010. pH induced polychromatic UV treatment for the removal of a mixture of SMX, OTC and CIP from water. *Journal of Hazardous Materials*, 175(1–3), pp. 1068–1074. https://doi.org/10.1016/j.jhazmat.2009.10.122

Azzouz, A. and Ballesteros, E., 2013. Influence of seasonal climate differences on the pharmaceutical, hormone and personal care product removal efficiency of a drinking water treatment plant. *Chemosphere*, 93(9), pp. 2046–2054. https://doi.org/10.1016/j.chemosphere.2013.07.037

Baccar, R., Sarrà, M., Bouzid, J., Feki, M. and Blánquez, P., 2012. Removal of pharmaceutical compounds by activated carbon prepared from agricultural by-product. *Chemical Engineering Journal*, 211, pp. 310–317. https://doi.org/10.1016/j.cej.2012.09.099

Bártíková, H., Podlipná, R. and Skálová, L., 2016. Veterinary drugs in the environment and their toxicity to plants. *Chemosphere*, 144, pp. 2290–2301. https://doi.org/10.1016/j.chemosphere.2015.10.137

Behera, S.K., Kim, H.W., Oh, J.E. and Park, H.S., 2011. Occurrence and removal of antibiotics, hormones and several other pharmaceuticals in wastewater treatment plants of the largest industrial city of Korea. *Science of the Total Environment*, 409(20), pp. 4351–4360. https://doi.org/10.1016/j.scitotenv.2011.07.015

Braeken, L. and Van der Bruggen, B., 2009. Feasibility of nanofiltration for the removal of endocrine disrupting compounds. *Desalination*, 240(1–3), pp. 127–131. https://doi.org/10.1016/j.desal.2007.11.069

Cai, M.Q., Wang, R., Feng, L. and Zhang, L.Q., 2015. Determination of selected pharmaceuticals in tap water and drinking water treatment plant by high-performance liquid chromatography-triple quadrupole mass spectrometer in Beijing, China. *Environmental Science and Pollution Research*, 22(3), pp. 1854–1867. https://doi.org/10.1007/s11356-014-3473–8

Cincinelli, A., Martellini, T., Coppini, E., Fibbi, D. and Katsoyiannis, A., 2015. Nanotechnologies for removal of pharmaceuticals and personal care products from water and wastewater. A review. *Journal of Nanoscience and Nanotechnology*, 15(5), pp. 3333–3347. https://doi.org/10.1166/jnn.2015.10036

Carlsson, C., Johansson, A. K., Alvan, G., Bergman, K., and Kühler, T., 2006. Are pharmaceuticals potent environmental pollutants?: Part I: Environmental risk assessments of selected active pharmaceutical ingredients. *Science of the Total Environment*, 364(1–3), pp. 67–87.

Clara, M., Strenn, B., Gans, O., Martinez, E., Kreuzinger, N. and Kroiss, H., 2005. Removal of selected pharmaceuticals, fragrances and endocrine disrupting compounds in a membrane bioreactor and conventional wastewater treatment plants. *Water Research*, 39(19), pp. 4797–4807. https://doi.org/10.1016/j.watres.2005.09.015

Crini, G. and Lichtfouse, E., 2019. Advantages and disadvantages of techniques used for wastewater treatment. *Environmental Chemistry Letters*, 17(1), pp. 145–155. https://doi.org/10.1007/s10311-018-0785-9

Czech, B., Kończak, M., Rakowska, M. and Oleszczuk, P., 2021. Engineered biochars from organic wastes for the adsorption of diclofenac, naproxen and triclosan from water systems. *Journal of Cleaner Production*, 288, p. 125686.

Dai, G., Wang, B., Fu, C., Dong, R., Huang, J., Deng, S., Wang, Y. and Yu, G., 2016. Pharmaceuticals and personal care products (PPCPs) in urban and suburban rivers of Beijing, China: Occurrence, source apportionment and potential ecological risk. *Environmental Science: Processes & Impacts*, 18(4), pp. 445–455. https://doi.org/10.1039/C6EM00018E

Dai, G., Wang, B., Huang, J., Dong, R., Deng, S. and Yu, G., 2015. Occurrence and source apportionment of pharmaceuticals and personal care products in the Beiyun River of Beijing, China. *Chemosphere*, 119, pp. 1033–1039.

de Lima Perini, J.A., e Silva, B.C., Tonetti, A.L. and Nogueira, R.F.P., 2017. Photo-Fenton degradation of the pharmaceuticals ciprofloxacin and fluoxetine after anaerobic pre-treatment of hospital effluent. *Environmental Science and Pollution Research*, 24(7), pp. 6233–6240.

Doummar, J. and Aoun, M., 2018. Assessment of the origin and transport of four selected emerging micropollutants sucralose, Acesulfame-K, gemfibrozil, and iohexol in a karst spring during a multi-event spring response. *Journal of Contaminant Hydrology*, 215, pp. 11–20. https://doi.org/10.1016/j.jconhyd.2018.06.003

Du, Z., Feng, L., Guo, Z., Yan, T., Hu, Q., Lin, J., Huang, Y., Tang, C. and Fang, Y., 2021. Ultrathin h-BN/Bi2MoO6 heterojunction with synergetic effect for visible-light photocatalytic tetracycline degradation. *Journal of Colloid and Interface Science*, 589, pp. 545–555.

Dubey, M., Mohapatra, S., Tyagi, V. K., Suthar, S., & Kazmi, A. A., 2021. Occurrence, fate, and persistence of emerging micropollutants in sewage sludge treatment. *Environmental Pollution*, 273, pp. 116515.

Fanourakis, S.K., Peña-Bahamonde, J., Bandara, P.C. and Rodrigues, D.F., 2020. Nano-based adsorbent and photocatalyst use for pharmaceutical contaminant removal during indirect potable water reuse. *NPJ Clean Water*, 3(1), pp. 1–15.

Ferre-Aracil, J., Valcárcel, Y., Negreira, N., de Alda, M.L., Barceló, D., Cardona, S.C. and Navarro-Laboulais, J., 2016. Ozonation of hospital raw wastewaters for cytostatic compounds removal. Kinetic modelling and economic assessment of the process. *Science of the Total Environment*, 556, pp. 70–79. https://doi.org/10.1016/j.scitotenv.2016.02.202

Ganiyu, S.O., Van Hullebusch, E.D., Cretin, M., Esposito, G. and Oturan, M.A., 2015. Coupling of membrane filtration and advanced oxidation processes for removal of pharmaceutical residues: A critical review. *Separation and Purification Technology*, 156, pp. 891–914. https://doi.org/10.1016/j.seppur.2015.09.059

Gao, P., Ding, Y., Li, H. and Xagoraraki, I., 2012. Occurrence of pharmaceuticals in a municipal wastewater treatment plant: Mass balance and removal processes. *Chemosphere*, 88(1), pp. 17–24. https://doi.org/10.1016/j.chemosphere.2012.02.017

Gao, Y., Xia, J., Liu, D., Kang, R., Yu, G. and Deng, S., 2019. Synthesis of mixed-linker Zr-MOFs for emerging contaminant adsorption and photodegradation under visible light. *Chemical Engineering Journal*, 378, p. 122118. https://doi.org/10.1016/j.cej.2019.122118

García-Espinoza, J.D. and Nacheva, P.M., 2019. Degradation of pharmaceutical compounds in water by oxygenated electrochemical oxidation: Parametric optimization, kinetic studies and toxicity assessment. *Science of the Total Environment*, 691, pp. 417–429. https://doi.org/10.1016/j.scitotenv.2019.07.118

Gupta, V.K., Gupta, B., Rastogi, A., Agarwal, S. and Nayak, A., 2011. Pesticides removal from waste water by activated carbon prepared from waste rubber tire. *Water Research*, 45(13), pp. 4047–4055. https://doi.org/10.1016/j.watres.2011.05.016

Hao, R., Xiao, X., Zuo, X., Nan, J. and Zhang, W., 2012. Efficient adsorption and visible-light photocatalytic degradation of tetracycline hydrochloride using mesoporous BiOI microspheres. *Journal of Hazardous Materials*, 209, pp. 137–145. https://doi.org/10.1016/j.jhazmat.2012.01.006

Heberer, T., 2002. Occurrence, fate, and removal of pharmaceutical residues in the aquatic environment: A review of recent research data. *Toxicology Letters*, 131(1–2), pp. 5–17. https://doi.org/10.1016/S0378–4274(02)00041–3

Heo, J., Flora, J.R., Her, N., Park, Y.G., Cho, J., Son, A. and Yoon, Y., 2012. Removal of bisphenol A and 17β-estradiol in single walled carbon nanotubes–ultrafiltration (SWNTs–UF) membrane systems. *Separation and Purification Technology*, 90, pp. 39–52. https://doi.org/10.1016/j.seppur.2012.02.007

Horovitz, I., Avisar, D., Baker, M.A., Grilli, R., Lozzi, L., Di Camillo, D. and Mamane, H., 2016. Carbamazepine degradation using a N-doped TiO_2 coated photocatalytic membrane reactor: Influence of physical parameters. *Journal of Hazardous Materials*, 310, pp. 98–107. https://doi.org/10.1016/j.jhazmat.2016.02.008

Huang, Z.H., Yin, Y.N., Aikebaier, G.L.M.L. and Zhang, Y., 2016. Preparation of a novel positively charged nanofiltration composite membrane incorporated with silver nanoparticles for pharmaceuticals and personal care product rejection and antibacterial properties. *Water Science and Technology*, 73(8), pp. 1910–1919.

Huber, M.M., Canonica, S., Park, G.Y. and Von Gunten, U., 2003. Oxidation of pharmaceuticals during ozonation and advanced oxidation processes. *Environmental Science & Technology*, 37(5), pp. 1016–1024. https://doi.org/10.1021/es025896h

Irandost, M., Akbarzadeh, R., Pirsaheb, M., Asadi, A., Mohammadi, P. and Sillanpää, M., 2019. Fabrication of highly visible active N, S co-doped TiO_2@ MoS_2 heterojunction with synergistic effect for photocatalytic degradation of diclofenac: Mechanisms, modeling and degradation pathway. *Journal of Molecular Liquids*, 291, p. 111342. https://doi.org/10.1016/j.molliq.2019.111342

Jiang, J. and Zhou, Z., 2013. Removal of pharmaceutical residues by ferrate (VI). *PLoS One*, 8(2), p. e55729. https://doi.org/10.1371/journal.pone.0055729

Jin, P., Jin, X., Wang, X., Feng, Y. and Wang, X.C., 2013. Biological activated carbon treatment process for advanced water and wastewater treatment. In: Miodrag Darko Matovic (ed) Biomass Now-Cultivation and Utilization, pp. 153–192. https://www.intechopen.com/books/3138

Jones, O. A., Lester, J. N., and Voulvoulis, N., 2005. Pharmaceuticals: a threat to drinking water?. *Trends in Biotechnology*, 23(4), pp. 163–167.

Joss, A., Keller, E., Alder, A.C., Göbel, A., McArdell, C.S., Ternes, T. and Siegrist, H., 2005. Removal of pharmaceuticals and fragrances in biological wastewater treatment. *Water Research*, 39(14), pp. 3139–3152. https://doi.org/10.1016/j.watres.2005.05.031

Kårelid, V., Larsson, G., Björlenius, B., 2017. Pilot-scale removal of pharmaceuticals in municipal wastewater: Comparison of granular and powdered activated carbon treatment at three wastewater treatment plants. *Journal of Environmental Management*. 193, 491–502. https://doi.org/10.1016/j.jenvman.2017.02.042

Kaur, A., Anderson, W.A., Tanvir, S. and Kansal, S.K., 2019. Solar light active silver/iron oxide/zinc oxide heterostructure for photodegradation of ciprofloxacin, transformation products and antibacterial activity. *Journal of Colloid and Interface Science*, 557, pp. 236–253. https://doi.org/10.1016/j.jcis.2019.09.017

Kaur, A., Gupta, G., Ibhadon, A.O., Salunke, D.B., Sinha, A.S.K. and Kansal, S.K., 2018. A Facile synthesis of silver modified ZnO nanoplates for efficient removal of ofloxacin drug in aqueous phase under solar irradiation. *Journal of Environmental Chemical Engineering*, 6(3), pp. 3621–3630. https://doi.org/10.1016/j.jece.2017.05.032

Kimura K., Toshima S., Amy G. and Watanabe Y., 2004. Rejection of neutral endocrine disrupting compounds (EDCs) and pharmaceutical active compounds (PhACs) by RO membranes. *Journal of Membrane Science*, 245(1–2): 71–78. https://doi.org/10.1016/j.memsci.2003.09.005

Kumar, A., Sharma, S.K., Sharma, G., Guo, C., Vo, D.V.N., Iqbal, J., Naushad, M. and Stadler, F.J., 2021. Silicate glass matrix@ Cu2O/Cu2V2O7 pn heterojunction for enhanced visible light photo-degradation of sulfamethoxazole: High charge separation and interfacial transfer. *Journal of Hazardous Materials*, 402, p. 123790. https://doi.org/10.1016/j.jhazmat.2020.123790

Lan, Y., Coetsier, C., Causserand, C. and Serrano, K.G., 2017. On the role of salts for the treatment of wastewaters containing pharmaceuticals by electrochemical oxidation using a boron doped diamond anode. *Electrochimica Acta*, 231, pp. 309–318. https://doi.org/10.1016/j.electacta.2017.01.160

Lapworth, D.J., Das, P., Shaw, A., Mukherjee, A., Civil, W., Petersen, J.O., Gooddy, D.C., Wakefield, O., Finlayson, A., Krishan, G. and Sengupta, P., 2018. Deep urban groundwater vulnerability in India revealed through the use of emerging organic contaminants and residence time tracers. *Environmental Pollution*, 240, pp. 938–949. https://doi.org/10.1016/j.envpol.2018.04.053

Lee, H.J., Kim, K.Y., Hamm, S.Y., Kim, M., Kim, H.K. and Oh, J.E., 2019. Occurrence and distribution of pharmaceutical and personal care products, artificial sweeteners, and pesticides in groundwater from an agricultural area in Korea. *Science of the Total Environment*, 659, pp. 168–176. https://doi.org/10.1016/j.scitotenv.2018.12.258

Lee, Y., Kovalova, L., McArdell, C.S., von Gunten, U., 2014. Prediction of micropollutant elimination during ozonation of a hospital wastewater effluent. *Water Res.* 64, 134–148. https://doi.org/10.1016/j.watres.2014.06.027

Lester, Y., Avisar, D., Gnayem, H., Sasson, Y., Shavit, M. and Mamane, H., 2014. Demonstrating a new BiOCl 0.875 Br 0.125 photocatalyst to degrade pharmaceuticals under solar irradiation. *Water, Air, & Soil Pollution*, 225(9), pp. 1–11. https://doi.org/10.1007/s11270-014-2132–5

Li, Z., Xiang, X., Li, M., Ma, Y., Wang, J. and Liu, X., 2015. Occurrence and risk assessment of pharmaceuticals and personal care products and endocrine disrupting chemicals in reclaimed water and receiving groundwater in China. *Ecotoxicology and Environmental Safety*, 119, pp. 74–80. https://doi.org/10.1016/j.ecoenv.2015.04.031

Liu, Y.J., Hu, C.Y. and Lo, S.L., 2019. Direct and indirect electrochemical oxidation of amine-containing pharmaceuticals using graphite electrodes. *Journal of Hazardous Materials*, 366, pp. 592–605. https://doi.org/10.1016/j.jhazmat.2018.12.037

Luo, Y., Guo, W., Ngo, H.H., Nghiem, L.D., Hai, F.I., Zhang, J., Liang, S. and Wang, X.C., 2014. A review on the occurrence of micropollutants in the aquatic environment and their fate and removal during wastewater treatment. *Science of the Total Environment*, 473, pp. 619–641. https://doi.org/10.1016/j.scitotenv.2013.12.065

Malakootian, M., Mahdizadeh, H., Dehdarirad, A. and Amiri Gharghani, M., 2019. Photocatalytic ozonation degradation of ciprofloxacin using ZnO nanoparticles immobilized on the surface of stones. *Journal of Dispersion Science and Technology*, 40(6), pp. 846–854. https://doi.org/10.1080/01932691.2018.1485580

Mendez-Arriaga, F., Otsu, T., Oyama, T., Gimenez, J., Esplugas, S., Hidaka, H. and Serpone, N., 2011. Photooxidation of the antidepressant drug Fluoxetine (Prozac®) in aqueous media by hybrid catalytic/ozonation processes. *Water Research*, 45(9), pp. 2782–2794. https://doi.org/10.1016/j.watres.2011.02.030

Mo, W.Y., Chen, Z., Leung, H.M. and Leung, A.O.W., 2017. Application of veterinary antibiotics in China's aquaculture industry and their potential human health risks. *Environmental Science and Pollution Research*, 24(10), pp. 8978–8989. https://doi.org/10.1007/s11356-015-5607-z

Molinari, A., Sarti, E., Marchetti, N. and Pasti, L., 2017. Degradation of emerging concern contaminants in water by heterogeneous photocatalysis with Na4W10O32. *Applied Catalysis B: Environmental*, 203, pp. 9–17. https://doi.org/10.1016/j.apcatb.2016.09.031

Moussavi, G., Hosseini, H. and Alahabadi, A., 2013. The investigation of diazinon pesticide removal from contaminated water by adsorption onto NH4Cl-induced activated carbon. *Chemical Engineering Journal*, 214, pp. 172–179. https://doi.org/10.1016/j.cej.2012.10.034

Nakada, N., Tanishima, T., Shinohara, H., Kiri, K. and Takada, H., 2006. Pharmaceutical chemicals and endocrine disrupters in municipal wastewater in Tokyo and their removal during activated sludge treatment. *Water Research*, 40(17), pp. 3297–3303. https://doi.org/10.1016/j.watres.2006.06.039

Nasr, O., Mohamed, O., Al-Shirbini, A.S. and Abdel-Wahab, A.M., 2019. Photocatalytic degradation of acetaminophen over Ag, Au and Pt loaded TiO_2 using solar light. *Journal of Photochemistry and Photobiology A: Chemistry*, 374, pp. 185–193. https://doi.org/10.1016/j.jphotochem.2019.01.032

Nasseh, N., Taghavi, L., Barikbin, B. and Nasseri, M.A., 2018. Synthesis and characterizations of a novel $FeNi_3/SiO_2/CuS$ magnetic nanocomposite for photocatalytic degradation of tetracycline in simulated wastewater. *Journal of Cleaner Production*, 179, pp. 42–54.

Niedergall, K., Bach, M., Hirth, T., Tovar, G.E. and Schiestel, T., 2014. Removal of micropollutants from water by nanocomposite membrane adsorbers. *Separation and Purification Technology*, 131, pp. 60–68. https://doi.org/10.1016/j.seppur.2014.04.032

Packer, J.L., Werner, J.J., Latch, D.E., McNeill, K. and Arnold, W.A., 2003. Photochemical fate of pharmaceuticals in the environment: Naproxen, diclofenac, clofibric acid, and ibuprofen. *Aquatic Sciences*, 65(4), pp. 342–351.

Park, J.S., Her, N. and Yoon, Y., 2011. Ultrasonic degradation of bisphenol A, 17β-estradiol, and 17α-ethinyl. *Desalination and Water Treatment*, 30(1–3), pp. 300–309.

Patel, M., Kumar, R., Kishor, K., Mlsna, T., Pittman Jr, C.U. and Mohan, D., 2019. Pharmaceuticals of emerging concern in aquatic systems: chemistry, occurrence, effects, and removal methods. *Chemical Reviews*, 119(6), pp. 3510–3673.

Peng, B., Chen, L., Que, C., Yang, K., Deng, F., Deng, X., Shi, G., Xu, G. and Wu, M., 2016. Adsorption of antibiotics on graphene and biochar in aqueous solutions induced by π-π interactions. *Scientific Reports*, 6, 1–10. https://doi.org/10.1038/srep31920

Pereira, J.H., Reis, A.C., Queirós, D., Nunes, O.C., Borges, M.T., Vilar, V.J. and Boaventura, R.A., 2013. Insights into solar TiO_2-assisted photocatalytic oxidation of two antibiotics employed in aquatic animal production, oxolinic acid and oxytetracycline. *Science of the Total Environment*, 463, pp. 274–283.

Prauchner, M.J., Sapag, K. and Rodríguez-Reinoso, F., 2016. Tailoring biomass-based activated carbon for CH_4 storage by combining chemical activation with H_3PO_4 or $ZnCl_2$ and physical activation with CO_2. *Carbon*, 110, pp. 138–147.

Ravina, M., Campanella, L. and Kiwi, J., 2002. Accelerated mineralization of the drug diclofenac via Fenton reactions in a concentric photo-reactor. *Water Research*, 36(14), pp. 3553–3560.

Rizzo, L., Meric, S., Kassinos, D., Guida, M., Russo, F. and Belgiorno, V., 2009. Degradation of diclofenac by TiO_2 photocatalysis: UV absorbance kinetics and process evaluation through a set of toxicity bioassays. *Water Research*, 43(4), pp. 979–988.

Rovani, S., Censi, M.T., Pedrotti Jr, S.L., Lima, É.C., Cataluña, R. and Fernandes, A.N., 2014. Development of a new adsorbent from agro-industrial waste and its potential use in endocrine disruptor compound removal. *Journal of Hazardous Materials*, 271, pp. 311–320.

Sánchez-Polo, M., Rivera-Utrilla, J., Prados-Joya, G., Ferro-García, M.A. and Bautista-Toledo, I., 2008. Removal of pharmaceutical compounds, nitroimidazoles, from waters by using the ozone/carbon system. *Water Research*, 42(15), pp. 4163–4171.

Shah, A.D., Huang, C.H. and Kim, J.H., 2012. Mechanisms of antibiotic removal by nanofiltration membranes: Model development and application. *Journal of Membrane Science*, 389, pp. 234–244.

Sharma, B.M., Bečanová, J., Scheringer, M., Sharma, A., Bharat, G.K., Whitehead, P.G., Klánová, J. and Nizzetto, L., 2019. Health and ecological risk assessment of emerging contaminants (pharmaceuticals, personal care products, and artificial sweeteners) in surface and groundwater (drinking water) in the Ganges River Basin, India. *Science of the Total Environment*, 646, pp. 1459–1467. https://doi.org/10.1016/j.scitotenv.2018.07.235

Shemer, H., Kunukcu, Y.K. and Linden, K.G., 2006. Degradation of the pharmaceutical metronidazole via UV, Fenton and photo-Fenton processes. *Chemosphere*, 63(2), pp. 269–276.

Sirtori, C., Zapata, A., Oller, I., Gernjak, W., Agüera, A. and Malato, S., 2009. Decontamination industrial pharmaceutical wastewater by combining solar photo-Fenton and biological treatment. *Water Research*, 43(3), pp. 661–668.

Stamatis, N.K. and Konstantinou, I.K., 2013. Occurrence and removal of emerging pharmaceutical, personal care compounds and caffeine tracer in municipal sewage treatment plant in Western Greece. *Journal of Environmental Science and Health. Part. B, Pesticides, Food Contaminants, and Agricultural Wastes*, 48, pp. 800–813. https://doi.org/10.1080/03601234.2013.781359

Thanekar, P., Panda, M. and Gogate, P.R., 2018. Degradation of carbamazepine using hydrodynamic cavitation combined with advanced oxidation processes. *Ultrasonics Sonochemistry*, 40, pp. 567–576. https://doi.org/10.1016/j.ultsonch.2017.08.001

Umbreen, N., Sohni, S., Ahmad, I., Khattak, N.U. and Gul, K., 2018. Self-assembled three-dimensional reduced graphene oxide-based hydrogel for highly efficient and facile removal of pharmaceutical compounds from aqueous solution. *Journal of Colloid and Interface Science* 527, pp. 356–367. https://doi.org/10.1016/j.jcis.2018.05.010

Van Doorslaer, X., Dewulf, J., De Maerschalk, J., Van Langenhove, H. and Demeestere, K., 2015. Heterogeneous photocatalysis of moxifloxacin in hospital effluent: Effect of selected matrix constituents. *Chemical Engineering Journal*, 261, pp. 9–16.

Verlicchi, P., Galletti, A., Petrovic, M., and Barceló, D., 2010. Hospital effluents as a source of emerging pollutants: an overview of micropollutants and sustainable treatment options. *Journal of Hydrology, 389*(3–4), pp. 416–428.

Wang, D., Sui, Q., Lu, S.G., Zhao, W.T., Qiu, Z.F., Miao, Z.W. and Yu, G., 2014. Occurrence and removal of six pharmaceuticals and personal care products in a wastewater treatment plant employing anaerobic/anoxic/aerobic and UV processes in Shanghai, China. *Environmental Science and Pollution Research* 21, pp. 4276–4285. https://doi.org/10.1007/s11356-013-2363-9

Wang, L., Albasi, C., Faucet-Marquis, V., Pfohl-Leszkowicz, A., Dorandeu, C., Marion, B. and Causserand, C., 2009. Cyclophosphamide removal from water by nanofiltration and reverse osmosis membrane. *Water Research*, 43(17), pp. 4115–4122.

Wang, Y., Zhu, J., Huang, H. and Cho, H.H., 2015. Carbon nanotube composite membranes for microfiltration of pharmaceuticals and personal care products: Capabilities and potential mechanisms. *Journal of Membrane Science* 479, 165–174. https://doi.org/10.1016/j.memsci.2015.01.034

Wei, X., Shi, Y., Fei, Y., Chen, J., Lv, B., Chen, Y., Zheng, H., Shen, J. and Zhu, L., 2016. Removal of trace phthalate esters from water by thin-film composite nanofiltration hollow fiber membranes. *Chemical Engineering Journal*, 292, pp. 382–388.

Wu, X., Dodgen, L.K., Conkle, J.L. and Gan, J., 2015. Plant uptake of pharmaceutical and personal care products from recycled water and biosolids: A review. *Science of the Total Environment,* 536, pp. 655–666.

Yaghmaeian, K., Moussavi, G. and Alahabadi, A., 2014. Removal of amoxicillin from contaminated water using NH4Cl-activated carbon: Continuous flow fixed-bed adsorption and catalytic ozonation regeneration. *Chemical Engineering Journal*, 236, pp. 538–544.

Zaib, Q., Mansoor, B., Ahmad, F., 2013. Photo-regenerable multi-walled carbon nanotube membranes for the removal of pharmaceutical micropollutants from water. *Environmental Science: Processes & Impacts*, 15, pp. 1582–1589. https://doi.org/10.1039/c3em00150d

Zakersalehi, A., Nadagouda, M. and Choi, H., 2013. Suppressing NOM access to controlled porous TiO_2 particles enhances the decomposition of target water contaminants. *Catalysis Communications*, 41, pp. 79–82.

Zhang, G., Ji, S., and Xi, B., 2006. Feasibility study of treatment of amoxillin wastewater with a combination of extraction, Fenton oxidation and reverse osmosis. *Desalination*, 196(1–3), 32–42.

Zhang, Z., Wang, B., Yuan, H., Ma, R., Huang, J., Deng, S., Wang, Y. and Yu, G., 2016. Discharge inventory of pharmaceuticals and personal care products in Beijing, China. *Emerging Contaminants*, 2(3), pp. 148–156.

Zhao, C., Pelaez, M., Duan, X., Deng, H., O'Shea, K., Fatta-Kassinos, D. and Dionysiou, D.D., 2013. Role of pH on photolytic and photocatalytic degradation of antibiotic oxytetracycline in aqueous solution under visible/solar light: Kinetics and mechanism studies. *Applied Catalysis B: Environmental*, 134, pp.83–92.

Zucker, I., Lester, Y., Avisar, D., Hübner, U., Jekel, M., Weinberger, Y. and Mamane, H., 2015. Influence of wastewater particles on ozone degradation of trace organic contaminants. *Environmental Science & Technology*, 49(1), pp. 301–308.

Zwiener, C.F.F.H. and Frimmel, F.H., 2000. Oxidative treatment of pharmaceuticals in water. *Water Research*, 34(6), pp. 1881–1885.

Chapter 5

Advanced oxidation process for removal of emerging contaminants in water and sustainable approaches

Harikaranahalli Puttaiah Shivaraju
JSS Academy of Higher Education & Research
Center for Water, Food and Energy, GREENS trust

Shivamurthy Ravindra Yashas
JSS Academy of Higher Education & Research

David Jenkins
University of Plymouth

CONTENTS

5.1 INTRODUCTION

Today, mankind is living in a compromised environment because of the exponential growth in population, industrialization, mining, agriculture, and, other technological advancements, which have posed a serious threat to natural resources. Especially, the water resources available on earth are the most vulnerable natural resources which are exploited maximum. Approximately, two-thirds of the world's population face water-scarcity or non-availability for some period in a year. The addition of diverse chemicals and organic materials into water has made it unsafe for supply without proper treatment. Unlike conventional pollutants which could be easily treated, contaminants of emerging concern (CECs) are the new generation organic and inorganic compounds that are being continuously introduced into water matrices. CECs, by and large, comprise agrochemicals and pesticides, drugs and medicines (antibiotics, anti-inflammatory, beta-blockers, analgesics, etc.), personal care products (musk compounds, UV

DOI: 10.1201/9781003247869-6

screens, parabens, etc.), synthetic hormones, plasticizers (phthalic esters), flame retardants, nanomaterials (ZnO, TiO_2, etc.) dyes, and many more (Gavrilescu et al., 2015). CECs generally include synthetic chemicals and microorganisms that are not being monitored and regulated by most of the countries but are likely to impact ecological and human health (Xu et al., 2019). Most of the CECs could be traced to the day-to-day items used to like medications, cleansers, detergents, and plastics. CECs are detected in all water sources (rivers, lakes, ponds, oceans, groundwater, drinking water supplies, and, wastewaters) in concentrations ranging from few parts-per-trillion to parts-per-billion. Out of large set of CECs, pharmaceutical and personal care products (PPCPs) constitute a significant subset inducing serious detrimental effects (Menon et al., 2020; Mohapatra et al., 2018, 2021; Priyanka and Mohapatra, 2020). For example, PPCPs entering the water sources are known to be persistent in the environment even after rigorous treatments. PPCPs are toxic, carcinogenic, and mutagenic and impart several ecotoxicological effects like growth inhibition in fishes, development of resistant bacterial strains, and imbalances in the nutrition cycle (Grenni et al., 2018) (Figure 5.1).

Further, the scientific community has invested lot of time and money in developing a sustainable water/wastewater treatment to address these complex and toxic CECs over the years. Generally, all existing treatment plants are designed to remove nutrients (nitrogen and phosphorous), organic load (simple carbonaceous compounds), and pathogens (microorganisms). This type of conventional treatment process (e.g.,

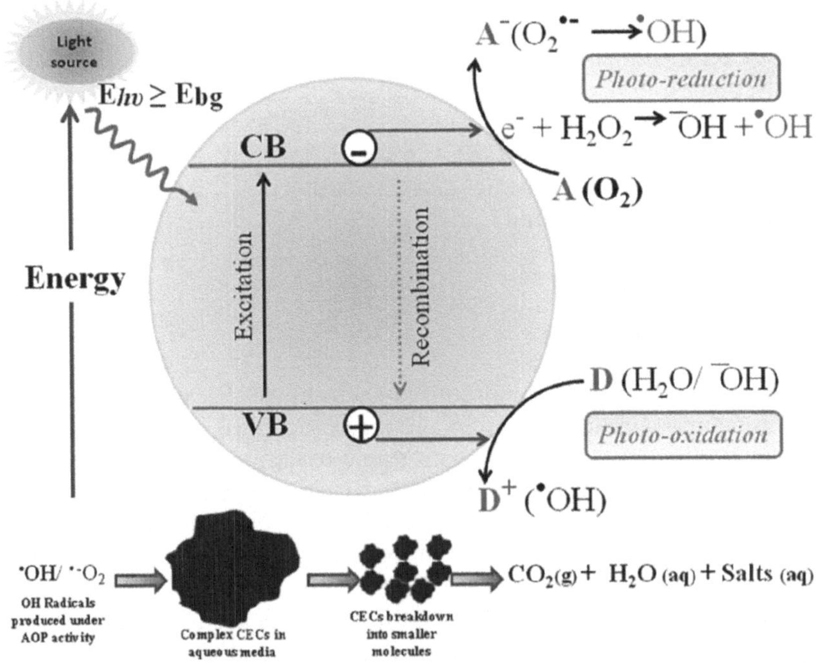

Figure 5.1 Schematic of AOP for removal of concern (CECs). CB, conduction band; CO_2, carbon dioxide; e^-, electron; E_{bg}, band gap energy; H_2O, water molecule; O_2^{-}, superoxide radical; •OH, hydroxyl radical; VB, valance band.

coagulation/flocculation, aeration, sequencing batch reactors, and many other biological reactors), when applied individually, turns out to be obsolete, as they are inefficient to mineralize the CECs (Liu et al., 2020). Alternatively, the advanced oxidation processes (AOPs) are reliable, robust, and sustainable to treat complex and trace CECs (Eregowda and Mohapatra, 2020; Priyanka et al., 2021). AOPs involves the in-situ generation of strong oxidants, hydroxyl radical (OH) or sulfate radical ($SO_4^{\bullet-}$) in some cases. Both of these oxidizing radicals attack the CECs/complex organics (pollutants) and break down into smaller organic molecules. In 1987, Glaze and the team were first to introduce the term AOP for the treatment process that was performed at ambient temperature, where there was in-situ production of •OH radical (oxidation potential of 1.9–2.7 V) (Glaze et al., 1987). Thereafter, the AOPs have evolved significantly. Today, catalyst (nanoscale)-based AOPs are the most readily tested mechanism for the treatment of most of the CECs. Specifically, photocatalysis (PC) and photoelectrocatalysis (PEC) are sought after methods with huge potential. Also, the integration of physico-chemical processes with AOPs is studied extensively for its efficient and economical operation.

Hence, the present chapter aims to provide brief description on CECs and the challenges in treating them. Specifically, the recent advances in nanocatalyst-driven AOPs, PC, and PEC are discussed in length. There is a special focus on solar PC as a sustainable approach in treatment of CECs. Finally, the prospective future and challenges of AOPs in addressing CECs are presented.

5.1.1 CECs: an overview

CECs could be described as "any anthropogenic or naturally occurring chemicals or microorganisms that are detected in very low concentrations, not regularly supervised in the environment, but potentially affect the environment instigating adverse environmental and/or human health effects" (*Contaminants of Emerging Concern including Pharmaceuticals and Personal Care Products | Water Quality Criteria | US EPA*). Generally, CEC is the collective term for chemical substances used in agriculture (pesticides, fungicides, synthetic growth regulators, etc.), pharmaceuticals (veterinary use and for humans), personal care products, and household products (cleansers, detergents, chemicals, disinfectants, etc.). The two important criteria for grouping of a substance as a CEC are (1) its perseverance in the environment (air, water, or soil) and (2) prospective human and ecotoxicological effects. Uniquely, the "emerging" and "concerning" about CECs is the development of accurate analytical techniques capable of measurements as low as ng/L in water, and whose potential risks, though reported based on laboratory tests, essentially remain indefinite in the environment respectively. Further, the concern about the CECs is the unavailability of rapid analytical protocol and regulatory framework. Firstly, the advanced techniques such as liquid chromatography (LC) coupled mass spectrometry (MS), time-of-flight mass spectrometry (LC/TOFMS), and gas chromatography coupled with mass spectrometry (GC/MS) are used to estimate CECs in the ng/L concentration range. However, these techniques are laborious and time-consuming. Secondly, the national agencies and international policymakers have not come out with standard monitoring and treatment guidelines for CECs. The regulatory aspects of CECs are at its infancy. Besides, there are numerous

Table 5.1 Examples of common classes of CECs detected in water samples

Class	*Compound*
Antibiotics	Sulfadiazine
	Sulfamethoxazole
	Norfloxacin
	Ofloxacin
	Ciprofloxacin
	Tetracycline
	Doxycycline
	Azithromycin
NSAIDs	Paracetamol
	Naproxen
	Ketoprofen
	Ibuprofen
Psychiatric drugs	Cotinine
	Carbamazepine
Beta-blockers	Atenolol
	Metoprolol
Pesticides	Isoprocarb
	Atrazine
	Imidacloprid
	Acetochlor
	Dichloro-diphenyl-trichloroethane (DDT)
Antibacterial agents	Triclosan
	Triclocarbon
Endocrine disruptors	Estriol
	Progesterone
	17-beta-Estradiol
	Bisphenol A
Personal care products	Benzophenone-3
	Climbazole
	Musk xylene, musk ketone
Nanomaterials	Zinc oxide
	Silver nanoparticles
	Titania nanoparticles

scientific reports on occurrence, fate, transport, and effects caused by selected CECs. For example, the US Environmental Protection Agency (EPA) has marked 88 emerging chemicals under National Primary Drinking Water Regulations (NPDWRs) that are monitored. A brief classification of CECs under different categories is listed in Table 5.1. Remarkably, all of these listed CECs are detected in water samples from effluent treatment plants, sewage treatment plants, surface water (lakes, streams, rivers, etc.), and groundwater (Lapworth et al., 2012).

Another essential threat from CECs is their entry into drinking supplies or disrupting food-chains. Unlike other pollutants, CECs are released from domestic, commercial, industrial, agricultural, and institutional set-ups. Because of their persistence and stability, CECs are even detected in environmental samples at far off. Usually, PPCPs are released from households, institutions, and healthcare facilities. Most of the CECs arising from pesticides and veterinary medications are released from agricultural practices and animal sheds, respectively. Therefore, once the CECs are released, they mostly leach into the ground (groundwater) or enter surface water sources (sewage or nearby lakes or ponds). The CECs leaching from cultivable lands seep into the ground and directly contaminate the soil and groundwater (Xu et al., 2019). However, the exact pathways of CEC transport is still a challenge as the definite properties of CECs are unknown.

As stated above, even the ecotoxicological effects of CECs are not completely known. There are countable reports that reveal the health impacts on aquatic species and human health. Most of the pharmaceutical residues (analgesics, antibiotics, beta-blockers) have adverse effects on the growth of embryonic kidney cells. The presence of UV filters (Benzophenone) is toxic to aquatic microorganisms and causes bleaching of coral reefs (Tsui et al., 2017). The CECs under the banner of endocrine-disrupting compounds (EDCs) directly target the hormonal changes in the species. These changes may be accompanied with impairments in the reproductive health of humans and aquatic organisms. The use of artificial musk compounds in perfumes, shampoos, and detergents can pose threat of high toxicity. Due to their high lipophilic property, they accumulate in living aquatic species, sludge, and sediments (Xu et al., 2019). Another very important class of CECs is the disinfection by-products (Sharma et al., 2021). These are the complex and unknown chemical molecules generated during the disinfection process (use of UV, chlorine, O_3, chloramines, etc.). Unfortunately, out of 700–800 by-products identified, very few are being monitored. The health risks from these chemicals may be a miscarriage, birth defects, and mainly bladder cancer. Hence, determining an accurate picture of health risks and ecotoxicological risks from CECs is a potential area for advanced research.

5.1.2 Treatment of CECs: conventional treatment systems vs. AOPs

With the brief background on the classification, sources, transport, and health impacts of CECs, the challenges encountered to treat the CECs with conventional treatment processes and justification for considering AOPs as a sustainable option are discussed. To date, several conventional physico-chemical processes and simple advanced treatments are investigated. The traditional sedimentation or/and flocculation or clariflocculation and the physical processes like adsorption and membrane processes are widely studied. However, most of the CECs escape every conventional process or partially removed owning their minimal octanol coefficients (log P) (Stackelberg et al., 2007). Out of many conventional treatment options available, biological water/wastewater is used to remediate the effluents. Firstly, when physical treatments like adsorption or membrane processes are employed, there is a problem of the generation of diluted or concentrated residues of target CECs as they work on the phase transfer mechanism. The disposal of such secondary wastes is the resulting challenge. Next, the most widely adopted treatments are biodegradation technologies. Here, the microorganisms consume organic compounds for their cell growth; however, the toxic nature of CECs impedes the biodegradation by hindering microbial development. In such scenarios, the microbes need additional substrates (nitrogen and phosphate sources) to switch-over to a co-metabolism way of biodegradation (Tran et al., 2013). The biodegradation of CECs has several limitations. The most popular biological treatment processes such as activated sludge process, attached-growth filters, and aerobic/anaerobic/facultative bioprocesses have been reported to be efficient to remove a few of the CECs, but they become obsolete for pesticides and PPCPs (Matamoros et al., 2016). For example, when advanced/hybrid biological treatment processes like membrane bioreactors are applied, the pollutant removal is by means of physical retention and microbial degradation on the surface. This process is efficient to remove few PPCPs (up to 100%), antibacterial agent (triclosan up to 99%), and beta-blockers (atenolol up to 97%). On the other hand, membrane bioreactors fail to mineralize pesticides, antibiotics because of clogging, biofouling, operational failure, and high cost. The probable order of treatment efficiency of membrane bioprocess may be pesticides < antibiotics < beta-blockers < personal care products < EDCs (Nguyen et al., 2013).

Therefore, the biological treatments were found efficient for the removal of many CECs, but each of which was selective toward a particular CECs/class of CECs. So, to achieve comprehensive performance, the biological processes were integrated with other physico-chemical treatments (membrane processes, ozonation, and AOPs). Usually, the chemical treatments break down the complex pollutants to decomposable compounds, either by thorough disassociation or by transformation to inorganic compounds like water, carbon dioxide, and smaller nitrogenous molecules. Here, various chemical agents, such as ozone, hydrogen peroxide, persulfate, chlorine, metal oxides, and nanostructured metallic catalysts (nanoscale), and some photon/energy sources such as ultrasonic, voltage, gamma (natural sunlight), and UV-radiation are used to support the reaction. Some of the traditional oxidation processes like photolysis, chlorination, ozonation, and Fenton reaction are based on the generation of oxidants from chlorine or bromine or hydrogen to target the pollutants. For instance, Noutsopoulos and team (2015) applied the chlorination technique to target CECs by the action of chlorine and bromine radicals. After 60 min of treatment, 95% of naproxen, 100% of diclofenac, and estradiol were eliminated, whereas other CECs like bisphenol A, ketoprofen, ibuprofen, triclosan, and nonylphenol were removed up to 34%–83% (Noutsopoulos et al., 2015). Likewise, the application of ozonation is very attractive as it assures almost complete degradation of CECs (pesticides, PPCPs, and EDCs) by the action of ozone on double bonds and aromatic rings of CECs. Nevertheless, there are many unintended intermediates formed during chlorination and ozonation that may be toxic but reduce the rate of mineralization. Therefore, any of the said conventional AOPs cannot be advocated readily against all the CECs.

To overcome the shortcomings of such oxidative treatments, AOPs were developed and evaluated against most of the identified CECs. Here, the organic pollutants are completely oxidized into environmentally benign products like carbon dioxide, water, and mineral acids. Mostly, AOPs engage strong oxidants like hydroxyl radical (OH), superoxide radical ($O_2^{\cdot-}$), ozone (O_3) which attack the organic compounds because of unpaired electrons. Also, AOPs mineralize the pollutants, not merely phase transfer or partial breakdown, theoretically, no further treatment is required after AOPs. AOPs are better option over conventional biological and oxidation processes for its reduced operational costs as no secondary wastes/contaminants are produced unlike adsorption, ion exchange, and stripping processes (Dhangar and Kumar, 2020). AOPs include both catalytic and non-catalytic processes that rely on reactive oxygen species (ROS) but differ in the way of radical generation. Usually, AOPs may be homogeneous or heterogeneous in a process where the latter usually operates with an external source of energy (light energy, ultrasonic energy, and electrical energy) (Garcia-Segura et al.,

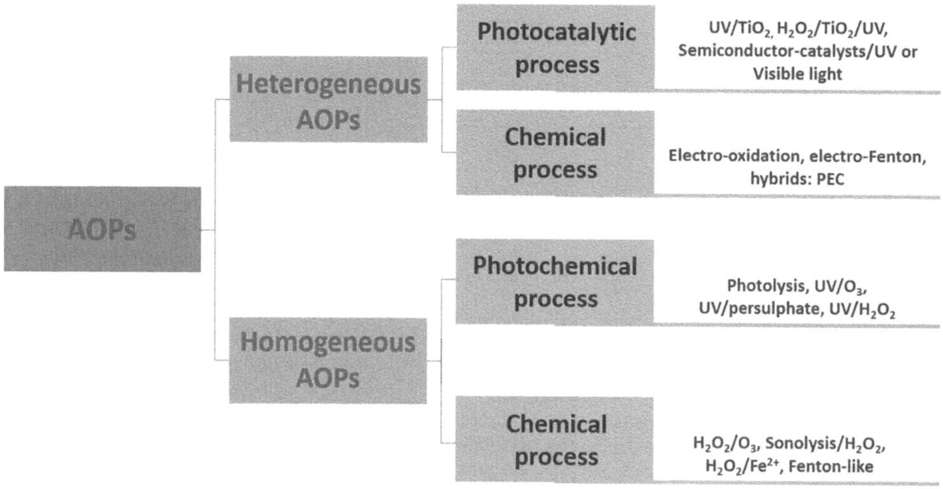

Figure 5.2 Types of AOPs applied for the treatment of CECs.

2012). Interestingly, catalyst-driven AOPs are gaining interest as they are relatively fast and efficient. Most common AOPs are depicted in Figure 5.2. This chapter discusses PC and a hybrid AOP, PEC from fundamentals to their application in pollutant remediation. Table 5.2 provides an overview of the advantages and disadvantages of conventional treatments and AOPs.

5.1.3 Factors influencing the performance of AOPs

As perceived, the efficiency of AOPs is related to the interaction of ROS with the organic contaminants, several key factors that influence the phenomenon. First, the intensity of light energy supplied in the case of PC, PEC, photo-Fenton, solar-driven PC, and many other hybrid AOPs impact on the generation of ROS. Usually, irradiation intensity is directly proportional to the rate of OH and $SO4^{\bullet -}$ formation that results in efficient pollutant degradation (Yang et al., 2008). Tobajas et al., (2017) revealed the photocatalytic degradation of acetaminophen and antipyrine using solar light with TiO_2@ZnO/clay catalyst. They concluded that by increasing the power of light (from 250 to 450 W/m^2), the decay rate constant (k) escalated from 01.9×10^3 to 05.4×10^3/min for acetaminophen with a photocatalyst dose at 250 mg/L (Tobajas et al., 2017). Another work by Mazierski et al., (2017) described the influence of light intensity on the photodegradation of toluene with TiO_2 catalyst. The study established the linearity between the toluene removal and light intensity from 0.50 to 20.0 mW/cm^2.

Second, the stock concentration of the analyte/pollutant is another important parameter that influences the AOPs. Mostly, as the concentration increases the degradation decreases keeping the operating parameters constant. Ravindra and team (2020) worked on the photodegradation of tetracycline antibiotics using ZnV_2O_6/g-C_3N_4 catalyst under LED irradiation. They observed the highest degradation for 20 mg/L, then followed by 30 mg/L, and 50 mg/L of tetracycline corresponding to 20 mg of catalyst

Table 5.2 Overview of pros and cons of available treatments for CECs

Treatment	*Pros*	*Cons*
Physical and physico-chemical treatments		
Adsorption	Wide range of natural and engineered adsorbents available, effective preparation and application of the process on wide range of CECs, low operating costs	Generation of concentrated spent adsorbent (pollutant accumulation), disposal of spent adsorbent, residues of CECs, needs polishing treatment for complete removal
Filtration	Processes like ultra-, nano-, and micro-filtration remove most of the organic CECs, effective for holistic depollution of water and wastewater, concomitant pathogen removal	High operational and energy costs, issues of membrane biofouling and dumping of used membranes, less effective in removal of PPCPs
Biological treatments		
Activated sludge process	Environmentally benign, low energy and operational costs, no harmful/toxic by-products, effective in removal of simpler CECs like dyes and pathogens	Unsuitable for water/wastewater with high strength (COD >4000 ppm), ineffective for removal of EDCs, generation of voluminous sludge, high sludge processing costs, susceptible to antibiotics/EDCs
Membrane bioreactors	Effective in removal of many CECs, less carbon footprint, volatile emissions during biodegradation processes	Needs high aeration, biofouling of membrane, cannot resist high pathogen/CECs load, non-adaptable for wide range of CECs, high operating costs
Oxidative treatments		
Coagulation and/or flocculation	Reduce turbidity, increase sedimentation rate, wide range of bio-/inorganic coagulants	High sludge production, not effective against CECs, complex maintenance
Chlorination	Easy inactivation of wide range of pathogens and mineralizes complex organic molecules, suitable for both water and wastewater treatments	Risk in handling the chlorine sources, issue of residual chlorine, affects palatability of the drinking water
Ozonation	O_3 is a strong oxidant against most of the CECs, presence of peroxide radicals enhances the process efficiency, simultaneous disinfection and sterilization	High energy requirement, generation of unknown and toxic by-products during O_3 reactions, radical quenchers inhibit the rate of degradation
AOPs (PC, PEC, photo-Fenton, etc.)	High removal efficiency of most of the CECs, different ROS can be engaged, nanoscale catalysts improve the efficiency, short run time, sustainable in terms of by-product formations	Laborious catalyst recovery process, energy requirement is high, unable to treat large volume of water/wastewater, rapid recombination of ROS during the process

throughout (Ravindra et al., 2020). At higher concentrations, the active adsorption of organic contaminants is restricted from absorbing light energy to generate photo-induced charges, thereby decreasing the degradation efficiency.

Third, pH is another deciding parameter that is likely to affect the surface charge on the catalysts, redox potential of ROS, and hence, the overall system efficiency of AOPs. An interesting study by Shah and team (2019) worked on the degradation of an antibiotic, ciprofloxacin, employing zero-valent manganese (Mn^0) with peroxydisulfate ($S_2O_8^{2-}$) as an activator. The authors noted that the maximum ciprofloxacin degradation was at pH 2.0 (95%) and least at pH 11.0 (27%) within 80 minutes of run under the same operating conditions. Here, the pollutant is positively charged below pH 6.0, so the $SO4^-$ being negatively charged efficiently reacts with remarkable removal at pH 2.0 and 5.7 (Shah et al., 2019).

Finally, the presence of interferents (competing ions and organic matters) in the reaction mixture significantly affects the performance of AOPs. The other anions and cations along with natural organic matters interfere very fast with the generated ROS over the target pollutant. Especially, in PC and PEC, these interferents compete to adsorb on the surfaces of catalyst resulting in less degradation of CECs. In support of this claim, Sayed et al., (2018) who studied peroxy-monosulfate activated TiO_2-based catalyst for the treatment of ciprofloxacin found lesser removal of ciprofloxacin in the presence of interferents. Besides, Chen et al., (2017a) fabricated a Z-scheme photocatalyst to target the tetracycline drug during which, they ascertained the reduced degradation efficiencies in different water sources. Further, the lesser degradation in lake water and river water was regarded to the presence of viable pollutants or interferents (organic/inorganic or natural). Hence, the aforementioned factors should be optimized for developing an efficient AOP.

5.2 CATALYST-BASED AOPs

As discussed above, the AOPs have been applied as a sustainable approach against most of the CECs. However, to increase the efficiency of ROS production and rapid interactions with targets, numerous inorganic/organic catalysts have been used. These catalysts fasten the generation of ROS and prolong their recombination to facilitate effective and complete mineralization of the pollutant. Interestingly, over the years, nanoscale electro-/photocatalysts have emerged with the highest potential in the AOPs. This chapter discusses two of the important catalyst-driven AOPs, namely, PC and PEC.

5.2.1 Photocatalysis (PC)

PC is one of the most sustainable AOPs that is being studied extensively for the degradation of CECs. Fundamentally, the PC process is inspired by the natural phenomenon of photosynthesis. Specifically, the word "PC" is being used in the following two meanings. First, if a material utilizes light energy to promote thermodynamic reactions ($\Delta G > 0$), such process is regarded as photosynthesis. Here, the photons constitute the reactants and such material is termed as photocatalyst. Classical examples of such systems are water splitting (Chen et al., 2017b) and carbon dioxide conversions

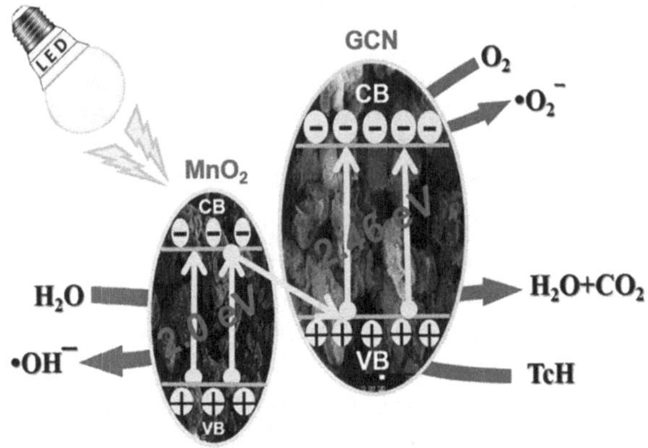

Figure 5.3 Schematic of PC using GCNMnO$_2$ against tetracycline hydrochloride (TcH). (From our previous work, Yashas et al., (2020), Copyright, 2020, Springer Nature.)

(Shivaraju et al., 2020a). On the other hand, if a material uses light energy to enable thermodynamically downhill reactions ($\Delta G<0$) by changing the reaction-kinetics through absorption of optical energy, such a material is called "photocatalyst". This section adopts the latter definition for a photocatalyst. PC was first introduced by Fujishima and Honda in 1972 who studied water splitting reaction using semiconductor photocatalyst, TiO_2 (Fujishima and Honda, 1972). The use of TiO_2 as photocatalyst remains as the best option till date. TiO_2 possesses certain key superiorities like broad band energy (~3.3 eV), chemical stability, non-toxicity, ease of synthesis, and surplus availability (Shivaraju et al., 2020b). Thereafter, the fabrication of photocatalysts has evolved to fit the purpose and target a specific pollutant. In addition, nanotechnology is very handy to engineer composite photocatalyst. Mainly, six general categories of photocatalysts are used: (1) molecular photocatalysts, (2) semiconducting photocatalysts, (3) quantum dot photocatalysts, (4) 2D photocatalysts, (5) plasmonic metal photocatalysts, and (6) photovoltaic-coupled catalysts (Yang and Wang, 2018). In a typical heterogeneous PC, semiconductors are illuminated by light energy (UV/Visible) with incident energy greater than its bandwidth; as a result, valance electrons are excited to the conduction band generating holes (in valance band). These generated photo-induced holes show strong oxidation potential and readily combine with OH− in an aqueous medium to form OH radicals. On the other hand, electrons in the conduction band exhibit reductive nature and react with dissolved oxygen to generate $O_2^{\cdot-}$ radicals. Therefore, both •OH and $O_2^{\cdot-}$ radicals efficiently attack the pollutant molecules in the photocatalytic process. The equations representing the above process are presented through Equations (5.1)–(5.4) (Shahmoradi et al., 2018), also the schematic of electron transition could be captured from Figure 5.3 of our previous study (Yashas et al., 2020). Here, the photocatalyst, GCN/MnO$_2$ (graphitic carbon nitride/manganese oxide), is applied to remove CECs, tetracycline hydrochloride under the visible light irradiation (LED).

$$GCN / MnO_2 + h\nu \rightarrow GCN / MnO_2 + h^+ + e^- \tag{5.1}$$

$$h^+ + H_2O \rightarrow \cdot OH + H^+ \tag{5.2}$$

$$e^- + O_2 \rightarrow \cdot O_2^- \rightarrow OH^- \tag{5.3}$$

$$OH / O_2^- / h^+ + TcH \rightarrow \text{Degradation products } (CO_2 + H_2O) \tag{5.4}$$

The application of a wide range of semiconductor-based photocatalysts has few limitations such as (1) difficulty in product separation or identification, (2) rapid recombination of photo-induced charges (electron-hole pairs), (3) poor chemical stability (poor dispersibility), (4) inability to harness a wide range of the solar spectrum, and (5) catalyst recovery and reuse (Yang and Wang 2018). These limitations are being addressed by the construction of composites like semiconductor/metallic nanoparticles, carbon/polymer with semiconductors, combinations of 1D, 2D, and 3D architectures, and many more. Few examples of emerging photocatalysts against the CECs are listed in Table 5.3.

Zinc oxide (ZnO) is a classic semiconductor photocatalyst employed in most of photo-driven processes like pollutant degradation, pathogen inactivation, photosensitization, optoelectronics, lasers, etc. due to its flexible/tunable band energy. Nagaraju and team (2020) have synthesized ZnO nanoparticles via sol-gel technique. Also,

Table 5.3 List of emerging photocatalysts applied to treat CECs

AOP	*Photocatalyst*	*Pollutant/class of CECs*	*Efficiency(%)/ duration*	*Reference*
PC	$Bi_{12}O_{17}C_{12}/Ag/AgFeO_2$	Tetracycline/ antibiotic	94.1/60 min	Guo et al., (2020)
	$Ag_2O/CeO_2/g\text{-}C_3N_4$	Ciprofloxacin/ antibiotic	~65/2.5 h	Pinheiro et al., (2020)
	TiO_2/natural zeolite	Metoprolol/ beta-blocker	~100/360 min	Piedra López et al., (2021)
	graphene/$TiO_2/g\text{–}C_3N_4$	Tetracycline/ antibiotics	83.5/80 min	Wu et al., (2020)
		Ciprofloxacin/ antibiotics	61.7/60 min	
		Bisphenol A/EDCs	79.5/70 min	
	ZnO/reduced graphene oxide	Metalaxyl/pesticide	90.25/120 min	Dehghan et al., (2020)
PEC	Cu_2O film on TiO_2 nanotube array	Ciprofloxacin/ antibiotics	73/4 h	Koiki et al., (2020)
	Ni-doped ZnO	Norfloxacin/ antibiotics	98/UV-A and 64/ visible light	Gholami et al., (2020)
	$ZnO/TiO_2/Ag_2Se$	Oxytetracycline/ antibiotics	96.5/360 min	Changanaqui et al., (2020)
	Co-$BiVO_4$ (peroxymonosulfate activator)	Bisphenol A/EDCs	~90/60 min	Bacha et al., (2020)
	$BiVO_4$/CuO	4-Nitrophenol/(by-product) pesticide	100/450 min	Martimiano do Prado et al., (2020)

catalyst based on heteroatom doping is prepared to target an emerging contaminant, chlorobenzene. Here, the authors study the effect of photodegradation with pristine ZnO, Ag-, Pb-, and Cd-ZnO with a visible light source (LED). Pb/ZnO was reported effective over other catalysts with ~100% removal in a shorter duration (<120 min) of run (Nagaraju et al., 2020). Similarly, Mei et al., (2020) targeted an EDC, bisphenol A, via simulated solar light with TiO_2/C_3N_4 composite. They reported nearly complete degradation of bisphenol A within 60 min of run accounting for the extended light-harvesting capability and rapid photoelectron production of TiO_2/C_3N_4 composite. They claim that the O^{2-} and $^1O_2$ (singlet oxygen) radicals were key ROS involved (Mei et al., 2020). Next, Alothman and team (2020) fabricated cobalt doped TiO_2 nanoparticles to degrade a beta-blocker, atenolol, under a UV light source. Ninety percent of atenolol was degraded in 40 min with atenolol concentration of 15 ppm, 2 g/L of catalyst, at a pH of 2.0, and hydrogen peroxide (substrate) amount of 2.0 mL/L at 30°C temperature (Alothman et al., 2020). Recently, Padilla Villavicencio et al., (2020) demonstrated photodegradation of a non-steroidal anti-inflammatory drug (NSAIDs), Ibuprofen, using Ag_2O and Ag/Ag_2O composites. Authors bring out interesting claims on introducing metallic silver to constitute the photocatalyst. Here, silver acts as an electron quencher which inhibits the recombination of photo-charges but promotes photocatalytic reactions (Padilla Villavicencio et al., 2020). Hence, all the information and examples reported will inspire the researchers to evaluate exhaustive combinations of a photocatalyst to make them cost-effective, easily recoverable, be able to absorb a wide band of energy, and aim at multi-pollutant degradation (non-selectivity).

5.2.2 Photoelectrocatalysis (PEC)

PEC is an integrated AOP which is a mixture of heterogeneous PC and electrochemical oxidation. Besides, PEC is an improvised version of anodic oxidation (water treatment technique). PEC process is administered by the deposition of photoactive materials on a suitable conducting substrate, exposing it to the appropriate light source and applying an optimum bias voltage to result in the generation of ROS (Park et al., 2012). PEC can be superior to the PC because of the boosted generation of ROS by the action of photoactive as well as conducting substrate. The efficiency of the PEC system accounts for the type and intrinsic properties of conducting substrate and photoactive material. The principle of generation of electron-hole pairs and the occurring of photochemical reactions are almost similar to that of PC. Here, the UV/visible light is irradiated on the photoanode which enables the excitation of valance electrons to the conduction band leaving behind the photogenerated holes. In turn, these holes react with water molecules to generate OH radicals. Typically, this phenomenon occurs on the electrode surface, but the radicals diffuse into the bulk phase solution to react with pollutants. As well, applied voltage diverts the electrons toward the cathode thereby producing $\bullet O^{2-}$ radicals which attack the target molecules. The typical reactions at the photoanode in a PEC system is represented through Equations (5.5)–(5.9) (Garcia-Segura and Brillas, 2017). Also, the schematic representation of PEC is depicted in Figure 5.4.

$$\text{Semiconductor} + h\nu \rightarrow h^+ + e^- \tag{5.5}$$

$$h^+ + H_2O \rightarrow \cdot OH + H^+ \tag{5.6}$$

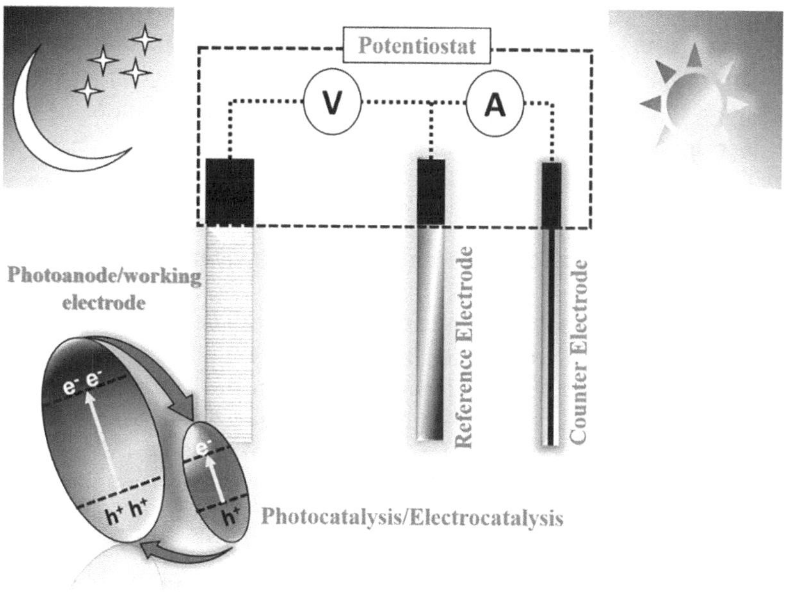

Figure 5.4 Schematic representation of PEC.

$$e^- + O_2 \rightarrow \cdot O_2^- \tag{5.7}$$

$$\cdot O_2^- + H^+ \rightarrow \cdot HO_2 \tag{5.8}$$

$$2 \cdot HO_2 \rightarrow H_2O_2 + O_2 \tag{5.9}$$

One of the major advantages of PEC over PC is the easy reusability of PEC materials since the photocatalysts are deposited over the photoanode (electrode) which could be washed/rinsed, dried, and reused. Contrarily, photocatalysts are usually in amorphous form, whose recovery is laborious and may involve the use of toxic reagents (Peleyeju and Arotiba 2018). One of the most studied PEC materials against the removal of dyes, phenolic derivatives, pharmaceuticals, etc. is TiO_2. However, the application of TiO_2 becomes expensive because of its inability to harness visible light. UV constitutes only 5% of the solar spectrum, for which TiO_2 needs to be tailored by doping, composite formation, metal loading, controlled synthesis, etc. (Sirés et al., 2014). Hence, the recent focus is toward the fabrication of visible light-responsive PEC materials.

Recently, Mafa and others (2019) evaluated the g-C_3N_4/BiOI/exfoliated graphite photoanode under visible light against an antibiotic, sulfamethoxazole. Here, the as-prepared anodic catalyst is characterized for structural, optical, photocatalytic, and electrocatalytic (cyclic voltammetry, electrochemical impedance, transient photocurrent) properties. The authors ascertain that the charge recombination is inhibited because of the p-n heterojunction of anodic material. Here, sulfamethoxazole was removed up to 88% in 180 min of PEC operation under the following conditions, 30 ppm of the target, pH unadjusted, electrolyte-0.1 M sodium sulfate, and current density-5 mA/cm^2 (Mafa et al., 2019). Similarly, Zheng et al., (2021) fabricated visible-light active MoS_2@$BiVO_4$ photoanode driven by chlorine to target an EDC, bisphenol A. Authors report complete degradation of bisphenol A (10 ppm) within 75 min of PEC run (with an application of 1.5 V against Ag/AgCl and 100 mM NaCl as electrolyte). They validated the role of chlorine radical and chlorine oxide radical in the mineralization of the target (Zheng et al., 2021). A few of the recent advances in PEC are listed in Table 5.3.

5.2.3 Solar PC: toward sustainable remediation

A sustainable approach in the treatment of CECs is very necessary to combat the large footprint of conventional treatments. Though the PC and PEC degrade most of the CECs, they are energy-intensive. Hence, the demand for utilizing renewable energy sources is high. Recently, research on employing sunlight to address many environmental problems have emerged. For example, solar irradiated PC systems and solar disinfection have emerged. The combination of sunlight and nanoscaled photocatalyst is a promising alternative for water/wastewater treatment. The major challenge in solar PC is the constant availability of a particular wavelength of radiation (Byrne et al., 2018). However, the hot regions within a country can effectively test solar PC. To complement the design of a solar PC system, the nanoscale photocatalysts are readily tunable according to the select CECs and wavelength of irradiation. For example, Shaban and team (2019) have used carbon-doped titanium dioxide (C-TiO_2) catalyst to target acetaminophen under natural sunlight. Here, the C-TiO_2 possessed a band gap of 1.78eV, and the system was operated under neutral pH. They have reported 100% degradation of acetaminophen within 1.5 h and 2 g/L of catalyst (Shaban and Fallata 2019). Similarly, Khedr et al., (2019) used tunable anatase/brookite heterojunctions of TiO_2 to treat ibuprofen drugs. Here, there is a complete removal of the drug within 5 h of solar PC (Khedr et al., 2019). Hence, recent reports on solar PC authenticate its potency in the treatment of CECs. The maximum utilization of solar spectrum needs to be addressed by careful design of photocatalysts. Applications of solar light-driven PC could be protracted from water treatment to CO_2 conversions and nitrogen fixation.

5.3 PERSPECTIVE AND FUTURE SCOPE

The establishment of various AOPs has opened new horizons in water/wastewater treatment. Both conventional and integrated/hybrid AOPs are efficient to treat complex organic contaminants like CECs. Considering the process flexibility, effective production of ROS, adaptability, cost, and sustainability, PC and PEC hold better status. In addition, PC and PEC are very potent in terms of the development of new catalysts. After a rigorous literature review, the following points may be projected as current challenges/research gaps related to the application of AOPs to address CECs.

i. The operational costs incurred on working of full-scale AOPs are lesser-known. However, literature have informed on the economics of a lab and pilot-scale AOP systems. The trend is that, if treatment costs decrease, then the pollutant concentration in water would increases. For instance, ozonation and Fenton oxidation are economical over the UV coupled AOPs as the operation with UV light is expensive. Also, people regard PC (with TiO_2) as feasible over other catalyst-based AOPs because of its maximum efficiency.

ii. It is claimed that AOPs completely mineralize the complex organics forming neutral products like water, carbon dioxide, and smaller amino groups, but it is irregular that the oxidation leads to different by-products or intermediates possessing similar or even higher toxicity than the parent CECs themselves. For example, the presence of excess nitrate in the effluents and subsequent AOP-based treatment will result in the formation of nitrated/nitrosated molecules, which are mutagenic. Though AOPs can remove CECs, they sometimes end up producing unsafe water.

iii. In a way to generate fair amount of OH radicals, PC and PEC are potential AOPs. Here, a semiconductor catalyst could be prepared according to the purpose. Catalysts like TiO_2 and ZnO are extensively applied in PC and PEC but suffer because of wide band gap, faster recombination of photo-charges, sensitivity to photo corrosion, unable to absorb visible light, etc. Though a vast number of photo and electroactive materials are reported, they cannot be readily engaged because less information is available on their stability, cost, and safety.

iv. Every AOP has performed effectively in remediating one or the other CECs, but there is no exact information on reaction. The formation of by-products, the chemistry of pollutant interactions, and to a certain extent, the kinetics of adsorption and degradation is not well established.

Therefore, it is important to orient the research on AOPs toward tackling the above-mentioned challenges. It is very important to design the AOPs which are non-selective to a particular class of pollutants. For this, PC and PEC can be tested as the catalyst synthesis could be controlled. Particularly, PEC is a promising AOP for the future because of its synergistic principle. Here, a multifunctional catalyst can keep the system functioning round the clock by switching between electro- and photo-oxidative removals of pollutants. However, looking at the ground reality, the initial move in terms of enhancing the treatment efficiencies, the conventional treatment processes could be retrofitted with either of AOPs to target CECs. Later on, integrated AOPs could be adopted as polishing treatment which will abet CECs removal. All in all, the AOPs to treat CECs are need of the hour from an environmental sustainability perspective.

REFERENCES

Alothman, Z.A., Badjah, A.Y., Alharbi, O.M.L., Ali, I. 2020. Cobalt doping of titanium oxide nanoparticles for atenolol photodegradation in water. *Environmental Science and Pollution Research*, 28, 7423–7430. https://doi.org/10.1007/s11356-020-11071–w

Bacha, A.-U.-R, Nabi, I., Cheng, H., Li, K., Ajmal, S., Wang, T., Zhang, L. 2020. Photoelectrocatalytic degradation of endocrine-disruptor bisphenol – A with significantly activated peroxymonosulfate by Co-$BiVO_4$ photoanode, *Chemical Engineering Journal*, 389, 124482. https://doi.org/10.1016/j.cej.2020.124482

Byrne, C., Subramanian, G., Pillai, S.C. 2018. Recent advances in photocatalysis for environmental applications. *Journal of Environmental Chemical Engineering*, 6, 3531–3555.

Changanaqui, K., Brillas, E., Alarcón, H., Sirés, I. 2020. ZnO/TiO_2/Ag_2Se nanostructures as photoelectrocatalysts for the degradation of oxytetracycline in water. *Electrochimica Acta*, 331, 135194.

Chen, F., Yang, Q., Li, X., et al., 2017a. Hierarchical assembly of graphene-bridged Ag_3PO_4/Ag/$BiVO_4$ (040) Z-scheme photocatalyst: An efficient, sustainable and heterogeneous catalyst with enhanced visible-light photoactivity towards tetracycline degradation under visible light irradiation. *Applied Catalysis B: Environmental*, 200, 330–342.

Chen, S., Takata, T., Domen, K. 2017b. Particulate photocatalysts for overall water splitting. *Nature Reviews Materials*, 2, 17050.

Contaminants of Emerging Concern including Pharmaceuticals and Personal Care Products | Water Quality Criteria | US EPA. n.d. https://www.epa.gov/wqc/contaminants-emerging-concern-including-pharmaceuticals-and-personal-care-products. Accessed 22 Oct 2020

Dehghan, S., Tahergorabi, M., Norzaee, S., et al., 2020. Preparation and photocatalytic performance of reduced graphene oxide/ZnO nanocatalyst for degradation of metalaxyl from aqueous solution: Effect of operational parameters, mineralisation and toxicity bioassay. *International Journal of Environmental Analytical Chemistry,* 1–23.

Dhangar, K., Kumar, M. 2020. Tricks and tracks in removal of emerging contaminants from the wastewater through hybrid treatment systems: A review. *Science of the Total Environment*, 738, 140320.

Eregowda, T., Mohapatra, S. 2020. Fate of micropollutants in engineered and natural environment. In: Kumar, M., Munoz-Arriola, F., Furumai, H., Chaminda, T. (eds) *Resilience, response, and risk in water systems. Springer transactions in civil and environmental engineering.* Springer, Singapore. https://doi.org/10.1007/978-981-15-4668-6_15

Fujishima, A., Honda, K. 1972. Electrochemical photolysis of water at a semiconductor electrode. *Nature*, 238, 37–38.

Garcia-Segura, S., Brillas, E. 2017. Applied photoelectrocatalysis on the degradation of organic pollutants in wastewaters. *Journal of Photochemistry and Photobiology C: Photochemistry Reviews*, 31, 1–35.

Garcia-Segura, S., El-Ghenymy, A., Centellas, F., et al., 2012. Comparative degradation of the diazo dye Direct Yellow 4 by electro-Fenton, photoelectro-Fenton and photo-assisted electro-Fenton. *Journal of Electroanalytical Chemistry*, 681, 36–43.

Gavrilescu, M., Demnerová, K., Aamand, J., et al., 2015. Emerging pollutants in the environment: Present and future challenges in biomonitoring, ecological risks and bioremediation. *New Biotechnology*, 32, 147–156.

Gholami, M., Rasoulzadeh, H., Ahmadi, T., Hosseini, M. 2020. Synthesis, characterization of nickel doped zinc oxide by radio-frequency sputtering and application in photo-electrocatalysis degradation of norfloxacin. *Materials Letters*, 269, 127647.

Glaze, W.H., Kang, J.-W., Chapin, D.H. 1987. The chemistry of water treatment processes involving ozone, hydrogen peroxide and ultraviolet radiation. *Ozone: Science & Engineering*, 9, 335–352.

Grenni, P., Ancona, V., Barra Caracciolo, A. 2018. Ecological effects of antibiotics on natural ecosystems: A review. *Microchemical Journal*, 136, 25–39.

Guo, J., Jiang, L., Liang, J., et al., 2020. Photocatalytic degradation of tetracycline antibiotics using delafossite silver ferrite-based Z-scheme photocatalyst: Pathways and mechanism insight. *Chemosphere*, 270, 128651.

Khedr, T.M., El-Sheikh, S.M., Ismail, A.A., Bahnemann, D.W. 2019. Highly efficient solar light-assisted TiO_2 nanocrystalline for photodegradation of ibuprofen drug. *Optical Materials*, 88, 117–127.

Koiki, B.A., Orimolade, B.O., Zwane, B.N., et al., 2020. Cu_2O on anodised TiO_2 nanotube arrays: A heterojunction photoanode for visible light assisted electrochemical degradation of pharmaceuticals in water. *Electrochimica Acta*, 340, 135944.

Lapworth, D.J., Baran, N., Stuart, M.E., Ward, R.S. 2012. Emerging organic contaminants in groundwater: A review of sources, fate and occurrence. *Environmental Pollution*, 163, 287–303.

Liu, W., Song, X., Huda, N., et al., 2020. Comparison between aerobic and anaerobic membrane bioreactors for trace organic contaminant removal in wastewater treatment. *Environmental Technology & Innovation*, 17, 100564.

Mafa, P.J., Kuvarega, A.T., Mamba, B.B., Ntsendwana, B. 2019. Photoelectrocatalytic degradation of sulfamethoxazole on g-C_3N_4/BiOI/EG p-n heterojunction photoanode under visible light irradiation. *Applied Surface Science*, 483, 506–520.

Martimiano do Prado, T., Lindo Silva, F., Grosseli, G., et al., 2020. Using $BiVO_4$/CuO-based photoelectrocatalyzer for 4-nitrophenol degradation. *Materials*, 13, 1322.

Matamoros, V., Uggetti, E., García, J., Bayona, J.M. 2016. Assessment of the mechanisms involved in the removal of emerging contaminants by microalgae from wastewater: A laboratory scale study. *Journal of Hazardous Materials*, 301, 197–205.

Mazierski, P., Nadolna, J., Lisowski, W., Winiarski, M.J. et al., 2020. Effect of irradiation intensity and initial pollutant concentration on gas phase photocatalytic activity of TiO_2 nanotube arrays, *Catalysis Today*, 284, 19–26. https://doi.org/10.1016/j.cattod.2016.09.004

Mei, P., Wang, H., Guo, H., et al., 2020. The enhanced photodegradation of bisphenol A by TiO_2/C_3N_4 composites. *Environmental Research*, 182, 109090.

Menon, N.G., Mohapatra, S., Padhye, L.P., Tatiparti, S.S.V., Mukherji, S. 2020. Review on occurrence and toxicity of pharmaceutical contamination in Southeast Asia. In: Kumar, M., Snow, D., Honda, R. (eds) *Emerging issues in the water environment during anthropocene. Springer transactions in civil and environmental engineering*. Springer, Singapore. https://doi.org/10.1007/978-981-32-9771-5_4

Mohapatra, S., Menon, N., Padhye, L.P., Tatiparti, S.S.V., Mukherji, S. 2021. Natural attenuation of pharmaceuticals in the aquatic environment and role of phototransformation. In: Kumar, M., Snow, D., Honda, R., Mukherjee, S. (eds) *Contaminants in drinking and wastewater sources. Springer transactions in civil and environmental engineering*. Springer, Singapore. https://doi.org/10.1007/978-981-15-4599-3_3

Mohapatra, S., Padhye, L.P., Mukherji, S. 2018. Challenges in detection of antibiotics in wastewater matrix. In: Gupta, T., Agarwal, A., Agarwal, R., Labhsetwar, N. (eds) *Environmental contaminants. Energy, environment, and sustainability*. Springer, Singapore. https://doi.org/10.1007/978-981-10-7332-8_1

M.S. Priyanka, Sanjeeb Mohapatra, N. Gayathri Menon, Hybrid membrane technology: an alternative to industrial wastewater treatment, in Membrane-Based Hybrid Processes for Wastewater Treatment, Eds: Maulin P. Shah, Susana Rodriguez-Couto,2021, 481-501, doi: https://doi.org/10.1016/B978-0-12-823804-2.00001-X

Nagaraju, P., Puttaiah, S.H., Wantala, K., Shahmoradi, B. 2020. Preparation of modified ZnO nanoparticles for photocatalytic degradation of chlorobenzene. *Applied Water Science*, 10, 1–15. https://doi.org/10.1007/s13201-020-01228–w

Nguyen L.N., Hai, F.I., Kang, J., et al., 2013. Removal of emerging trace organic contaminants by MBR-based hybrid treatment processes. *Biodeterioration & Biodegradation*, 85, 474–482.

Noutsopoulos, C., Koumaki, E., Mamais, D., et al., 2015. Removal of endocrine disruptors and non-steroidal anti-inflammatory drugs through wastewater chlorination: The effect of pH, total suspended solids and humic acids and identification of degradation by-products. *Chemosphere*, 119, S109–S114.

Padilla Villavicencio, M., Escobedo Morales, A., Ruiz Peralta M., et al., 2020. Ibuprofen photodegradation by Ag_2O and Ag/Ag_2O composites under simulated visible light irradiation. *Catalysis Letters*, 150, 2385–2399.

Park, H., Bak, A., Ahn, Y.Y., et al., 2012. Photoelectrochemical performance of multi-layered BiO_x–TiO_2/Ti electrodes for degradation of phenol and production of molecular hydrogen in water. *Journal of Hazardous Materials*, 211, 47–54.

Peleyeju, M.G., Arotiba, O.A. 2018. Recent trend in visible-light photoelectrocatalytic systems for degradation of organic contaminants in water/wastewater. *Environmental Science: Water Research & Technology*, 4, 1389–1411.

Piedra López, J.G., González Pichardo, O.H., Pinedo Escobar, J.A., et al., 2021. Photocatalytic degradation of metoprolol in aqueous medium using a TiO_2/natural zeolite composite. *Fuel*, 284, 119030.

Pinheiro, D., Sunaja Devi K.R., Jose, A., Karthik, K. 2020. Box–Behnken design and experimental study of ciprofloxacin degradation over $Ag_2O/CeO_2/g\text{-}C_3N_4$ nanocomposites. *International Journal of Environmental Science and Technology*, 18, 2303–2324. https://doi.org/10.1007/s13762-020-02971–y

Priyanka, M.S., Mohapatra, S. (2020) Shifts and trends in analysis of contaminants of emerging concern: Sulfonamides. In: Kumar, M., Munoz-Arriola, F., Furumai, H., Chaminda, T. (eds) *Resilience, response, and risk in water systems. Springer transactions in civil and environmental engineering.* Springer, Singapore. https://doi.org/10.1007/978-981-15-4668-6_14

Ravindra, Y.S., Puttaiah, S.H., Yadav, S., Prabagar, J.S. 2020. Evaluation of polymeric $g\text{-}C_3N_4$ contained nonhierarchical ZnV_2O_6 composite for energy-efficient LED assisted photocatalytic mineralization of organic pollutant. *Journal of Materials Science: Materials in Electronics*, 31, 16806–16818.

Sayed, M., Ali Khan, J., Ali Shah, L. Noor S. Shah, et al., 2018. Solar light responsive poly(vinyl alcohol)-assisted hydrothermal synthesis of immobilized TiO_2/Ti film with the addition of peroxymonosulfate for photocatalytic degradation of ciprofloxacin in aqueous media: A mechanistic approach. *The Journal of Physical Chemistry C*, 122 (1), 406–421. https://doi.org/10.1021/acs.jpcc.7b09169

Shaban, Y.A., Fallata, H.M. 2019. Sunlight-induced photocatalytic degradation of acetaminophen over efficient carbon doped TiO_2 ($CTiO_2$) nanoparticles. *Research on Chemical Intermediates*, 45, 2529–2547.

Shah, N.S., Ali Khan, J., Sayed, M., et al., 2019. Hydroxyl and sulfate radical mediated degradation of ciprofloxacin using nano zerovalent manganese catalyzed $S_2O_8^{2-}$. *Chemical Engineering Journal*, 356, 199–209.

Shahmoradi, B., Yavari, S., Zandsalimi, Y., et al., 2018. Optimization of solar degradation efficiency of bio-composting leachate using Nd: ZnO nanoparticles. *Journal of Photochemistry and Photobiology A: Chemistry*, 356, 201–211.

Sharma, N., Mohapatra, S., Padhye, L.P. and Mukherji, S., 2021. Role of precursors in the formation of trihalomethanes during chlorination of drinking water and wastewater effluents from a metropolitan region in western India. *Journal of Water Process Engineering*, 40, 101928.

Shivaraju, H.P., Anilkumar, K.M., Yashas, S.R., et al., 2020a. Facile synthesis of Mn-Ce/N-TiO_2 composite for CO_2 hydrogenation into methane and intensifying methane yield in biomethanation. *Biofuels, Bioproducts and Biorefining*, 15, 189–201. https://doi.org/10.1002/bbb.2150

Shivaraju, H.P., Yashas, S.R., Harini, R. 2020b. Application of Mg-doped TiO_2 coated buoyant clay hollow-spheres for photodegradation of organic pollutants in wastewater. *Materials Today: Proceedings*, 27, 1369–1374.

Sirés, I., Brillas, E., Oturan, M.A., et al., 2014. Electrochemical advanced oxidation processes: today and tomorrow. A review. *Environmental Science and Pollution Research*, 21, 8336–8367.

Stackelberg, P.E., Gibs, J., Furlong, E.T., et al., 2007. Efficiency of conventional drinking-water-treatment processes in removal of pharmaceuticals and other organic compounds. *Science of the Total Environment*, 377, 255–272.

Tobajas, M., Belver, C., Rodriguez, J.J. 2017. Degradation of emerging pollutants in water under solar irradiation using novel TiO_2-ZnO/clay nanoarchitectures. *Chemical Engineering Journal*, 309, 596–606.

Tran, N.H., Urase, T., Ngo, H.H., et al., 2013. Insight into metabolic and cometabolic activities of autotrophic and heterotrophic microorganisms in the biodegradation of emerging trace organic contaminants. *Bioresource Technology*, 146, 721–731.

Tsui, M.M.P., Lam, J.C.W., Ng, T.Y., et al., 2017. Occurrence, distribution, and fate of organic UV filters in coral communities. *Environmental Science & Technology*, 51, 4182–4190.

Wu, Z., Liang, Y., Yuan, X., et al., 2020. MXene Ti_3C_2 derived Z–scheme photocatalyst of graphene layers anchored TiO_2/g–C_3N_4 for visible light photocatalytic degradation of refractory organic pollutants. *Chemical Engineering Journal*, 394, 124921.

Xu, M., Huang, H., Li, N., et al., 2019. Occurrence and ecological risk of pharmaceuticals and personal care products (PPCPs) and pesticides in typical surface watersheds, China. *Ecotoxicology and Environmental Safety*, 175, 289–298.

Yang, L., Yu, L.E., Ray, M.B. 2008. Degradation of paracetamol in aqueous solutions by TiO_2 photocatalysis. *Water Research*, 42, 3480–3488.

Yang, X., Wang, D. 2018. Photocatalysis: From fundamental principles to materials and applications. *Applied Energy Materials*, 1, 6657–6693.

Yashas, S.R., Shivaraju, H.P., Pema, G., et al., 2020. Sonochemical synthesis of graphitic carbon nitride-manganese oxide interfaces for enhanced photocatalytic degradation of tetracycline hydrochloride. *Environmental Science and Pollution Research*, 28, 4778–4789. https://doi.org/10.1007/s11356-020-10813-0

Zheng, Z., Ng, Y.H., Tang, Y., et al., 2021. Visible-light-driven photoelectrocatalytic activation of chloride by nanoporous MoS_2@$BiVO_4$ photoanode for enhanced degradation of bisphenol A. *Chemosphere*, 263, 128279.

Section 2

Removal of geo- and anthropo-genic contaminants

Chapter 6

Solid waste and landfill leachate

A transient source of emerging microbes and legacy contaminants for groundwater pollution

Sasmita Chand
Centre of Sustainable Built Environment, Manipal School of Architecture and Planning, Manipal Academy of Higher Education

Bhubaneswar Pradhan
IRDM Faculty Centre, Ramakrishna Mission Vivekananda Educational and Research Institute, Narendrapur

Sujata Chand
Cluster Development Executive, Padampur Block Level Cluster, National Handloom Development Program, Ministry of Textile

Sushanta Kumar Naik
ICAR-RCER, Farming System Research Centre for Hill and Plateau Region, Plandu, Namkum

CONTENTS

DOI: 10.1201/9781003247869-8

6.1 INTRODUCTION

The wastes are the materials coming from anthropogenic and animal activities that have no or marginal value for the owner and are discarded. Solid waste is either solid or liquid as sludge or exists as a free chemical phase. They serve no further purpose for production, transformation, or consumption. Solid waste is generated by human activities during the manufacturing of raw materials to intermediates and final products. These wastes include commercial waste, institutional waste, domestic waste, sanitary waste, and other non-residential wastes. They also include street sweepings, farming, and animal husbandry waste excluding electronic waste, biomedical waste, nuclear waste, industrial waste, battery waste, etc. These wastes are wide-ranging and include all sources, classification, composition, and characteristics. Therefore, solid waste management has become one of the most significant problems of the present time. However, these solid waste materials can be reused and recycled following some pretreatment procedure. These waste materials can become a resource for industrial production or energy generation if managed properly (Solid Waste Management Rules, 2016, Ministry of Environment, Forest and Climate Change [MoEF & CC, 2016a]).

India is the sixth largest economy and the fastest developing nation in the world, with an area of 3.287 million sq. km and a population 1.38 billion in 2021. Over time, the number of towns has increased from 5,161 in 2001 to 7,935 in 2011 (Census 2011). Increasing population and a growing number of cities and towns generate a large amount of solid wastes. With rapid population growth and urbanization in India, it is expected that annual waste generation may reach to 3.4 billion tonnes in 2050 (Kaza et al., 2018). In India, municipalities take the sole responsibility to manage solid wastes and usually dump them in open landfills. Ineffective collection system and scientific management of solid wastes, many political, legal, financial and socioeconomic constraints together with increased urban population and waste generation greatly affect the solid waste management practices (Srivastava et al., 2014). For the best management of solid waste and efficient recycling, the waste has to be separated at the source point. The recycling industry will play a major role in the management of solid waste.

6.2 SOURCE, TYPES, AND DIFFERENT SECTOR OF SOLID WASTE

Gaining knowledge on the sources and types of solid waste generation is basic to engineering solid waste management. There are several ways to classify solid waste and its sources. The key sources of solid waste generation are the medical centers, manufacture and production industries, the mining sector, fertilizers, pesticides and biocides, electronics, radioactive materials, warfare, natural disasters, food stores and distribution centers, feeding centers, slaughter areas, warehouses, domestic areas, and markets. These wastes are defined by their types, composition, and different important properties. Wastes are characterized on the basis of physical, chemical, biological, reactive, toxic, flammable, explosive, or corrosive properties (Management and Transboundary Movement Rules, 2016, (MoEF & CC, 2016b)).

Wastes generated from medicals by identifying, treating, or immunizing human beings or animals from health camps or from research activities are defined as biomedical wastes (Bio Medical Waste Management Rules, 2016 [MoEF & CC, 2016c]). Wastes generated from electrical and electronic equipment during manufacturing, renovation, and repair processes are defined as electronic wastes (E Waste Management Rules, 2016 [MoEF & CC, 2016d]). Wastes from construction, remodeling, and demolition of building materials and any other civil structure are defined as demolition and construction wastes (Construction and Demolition Waste Management Rules, 2016 [MoEF & CC, 2016e]). Solid wastes generated from lead-acid batteries after use are defined as battery wastes (Batteries Management and Handling Rules, 2011 [MoEF, 2011]). Waste plastic products from plastic manufacturing and abandoned plastic are defined as plastic wastes (Plastic Waste Management Rules, 2016 [MoEF & CC, 2016f]).

6.2.1 Municipal wastes

Wastes generated from different sources, which include household waste, waste from streets, office buildings, hotels, hospitals, institutions, etc., are generally collected by municipality vehicles and are dumped over a large area of land. These wastes are generally termed as municipal solid wastes (MSW), which would cause various environmental problems near the dumped site. In general, the MSW can be classified as residential, commercial, or industrial types based on the sources of waste generation. The chief components of these wastes are food wastes arising from residential flats and restaurants, demolition and construction wastes from residential flats and other civil works, hospitals and biomedical wastes, solid and semisolid wastes and residual sludge of waste water treatment processes and other industrial sites, as well as the miscellaneous wastes such as wastes from street cleaning activities, roadside litter, dead animals, abandoned materials arising from highways, streets, parks, playgrounds, and beaches. (Charles et al., 2019; Peavy et al., 1985; Ghosh, 2017).

6.2.2 Industrial wastes

Industrial wastes are those solid wastes that arise from manufacturing industries and usually include ashes, rubbish, demolition and construction waste, special wastes, and hazardous wastes from different sectors. The industrial waste can be categorized into biodegradable and non-biodegradable wastes based on their decomposing ability. Many kinds of wastes generated from food processing industries, dairy, textile mills, house wastes, slaughterhouses, etc. can be decomposed to harmless products by the action of microbes are referred as biodegradable wastes. The treatment processes for these wastes include combustion, composting, gasification, bio-methanation, etc. On the other hand, 10%–15% of the total industrial wastes are non-biodegradable and hazardous, and these wastes are resistant to microbial activity, hence cannot be decomposed into non-poisonous substances (Junankar, 2018). These wastes are generated from iron and steel plants, fertilizer industries, chemicals, drugs, and dyes like plastics, fly ash, synthetic fibers, gypsum, silver foil, glass objects, radioactive wastes, etc. (www.byjus.com/chemistry/industrial-waste-types). This category of waste poses serious threat to living organisms, as their concentration is increasing year after year and in the long run enters the bodies of animals and plants, causing diseases and environmental pollution (Junankar, 2018; Ghosh, 2017; Pariatamby, 2017; Srivastava et al., 2016).

6.2.3 Hazardous wastes

Any waste that causes harm, instantly or over a period of time, to the life of any living organism can be defined as hazardous waste (Misra and Pandey, 2005). These wastes are classified by their characteristics such as reactivity, toxicity, ignitability, and corrosivity. Hazardous waste shows the following four attributes as stated below.

a. **Ignitability:** These wastes have low flash points (less than 60°C) and can catch fire easily. Waste oils and used solvents are an example of this category.
b. **Corrosivity:** These wastes are either highly acidic or basic in nature. They readily corrode metal containers like storage tanks, barrels, and drums.
c. **Reactivity:** These wastes are highly volatile and may explode and produce toxic fumes, upon heating or if compressed (TERI, 2014).
d. **Toxicity:** Wastes containing lead or mercury may prove to be harmful or fatal when consumed or absorbed, like contamination of groundwater. Toxicity Characteristic Leaching Procedure is a procedure used to define toxicity and helps to identify wastes that are harmful to human health (TERI, 2014).

The principal sources of hazardous wastes are hospitals and biological research laboratories. Hazardous wastes are grouped by their categories like radioactive, biological, chemicals (reactive, toxic, and flammable), and explosives wastes. This waste includes waste specified under column Schedule I, Schedule II, and Schedule III of the Rules (Hazardous and Other Wastes (Management and Transboundary Movement) Rules, 2016 (MoEF & CC, 2016b)).

6.3 SOLID WASTE GENERATION, COMPOSITION, AND CHARACTERIZATION

Increasing population improved living standards, and rapid urbanization has led to the generation of huge amount of MSW every day. According to the Union Housing and Urban Affairs Ministry, India generated nearly 1.45 lakh MT of waste per day in 2019, and nearly half of the waste was processed to achieve overall scientific management of solid waste. It is expected that the total urban MSW rate will be 165, 230, and 436 MT by 2030, 2041, and 2050, respectively (Annepu, 2012; Sharma and Jain, 2019). Small towns with a population of 0.2 million generate solid waste between 200 and 300 g/capita/day (Sharma and Jain, 2019). As the population ranges increase from 0.2 to 0.5 million, 0.5 to 1 million, and above, the solid waste generated also increases to 300–350, 350–400, and 400–600 g/capita/day, respectively. Approximately 82% (50 MT) of MSW was collected, and only 18% (12 MT) comprised of litter. About 14 MT (28%) of the total waste collected was treated, and the remaining are openly dumped (CPCB 2016). According to the Ministry of Finance, if these solid wastes are not handled properly, then the rate of the waste generation by 2021 will need 590 km^2 of a projected area like Hyderabad city (CPCB 2016; www.statista.com, 2017; www.newindianexpress.com, 2019; Sharma and Jain, 2019; Ghosh, 2017).

The knowledge of physical and chemical constituents of solid waste is important to improve the solid waste management practices such as composting, recycling of economically usable materials and converting waste to energy. The physical composition of waste involves identification of individual components, analysis of particle size, moisture content, density, and calorific value (CV) and the chemical composition includes temperature, pH, C/N ratio, organic and inorganic materials, etc. For example, useful materials can be recovered from wastes by identifying individual components and analyzing the particle size; gathering information on moisture content, C/N ratio, and organic component would be helpful for composting; CV and chemical composition analysis are desirable to recover energy from wastes (Luz et al., 2015; Srivastava et al., 2016; Kumar and Samadder, 2017; Awasthi et al., 2019). The main composition of MSW in India consists of nearly 40%–60% compostable, 30%–50% inert, and 10%–30% recyclable wastes. Additionally, a report by National Environmental Engineering Research Institute suggested waste in India primarily consists of 0.68%±0.15% K, 0.67%±0.15% P, 0.64%±0.8% N, and a 26±5 C:N ratio (Joshi and Ahmed, 2016). A considerable amount (nearly 40%) of urban waste is originated from the Indian metropolitan cities followed by the small cities of North India, South India, West India, and East India. The MSW generated over time in the four parts of India and the urban areas are shown in Figure 6.1. Nearly 50% of the wastes are compostable in nature and the moisture contents were ranged 45%–50% (Sharma and Jain, 2019). Earlier studies reported that the waste composition and characteristics such as high CV have changed promptly from 1996 to 2011 (Ghosh, 2017; Sharma and Jain, 2019). It has also been reported that there was a substantial increase in paper and plastic wastes produced and the chemical composition of solid wastes in India during 1995, according to population range by National Environmental Engineering Research Institute. Population ranges from 0.1 to 0.5 and 5 to above the total nitrogen, phosphorus, potassium, C/N ratio, and CV (kcal/kg) are in the range 0.71–0.56, 0.63–0.52, 0.83–0.52, 30.94–30.11, and 1,009.89–800.70, respectively (Nandan et al., 2017). The variations in the physical and chemical composition of MSW over the years are depicted in Figure 6.2. The overall impact and the expected MSW generation rate per year in India are depicted in Figure 6.3. It is imperative to understand the trend of the solid waste composition in MSW, and for their better management, it is essential to characterize the solid waste along with the geotechnical and geological features, the surrounding area stability, groundwater composition, soil condition, recyclable waste materials, and hazardous waste of the municipal landfill site. Thus, it helps in the development of a robust and locality-specific management system (Naveen and Sivapullaiah, 2020).

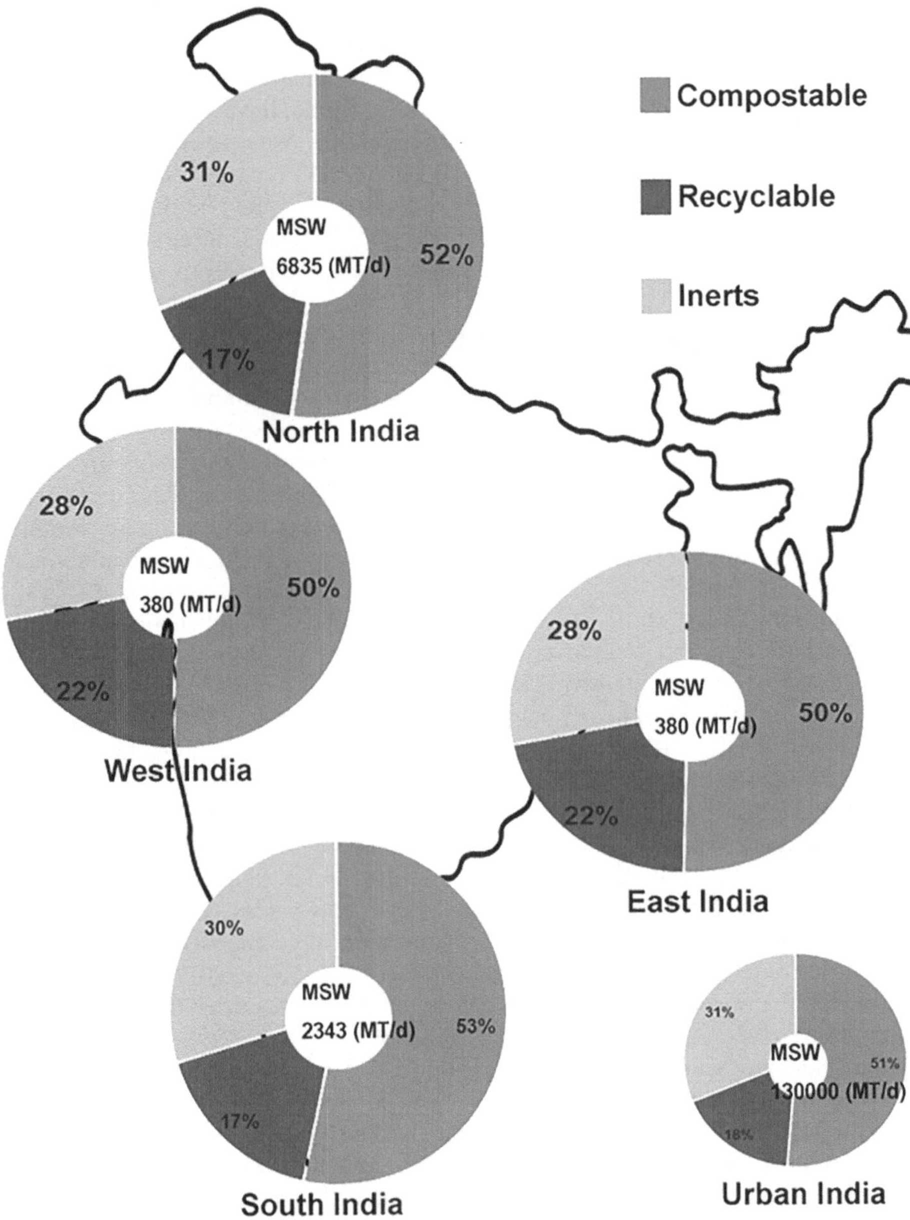

Figure 6.1 Municipal solid waste (MSW) generated over time in India. (Modified from Sharma and Jain, 2019.) The percentage composition of compostable, recyclable, and inert materials with the MSW generated per day from East India, South India, West India, North India, and Urban India are highlighted in pie charts.

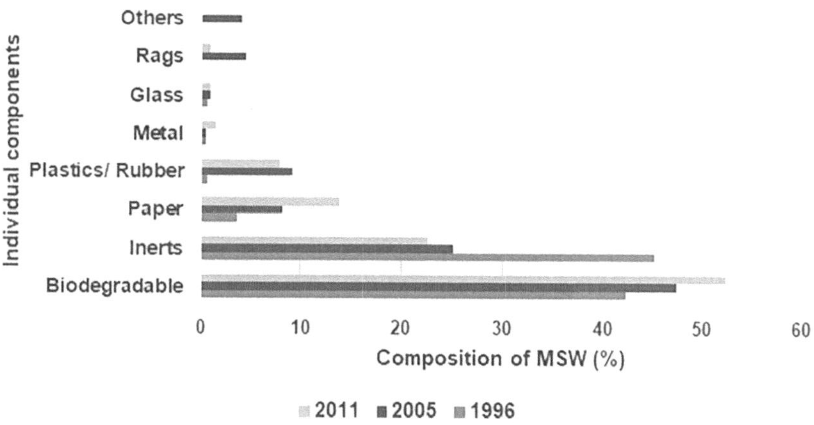

Figure 6.2 The variation in the physical and chemical composition of MSW in India over the years. (Modified from Charles et al., 2019; Sharma and Jain, 2019.) The percentage composition of physical and chemical materials generated in 1996, 2005, and 2011 from India is represented in the form of bar diagram.

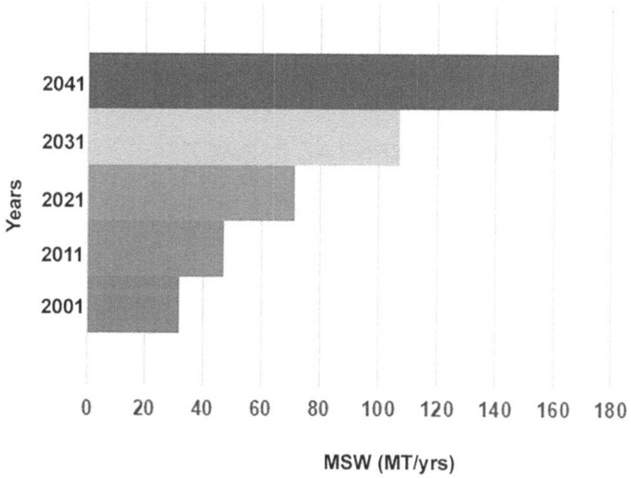

Figure 6.3 The overall impact and the expected MSW generation rate per year in India (Modified from Sharma and Jain, 2019) The anticipated MSW generated in million ton/year (MT/year) from 2001 to 2041 from India are represented in the form of bar diagram.

6.4 EMERGING AND LEGACY CONTAMINANTS IN SOLID WASTES

Contaminants that cause adverse ecological or human health effects and not undergoing the current environmental regulations and guidelines are referred to as emerging contaminants (ECs). These are pollutants of significant concentrations (usually in the range of ppb to ppm) found in the environment such as pharmaceuticals and personal care products (PPCPs), artificial sweeteners (ASs), pesticides, and industrial chemicals of a diverse group such as fungicides, insecticides, organo-chlorine and organo-phosphate pesticides, surfactants and their degradation products, plasticizers, pesticides, insect repellents, and fire retardants which previously remained undetectable, now being detected frequently (Terzić et al., 2008; Kapelewska et al., 2018; Sharma et al., 2019; Sardiña et al., 2019). On the other hand, legacy contaminants mostly include many industrial and agricultural chemicals, such as polyaromatic hydrocarbons, polychlorinated biphenyls (PCBs), and a range of pesticides and heavy metals (lead, mercury, etc.), which persist in the environment in higher concentrations (Sardiña et al., 2019). In other words, legacy contaminants have traditionally been monitored, whereas ECs are pollutants of recent origin, and their presence was not monitored in the pasts (Hildebrandt et al., 2008; Houtman, 2010; Sardiña et al., 2019).

A major portion of MSW is biodegradable due to their organic nature; however, some wastes either do not undergo degradation such as heavy metals or can also be modified to toxic/nontoxic by-products. These are mostly the ECs such as food wastes, batteries, pharmaceuticals and personal care products, and plasticizers (Jouhara et al., 2017). Antibiotics, hormones, other chemicals and their containers used in agriculture and allied sources such as animal husbandry and fish farming also contribute ECs. Since these ECs reach the municipal landfills via MSW, they might pollute groundwater through the water percolation (Nawaz and Sengupta, 2019). Pesticides, heavy metals, and many other toxic and hazardous chemicals are the major ECs from industrial sectors, which could also possibly deteriorate the water quality through accidental spillage and if improperly disposal.

6.5 POTENTIAL ENVIRONMENTAL IMPACTS

In India, rapid industrialization, urbanization, developed living standards, and exponential population growth are the foremost reasons for the generation of solid waste, which become an acute problem. The MSW contains wastes of heterogeneous composition from diversified origin such as household waste, recyclable waste, hazardous wastes from biomedical and industrial origin, and debris from construction and demolition of civil structures (Liyala, 2011). The population of India is around 1.3 billion and each day a single person is generating 0.5 kg of waste. Over the past decades, the population growth rate and per capita waste generation in India has also risen. This has led to various environmental problems, which is evident that there has been no separation of waste at source. It becomes exceedingly difficult to separate waste at the transfer station, inappropriate siting, design, operation, or maintenance of dumps in landfill sites. Adverse environmental impacts from solid waste are embedded in inadequate collection and recovery of recyclable and disposal of hazardous wastes, which is the critical phase. An enormous pile of garbage is landfilled in an unsanitary manner.

Improper waste management causes many environmental problems like an increase in disease transmission, contamination of ground and surface water, destruction of ecosystems, damage to the society and loss of property, decrease in tourism and other business, air pollution, vectors of diseases, and also produce greenhouse gas (GHGs) emissions. The organic constituents pose a great public health risk of poisoning, cancer, birth defects, and other ailments. Leachates in the landfill are composed of extracts of soluble waste, liquid waste, decomposed organic waste, infiltrated rain, etc., and MSW may add toxic materials and pathogenic organisms to the leachates. The ground and surface water will get contaminated if the landfills are unlined. In landfills, disposed organic wastes can undergo anaerobic degradation or biomethanation, which becomes a significant source of CH_4, a GHG gas with 21 times the effect of CO in trapping heat, unpleasant odor, and appearance and could also degrade the atmospheric air quality. Solid wastes dumped into water bodies affect aquatic habitats. A high nutrient content in organic wastes can reduce dissolved oxygen and affect aquatic life forms. The accumulation of solid waste results in the clogging of drains and flooding (www.waterfund.go.ke; Charles et al., 2019; Ferronato and Torretta, 2019).

6.6 SOLID WASTE TREATMENT TECHNIQUES

Owing to India's ever-growing population, huge amounts of MSW are produced due to various domestic, commercial, and institutional activities. Such huge amounts of waste generation are extremely dangerous to living organisms. Leachate filtration affects the groundwater quality and pollutes the air by the emission of GHGs. Apart from MSW, electronic waste and radioactive waste also deserve due to prudence due to fast-growing electronics and nuclear divisions. Thus, efficient solid waste management is the need of the hour to reduce environmental problems. For this, the U.S. Environmental Protection Agency has adopted the hierarchy of waste management practices such as reduction at the source level by recycling of materials, and finally burning, and landfilling.

6.6.1 Landfill disposal

In India, the state of solid waste management was not prioritized initially. But, during recent days much attention is given for proper scientific management of the solid waste disposal. Based on the type and characteristics of solid waste, different types of treatment methods are practiced in India. The treatment methods accomplished for MSW are composting, landfills, thermal processes including incineration and pyrolysis and chemical fixation, volume reduction, detoxification, degradation, encapsulation, etc. (Nandan et al., 2017; Ahluwalia and Patel, 2018). Apart from these techniques, the biodegradable (kitchen, garden, and food wastes), recyclable (plastic, paper, and glass), and inert (sand, pebbles, and gravels) wastes are treated by various treatments and processes. The biodegradable waste treatment is treated by aerobic and anaerobic biological treatment, thermal treatment by pyrolysis, incineration and gasification, and transformation by a mechanical and thermal process. The recyclable waste is treated by plasma pyrolysis technology (PPT), dissolution, screening, de-inking, sterilization, bleaching process, and vitrification technology. Inert wastes are treated by the landfilling procedure. Both organic and inorganic matter constitute MSW, and they are recovered by the adoption of suitable waste processing and treatment technologies options, such as sanitary landfill, aerobic and anaerobic digestion, biomethanation, composting, incineration, pyrolysis, and gasification (www.cbs.teriin.org, 2014; Ahluwalia and Patel, 2018; Nandan et al., 2017; Charles et al., 2019; Swati et al., 2018).

6.6.2 Sanitary landfill

In India, MSW is disposed of in an open area in a haphazard manner, which leads to environmental worsening, pollution of ground and surface water bodies by the percolation of leachate, and air pollution by the emission of harmful gases. Previous reports state that many landfill sites are operating even though their pollution reduction efficiency is a matter of concern. Owing to fast industrialization and urbanization, land availability is shrinking but waste generation is ever increasing. To address this issue, a secure and sanitary landfill is needed in landfilling practices for MSW management. The merits and demerits of this sanitary landfill technique have been reported by The Energy and Resources Institute in 2014. This technique decreases the emission of CH_4 and other toxic gases, lessens water and land pollution, decreases the threat to birds and rodents, controls fire hazards, and takes care of the issue of bad odor in the nearby settlement. Shortcomings of the technique include high initial investment and emission of insufferable gases. Land acquisition and availability is an additional issue associated with the technique.

6.7 EMERGING AND LEGACY CONTAMINANTS IN LANDFILL LEACHATES

Landfills are the final destination of MSW management practices worldwide, hence can be regarded as the general reservoirs of ECs. The landfills are also a major hub for the hazardous materials originated from the hospital and biomedical wastes and the ash generated from incinerated biomedical wastes (Zhao et al., 2009; Anitha and Jayraaj, 2012). Photographic materials such as X-ray film and many other medical equipment and medicines contribute to heavy metal wastes. The leachate analysis indicates a higher concentration of heavy metals and many ECs containing cosmetics, industrial compounds, plasticizers, and pharmaceuticals (Eggen et al., 2010; Anitha and Jayraaj, 2012; Ramakrishnan et al., 2015). Wastes containing ECs in the landfills can undergo the following physical, chemical, and biological conversion. Due to their varying chemical nature, they could adsorb to other solid wastes in the landfill, soil matrices, and even mix into the nearby aquatic bodies. Biochemical degradation in the presence of microbes is also another potential transformation process releasing toxic, nontoxic by-products (Samadder et al., 2017). ECs adsorbed to soil matrices, their transformed by-products from the landfills along with the leachate, thus provoke groundwater contamination (Taheran et al., 2018).

6.8 FATE OF LANDFILL LEACHATE

In recent days, inadequate solid waste management is a global concern with reference to social inclusion, environmental contamination, and economic sustainability, which requires an integrated waste management approach for its solution. The landfill is the most popular method of MSW disposal in many developing nations including India, because of less cost and maintenance. But this has led to many environmental worsening situations. The wastes in the landfills undergo physicochemical and biological transformation to watery liquids, which enters to nearby water bodies in the form of leachate. Due to the exponential growth of population and development of the industry, the landfill leachate problem becomes increasingly serious. If the leachate enters the water bodies, it is very difficult and expensive to clean up the contaminated water. However, these countries are witness to a huge surge in MSW resulting in waste management problems. The landfill method as sanitary landfill has been extensively used to describe solid waste disposal process constructed based on of dump and cover, without proper protection against groundwater pollution. Most of the landfills in India are usually open dumps without any specific design of scientific solid waste management principles, and a small fraction can be regarded as engineered landfills indicating designed and constructed according to engineering specifications. Therefore, landfill sites are vulnerable to the local inhabitants and other animals including high risk of environmental pollution (Bhalla et al., 2014). The risk depends on the concentration and types of toxic compounds in the leachate. Additionally, the groundwater is also contaminated by the leachates depending on the permeability of geological strata, depth of water table, and groundwater flow direction. The solid wastes in the landfill and their decomposition by-products move into the waterbodies during rainfall which eventually pollute groundwater through surface water-groundwater exchange.

6.9 PROCESS OF DECOMPOSITION

Leachate production in sanitary landfills is a complex combination of physical, chemical, and biological processes. In the landfills, decomposition process experiences the following four phases—aerobic phase, anaerobic acid phase, methanogenic phase, and stable methanogenic phase. In aerobic degradation process, the organic matters undergo hydrolysis which is the earliest event in order and spans for a short period followed by anaerobic and methanogenic phase consisting of formation of organic acids, acetic acids, and methane (Sikora et al., 2017).

1. **Hydrolysis:** The chemical bonds present in high molecular substances are broken down into low molecular compounds in the presence of water in the process of decomposition, which is relatively a slow process in acid digestion. The hydrolysis process is summarized in Equation (6.1).

$$C_6H_{10}O_5)\,n + nH_2O \rightarrow nC_6H_{12}O_6 + n\,H_2 \quad (6.1)$$

2. **Acidification:** It is the second stage of decomposition in which organic acids, such as propionic acid, butyric acid, lactic acid, acetic acid, and ethanol, H_2, and CO_2 are formed with the help of microorganisms. Acetic acid is the important organic acid formed as by-product in this stage, and it is also utilized as substrate in subsequent phase by methane-forming microbes (Rabii et al., 2019; Anukam et al., 2019). The following equations summarize the reaction sequence of the acidogenic stage.

$$C_6H_{12}O_6 \leftrightarrow 2CH_3CH_2OH + 2CO_2 \quad (6.2)$$

$$C_6H_{12}O_6 + 2H_2 \leftrightarrow 2CH_3CH_2COOH + 2H_2O \quad (6.3)$$

$$C_6H_{12}O_6 \rightarrow 3CH_3COOH \quad (6.4)$$

3. **Acetogenesis:** It is the penultimate stage in which acetic acid is formed along with H_2 and CO_2 of the anaerobic digestion process. This is also known as the dehydrogenation stage as H_2 is released in this process. In presence of H_2, the growth and metabolism of acetogenic bacteria is inhibited. Thus, acetic acid is consumed by CH_4 producing bacteria in the later stages of decomposition. The reaction series accompanying acetogenesis is denoted by Equations (6.5–6.7).

$$CH_3CH_2COO^- + 3H_2O \leftrightarrow CH_3COO^- + H + HCO_3^- + 3H_2 \tag{6.5}$$

$$C_6H_{12}O_6 + 2H_2O \leftrightarrow 2CH_3COOH + 2\ CO_2 + 4H_2 \tag{6.6}$$

$$CH_3CH_2OH + 2H_2O \leftrightarrow CH_3COO^- + 3H_2 + H^+ \tag{6.7}$$

4. **Methanogenesis:** It is the formation of CH_4 and CO_2. This is the final stage of the anaerobic digestion process carried out by methanogenic bacteria. Due to their extreme sensitivity to environmental fluctuations, especially to O_2 concentration, the methanogens are highly essential for anaerobic digestion process in the landfills. Few prominent species of methanogens are *Methanobrevibacter ruminantium, M. bryantic, M. thermoautotrophicum, Methanogenium cariaci*, and *M. marinsnigri*. They can absorb and digest the simplest of substrates. The following reactions show the methanogenic stage of the anaerobic digestion process:

$$CH_3COOH \rightarrow CH_4 + CO_2 \tag{6.8}$$

$$CO_2 + 4H_2 \rightarrow CH_4 + 2H_2O \tag{6.9}$$

$$2CH_3CH_2OH + CO_2 \rightarrow CH_4 + 2CH_3COOH \tag{6.10}$$

Anaerobic degradation of organic substances is the major degradation pathway in landfills (Anukam et al., 2019).

6.10 ENVIRONMENTAL CONCERNS TO LANDFILL DISPOSAL AND LEACHATE

In India, the past decades have witnessed a huge quantity of waste generation with increasing trend in every year (Swati et al., 2018). This problem is accompanied by the exponential increase in population, overall improvement of living standard and rapid growth of global manufacturing. Out of 62 million t of solid wastes collected each year in India, 28% undergo scientific treatment technologies and the rest being dumped in the municipality landfills (Swati et al., 2018). If the current rate of garbage dumping continues, then there will be a requirement of 3,000 acres of land for landfill disposal. It is currently predicted that waste generation will touch 165 million t in 2030 and would require a landfill area of 700 km^2 equal to the size of Bengaluru city (Swati et al., 2018). It has also been found that Mumbai, Hyderabad, Kolkata, Chennai, and Delhi are the top five solid waste generating cities in India. These cities generate solid waste ranging from biomedical, e-waste, plastic waste to hazardous waste. Anaerobic fermentation of biodegradable organic waste found in landfill leads to the formation of volatile fatty acid (VFA) and the methanogenic microbes present in many mature landfills convert VFA into biogas as CH_4 and CO_2 (Karthikeyan and Visvanathan, 2013). The quantity of leachate generation is affected by landfill moisture distribution influencers, namely refuse age, pretreatment, permeability, compaction, particle size and density, and direct landfill moisture contributors like rainfall, snowmelt, groundwater intrusion, initial moisture, and leachate recirculation. The quantity of leachate and the complex variety of substances and organic compounds including heavy metals, and many other hazardous chemicals can affect the groundwater quality (Wang et al., 2003). The following are the associated environmental concerns:

- Major health complications associated with emissions from landfills include difficulty in breathing, atmospheric air pollution, asthma, bacterial infections, and increased cardiovascular risk.
- The organic wastes in the landfills produce GHGs due to the decomposition and cause climate change.
- Percolation of leachate to groundwater and lower soil strata leads to groundwater contamination. The contaminated surface waterbodies act as breeding ground of pests (mosquitoes, flies, rats, cockroaches, etc.) and cause many diseases such as West Nile fever, dengue, malaria, murine typhus, histoplasmosis, and plague. During summer season, due to heavy odor, nearby settlements face lots of difficulties.
- Major issues associated with landfills that adversely affect the environment and human health are leachate, greenhouse, and poisonous gases and toxic substances.

6.11 OCCURRENCE OF POLLUTANTS IN GROUNDWATER

When the waste in the landfills decomposes in the presence of moisture, leachate is produced. If leachate penetrates the landfill layers and reaches the groundwater in aquifers, it pollutes water, as it contains innumerable organic and inorganic matters, humic and fulvic substances, xenobiotic substances, and heavy metals (Baghanam et al., 2020). For example, the Suchi landfill in India, an improper and unusual landfill, has resulted in groundwater contamination. The groundwater contamination problems caused by landfills have been an environmental concern in many developed countries as well since 1970 (www. mohua.gov.in, 2016; Han et al., 2016; Nagarajan et al., 2012, Bhalla et al., 2014; Zahari et al., 2016, Swati et al., 2018; Ferronato and Torretta, 2019).

6.12 EMERGING CONTAMINANTS IN GROUNDWATER (ORGANIC, INORGANIC, AND BIOLOGICAL CONTAMINANTS)

In recent times, groundwater contamination has become an emergent concern and is relatively poorly understood. The huge production of solid waste by anthropogenic activities has increased the ECs in the environment, and the subsequent groundwater pollution has made the situation alarming.

6.12.1 Sources and pathways

The key source and pathways of EC to groundwater include waste water discharge from municipal sewage treatment plant, industrial spillage, combined sewage-storm-water overflows, mining, waste disposal sites, and buried septic tanks. The point and diffusion sources are the two main sources of ECs in the environment. The point source and pathways for groundwater pollution by ECs mainly include industrial solid waste disposal in landfill sites, industrial impoundments, and buried septic tanks, whereas the diffuse pollution originates from poorly defined areas such as runoff from agricultural contaminants through irrigation water and rainfall, storm-water and urban runoff, leakage from reticulated urban sewerage systems, and diffuse aerial deposition. These are the major sources of ECs in the environment, and are mostly found in places where wastewater is not properly treated and remain in groundwater for a long time and contaminate it (Lapworth et al., 2012, Lamastra et al., 2016; Menon et al., 2020; Mohapatra et al., 2018, 2021; Priyanka and Mohapatra, 2020).

6.12.2 Organic and inorganic pollutants

The waste from landfills containing varied substances mostly consists of organic and inorganic pollutants which are extremely toxic to the environment (Stefanakis et al., 2014). As discussed earlier, the major organic pollutants are many household wastes, pharmaceuticals, industrials waste, pesticides, fungicide and insecticide, personal care products (PCPs), etc. The organic pollutants in landfill also undergo transformation processes with the help of bacteria producing carbonic and organic acids. In the absence of oxygen, the complex organic compounds undergo fermentation yielding soluble organic acids such as free VFAs, amino acids, other low molecular weight compounds, and gases like H_2 and CO_2 (Bhalla et al., 2013). High organic strength in the leachates found at the landfill locations (Delhi, Hyderabad and Kadapa) could be due to high values of total dissolved solids, alkalinity, biological oxygen demand, chemical oxygen demand, and chloride (Somani et al., 2019). The leachate from landfills contains high concentration of inorganic nutrients, heavy metals, reduced sulfate pollutants, and ammonium and nitrate ions from the bacterial transformation, which are the major inorganic pollutants. The leachates, which undergo the oxidation process in landfills, turn brown, which is due to the formation of ferric hydroxide colloids and the presence of humic substances (Gamar et al., 2017).

The concentrations of these organic and inorganic pollutants vary seasonally from landfill sites which are characterized by pH, biological oxygen demand, chemical oxygen demand, ammonia, microorganisms, heavy metals, and phosphates. The different parameters of leachate characteristics at different landfill sites in major cities like Tamil Nadu, Bangalore, Kolkata, Delhi, and Mumbai are given in Table 6.1 (Nagarajan et al., 2012; Naveen et al., 2013; Maiti et al., 2016; Yadav and Dikshit, 2017; Somani et al., 2019). The data from Table 6.1 reveal that the physicochemical parameters in landfill leachate are higher than the Indian Standard for Drinking Water as per BIS specifications (IS-10500-2012, n.d.) (Somani et al., 2019). Leachates from landfills containing these organic and inorganic pollutants are released into the natural

Table 6.1 Characteristics of leachates from Indian landfills

Sl. No.	*Parameters*	*Tamil Nadu*	*Bangalore*	*Kolkata*	*New Delhi*	*Mumbai*	*Standards for disposal*			*Typical range in leachate (as per the MoEF, 2016)*	*References*
							Inland surface water	*Public sewer*	*Land disposal*		
1.	pH	6.9	8.1	7.4	8.1	8.5	5.5–9.0	5.5–9.0	5.5–9.0	3.7–8.9	Nagarajan et al., (2012); Naveen et al., (2013); Maiti et al., (2016); Naveen and Malik, (2017); Yadav and Dikshit, (2017); Somani et al., (2019)
2.	TDS (mg/L)	24,123	1250	920	11,135	3,400	21,002	2,100	2,100	725–55,000	Nagarajan et al., (2012); Naveen et al., (2013); Maiti et al., (2016); Naveen and Malik, (2017); Yadav and Dikshit, (2017); Somani et al., (2019)
3.	COD (mg/L)	25,102	1080	48	4,560	2,304	250	-	-	50–99,000	Nagarajan et al., (2012); Naveen et al., (2013); Maiti et al., (2016); Naveen and Malik, (2017); Yadav and Dikshit, (2017); Somani et al., (2019)
4.	BOD (mg/L)	17,552	105	0.32	2,825	238	-	-	-	-	Nagarajan et al., (2012); Naveen et al., (2013); Maiti et al., (2016); Naveen and Malik, (2017); Yadav and Dikshit, (2017); Somani et al., (2019)
5.	Na^+ (mg/L)	532	1,676	-	-	1,412.43	-	-	-	-	Nagarajan et al., (2012); Naveen et al., (2013); Yadav and Dikshit (2017); Somani et al., (2019)
6.	K^+ (mg/L)	1,392	1,078	-	-	-	-	-	-	-	Nagarajan et al., (2012); Naveen et al., (2013); Somani et al., (2019)
7.	NH_4^+ (mg/L)	1,932	-	0.018	29.50	-	-	-	-	-	Nagarajan et al., (2012); Maiti et al., (2016); Naveen and Malik (2017); Somani et al., (2019)

(Continued)

Table 6.1 (Continued) Characteristics of leachates from Indian landfills

Sl. No.	*Parameters*	*Tamil Nadu*	*Bangalore*	*Kolkata*	*New Delhi*	*Mumbai*	*Standards for disposal*			*Typical range in leachate (as per the MoEF, 2016)*	*References*
							Inland surface water	*Public sewer*	*Land disposal*		
8.	Cd (mg/L)	0.05	0.05	-	-	0.09	2	1	-	-	Nagarajan et al., (2012); Naveen et al., (2013); Yadav and Dikshit (2017); Somani et al., (2019)
9.	Cr (mg/L)	0.28	0.03	-	0.75	0.024	2	2	-	-	Nagarajan et al., (2012); Naveen et al., (2013); Naveen and Malik (2017; Yadav and Dikshit (2017); Somani et al., (2019)
10.	Cu (mg/L)	0.71	0.15	0.004	0.54	-	3	3	-	0–9.0	Nagarajan et al., (2012); Naveen et al., (2013); Maiti et al., (2016); Naveen and Malik (2017); Somani et al., (2019)
11.	Fe (mg/L)	63.41	11.16	-	10.20	1.06	-	-	-	-	Nagarajan et al., (2012); Naveen et al., (2013); Naveen and Malik (2017); Yadav and Dikshit (2017); Somani et al., (2019)
12.	Ni (mg/L)	0.38	1.33	-	0.31	-	3	3	-	-	Nagarajan et al., (2012); Naveen et al., (2013); Naveen and Malik (2017); Somani et al., (2019)
13.	Pb (mg/L)	1.20	0.30	-	-	-	0.1	1	-	0–17.2	Nagarajan et al., (2012); Naveen et al., (2013); Somani et al., (2019)
14.	Zn (mg/L)	2.10	3.0	0.02	0.56	-	5	5	15	-	Nagarajan et al., (2012); Naveen et al., (2013); Maiti et al., (2016); Naveen and Malik (2017); Somani et al., (2019)
15.	Cl^- (mg/L)	-	250	799	56.80	349.53	1,000	1,000	600	2–11,375	Naveen et al., (2013); Maiti et al., (2016); Naveen and Malik (2017), Yadav and Dikshit (2017), Somani et al., 2019)
16.	SO_4^{2-} (mg/L)	-	10	31.94	-	-	-	-	-	-	Naveen et al., (2013); Maiti et al., (2016); Somani et al., (2019)

ecosystem that can cause eutrophication, threats to natural biodiversity, and can also cause ground water pollution.

6.12.3 Biological contaminants

Among the biological contaminants, harmful bacteria are most commonly seen in the groundwater. The quality of groundwater findings has been stated by previous researchers that, in comparison to surface and groundwater, contaminants from leachate showed significant microbial populations such as *Escherichia coli*, *Shigella* sp., and *Shalmonella* sp. These microbial contaminants might lead to serious health hazards in human and farm animals (Naveen and Malik, 2017; Naveen et al., 2018; Mukherjee et al., 2015).

6.12.4 Processes of remediation of organic and inorganic wastes

Organic and inorganic portion of solid wastes undergoing different treatment procedures based on their chemical composition may be recycled back as a significant energy source. For instance, composting is viable only for the organic portions of the solid wastes, but to reduce or recycle the inorganic portions, thermal degradation or incineration technologies might be helpful, which are discussed below.

6.12.4.1 Composting

This process involves the use of microorganisms and environmental heat and moisture to decompose organic solid waste. Composting can be classified into aerobic and anaerobic types. Anaerobic composting is also known as biomethanation. In general, bacteria, fungi, and actinomycetes are involved in the process of composting and the wastes are converted into sugars, starch, and organic acids by the action of the microorganisms (TERI, 2014). Solid waste in India is mostly treated in this method, and the efficiency of the method largely depends on temperature. Advantages associated with composting technique include production of organic manure, a decline in the volume of waste that eventually decreases the burdens of municipal landfills and easy to practice. Some technologies for composting waste are:

a. Window compositing: this process of compost production involves piling of the organic waste in long rows. This helps easy air flow to the deposited matter for efficient aerobic microbial degradation.
b. Aerated static pile composting: This method is relatively quicker (3–6 months) in producing compost. It involves mechanical aeration of the solid waste by use of a network of pipelines connected to a blower.
c. In-vessel composting: here, the solid waste is stored in a vessel and required aeration, moisture, and temperature is maintained in it.
d. Vermicomposting: food material, kitchen waste papers, etc. containing organic wastes can be converted into compost by culturing special types of earthworms such as *Eudrilus eugeniae*, *Eisenia foetida*, and *Perionyx excavatus*.

6.12.4.2 *Aerobic digestion and anaerobic digestion*

Aerobic composting involves the conversion of organic solid waste into compost by different bacteria in the presence of air. A reduction of 50%–85% in volume can be achieved by this process. The compost yielded is mainly used as fertilizer (Srivastava et al., 2017). Final product is free from odor and pathogens. In anaerobic digestion, organic waste is anaerobically digested to produce biogas (CH_4 and CO_2) and organic fertilizer. The optimal conditions required for the process are a temperature of 37°C and pH 7. Anaerobic biodegradation provides clean fuel and digested sludge, which can be used as manure. Different types of anaerobic digestion processes consist of conventional slurry digestion (Ref Com process: refuse conversion to methane, WMC process, Cal recovery process, and waste biogas process), dry digestion (VALOGRA and BIOCEL process), and two-phase digestion (hydrolysis, acidification, and methanation phases). Segregating the biodegradable fractions from that of non-biodegradable substances and their pretreatment processing is the foremost step in almost all categories of the anaerobic digestion processes.

6.12.4.3 *Biomethanation*

This method involves the anaerobic degradation of organic waste by microorganisms. Biogas containing CH_4 is produced in this process. Many researchers have reported approximately 16 million metric (CO equivalents per annum) of CH_4 is emitted from Indian landfill sites, and the annual energy potential from landfill gas of different landfills sites like Mumbai (Deonar and Gorai) has the potential to produce 5.6 MW, Ahmedabad (Pirana) 1.3 MW, Delhi (Bhalswa, Ghazipur, and Okhla) 8.4 MW, and Pune (Uruli) 0.7 MW. Biogas produced in the process is used to generate electricity, and the inert slurry can be used as an organic fertilizer (Vaish et al., 2020). The gas can be used for conventional fuels like liquid petroleum gas or compressed natural gas, compressed biogas, which can be converted into electricity with use of generators yielding 30% electricity. However, the demerits of this method are almost 70% of the energy is lost as heat in the process of energy conversion. The biomethanation slurry is used as excellent liquid manure for agriculture. Thus, biomethanation not only produces energy but also distributes nutrients for the soil.

6.12.4.4 *Incineration*

This is a direct burning process of waste in the presence of air, at temperatures of about above 800°C to 2,000°C, liberating heat energy, inert gases, and ash. The final products of the process are CO_2 and H_2O. In India, incineration is not prevalent because of poor segregation of solid waste at the source. The presence of high organic matter (40%–60%), moisture content (40%–60%), and inert material (30%–50%) and low CV (800 to 1,100 kcal/kg) make the incineration process difficult. CV of the solid waste must fall in the range of 1,200–1,400 kcal/kg for energy production. If carried out properly, there is a scope to harness about 65%–80% of the energy content of the organic matter as heat energy, which can be further utilized for thermal application. Deterioration of air quality due to emission of sulfur, nitrogen, and halogen-containing compounds is the disadvantage associated with the process.

6.12.4.5 *Pyrolysis and gasification*

Pyrolysis is a process of thermal degradation of solid waste with low moisture content like paper, cloth, plastic, yard wastes, etc. in the absence of air. The process yields recyclable products such as char, oil, and combustible gases (H_2, CH_4, CO, CO_2, etc.). The waste is treated in a reactor at 600°C–1,000°C. When compared with incineration, pyrolysis is a costly process but has little effect on the environment. Gasification involves high-temperature degradation of solid organic waste in the presence of limited amount of air. This process yields gases like CO and CO_2. The merits of this process are like the handling of gas is easier than solid fuel, reduced volume leads to a reduction in dumping, lesser ash production to incineration, and no emission of hazardous gases. Disadvantages associated with the process include the cost factor and release of tar and poisonous outputs into the environment.

6.13 TREATMENT TECHNOLOGIES FOR LANDFILL LEACHATE

Leachate generated from a landfill site is a concentrated stream rich in heavy metals, organic matter, xenobiotics, and inorganic macro constituents. The presence of all these components of pollutants makes it inappropriate for its discharge into water bodies. In India, most of the solid waste is disposed of in the open, in an unscientific and haphazard manner, which creates problems for public health and the environment. Considering the leachate composition and many physicochemical factors such as temperature, pH, solubility, chemical composition, etc., treatment procedures are chosen to minimize the pollutant level. Thus, several conventional and advanced treatment technologies are used to treat organic and inorganic constituents of leachate found in landfills. In principle, the leachates can undergo the following treatment processes, broadly categorized as (1) physicochemical, (2) biological (aerobic or anaerobic), and (3) combination of physicochemical and biological.

1. Physicochemical treatment:

   The physicochemical treatments are carried out to remove refractory substances from stabilized leachate and carried out by process like flocculation-coagulation, separation treatments with membrane filtration, air stripping, adsorption by activated carbon, chemical precipitation, ion exchange, chemical oxidation, and advanced oxidation processes, Fenton process, photocatalysis, and electrochemical process.
2. Biological (aerobic or anaerobic) treatment:

   Landfill leachate treatment using microorganism is the most common practice followed everywhere and can be classified as aerobic or anaerobic type. Leachate with larger quantities of organic substances is most suited to this type of treatment. Aerobic treatment shows rapid removal kinetics and succeeds in reducing organic pollutants, and the nitrification process is accomplished. The demerits of this treatment are high cost, energy-intensive, and excess sludge production. The aerobic treatment of landfill leachate applied in both developed and developing countries are aerated lagoons or stabilization ponds, constructed wetlands, aerated reactors, rotating biological contactors, sequencing batch reactor (SBR), trickling filters, moving bed bioreactor, fluidized bed bioreactors, membrane biological reactor, membrane aerated biofilm reactor, up flow anaerobic sludge blanket, and anaerobic ammonium oxidation. On the other hand, in the case of anaerobic and anoxic processes, the organic matter is broken down in the absence of oxygen by microbial action. Anaerobic treatment also shows low sensitivity for the presence of toxic substances and considerable efficiency in ammonia stripping. Advantages associated with anaerobic treatment are limited due to poor growth rate of anaerobic microorganisms, ineffective NH_4-N removal, and poor retention of biomass.

3. Combination of physicochemical and biological treatment:

   The combination of physical-chemical and biological processes is an efficient process to treat for dissimilar leachate pollutant concentrations. These are taken by different process like photo-electro-Fenton process–membrane bioreactor (India), submerged anaerobic membrane bioreactor-membrane bioreactor (London), sequencing batch biofilter granular reactor (Italy), sequencing batch reactor-SBR-Fenton-Like-SBR Post-Oxidation (Estonia), trickling filters-electro-coagulation (Magnesium-Based Anode) (Canada), Fenton process–passive aerated immobilized biomass (Egypt), and aerobic SBR-zeolite adsorption (Malaysia) (Torretta et al., 2017; Show et al., 2019; Durai et al., 2020).

6.14 ENVIRONMENTAL MONITORING, WASTE MANAGEMENT PRACTICES, AND CHALLENGES

Recently, solid waste management in India has become better with new and revised legislation of solid waste management in 2016 (Solid Waste Management Rules, 2016), new schemes for cleanliness and waste management like SBM, Atal Mission for Rejuvenation and Urban Transformation, Smart Cities, Jawaharlal Nehru National Urban Renewal Mission, etc. (Ghosh, 2017). The main aim of MSW management strategies is to address the human health, environmental, aesthetic, land-use, resource, and economic concerns related to inadequate waste disposal. Unscientific disposal of MSW in open dumps or low-lying areas without adequate precautions or controls affects both human health and environment (Srivastava et al., 2015; Kumar and Samadder, 2017; Sharma and Jain, 2019). The effective solid waste management in new legislation is to minimize the quantum of waste by adopting integrated waste management and the concept of 5R like Reduce, Reuse, Recover, Recycle, and Remanufacture. The eventual objective is zero waste to landfills to reduce the soil contamination, water and air, minimization of adverse environment and health impacts, resource recovery, and recycling to reduce the GHGs. The system should also be cost-effective, easy to operate, and acceptable to the local community (Ghosh, 2017).

Thus, the potential environmental impacts of MSW management programs cover:

- The collection of waste from households and commercial institutions to acceptable final disposal to minimize the quantity of waste that must be placed in landfills through elimination, recovery, reuse, and recycling, remanufacturing, composting, and similar methods.

- Effective and efficient collection and transportation of all waste, separating the hazardous and non-hazardous waste.
- Designing the sanitary landfills to confirm proper siting, operation, monitoring, and closure.

Three main functional components of the MSW management plans in India are: (1) On-site handling, storage, separation, and processing of wastes, (2) waste collection, processing, and transport, and (3) landfill disposal. The major aim is to reduce waste production, reuse, and recycle wherever possible that helps in conservation of resources and revenue generation (Charles et al., 2019; Ghosh, 2017; Hanrahan et al., 2006; Sharma and Jain, 2019). According to Central Pollution Control Board, India (CPCB, 2013), only 12.45% solid waste is methodically processed, and the rest are disposed in a haphazard manner in open dumps. The major challenges include:

- improper segregation of the waste due to inadequate public support, absence of proper waste processing facilities, and lack of door-to-door collection systems,
- lack of scientific methods in collection, transportation, treatment, disposal,
- lack of participation from the urban population, and
- lack of a complete integrated system for efficient waste management (Ghosh, 2017).

6.15 SUSTAINABLE SOLID WASTE MANAGEMENT

Sustainable solid waste management relies on two principles, i.e., waste reduction and waste separation (Permana et al., 2015). Effective solid waste management systems are environmentally and financially sustainable. Sustainability can no longer be sidelined. It must be an integrated aspect of different elements of the system and it also quantify the adverse impact of solid waste, its generation on the environment. There are social, economic, environmental, and technical challenges for implementing solid waste management, which needs to be established for a sustainable solid waste management plan (Bui et al., 2020).

- The societal challenge is the lack of active public participation in solid waste management. Hence, creating public awareness can eliminate the negative impact (Ikhlayel, 2018).
- The Ministry of Environment, Forest and Climate Change revised solid waste management Rules in 2016, in supersession of MSW Rules 2000, to improve the collection, segregation, recycling, treatment, and disposal of solid waste in an environmentally sound manner. Solid waste management Rules 2016 mandate local authorities to frame laws to levy user charges and also impose fines for littering and noncompliance. This would be a useful source of financial empowerment for the cities and the state governments would approve the local authorities to levy user charges.
- A major transformation is required in the current practice of solid waste handling and management at the source of generation, at one end and at the landfill disposal on the other. Implementing an adequate solid waste management technique can considerably diminish the various environmental problems.

In India, the sustainability of solid waste management lies in its integrated sustainable approach that includes technological, institutional, and financial aspects. The basic issue faced by urban local bodies in India is their overdependence on higher-level governments for cash inflow. The Government of India has allotted Rs. 14,620 crores under the SBM over next 5-year period for all the existing 4,041 statutory towns in the country. The mission will be mostly funded by the state/urban local governments, amplified by user charges, unlocking land value, and private sector contributions. A sustainability model prepared by the National Solid Waste Association of India can solve the problems associated with solid waste management in Indian cities as it has utilized all the stake holders and also given considerations to different factors such as technical, environmental, financial, political, sociocultural, legal, and policy-related issues associated with the MSW management (Sharma and Jain, 2019). After the introduction of Jawaharlal Nehru National Urban Renewal Mission by the Government of India, the solid waste management sector in India has progressed in the right direction during the last few years. In recent times, there has been an increasing demand for sustainable approaches for solid waste management and steps are being taken by the government. However, a more systematic approach is required along with the practice of the latest and cost-effective technologies for appropriate solid waste management (Mani and Singh, 2016; Ahluwalia and Patel, 2018).

6.16 CONCLUSIONS

Exponential growth in population, rapid urbanization, industrialization, economic development, and an overall improvement in living standard has led to the generation of huge amounts of waste production in India. The waste management systems in India are not adequate due to their dependence on insufficient waste infrastructure, the informal sector, and open dumping when compared to developed countries. Lack of effective solid waste management policies, appropriate technology, and sufficiency of trained manpower for waste management are some of the major challenges. Therefore, community awareness and changes in mindset among the public is the need of hour for developing an appropriate and sustainable solid waste management strategy. Extensive techniques of waste management and alternatives such as source reduction and increase in recycling rate and increase in the number of recycling industries should be implemented. This will also provide new employment opportunities, increase the proper disposal, and bring awareness about the service. Additionally, there should be research collaboration between production plants and the research institutions to develop the waste-to-energy plants for the development of effective waste management techniques and technology and study of environmental impacts.

REFERENCES

Ahluwalia, I.J., Patel, U., 2018. Solid waste management in India: an assessment of resource recovery and environmental impact. https://think-asia.org/handle/11540/8143

Anitha, J., Jayraaj, I.A., 2012. Isolation and identification of bacteria in biomedical wastes (BMW). *Int. J. Pharm. Pharm. Sci.* 4, 386–388.

Annepu, R.K., 2012. *Sustainable solid waste management in India*. Columbia University, New York 2.

Anukam, A., Mohammadi, A., Naqvi, M., Granström, K., 2019. A review of the chemistry of anaerobic digestion: methods of accelerating and optimizing process efficiency. *Processes* 7, 504.

Awasthi, M.K., Zhao, J., Soundari, P.G., Kumar, S., Chen, H., Awasthi, S.K., Duan, Y., Liu, T., Pandey, A., Zhang, Z., 2019. Sustainable management of solid waste. In: Taherzadeh M., Bolton K., Wong J., Pandey A. (eds) *Sustainable Resource Recovery and Zero Waste Approaches*. Elsevier, pp. 79–99.

Baghanam, A.H., Nourani, V., Aslani, H., Taghipour, H., 2020. Spatiotemporal variation of water pollution near landfill site: application of clustering methods to assess the admissibility of LWPI. *J. Hydrol.* 591, 125581.

Bhalla, B., Saini, M.S., Jha, M.K., 2013. Effect of age and seasonal variations on leachate characteristics of municipal solid waste landfill. *Int. J. Res. Eng. Technol.* 2, 223–232.

Bhalla, B., Saini, M.S., Jha, M.K., 2014. Assessment of municipal solid waste landfill leachate treatment efficiency by leachate pollution index. *Assessment* 3, 8447–8454.

Bui, T.D., Tsai, F.M., Tseng, M.-L., Ali, M.H., 2020. Identifying sustainable solid waste management barriers in practice using the fuzzy Delphi method. *Resour. Conserv. Recycl.* 154, 104625.

CPCB (Central Pollution Control Board). 2016. Central Pollution Control Board (CPCB) Bulletin. Government of India.

Charles, R.K.J., Bernard, M.A., Raju, J., Abdulmajid, M., 2019. Sustainable waste management through waste to energy technologies in India-opportunities and environmental impacts. *Int. J. Renew. Energy Res.* 9, 309–342.

Durai, N.J., Gopalakrishna, G.V.T., Padmanaban, V.C., Selvaraju, N., 2020. Oxidative removal of stabilized landfill leachate by Fenton's process: Process modeling, optimization & analysis of degraded products. *RSC Adv.* 10, 3916–3925.

Eggen, T., Moeder, M., Arukwe, A., 2010. Municipal landfill leachates: a significant source for new and emerging pollutants. *Sci. Total Environ.* 408, 5147–5157.

Ferronato, N., Torretta, V., 2019. Waste mismanagement in developing countries: a review of global issues. *Int. J. Environ. Res. Public Health* 16, 1060.

Gamar, A., Khiya, Z., Zair, T., El Kabriti, M., Bouhlal, A., El Hilali, F., 2017. Assessment of physicochemical quality of the polluting load of leachates from the wild dump of el hajeb city (Morocco). *Int. J. Res.* 5, 63–71.

Ghosh, S.K., 2017. Country Chapter.

Han, Z., Ma, H., Shi, G., He, L., Wei, L., Shi, Q., 2016. A review of groundwater contamination near municipal solid waste landfill sites in China. *Sci. Total Environ.* 569, 1255–1264.

Hanrahan, D., Srivastava, S., Ramakrishna, A.S., 2006. *Improving Management of Municipal Solid Waste in India: Overview and Challenges*. Environment and Social Development Unit, South Asia Region. The World Bank (India Country Office), New Delhi.

Hildebrandt, A., Guillamón, M., Lacorte, S., Tauler, R., Barceló, D., 2008. Impact of pesticides used in agriculture and vineyards to surface and groundwater quality (North Spain). *Water Res.* 42, 3315–3326.

Houtman, C.J., 2010. Emerging contaminants in surface waters and their relevance for the production of drinking water in Europe. *J. Integr. Environ. Sci.* 7, 271–295.

Ikhlayel, M., 2018. Indicators for establishing and assessing waste management systems in developing countries: a holistic approach to sustainability and business opportunities. *Bus. Strategy Dev.* 1, 31–42.

IS-10500-2012, n.d. Drinking water—Specification second revision, Gr 6. http://cgwb.gov.in/Documents/WQ-standards.pdf

Joshi, R., Ahmed, S., 2016. Status and challenges of municipal solid waste management in India: a review. *Cogent Environ. Sci.* 2, 1139434.

Jouhara, H., Czajczyńska, D., Ghazal, H., Krzyżyńska, R., Anguilano, L., Reynolds, A.J., Spencer, N., 2017. Municipal waste management systems for domestic use. *Energy* 139, 485–506.

Junankar, A.K.N., 2018. Use of industrial waste in the field of construction. *Int. J. Eng. Res.* 7, 6.

Kapelewska, J., Kotowska, U., Karpińska, J., Kowalczuk, D., Arciszewska, A., Świrydo, A., 2018. Occurrence, removal, mass loading and environmental risk assessment of emerging organic contaminants in leachates, groundwaters and wastewaters. *Microchem. J.* 137, 292–301.

Karthikeyan, O.P., Visvanathan, C., 2013. Bio-energy recovery from high-solid organic substrates by dry anaerobic bio-conversion processes: a review. *Rev. Environ. Sci. Biotechnol.* 12, 257–284.

Kaza, S., Yao, L., Bhada-Tata, P., Van Woerden, F., 2018. *What a Waste 2.0: A Global Snapshot of Solid Waste Management to 2050.* World Bank Publications.

Kumar, A., Samadder, S.R., 2017. A review on technological options of waste to energy for effective management of municipal solid waste. *Waste Manag.* 69, 407–422.

Lamastra, L., Balderacchi, M., Trevisan, M., 2016. Inclusion of emerging organic contaminants in groundwater monitoring plans. *MethodsX* 3, 459–476.

Lapworth, D.J., Baran, N., Stuart, M.E., Ward, R.S., 2012. Emerging organic contaminants in groundwater: a review of sources, fate and occurrence. *Environ. Pollut.* 163, 287–303.

Liyala, C.M., 2011. Modernising solid waste management at municipal level: Institutional arrangements in urban centres of East Africa. Thesis, Wageningen University, Wageningen, Netherlands.

Luz, F.C., Rocha, M.H., Lora, E.E.S., Venturini, O.J., Andrade, R.V., Leme, M.M.V., del Olmo, O.A., 2015. Techno-economic analysis of municipal solid waste gasification for electricity generation in Brazil. *Energy Convers. Manag.* 103, 321–337.

Maiti, S.K., De, S., Hazra, T., Debsarkar, A., Dutta, A., 2016. Characterization of leachate and its impact on surface and groundwater quality of a closed dumpsite–a case study at Dhapa, Kolkata, India. *Procedia Environ. Sci.* 35, 391–399.

Mani, S., Singh, S., 2016. Sustainable municipal solid waste management in India: a policy agenda. *Procedia Environ. Sci.* 35, 150–157.

Menon N.G., Mohapatra S., Padhye L.P., Tatiparti S.S.V., Mukherji S. (2020) Review on occurrence and toxicity of pharmaceutical contamination in Southeast Asia. In: Kumar M., Snow D., Honda R. (eds) *Emerging Issues in the Water Environment during Anthropocene. Springer Transactions in Civil and Environmental Engineering.* Springer, Singapore. https://doi.org/10.1007/978-981-32-9771-5_4

Mohapatra S., Menon N., Padhye L.P., Tatiparti S.S.V., Mukherji S. (2021) Natural attenuation of pharmaceuticals in the aquatic environment and role of phototransformation. In: Kumar M., Snow D., Honda R., Mukherjee S. (eds) *Contaminants in Drinking and Wastewater Sources. Springer Transactions in Civil and Environmental Engineering.* Springer, Singapore. https://doi.org/10.1007/978-981-15-4599-3_3

Mohapatra S., Padhye L.P., Mukherji S. (2018) Challenges in detection of antibiotics in wastewater matrix. In: Gupta T., Agarwal A., Agarwal R., Labhsetwar N. (eds) *Environmental Contaminants. Energy, Environment, and Sustainability.* Springer, Singapore. https://doi.org/10.1007/978-981-10-7332-8_1

Misra, V., Pandey, S.D., 2005. Hazardous waste, impact on health and environment for development of better waste management strategies in future in India. *Environ. Int.* 31, 417–431.

MoEF & CC. 2016a. Solid Wastes (Handling and Management) Rules, Ministry of Environment, Forests and Climate Change, Government of India.

MoEF & CC. 2016. Notification on C & D waste, dated 29th March 2016. https://dpccocmms.nic.in/SPCB_DOCUMENTS/MSW.pdf

MoEF & CC. 2016b. Hazardous and Other Wastes (Management and Transboundary Movement) Rules, Ministry of Environment, Forests and Climate Change.

MoEF & CC. 2016c. Biomedical Wastes (Handling and Management) Rules, Ministry of Environment, Forests and Climate Change, Government of India.

MoEF & CC. 2016d. E Wastes Management Rules, Ministry of Environment, Forests and Climate Change, Government of India.

MoEF & CC. 2016e. Construction and Demolition Waste Management Rules, Ministry of Environment, Forests and Climate Change, Government of India.

MoEF & CC. 2016f. Plastic Wastes Management Rules, Ministry of Environment, Forests and Climate Change, Government of India.

MoEF, 2016. Municipal Solid Waste Management Rules, 2016. Available at The government of India. http://www.moef.nic.in/sites/default/files/SWM%202016_0.pdf

MoEF. 2011. Batteries (Management and Handling) Rules, Ministry of Environment and Forests, Government of India.

Mukherjee, S., Mukhopadhyay, S., Hashim, M.A., Sen Gupta, B., 2015. Contemporary environmental issues of landfill leachate: Assessment and remedies. *Crit. Rev. Environ. Sci. Technol.* 45, 472–590.

Nagarajan, R., Thirumalaisamy, S., Lakshumanan, E., 2012. Impact of leachate on groundwater pollution due to non-engineered municipal solid waste landfill sites of Erode City, Tamil Nadu, India. *Iran. J. Environ. Health Sci. Eng.* 9, 1–12.

Nandan, A., Yadav, B.P., Baksi, S., Bose, D., 2017. Recent scenario of solid waste management in India. *World Sci. News* 66, 56–74.

Naveen, B.P., Malik, R.K., 2017. Assessment of leachate pollution index for Delhi landfill sites, India. *Open Access Int. J. Sci. Eng.* 2, 98–101.

Naveen, B.P., Sharma, A.K., Sivapullaiah, P.V., Sitharam, T.G., Narayana, M.A., 2013. Characteristics of the leachate from MSW landfill, Bangalore. In: *Silver Jubliee Celebrations of Indian Chapter of IGS-Interenational Symposium"Geosynthetics India-2013"23–25 October 2013, NewDelhi, India.* pp. 23–25.

Naveen, B.P., Sivapullaiah, P. V., 2020. Solid waste management: current scenario and challenges in Bengaluru. In: *Sustainable Sewage Sludge Management and Resource Efficiency.* IntechOpen.

Naveen, B.P., Sumalatha, J., Malik, R.K., 2018. A study on contamination of ground and surface water bodies by leachate leakage from a landfill in Bangalore, India. *Int. J. Geotech. Eng.* 9, 1–20.

Nawaz, T., Sengupta, S., 2019. Contaminants of emerging concern: occurrence, fate, and remediation. In: *Advances in Water Purification techniques.* Elsevier, pp. 67–114.

Pariatamby, A., 2017. *Country Chapter: State of the 3Rs in Asia and the Pacific.* United Nations Centre for Regional Development, Nagoya.

Peavy, H.S., Rowe, D.R., Tchobanoglous, G., 1985. *Environmental Engineering.* McGraw-Hill Book Co., New York.

Permana A.S., Towolioe, S. Aziz, N.A., Ho, C.S., 2015. Sustainable solid waste management practices and perceived cleanliness in low-income city. *Habitat Int.* 49, 197–205.

Priyanka M.S., Mohapatra S. 2020. Shifts and trends in analysis of contaminants of emerging concern: Sulfonamides. In: Kumar M., Munoz-Arriola F., Furumai H., Chaminda T. (eds) *Resilience, Response, and Risk in Water Systems. Springer Transactions in Civil and Environmental Engineering.* Springer, Singapore. https://doi.org/10.1007/978-981-15-4668-6_14

Rabii, A., Aldin, S., Dahman, Y., Elbeshbishy, E., 2019. A review on anaerobic co-digestion with a focus on the microbial populations and the effect of multi-stage digester configuration. *Energies* 12, 1106.

Ramakrishnan, A., Blaney, L., Kao, J., Tyagi, R.D., Zhang, T.C., Surampalli, R.Y., 2015. Emerging contaminants in landfill leachate and their sustainable management. *Environ. Earth Sci.* 73, 1357–1368.

Samadder, S.R., Prabhakar, R., Khan, D., Kishan, D., Chauhan, M.S., 2017. Analysis of the contaminants released from municipal solid waste landfill site: A case study. *Sci. Total Environ.* 580, 593–601.

Sardiña, P., Leahy, P., Metzeling, L., Stevenson, G., Hinwood, A., 2019. Emerging and legacy contaminants across land-use gradients and the risk to aquatic ecosystems. *Sci. Total Environ.* 695, 133842.

Sharma, B.M., Bečanová, J., Scheringer, M., Sharma, A., Bharat, G.K., Whitehead, P.G., Klánová, J., Nizzetto, L., 2019. Health and ecological risk assessment of emerging contaminants (pharmaceuticals, personal care products, and artificial sweeteners) in surface and groundwater (drinking water) in the Ganges River Basin, India. *Sci. Total Environ.* 646, 1459–1467.

Sharma, K.D., Jain, S., 2019. Overview of municipal solid waste generation, composition, and management in India. *J. Environ. Eng.* 145, 4018143.

Show, P.L., Pal, P., Leong, H.Y., Juan, J.C., Ling, T.C., 2019. A review on the advanced leachate treatment technologies and their performance comparison: an opportunity to keep the environment safe. *Environ. Monit. Assess.* 191, 1–28.

Sikora, A., Detman, A., Chojnacka, A., Błaszczyk, M.K., 2017. Anaerobic digestion: I. A common process ensuring energy flow and the circulation of matter in ecosystems. II. A tool for the production of gaseous biofuels. *Ferment. Process.* 14, 271.

Somani, M., Datta, M., Gupta, S.K., Sreekrishnan, T.R., Ramana, G.V., 2019. Comprehensive assessment of the leachate quality and its pollution potential from six municipal waste dumpsites of India. *Bioresour. Technol. Rep.* 6, 198–206.

Srivastava, V., De Araujo, A.S.F., Vaish, B., Bartelt-Hunt, S., Singh, P., Singh, R.P., 2016. Biological response of using municipal solid waste compost in agriculture as fertilizer supplement. *Rev. Environ. Sci. Biotechnol.* 15, 677–696.

Srivastava, V., Ismail, S.A., Singh, P., Singh, R.P., 2015. Urban solid waste management in the developing world with emphasis on India: challenges and opportunities. *Rev. Environ. Sci. Biotechnol.* 14, 317–337.

Srivastava, V., Sarkar, A., Singh, S., Singh, P., de Araujo, A. S., & Singh, R. P., 2017. Agroecological responses of heavy metal pollution with special emphasis on soil health and plant performances. *Front. Environ. Sci.*, 5, 64.

Stefanakis, A., Akratos, C.S., Tsihrintzis, V.A., 2014. Treatment of special wastewaters in VFCWs. In: Stefanakis A., Akaratos C.S., Tsihrintzis V.A., editors. *Vertical Flow Constructed Wetlands*. Elsevier, Amsterdam, pp. 145–164.

Stuart, M., Lapworth, D. 2013. Emerging organic contaminants in groundwater. In *Smart Sensors for Real-Time Water Quality Monitoring*. Springer, Berlin, Heidelberg, pp. 259–284.

Swati, T.I.S., Vijay, V.K., Ghosh, P., 2018. Scenario of landfilling in India: problems, challenges, and recommendations. In: Chaudhery M. H. (ed) *Handbook of Environmental Materials Management*. Springer, Cham, pp. 1–16.

Taheran, M., Naghdi, M., Brar, S.K., Verma, M., Surampalli, R.Y., 2018. Emerging contaminants: here today, there tomorrow! *Environ. Nanotechnol. Monit. Manag.* 10, 122–126.

TERI, 2014. *Waste to Resources: A Waste Management Handbook*. The Energy and Resources Institute, India.

Terzić, S., Senta, I., Ahel, M., Gros, M., Petrović, M., Barcelo, D., Müller, J., Knepper, T., Martí, I., Ventura, F., 2008. Occurrence and fate of emerging wastewater contaminants in Western Balkan Region. *Sci. Total Environ.* 399, 66–77.

Torretta, V., Ferronato, N., Katsoyiannis, I.A., Tolkou, A.K., Airoldi, M., 2017. Novel and conventional technologies for landfill leachates treatment: A review. *Sustainability* 9, 9.

Vaish, B., Srivastava, V., Singh, P.K., Singh, P., Singh, R.P., 2020. Energy and nutrient recovery from agro-wastes: rethinking their potential possibilities.

Wang, F., Smith, D.W., El-Din, M.G., 2003. Application of advanced oxidation methods for landfill leachate treatment–a review. *J. Environ. Eng. Sci.* 2, 413–427.
www.newindianexpress.com. 2019. India processed 53 per cent of solid waste generated till February 2019: RTI- The New Indian Express. 5/5/2019.
www.statista.com. 2017. Leading solid waste generating urban areas across India as on November 2017, by state. https://www.statista.com/statistics/1009071/india-major-solid-waste-generating-areas-by-state/
Yadav, J.S., Dikshit, A.K., 2017. Stabilized old landfill leachate treatment using electrocoagulation. *Environment Asia* 10, 25–33.
Zahari, M. S. M., Ismail, S., & Jaafar, I., 2016. Leachate generation, transport, and control. In: Hamidi A. A., Salem A. A. (eds) *Control and Treatment of Landfill Leachate for Sanitary Waste Disposal*. IGI Global, pp. 115–144.
Zhao, L., Zhang, F.-S., Wang, K., Zhu, J., 2009. Chemical properties of heavy metals in typical hospital waste incinerator ashes in China. *Waste Manag.* 29, 1114–1121.

Chapter 7

Synthesis of hydroxyapatite nanoparticles by modified co-precipitation technique and investigation of the ceramic characteristics upon thermal treatment for their potential applications for water treatment

Shruti Sharma[1] *and Ajai Singh*
Central University of Jharkhand

Rupesh M. Das and Mahesh Kumar
CSIR-National Physical Laboratory

CONTENTS

7.1 INTRODUCTION

Water is one of the inevitable constituents of the ecosystem and is the fundamental necessity for supporting all life forms on earth. However, it is under extreme menace in the past few decades due to ever-increasing anthropogenic pollution caused by industrial and non-industrial sectors (Mohapatra et al., 2014; 2021). There are many facets of environmental pollution. Nonetheless, water pollution has been regarded as the top-priority issue due to the associated risk of health hazards and ecological imbalance (Leung et al., 2020). According to the United Nations, around 1.5×10^{16} L of wastewater is being produced worldwide annually, which approximately comes out six

DOI: 10.1201/9781003247869-9

times the water bodies (Geissen et al., 2015). The accelerated pace of industrialization, urbanization, and unplanned settlements aided with a lack of monitoring, treatment, and disposal systems are the key players that have exacerbated water pollution and have limited the availability of clean water by increasing water complexity and sanitation issues.

Various classes of water pollutants have been identified by regulatory agencies that pose a risk to the environment and human health when present in surface and groundwater systems. Some of them include, but are not limited to, heavy metals, nutrients, organics, pharmaceuticals, dyes, pesticides, and pharmaceuticals (Menon et al., 2020; Mohapatra et al., 2018; Priyanka and Mohapatra, 2020). Phosphorus and nitrogen are the commonly encountered pollutants that find their way to water sources through agricultural activities, sewage, and industrial discharge, thus causing eutrophication of lakes, rivers, and estuaries. In developing nations like India, the discharge of poorly treated or untreated wastewater in local bodies has increased the jeopardy of public health hazards (Kookana et al., 2020). Besides, the severe environmental implications and health hazards caused due to heavy metal contamination of the aquatic environment represent one of the major environmental challenges and are of great concern owing to their toxicity, bio-accumulative characteristics, and carcinogenic nature (Ali et al., 2019). There are several routes through which heavy metals find their ways to waterbodies; however, few primary anthropogenic sources include discharge of unregulated or poorly treated industrial effluents, agricultural activities, mining, urban, and runoffs (Obasi and Akudinobi, 2020). A broad and diverse set of synthetic compounds have been recently identified in the aquatic environment as significant water pollutants despite being present in water bodies for a long time and have been termed as emerging contaminants (Rasheed et al., 2019). Their occurrence has received noteworthy consideration from scientific communities, innovators, NGOs, and governments worldwide owing to their potential risk to human health and ecological balance even at a minor concentration (Patel et al., 2019). Recently, World Health Organization (WHO) has published a detailed and comprehensive report on the monitoring, detection, and the associated risk of the emerging contaminants in surface water bodies, groundwater, drinking water, and wastewater and emphasized the need for cost-effective and eco-friendly technological innovations for water and wastewater treatment to alleviate the potential health hazards caused due to emerging contaminants (Maycock and Watts, 2011).

The challenges associated with water treatment can be addressed by designing and developing materials with specific and tunable structural and morphological characteristics. Thus, this study presents an attempt to develop an eco-friendly adsorbent through a facile synthesis procedure and low-cost precursors for its potential applications in water treatment.

Hydroxyapatite (HA) ($Ca_{10}(PO_4)_6(OH)_2$) is an inorganic compound from the family of apatite group and is one of the major constituents of human bones and teeth (Wang et al., 2019). The structural and compositional similarity with natural bone, along with the excellent attributes including bioactivity, biocompatibility, and high stability, has led to several applications of HA in biomedicals ranging from bone tissue regeneration to controlled delivery of drugs at the targeted sites (Hubadillah et al., 2020). The versatility of hydroxyapatite is not only limited to biomedical applications but has also been recognized in the field of water and wastewater treatment (Hubadillah et al., 2020). It is worth highlighting that recent years have witnessed remarkable progress in the applications of hydroxyapatite for the elimination of heavy metals, dyes, pharmaceuticals, endocrine disruptors, nutrients, biomolecules, anions, etc. (Pereira et al., 2020) (Martínez-Olivas et al., 2020). Such applications of HA could be credited to its higher reactivity, ion-exchange capacity, sorption affinity, stability under reducing and oxidizing conditions, low water solubility, and non-toxicity (Mobasherpour et al., 2011). The stability of HA within the wide pH range of 4–14 at room temperature is also one of the important factors behind its utilization as novel materials for water and wastewater treatment (Yelten-Yilmaz and Yilmaz, 2018). Furthermore, the suitability of utilization of HA as a matrix for wastewater treatment has also been credited to its flexible lattice structure, which allows the immobilization of a wide range of contaminants through various pathways such as complexation formation, ion-exchange, and dissolution-precipitation (Ibrahim et al., 2020). Reportedly, HA has two different binding sites (Ca and PO_4^{-3} or OH^-) at its surface, allowing substitution of contaminants in the lattice structure (Wang et al., 2019). The structural substitutions might occur either at the cationic or the anionic site and are found to be the function of ionic charge, size, and electronegativity of the substituting species.

There are various techniques available for the synthesis of hydroxyapatite nanoparticles and could be broadly categorized into three types: dry, wet, and high-temperature method (Mohd Pu'ad et al., 2019). However, wet-chemical methods, mainly chemical precipitation technique, are widely employed for the synthesis of HA nanoparticles owing to their simplicity, low-cost, minimal use of chemical reagents, and ability to tune the morphology of HA nanoparticles (Sadat-Shojai et al., 2013).

In the present study, HA nanoparticles were synthesized by modifying the wet-chemical precipitation technique and sintered at four different temperatures for exploring their potential applications for water treatment. The study also investigated the effect of sintering on phase composition, surface chemistry, morphology, thermal stability, and pore volume of the prepared ceramic for exploring its potential to develop a stable and porous ceramic material for water treatment applications. The structural composition, crystallinity, and phase transformations of the prepared hydroxyapatite nanoparticles were examined using X-Ray Diffraction (XRD). Fourier Transform Infrared Spectroscopy (FTIR) elucidated the presence of surface functional moieties. The morphology, thermal behavior, pore volume, and specific surface area (SSA) of nano-hydroxyapatite (nHA) were studied using Field Emission Secondary Electron Microscopy (FESEM), Thermogravimetric Analysis (TGA), and Brunauer, Emmett, and Teller particle size analyzer, respectively.

7.2 EXPERIMENTAL

7.2.1 Materials

The wet synthesis of hydroxyapatite involved the usage of analytical grade calcium hydroxide [$Ca(OH)_2$, >95%], orthophosphoric acid [H_3PO_4, ≥85%], and ammonium hydroxide [NH_4OH, aqueous]. All reagents were purchased from Merck Pvt. Ltd., Mumbai, India. To ensure no interference of foreign elements during the synthesis, double-distilled elements were used throughout.

7.2.2 Synthesis of nano-hydroxyapatite

The nano-hydroxyapatite (nHA) was synthesized by modifying the wet-chemical precipitation method, which is commonly reported in literature (Yelten-Yilmaz and Yilmaz, 2018). Here, calcium hydroxide served as a calcium source, while phosphorus was derived from orthophosphoric acid. Calcium and phosphorus precursors were precisely weighed to obtain a molar ratio of Ca/P = 1.67, crucial for the formation of stoichiometric HA. A white suspension of $Ca(OH)_2$ was prepared by mixing the weighed amount in 600 mL distilled water, and the obtained suspension was continuously stirred for 3 h at a temperature of 75°C. After that, 87 mL of an aqueous orthophosphoric acid solution was added dropwise with an addition rate maintained at 4.5 mL/min to the alkaline $Ca(OH)_2$. During the study, it was made sure that the solution pH lies in the range of 9–11 during the reaction and has been carried out by simultaneously adding the aqueous ammonia solution into the acidic slurry with a relatively slower rate of 2.5 mL/min. The system pH was monitored using the handheld pH meter. The whole mixture was subjected to continuous mixing at 1200 rpm using a magnetic stirrer. A dwelling time of 48 h was provided to the reaction mixture for aging at room temperature. The steps performed in the sequence mentioned above resulted in the milky white precipitate. Equation (7.1) shows the chemical reaction outlining the synthesis of hydroxyapatite (Anwar and Akbar, 2018).

$$10\,Ca(OH)_2(susp.) + 6H_3PO_4(1) \rightarrow Ca_{10}\left(PO_4\right)_6(OH)_2(s) + 18\,H_2\,O(1) \tag{7.1}$$

Later, after the maturation stage, the precipitate separation was carried out using vacuum filtration, and thereafter, to remove any impurities, the precipitate was washed using distilled water and ethanol till neutral pH. The obtained white precipitate was then oven-dried at 130°C for 24 h and subsequently ground into powder using mortar and pestle. To examine the effects of thermal treatment on morphology, phase composition, Ca/P ratio, functionalities, and other physicochemical properties of nHA, the oven-dried fine powder (nHA) was thermal- treated at four different temperatures from 200°C to 800°C with a heating rate of 5°C/min for 2 h. Figure 7.1 shows the process flow chart followed for the synthesis of nano-hydroxyapatite. The sintered samples were labeled as nHA-2, nHA-4, nHA-6, and nHA-8 corresponding to HA thermally treated at 200°C, 400°C, 600°C, and 800°C.

7.2.3 Characterizations

XRD (Miniflex, M/s. Rigaku Corporation, Japan) was performed to study the phase composition and transformations in the constituting phase of nHA on heat treatment. The diffraction pattern was monitored in the 2θ range of 20°–60° using monochromatic CuKα radiation with 0.02° as the step size. The mean crystallite size of synthesized n-HA was estimated using Scherrer's formula, which has been given in Equation (7.2) (Wu et al., 2015).

$$D = K\lambda / \beta Cos(\theta) \tag{7.2}$$

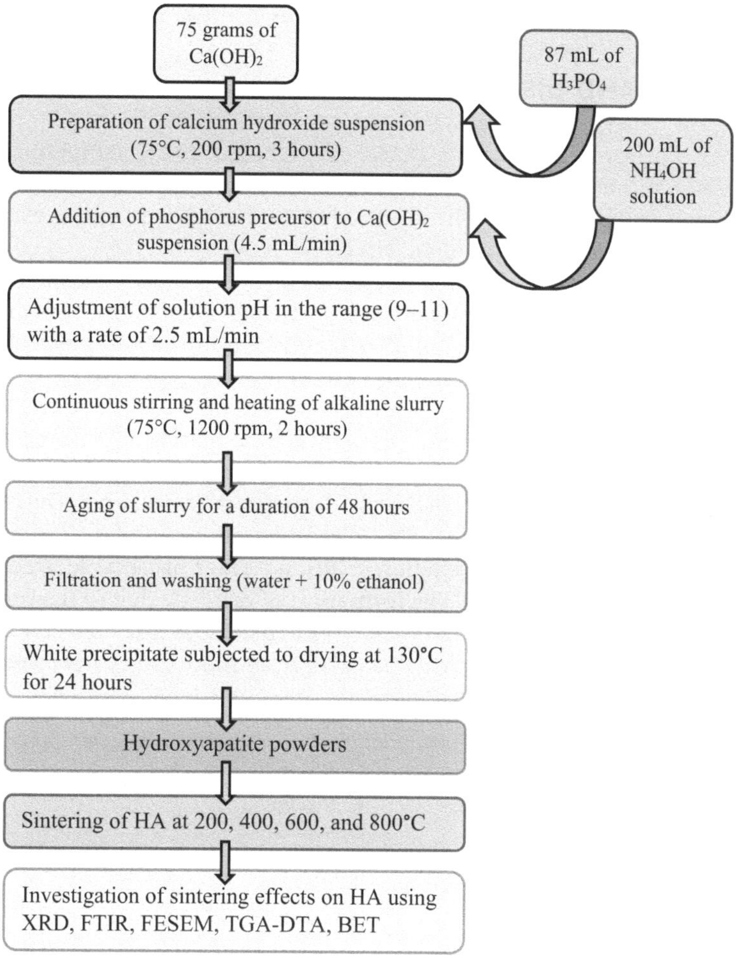

Figure 7.1 Flow chart of synthesis of nHA.

where β = full width at half maximum (FWHM) for the diffraction peak under consideration (radian), θ = Bragg's angle in degree (°), λ = 0.15406 nm wavelength for Cu Kα radiation, and K = Scherrer's constant (0.94). The specific surface area (SSA) and pore volume of the prepared nHA powders were estimated by the nitrogen adsorption-desorption isotherm using Micromeritics Gemini Model Number 2380.

While the surface functionalities were characterized using IR-spectroscopy (Perkin Elmer Spectrum 100) using the KBr method. The spectra were captured in the region of 400–4,000 cm^{-1}. The morphological analyses of the nHA powders were carried out using FESEM (Supra 40VP, M/s. Zeiss, UK) with the electron accelerating voltage of 300 kV. To avoid image distortion due to the charging effect, the samples were gold coated, having a thickness of approximately 200 Å before capturing the FESEM images. The quantitative elemental study was carried out using Energy Dispersive Spectroscopy (EDS), which comes with the FESEM system. The thermal behavior of nHA samples was explored using TGA (Mettler Toledo TGA/DSC 3+ HT) under the nitrogen atmosphere with a 50 mL/min flow rate and a heating rate of 10°C/min.

7.3 RESULTS AND DISCUSSIONS

The X-Ray Diffraction (XRD) spectra of prepared nHA samples are shown in Figure 7.2a. A minor phase of brushite ($CaHPO_4.2H_2O$) corresponding to JCPDS file (09-0077) was found to be present in the nHA and nHA-2 samples, while the major phase was comprised of calcium-deficient HA.

The formation of such a minor phase lies in the formation mechanism of nHA. The wet synthesis of hydroxyapatite involving calcium hydroxide and orthophosphoric acid as the starting materials is a complex reaction; however, simplistically, the entire process could be divided into five steps (Ferreira et al., 2003; Ma, 2019). Firstly, hydroxyapatite precipitation occurs in the initial stage at the cost of undissolved calcium hydroxide present in the suspension, as shown in Equation (7.3). Secondly, calcium hydroxide gets completely dissolved in the solution, followed by the growth of HA crystals. However, subsequently, the nuclei formation of brushite occurs and successively transforms HA into brushite. The lower crystallinity and decrease in the supersaturation degree of HA in the reaction solution trigger brushite formation (Ferreira et al., 2016; Oliveira et al., 2007). The formation of brushite due to the transformation of HA could be described by Equation (7.4). Furthermore, the conversion of brushite into calcium-deficient hydroxyapatite has also been reported by the researchers when

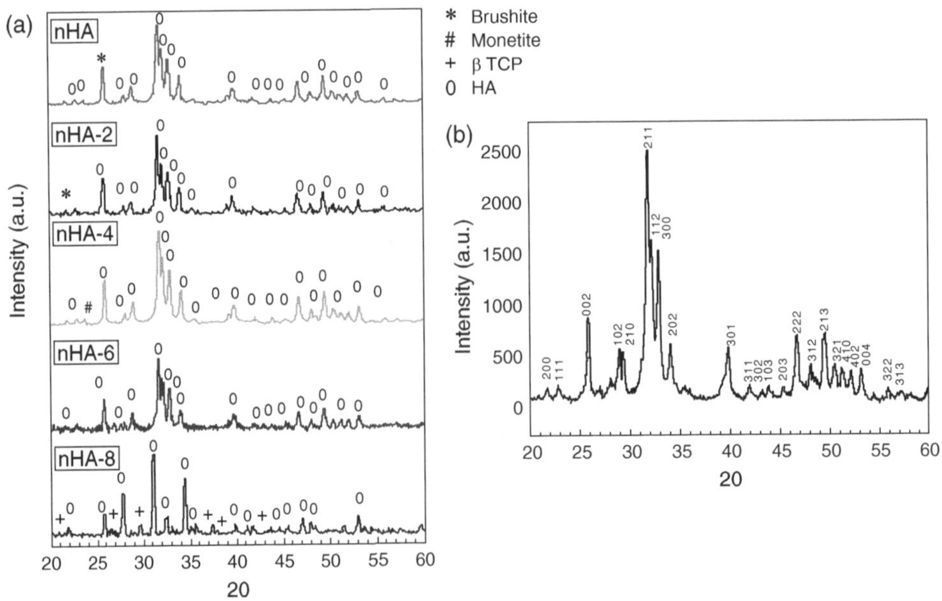

Figure 7.2 XRD spectra: (a) synthesized nHA powder and (b) HA sintered at 600°C (nHA-6).

the prolonged hydrolysis of brushite is carried out in the alkaline pH (Ferreira et al., 2003; Da Rocha et al., 2018).

$$5Ca(OH)_2 + 3H_3PO_4 \rightarrow Ca_5OH(PO_4)_3 + 9H_2O \tag{7.3}$$

(Ferreira et al., 2003)

$$Ca_5OH(PO_4)_3 + 2H_3PO_4 + 9H_2O \rightarrow 5CaHPO_4.2H_2O \tag{7.4}$$

(Ferreira et al., 2003)

In this study, the occurrence of a major phase of Ca-deficient HA in nHA and nHA-2 might be the result of hydrolysis of the formed brushite in the alkaline environment. The continuous addition of ammonia solution can be held responsible for maintaining the pH in the alkaline range. The appearance of monetite diffraction peak having JCPDS file (09-0080) in the HA sample (nHA-4) could be ascribed to the loss of structural water of brushite upon thermal treatment (Jmai et al., 2019; Dosen and Giese, 2011). However, the obtained nHA powder after thermal treatment at 600°C (nHA-6) showed good agreement with the characteristics peaks of stoichiometric HA when indexed with the JCPDS file (09-0432). The well-developed signature peaks (0 0 2), (2 1 0), (2 1 1), (1 1 2), (3 0 0), (2 0 2), (2 2 2), (2 1 3), etc. of stoichiometric HA are shown in Figure 7.2b. No other phase was observed for this sample.

The further increase in sintering temperature to 800°C (nHA-8) resulted in the partial decomposition of nHA into a biphasic mixture of β-TCP and nHA, which was confirmed by indexing the obtained pattern with the ICDD database (09-0169). The formation of β-TCP above 600°C could be due to the dehydroxylation of synthesized HA (Savino and Yates, 2015). Similar findings have been observed regarding the occurrence of a biphasic mixture of HA and β-TCP upon sintering below 1,000°C (Ghosh and Sarkar, 2016). The above XRD spectra showed that up to 600°C, the synthesized HA was stable, whereas the thermal treatment at higher temperatures caused partial decomposition of HA. The crystallinity of synthesized nano-hydroxyapatite samples was estimated using Equation (7.5).

$$X_c = \left(0.24/\beta_{002}\right)^3 \tag{7.5}$$

(Sa et al., 2017)

where β_{002} represents the FWHM (°) of (0 0 2) diffraction peak.

To elucidate the structural characteristics of hydroxyapatite, generally, the diffraction peak located at a 2θ value of 25.883° and between 30 and 35° is considered (Reyes-Gasga et al., 2013). However, the low resolution of diffraction peaks in the 2θ range of 30–35° makes it challenging to distinguish them clearly, resulting in erroneous estimation of FWHM (Bohic et al., 1998). Here, a similar condition was observed in the case of synthesized HA. Nevertheless, it is noteworthy that the peak with (h k l) (0 0 2) showed fair resolution as compared to (1 1 2) and (3 0 0). It also has an added advantage of determination of FWHM (0 0 2) with more accuracy and ease. Hence, in the present study, the diffraction peak (0 0 2) was selected for further calculations.

Meanwhile, it could also be observed from the above XRD spectra that with increasing temperature, the diffraction peak becomes narrower, thus indicating the enhancement in the crystallinity of synthesized HA. The variation in the crystallinity index with that of thermal treatment is shown in Figure 7.3. The sample nHA-4 showed a minor reduction upon thermal treatment. The thermal treatment of brushite results in the formation of monetite along with an amorphous phase, which is primarily attributed to the loss of water molecules trapped within the brushite structure (Dosen and Giese, 2011). The decrease in the crystallinity of as-synthesized nHA could be due to the occurrence of the amorphous phase during the transformation process.

The average crystallite size was determined using the diffraction peak at $2\theta = 25.92°$ with (h k l) value (002) as it was sharper and isolated compared to others in the obtained XRD spectra. Figure 7.4 shows the variation in crystallite size with thermal treatment.

The crystallite size was found to be increasing with an increase in sintering temperature, with a slight reduction in the case of sample nHA-4. This positive correlation between thermal treatment and crystallite size could be explained by the role of diffusion in the process of crystallization. Higher temperature treatment promotes the phenomenon of diffusion, and hence, there is an increase in crystallite size (Pratihar

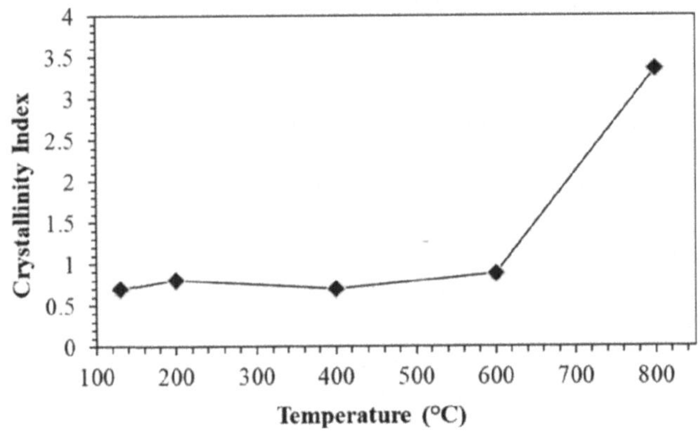

Figure 7.3 Variation in crystallinity of nHA samples with the thermal treatment.

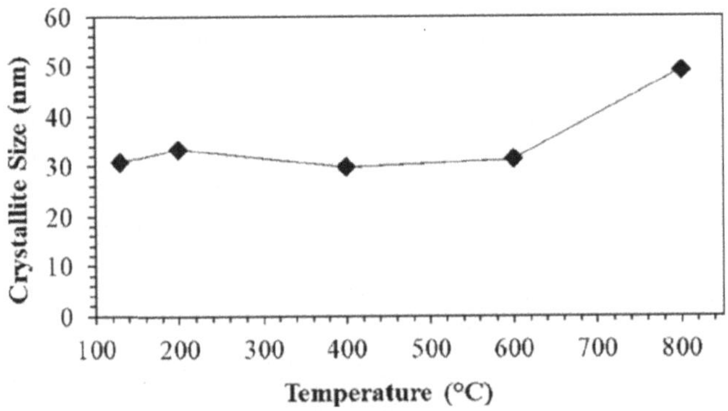

Figure 7.4 Variation in crystallite size of nHA samples with the thermal treatment.

et al., 2006). A similar trend has also been reported (Pang and Bao, 2003). Additionally, the average crystallite size of the minor phase of β-TCP was also determined using Scherrer's Equation and was found to be 57 nm.

The SSA of as-prepared nHA and sintered samples are shown in Table 7.1. The as-prepared sample nHA exhibited the highest SSA of 45.68 m^2/g, while it exhibited a decreasing trend with an increase in sintering temperature. The decline in SSA was attributed to the increase in the crystallinity of the samples, which has also been observed during XRD analysis.

The pore volume of prepared nHA samples exhibited a declining trend with an increase in sintering temperature and is shown in Figure 7.5. Furthermore, the prepared nHA samples showed a pore size distribution ranging from meso to macro pores. The study of pore volume and SSA is of great significance as it plays a vital role in the immobilization of contaminants on the adsorbent surface from an aqueous solution. The surface chemistry of nHA samples has been studied using FTIR spectra, and the plotted spectra of nHA samples are presented in Figure 7.6. nHA showed the characteristic vibrational modes of phosphate ions at 567, 600, 962, 1,020 and 1,090 cm^{-1}. The bands observed at 567 and 600 cm^{-1} correspond to the v4 bending vibrations produced by the hydroxyapatite phosphate groups (Yelten-Yilmaz and Yilmaz, 2018) (Destainville et al., 2003). Additionally, the band at 962 cm^{-1} indicates the v1 symmetric stretching vibrations, while 1020 and 1,090 cm^{-1} wavenumbers could be assigned to the v3 bending vibrations of PO_4^{3-} group (Paz et al., 2012). The bending and stretching vibrations of hydroxyl groups were observed at 630 and 3,571 cm^{-1}, respectively (Ramesh et al., 2013).

Table 7.1 Variation in specific surface area with thermal treatment

Temperature (°C)	*HA samples*	*Specific surface area (m^2/g)*
130	nHA-2	45.68
200	nHA-2	42.31
400	nHA-4	38.19
600	nHA-6	30.92
800	nHA-8	20.04

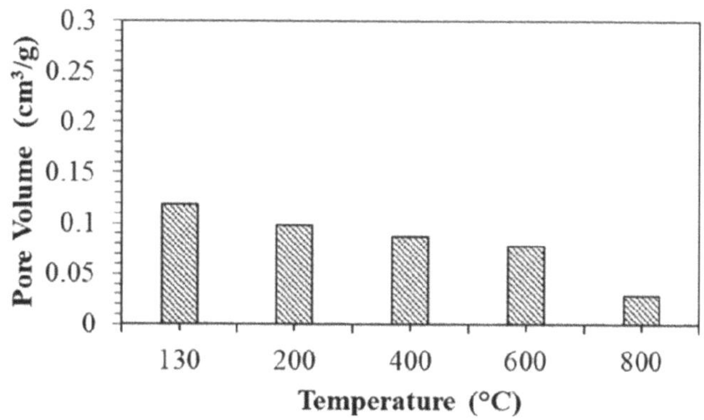

Figure 7.5 Variation in the pore volume of nHA samples with thermal treatment.

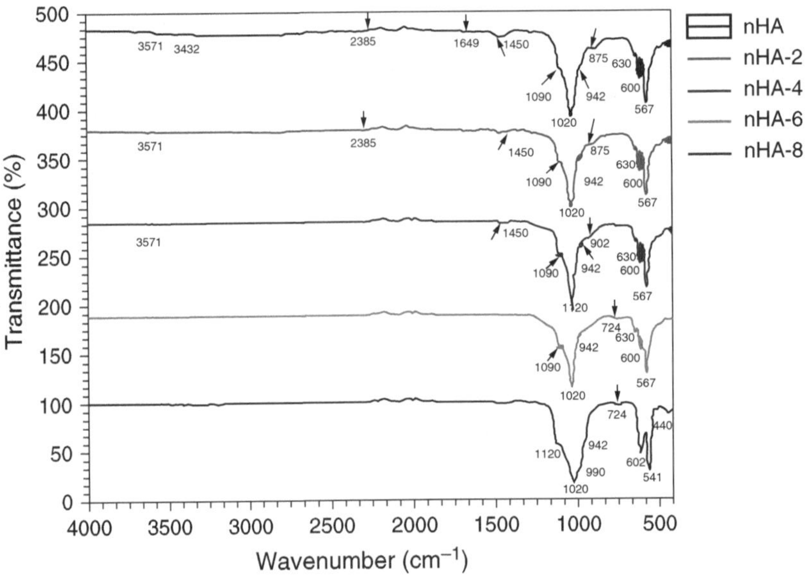

Figure 7.6 FTIR spectra of synthesized nHA samples.

The appearance of a stretching band at 3,432 cm^{-1} belongs to the adsorbed water, which might be due to the aqueous phase synthesis of nHA (Paz et al., 2012). The occurrence of the band at 875 cm^{-1} corresponds to the acidic phosphate groups, which is one of the key characteristics of hydroxyapatite deficient in calcium (Destainville et al., 2003). The result obtained from IR spectra supports the finding of X-ray analysis, as mentioned in the earlier part.

The low-transmittance band at 1,450 cm^{-1} indicates a minor carbonate fraction in the prepared nano-hydroxyapatite powder suggesting the substitution of PO_4^{3-} group by $CO_3{}^{2-}$ ions (Meejoo et al., 2006). The carbonate formation could be due to the interaction of carbon dioxide present in the surroundings with alkaline precursors used during the synthesis. The IR spectrum also witnessed the presence of brushite in the nHA sample as observed in the XRD spectra and could be realized by the occurrence of the absorption peak at 2,385 cm^{-1} corresponding to HPO_4^{2-}, whereas the weak band at 1,649 cm^{-1} was ascribed to the symmetric bending vibrations of O-H moieties (Suryawanshi and Chaudhari, 2014).

The hydroxyapatite sintered at 200°C (nHA-2) exhibited lower transmittance of CO_3^{2-} ions present at 1,450 cm^{-1} compared to nHA. Also, the band associated with carbonate groups (1,450 cm^{-1}) showed a diminishing trend with increasing sintering temperature, and ultimately it disappeared at a higher temperature of 800°C. The above findings agree with the available literature (Rodríguez-Lugo et al., 2018). The absence of a band located at 3,432 cm^{-1} indicates the removal of absorbed water. The weak absorption peak of brushite at 1,649 cm^{-1} due to bending vibrations of O-H groups was also found to be lacking upon thermal treatment at 200°C. Furthermore, it is worth highlighting the gradual weakening of the band at 875 cm^{-1} ascribed to HPO_4^{2-} upon an increase in the temperature of thermal treatment. Such changes in the IR spectra thus indicate the transformation processes occurring in the nano-HA due to thermal treatment.

The sample (nHA-4) showed the appearance of a shoulder at 902 cm^{-1} credited to the phosphate groups, thereby indicating the presence of monetite in the nHA sample sintered at 400°C (Salimi and Javadpour, 2012). Further increase in sintering temperature led to the complete elimination of carbonate fraction present in synthesized HA, which could be readily seen by the fading of the absorption peak at 1450 cm^{-1} in the heat-treated sample at 600°C. Upon sintering, the shoulder at 902 cm^{-1} corresponding to monetite phosphate groups was also found to be absent. Furthermore, the noteworthy effect of thermal treatment was observed on the absorption peak of OH^-. One such influence is the elimination of the adsorption band (3571 cm^{-1}) assigned to the OH^- group. Only characteristic peaks of phosphate and hydroxyl groups were observed in the case of nHA-6. However, a minor shoulder assigned to the pyrophosphate group appeared at 724 cm^{-1}.

The high-temperature thermal treatment of calcium-deficient hydroxyapatite powders resulted in the β-TCP phase formation, as evident from the obtained IR spectra. The sample sintered at 800°C showed intense vibrational peaks at 561 and 607 cm^{-1} corresponding to PO_4^{3-} groups present in β-TCP (Raynaud et al., 2002). The shoulder assigned to pyrophosphate groups also appeared at 724 cm^{-1} (Ito et al., 1996). The other featured stretching vibrations attributed to PO_4^{3-} were also found at 945, 980, 1,020, and 1,120 cm^{-1} (Meejoo et al., 2006) (Massit et al., 2014). Similar observations have also been discussed in the X-ray analysis.

The FESEM micrograph of nHA (Figure 7.7a) showed the agglomerated structures having a rough surface with no defined morphology and varying sizes. Similar morphology has also been observed in other reported studies (Manatunga et al., 2016). The high-temperature treatment promotes the growth of particles and hence results in a more condensed structure. In the current research, the increase in sintering temperature led to particle growth and the development of grain boundaries, which is readily evident in the FESEM micrographs. The increase in sintering temperature showed a more evolved and refined structure, which could be observed in Figure 7.7b–e. The sample nHA-2 showed small particles with irregular morphology. The thermal treatment showed a pronounced effect on the morphology of the prepared nHA, which could be readily observed in terms of the transformation of irregular morphology into HA crystal flakes in the case of nHA-4; the sample sintered at 400°C.

Moreover, the FESEM micrograph of the obtained nHA after heat treatment at 400°C showed flat-plate-like morphology, which could be attributed to the presence of monetite crystals. At 600°C, the sharp rod-like morphology of HA crystals appeared. According to the FESEM micrograph, the nHA heat-treated at 600°C appears to be monocrystalline. The formation of nanorods might be the result of self-arrangement reactions within the crystallites. Nevertheless, a further increase in the temperature led to a decrease in the shape factor of the HA crystallites and resulted in more regular and close to spherical-shaped particles when sintered at 800°C. A similar phenomenon was observed by Bose and Saha (2003). The sample was composed of two types of particles: one with spherical morphology and the other was of flakes types. The transformations observed in the morphology of the nHA nanoparticles upon sintering could be ascribed by the relative specific surface energies of the different planes of HA

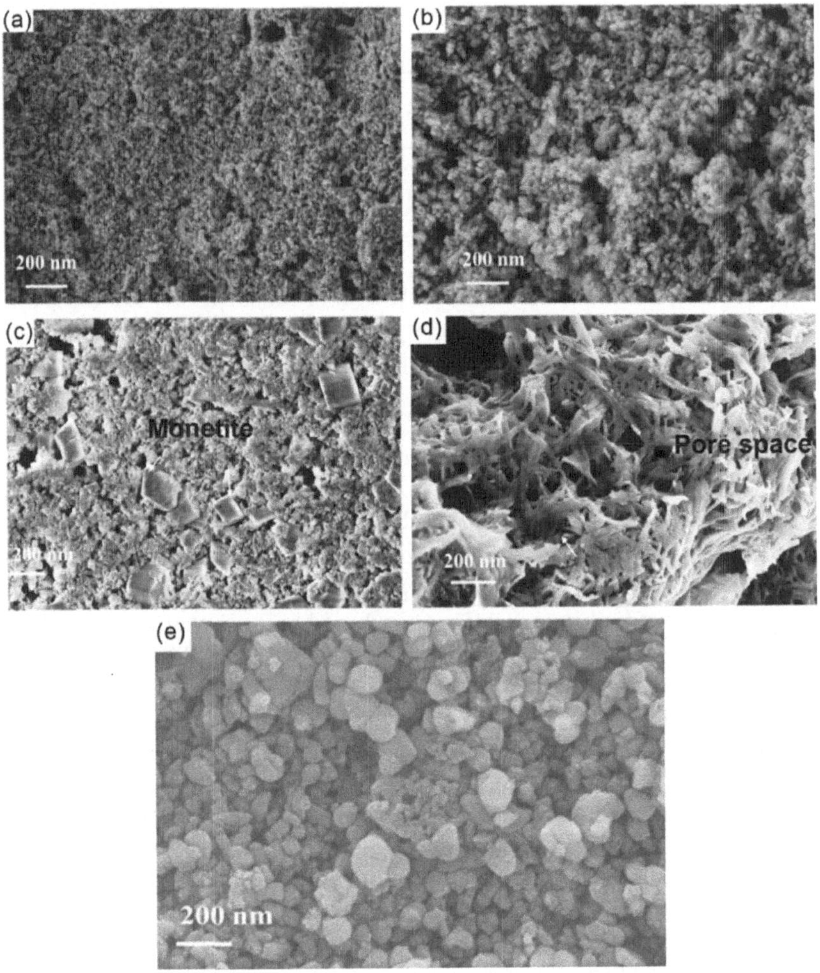

Figure 7.7 FESEM micrographs of synthesized nHA samples.

crystals. The difference in surface energies possessed by the different planes of nHA crystal plays a critical role in the absorption of hydroxyl ions and their associated shielding effects on the growth and morphology of HA nanoparticles (Sanosh et al., 2009). Certain facets preferably absorb hydroxyl ions and thus limit the movement of Ca^{2+} and PO_4^{3-} in the crystallites, which ultimately results in a reduced growth rate. The hydroxyl ion concentration associated with the HA planes greatly influences the morphology of HA nanoparticles (Khalil et al., 2011). The calculated Ca/P atomic ratio for sample nHA-6 was found to be as per the stoichiometric HA (Ca/P = 1.67), while for other HA samples, the estimated values refer to the sub-stoichiometric hydroxyapatite, corroborated with the results of in XRD and FTIR. Furthermore, with an increase in sintering temperature, the Ca/P ratio increased to 600°C; however, a sudden decrease was observed in the case of a biphasic mixture of HA and β-TCP, probably due to the decomposition of nHA.

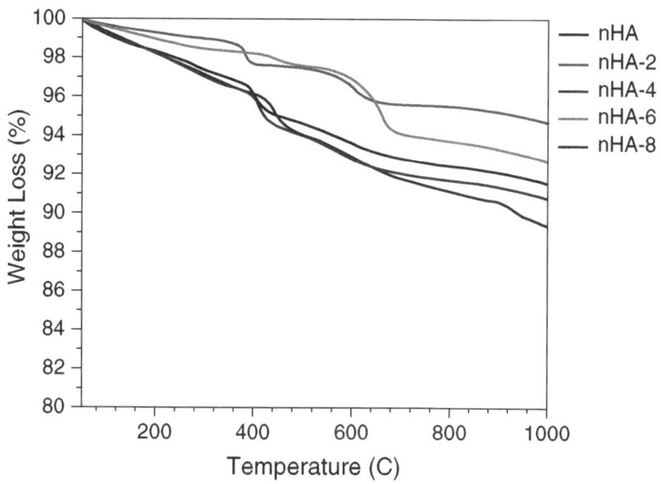

Figure 7.8 TGA plot of synthesized nHA samples.

The typical thermographs of prepared HA nanoparticles are shown in Figure 7.8, whereas the DTG curves for the synthesized nHA are shown in Figure 7.9. The sample nHA showed a total mass loss of 9.4% at the end of the heating cycle due to water loss and carbon dioxide liberation. The analysis of the thermograph reflects a three-step mass loss profile over the considered range of temperature. A small mass loss observed up to 250°C with an endothermic peak could be due to the loss of physically adsorbed water (Kalaiselvi et al., 2017). However, a pronounced mass loss occurred between 360°C and 430°C with a sharp endothermic peak at 408°C resulting from the loss of lattice water (Tõnsuaadu et al., 2012).

Further increase in temperature caused a minor loss in sample mass in the range of 570°C–630°C with an endothermic shoulder at 620°C, corresponding to the liberation of carbon dioxide due to the decomposition of the present carbonate fraction in the nHA (Costescu et al., 2010). The thermograph for nHA-2 showed a similar pattern concerning mass loss caused due to loss of structural water molecules and decarboxylation of prepared hydroxyapatite. However, the distinguishing feature of the TG profile lies in the mass loss of samples occurring in the range of 150°C–280°C. Such mass loss could be due to the loss of water of hydration associated with the brushite phase (Suryawanshi and Chaudhari, 2014). The obtained finding is concurrent with FTIR results, where the band at 1649 cm^{-1} corresponding to H-O-H gets diminished with sintering at higher temperatures.

The total mass, in this case, was found to be 9.2%. Upon further sintering, the endothermic peak corresponding to decarboxylation of HA persisted, but later, at elevated temperature, it got disappeared. A close examination of the thermograph has revealed a shift in the endothermic peak position from 408°C to 443°C. The sintered sample nHA-4 exhibited a mass loss between 398°C and 450°C, with an intense endothermic peak at 443°C and was assigned to monetite conversion to form calcium pyrophosphate owing to heat treatment (Dey, 2014). Moreover, the thermograph also revealed a minor mass loss in the temperature range of 800–1,000°C owing to the

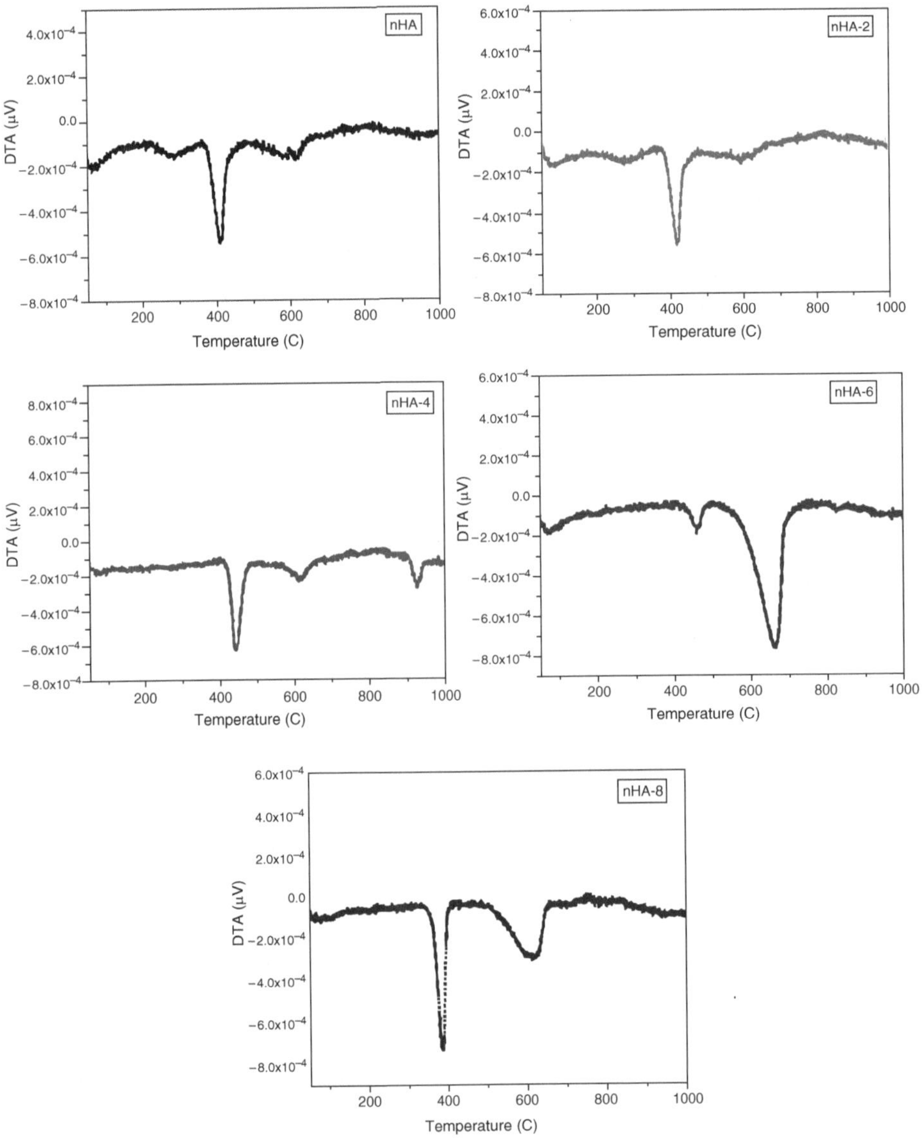

Figure 7.9 DTA plot of synthesized nHA samples.

gradual removal of hydroxyl groups from nHA (Eslami et al., 2010). A mass loss of 8.2% was observed after the heating cycle.

Thermal analysis of HAP heat-treated at 600°C showed a total mass loss of 7.2%. A two-stage mass loss was observed in the obtained thermograph. The first endothermic region ranges from 440°C to 480°C with an endothermic peak at 460°C, credited to the release of OH^-groups (Eslami et al., 2010). However, with an increase in temperature, a relatively pronounced mass loss has been encountered with a well-defined endothermic peak at 660°C in the TG curve. The dominating mass loss in the region 545°C–710°C could be assumed to result from the decomposition of phosphate ions (Granados-Correa et al., 2010).

Furthermore, the thermograph for nHA-8 reflected mass loss in two steps. The first endothermic region ranges from 355°C to 420°C with a sharp endothermic peak at 388°C. Probably, the reason behind this mass loss could be the disintegration of the pyrophosphate clusters. The mass loss has also been observed between 500°C and 650°C with an endothermic peak at 610°C. The resulting mass loss in this temperature range indicates the disintegration of HA into β-TCP, also been confirmed in XRD analysis (Othman et al., 2017). A mass loss of 6.3% was observed after heat treatment. The mass loss for the prepared samples followed a decreasing trend with respect to thermal treatment, as observed in TG analysis.

7.4 CONCLUSIONS

The wet-chemical method was employed to synthesize the hydroxyapatite nanoparticles aiming for its potential application for water treatment. The study investigated the effect of thermal treatment on composition, morphology, functional groups, and surface properties of prepared nHA. The XRD spectra revealed that the sample heat-treated at 600°C (nHA-6) consists of monophasic stoichiometric hydroxyapatite, while further heat treatment resulted in a biphasic mixture of nHA and β-TCP. The EDS analysis demonstrated the Ca/P ratio of 1.67 for nHA-6, thus confirming stoichiometric HA synthesis. The values were found to be in the range of 1.3–1.6, indicating the Ca-deficient hydroxyapatite in the case of other samples. A minor fraction of brushite and monetite were also observed in nHA, nHA-2, and nHA-4, respectively. These minor phases indicate the sensitivity of the wet-chemical precipitation method to the surrounding environment, and the reaction setup. The crystalline character of the prepared hydroxyapatite nanoparticles showed an increasing trend with increasing thermal treatment; however, a slight decrease in sample nHA-4 was observed, attributed to the transformation of brushite into monetite. Furthermore, the crystallite size was varied between 30.9 and 49.1 nm with an increase in sintering temperature, and the trend observed is similar to that of the crystallinity index. FESEM images showed remarkable transformations in the morphology of the hydroxyapatite nanoparticles owing to thermal treatment.

Interestingly, nHA-6 exhibited a porous structure with sharp-rod-like morphology having a SSA of 30.92 m^2/g. Additionally, FTIR spectra confirmed the presence of oxygen-containing functional groups on the adsorbent surface. The thermal stability of prepared samples investigated using thermogravimetric analysis showed a decreasing mass loss trend with increasing sintering temperatures. Based on the experimental findings, it could be said that the as-prepared monophasic stoichiometric nHA-6 could be potentially used for wastewater treatment. However, studies need to be conducted to study the interaction of prepared HA nanoparticles with real wastewater samples.

CONFLICTS OF INTEREST

The authors declare that there is no conflict of interest regarding the authorship and publication of this chapter.

ACKNOWLEDGMENTS

The authors would like to thank the Director of CSIR-National Physical Laboratory, New Delhi, India, for providing the facilities for material preparation and various state-of-the-art characterization techniques. The author Shruti Sharma also wants to express her thanks to the Central University of Jharkhand, Ranchi, India, for their support during the research work. She also wants to extend her thanks to Mr. Amritanshu for supporting her during the research work.

NOTE

1 Currently working as a research scholar in the Environmental Science and Engineering Department,at Indian Institute of Technology Bombay, Mumbai, India.

REFERENCES

Ali, H., Khan, E., Ilahi, I., 2019. Environmental chemistry and ecotoxicology of hazardous heavy metals: Environmental persistence, toxicity, and bioaccumulation. *J. Chem.* 2019, 6730305. https://doi.org/10.1155/2019/6730305

Anwar, A., Akbar, S., 2018. Continuous microwave assisted flow synthesis and characterization of calcium deficient hydroxyapatite nanorods. *Adv. Powder Technol.* 29, 1493–1498. https://doi.org/10.1016/j.apt.2018.03.014

Bohic, S., Heymann, D., Pouëzat, J.A., Gauthier, O., Daculsi, G., 1998. Transmission FT-IR microspectroscopy of mineral phases in calcified tissues. *Comptes Rendus l'Académie des Sci. - Ser. III - Sci. la Vie* 321, 865–876. https://doi.org/10.1016/s0764-4469(99)80027–4

Bose, S., Saha, S.K., 2003. Synthesis and characterization of hydroxyapatite nanopowders by emulsion technique. *Chem. Mater.* 15, 4464–4469. https://doi.org/10.1021/cm0303437

Costescu, A., Pasuk, I., Ungureanu, F., Dinischiotu, A., Costache, M., Huneau, F., Galaup, S., le Coustumer, P., Predoi, D., 2010. Physico-Chemical Properties Of Nano-Sized Hexagonal Hydroxyapatite Powder Synthesized By Sol-Gel. *Dig. J. Nanomater. Biostructures* 5, 989–1000.

Da Rocha, D.N., Da Silva, M.H.P., De Campos, J.B., Marçal, RLSB, Mijares, D.Q., Coelho, P.G., Cruz, L.R., 2018. Kinetics of conversion of brushite coatings to hydroxyapatite in alkaline solution. *J. Mater. Res. Technol.* 7, 479–486. https://doi.org/10.1016/j.jmrt.2018.02.002

Destainville, A., Champion, E., Bernache-Assollant, D., Laborde, E., 2003. Synthesis, characterization and thermal behavior of apatitic tricalcium phosphate. *Mater. Chem. Phys.* 80, 269–277. https://doi.org/10.1016/S0254–0584(02)00466–2

Dey, T., 2014. Thermal properties of a sustainable cement material: Effect of cure conditions. *Ceram. – Silik.* 58, 275–281.

Dosen, A., Giese, R.F., 2011. Thermal decomposition of brushite, CaHPO4·2H2O to monetite CaHPO4 and the formation of an amorphous phase. *Am. Mineral.* 96, 368–373. https://doi.org/10.2138/am.2011.3544

Eslami, H., Solati-Hashjin, M., Tahriri, M., Bakhshi, F., 2010. Synthesis and characterization of nanocrystalline hydroxyapatite obtained by the wet chemical technique. *Mater. Sci. Pol.* 28, 5–13.

Ferreira, A., Oliveira, C., Rocha, F., 2003. The different phases in the precipitation of dicalcium phosphate dihydrate. *J. Cryst. Growth* 252, 599–611. https://doi.org/10.1016/S0022–0248(03)00899–6

Ferreira, C.I.A., Calisto, V., Cuerda-Correa, E.M., Otero, M., Nadais, H., Esteves, V.I., 2016. Comparative valorisation of agricultural and industrial biowastes by combustion and pyrolysis. *Bioresour. Technol.* 218, 918–925. https://doi.org/10.1016/j.biortech.2016.07.047

Geissen, V., Mol, H., Klumpp, E., Umlauf, G., Nadal, M., van der Ploeg, M., van de Zee, SEATM, Ritsema, C.J., 2015. Emerging pollutants in the environment: A challenge for water resource management. *Int. Soil Water Conserv. Res.* 3, 57–65. https://doi.org/10.1016/j.iswcr.2015.03.002

Ghosh, R., Sarkar, R., 2016. Synthesis and characterization of sintered beta-tricalcium phosphate: A comparative study on the effect of preparation route. *Mater. Sci. Eng. C* 67, 345–352. https://doi.org/10.1016/j.msec.2016.05.029

Granados-Correa, F., Bonifacio-Martinez, J., Serrano-Gomez, J., 2010. Synthesis and characterization of calcium phosphate and its relation to. *Rev. Int. Contam. Ambient.* 26, 129–134.

Hubadillah, S.K., Othman, M.H.D., Tai, Z.S., Jamalludin, M.R., Yusuf, N.K., Ahmad, A., Rahman, M.A., Jaafar, J., Kadir, SHSA, Harun, Z., 2020. Novel hydroxyapatite-based bio-ceramic hollow fiber membrane derived from waste cow bone for textile wastewater treatment. *Chem. Eng. J.* 379. https://doi.org/10.1016/j.cej.2019.122396

Ibrahim, M., Labaki, M., Giraudon, J.M., Lamonier, J.F., 2020. Hydroxyapatite, a multifunctional material for air, water and soil pollution control: A review. *J. Hazard. Mater.* 383. https://doi.org/10.1016/j.jhazmat.2019.121139

Ito, A., Nakamura, S., Aoki, H., Akao, M., Teraoka, K., Tsutsumi, S., Onuma, K., Tateishi, T., 1996. Hydrothermal growth of carbonate-containing hydroxyapatite single crystals. *J. Cryst. Growth* 163, 311–317. https://doi.org/10.1016/0022-0248(95)00955-8

Jmai, S., Bagane, M., Queneudec-T'kint, M., 2019. Physico-chemical, thermal, thermodynamic and kinetic characterization of a porous material (Di-calcium phosphate). *Heat Mass Transf.* 55, 3589–3602. https://doi.org/10.1007/s00231-019-02625–x

Kalaiselvi, V., Mathammal, R., Vetrivel, V., 2017. An efficient and regioselective synthesis of tetrazoles under transition - metal free conditions. *J. Environ. Nanotechnol.* 6, 17–21. https://doi.org/10.13074/jent.2017.12.174289

Khalil, K.A., Almajid, A.A., Soliman, M.S., 2011. Effect of hydroxide ion concentration on the morphology of the hydroxyapatite nanorods synthesized using electrophoretic deposition. *Mater. Sci. Appl.* 02, 105–110. https://doi.org/10.4236/msa.2011.22014

Kookana, R.S., Drechsel, P., Jamwal, P., Vanderzalm, J., 2020. Urbanisation and emerging economies: Issues and potential solutions for water and food security. *Sci. Total Environ.* 732, 139057. https://doi.org/10.1016/j.scitotenv.2020.139057

Leung, K.M.Y., Yeung, K.W.Y., You, J., Choi, K., Zhang, X., Smith, R., Zhou, G.J., Yung, M.M.N., Arias-Barreiro, C., An, Y.J., Burket, S.R., Dwyer, R., Goodkin, N., Hii, Y.S., Hoang, T., Humphrey, C., Iwai, C.B., Jeong, S.W., Juhel, G., Karami, A., Kyriazi-Huber, K., Lee, K.C., Lin, B. Le, Lu, B., Martin, P., Nillos, M.G., Oginawati, K., Rathnayake, I.V.N., Risjani, Y., Shoeb, M., Tan, C.H., Tsuchiya, M.C., Ankley, G.T., Boxall, A.B.A., Rudd, M.A., Brooks, B.W., 2020. Toward sustainable environmental quality: priority research questions for Asia. *Environ. Toxicol. Chem.* 39, 1485–1505. https://doi.org/10.1002/etc.4788

Ma, G. (2019). Three common preparation methods of hydroxyapatite. In: *IOP Conference Series: Materials Science and Engineering* (Vol. 688, No. 3, p. 033057). IOP Publishing. doi:10.1088/1757–899X/688/3/033057

Manatunga, D.C., De Silva, R.M., De Silva, K.M.N., Ratnaweera, R., 2016. Natural polysaccharides leading to super adsorbent hydroxyapatite nanoparticles for the removal of heavy metals and dyes from aqueous solutions. *RSC Adv.* 6, 105618–105630. https://doi.org/10.1039/c6ra22662k

Martínez-Olivas, A., Torres-Pérez, J., Balderas-Hernández, P., Reyes-López, S.Y., 2020. Oxytetracycline sorption onto synthetized materials from hydroxyapatite and aluminosilicates. *Water. Air. Soil Pollut.* 231: 1–21. https://doi.org/10.1007/s11270-020-04638–3

Massit, A.E.Y.A., Idrissi, B.C.E., Yamni, K., 2014. Synthesis and characterization of nano-sized β-tricalcium phosphate: effects of the aging time. *IOSR J. Appl. Chem.* 7, 57–61. https://doi.org/10.9790/5736–07715761

Maycock, D.S., Watts, C.D., 2011. Pharmaceuticals in drinking water. *Encycl. Environ. Heal.* 472–484. https://doi.org/10.1016/B978–0–444–52272–6.00457–8

Meejoo, S., Maneeprakorn, W., Winotai, P., 2006. Phase and thermal stability of nanocrystalline hydroxyapatite prepared via microwave heating. *Thermochim. Acta* 447, 115–120. https://doi.org/10.1016/j.tca.2006.04.013

Menon N.G., Mohapatra S., Padhye L.P., Tatiparti S.S.V., Mukherji S. (2020) Review on occurrence and toxicity of pharmaceutical contamination in Southeast Asia. In: Kumar M., Snow D., Honda R. (eds) *Emerging Issues in the Water Environment during Anthropocene. Springer Transactions in Civil and Environmental Engineering.* Springer, Singapore. https://doi.org/10.1007/978-981-32-9771-5_4

Mohapatra, S., Sargaonkar, A. and Labhasetwar, P.K., 2014. Distribution network assessment using EPANET for intermittent and continuous water supply. *Water Resour. Manag.* 28(11), pp.3745–3759.

Mohapatra S., Padhye L.P., Mukherji S. (2018) Challenges in detection of antibiotics in wastewater matrix. In: Gupta T., Agarwal A., Agarwal R., Labhsetwar N. (eds) *Environmental Contaminants. Energy, Environment, and Sustainability.* Springer, Singapore. https://doi.org/10.1007/978-981-10-7332-8_1

Mohapatra, S., Sharma, N., Mohapatra, G., Padhye, L.P. and Mukherji, S., 2021. Seasonal variation in fluorescence characteristics of dissolved organic matter in wastewater and identification of proteins through HRLC-MS/MS. *J. Hazard. Mater.* 413, p.125453.

Mobasherpour, I., Salahi, E., Pazouki, M., 2011. Removal of divalent cadmium cations by means of synthetic nano crystallite hydroxyapatite. *Desalination* 266, 142–148. https://doi.org/10.1016/j.desal.2010.08.016

Mohd Pu'ad, N.A.S., Abdul Haq, R.H., Mohd Noh, H., Abdullah, H.Z., Idris, M.I., Lee, T.C., 2019. Synthesis method of hydroxyapatite: A review. *Mater. Today Proc.* 29, 233–239. https://doi.org/10.1016/j.matpr.2020.05.536

Obasi, P.N., Akudinobi, B.B., 2020. Potential health risk and levels of heavy metals in water resources of lead–zinc mining communities of Abakaliki, southeast Nigeria. *Appl. Water Sci.* 10, 1–23. https://doi.org/10.1007/s13201-020-01233–z

Oliveira, C., Ferreira, A., Rocha, F., 2007. Dicalcium phosphate dihydrate precipitation: Characterization and crystal growth. *Chem. Eng. Res. Des.* 85, 1655–1661. https://doi.org/10.1016/S0263–8762(07)73209–4

Othman, R., Mustafa, Z., Kien, P. T., Ishak, N. F., Shaaban, A., & Noor, A. F. (2017). Parameters affecting the synthesis of beta-tricalcium phosphate powder using a wet precipitation method. *J. Mech. Eng. Sci.* 11(4), 3197–3205.

Pang, Y. X., & Bao, X. (2003). Influence of temperature, ripening time and calcination on the morphology and crystallinity of hydroxyapatite nanoparticles. *J. Eur. Ceram. Soc.*, 23(10), 1697–1704. https://doi.org/10.1016/S0955–2219(02)00413–2

Patel, M., Kumar, R., Kishor, K., Mlsna, T., Pittman, C.U., Mohan, D., 2019. Pharmaceuticals of emerging concern in aquatic systems: Chemistry, occurrence, effects, and removal methods. *Chem. Rev.* 119, 3510–3673. https://doi.org/10.1021/acs.chemrev.8b00299

Paz, A., Guadarrama, D., López, M., E González, J., Brizuela, N., Aragón, J. (2012). A comparative study of hydroxyapatite nanoparticles synthesized by different routes. *Química Nova*, 35(9), 1724–1727. https://doi.org/10.1590/S0100–40422012000900004

Pereira, M.B.B., França, D.B., Araújo, R.C., Silva Filho, E.C., Rigaud, B., Fonseca, M.G., Jaber, M., 2020. Amino hydroxyapatite/chitosan hybrids reticulated with glutaraldehyde at different pH values and their use for diclofenac removal. *Carbohydr. Polym.* 236, 116036. https://doi.org/10.1016/j.carbpol.2020.116036

Pratihar, S.K., Garg, M., Mehra, S., Bhattacharyya, S., 2006. Phase evolution and sintering kinetics of hydroxyapatite synthesized by solution combustion technique. *J. Mater. Sci. Mater. Med.* 17, 501–507. https://doi.org/10.1007/s10856-006-8932-4

Priyanka M.S., Mohapatra S. (2020) Shifts and trends in analysis of contaminants of emerging concern: sulfonamides. In: Kumar M., Munoz-Arriola F., Furumai H., Chaminda T. (eds) *Resilience, response, and risk in water systems. Springer transactions in civil and environmental engineering.* Springer, Singapore. https://doi.org/10.1007/978-981-15-4668-6_14

Ramesh, S., Aw, K.L., Tolouei, R., Amiriyan, M., Tan, C.Y., Hamdi, M., Purbolaksono, J., Hassan, M.A., Teng, W.D., 2013. Sintering properties of hydroxyapatite powders prepared using different methods. *Ceram. Int.* 39, 111–119. https://doi.org/10.1016/j.ceramint.2012.05.103

Rasheed, T., Bilal, M., Nabeel, F., Adeel, M., Iqbal, H.M.N., 2019. Environmentally-related contaminants of high concern: Potential sources and analytical modalities for detection, quantification, and treatment. *Environ. Int.* 122, 52–66. https://doi.org/10.1016/j.envint.2018.11.038

Raynaud, S., Champion, E., Bernache-Assollant, D., Thomas, P., 2002. Calcium phosphate apatites with variable Ca/P atomic ratio I. Synthesis, characterisation and thermal stability of powders. *Biomaterials* 23, 1065–1072. https://doi.org/10.1016/S0142–9612(01)00218–6

Reyes-Gasga, J., Martínez-Piñeiro, E.L., Rodríguez-Álvarez, G., Tiznado-Orozco, G.E., García-García, R., Brès, E.F., 2013. XRD and FTIR crystallinity indices in sound human tooth enamel and synthetic hydroxyapatite. *Mater. Sci. Eng. C* 33, 4568–4574. https://doi.org/10.1016/j.msec.2013.07.014

Rodríguez-Lugo, V., Karthik, T.V.K., Mendoza-Anaya, D., Rubio-Rosas, E., Villaseñor Cerón, L.S., Reyes-Valderrama, M.I., Salinas-Rodríguez, E., 2018. Wet chemical synthesis of nanocrystalline hydroxyapatite flakes: Effect of pH and sintering temperature on structural and morphological properties. *R. Soc. Open Sci.* 5, 180962. https://doi.org/10.1098/rsos.180962

Sa, Y., Guo, Y., Feng, X., Wang, M., Li, P., Gao, Y., Yang, X., Jiang, T., 2017. Are different crystallinity-index-calculating methods of hydroxyapatite efficient and consistent? *New J. Chem.* 41, 5723–5731. https://doi.org/10.1039/c7nj00803a

Sadat-Shojai, M., Khorasani, M.T., Dinpanah-Khoshdargi, E., Jamshidi, A., 2013. Synthesis methods for nanosized hydroxyapatite with diverse structures. *Acta Biomater.* 9, 7591–7621. https://doi.org/10.1016/j.actbio.2013.04.012

Salimi, E., Javadpour, J., 2012. Synthesis and characterization of nanoporous monetite which can be applicable for drug carrier. *J. Nanomater.* 2012, 135. https://doi.org/10.1155/2012/931492

Sanosh, K.P., Chu, M.C., Balakrishnan, A., Lee, Y.J., Kim, T.N., Cho, S.J., 2009. Synthesis of nano hydroxyapatite powder that simulate teeth particle morphology and composition. *Curr. Appl. Phys.* 9, 1459–1462. https://doi.org/10.1016/j.cap.2009.03.024

Savino, K., Yates, M.Z., 2015. Thermal stability of electrochemical-hydrothermal hydroxyapatite coatings. *Ceram. Int.* 41, 8568–8577. https://doi.org/10.1016/j.ceramint.2015.03.065

Suryawanshi, V.B., Chaudhari, R.T., 2014. Growth and characterization of agar gel grown brushite crystals. *Indian J. Mater. Sci.* 2014, 1–6. https://doi.org/10.1155/2014/189839

Tõnsuaadu, K., Gross, K.A., Pluduma, L., Veiderma, M., 2012. A review on the thermal stability of calcium apatites. *J. Therm. Anal. Calorim.* 110, 647–659. https://doi.org/10.1007/s10973-011-1877–y

Wang, Z., Sun, K., He, Y., Song, P., Zhang, D., Wang, R., 2019. Preparation of hydroxyapatite-based porous materials for absorption of lead ions. *Water Sci. Technol.* 80, 1266–1275. https://doi.org/10.2166/wst.2019.370

Wu, S.C., Hsu, H.C., Hsu, S.K., Chang, Y.C., Ho, WF, 2015. Effects of heat treatment on the synthesis of hydroxyapatite from eggshell powders. *Ceram. Int.* 41, 10718–10724. https://doi.org/10.1016/j.ceramint.2015.05.006

Yelten-Yilmaz, A., Yilmaz, S., 2018. Wet chemical precipitation synthesis of hydroxyapatite (HA) powders. *Ceram. Int.* 44, 9703–9710. https://doi.org/10.1016/j.ceramint.2018.02.201

Chapter 8

Nanostructured adsorbents for uranium removal from drinking water

A review

Alok Kumar Meher
Central Pollution Control Board (CPCB), Ministry of Environment, Forest & Climate Change (MoEF & CC)

CONTENTS

DOI: 10.1201/9781003247869-10

8.1 INTRODUCTION

Water pollution due to the presence of elevated concentration of toxic elements is considered to be a serious global environmental problem (Tchounwou et al., 2012; Briffa et al., 2020). Uranium (U) is considered to be carcinogenic (Tirmarche et al., 2004) and richly available in environmental matrices originated from geogenic as well as anthropogenic sources namely nuclear waste disposal facilities, uranium mining, combustion of fossil fuels and the application of chemical fertilizers. These anthropogenic activities result in the entry of U into the environmental components, which poses a negative impact to the aquatic organism as well as human health (Barnett et al., 2000; Seidel et al., 2005). U predominantly enters through the ingestion route (drinking water) and accumulates in the body parts namely kidneys, liver and bones that trigger radiological activity, which can cause cancer with extensive exposures over a long period of time. The elevated concentration of uranium ingestion results in adverse human health effects namely nephritis, and kidney is reported to be the primary targeted organ. Prolonged exposure to uranium is also reported for the alteration of the concentration of neurochemicals in the brain and decrease in the fertility of some mammals (Limson Zamora et al., 1998; Hakonson-Hayes et al., 2002). Considering the toxicity of the uranium and uranyl compounds, several agencies have framed a strict guideline for safety concerns. For example, the World Health Organization (WHO) allows a Maximum Admissible Concentration (MAC) for uranium as 30 μg/L in drinking water (World Health Organization, 2008). Similarly, the United States Environmental Protection Agency (USEPA) has set a Maximum Contaminant Level (MCL) of 30 μg/L (United States Environmental Protection Agency, 2009). Elevated concentrations of U have been recorded from various parts of India, e.g. Hoshiarpur district, Punjab, in the range of 2.1–2.2 mg/L (Central Groundwater Water Board, 2020), which is almost 70-fold higher as compared to the USEPA prescribed limit for drinking water.

The chemical speciation of U in the aqueous matrices depends on the redox potential (Eh) and the solution pH, which has been presented in Table 8.1.

Uranium occurs in various valence states, i.e., +2, +3, +4, +5 and +6 in the nature. However, hexavalent form, i.e., U(VI), is the most dominating species and found to be commonly associated with oxygen as the uranyl cation (UO_2^{2+}). It also resides as the complex conjugation with other anions namely carbonate, phosphate, sulfate, etc., or as hydroxyl complexes, depending on the ionic composition of the water and the solution

Table 8.1 Chemical speciation of U ion in the aqueous matrices

pH of the aqueous matrix	<5.0	5.0	6.5	7.6	10.5	>10.5
Form of the U species	UO_2^{2+}	UO_2CO_3	$UO_2(CO_3)_2^{2-}$	$UO_2(CO_3)_3^{4-}$	$(UO_2)_3(OH)_5^{+}$	

pH. The presence of carbonate (alkalinity) significantly influences the chemical speciation of U and thus plays an important role in deciding the specific U removal process from the groundwater matrix (Alam and Cheng, 2014). Toxicity of the U species has a strong correlation with their solubility in water, and more soluble uranium ions are reported to be the more toxic forms. The most soluble and the toxic forms of the uranium compounds may be listed as UF_6, $UO_2(NO_3)_2$, UO_2Cl_2, UO_2F_2, uranyl acetates, sulfates and carbonates. Some of the moderately soluble ions are UO_2, UO_4, $(NH_4)_2U_2O_7$ and UO_3. The least soluble and low toxic forms are UO_3, $Na_2U_2O_7$ and $(NH_4)_2U_2O_7$. In general, hexavalent compounds observed to be more soluble and toxic as compared to the other valence state compounds of the uranium (Aly and Hamza, 2013).

Due to the serious negative impact of uranium ions, it's imperative to treat the uranium contaminated water up to a safe level or convert it to a less toxic form prior to discharge it to the environmental matrices, for the safeguard of the human health and the environment. In this aspect, various methods namely lime softening/coagulation, membrane separation, ion exchange resins, adsorption, reactive permeable membranes, biological detoxification, etc. have been reported to remove toxic U species from various aqueous matrices (Katsoyiannis and Zouboulis, 2013; Li and Zhang, 2012; Aly and Hamza, 2013).

The simplest method for uranium removal from water is by application of coagulating agents namely alum and iron sulfate salts. After mixing the coagulants with the contaminated water, the resultant flocks can be easily separated from the bulk solution by sedimentation and filtration process. The method performs well at the solution pH range of 6–10 (Lee and Bondietti, 1983). Membrane separation methods (reverse osmosis or nanofiltration) methods are based on the separation of the larger size of uranyl ions as compared to the smaller pore size of the applied membranes. The technology is well established and proven to be highly efficient (>99% removal capacity) regardless of the solution pH. However, high rejection rate of the wastewater is a significant challenge for field applications (Shen and Schäfer, 2014). Removal of uranium by ion exchange mechanism is significantly affected by the composition of the water and highly sensitive to the solution pH. At the solution pH > 6, uranium exists as the anionic uranyl carbonate complex, which has a high affinity for strong anion resins. However, in acidic matrices, the cationic resins in the H_2 form reported being performed well, due to the conversion of uranium complexes to simple uranyl cations (Hanson et al., 1987; Lee and Bondietti, 1983; Rosenberg et al., 2016). Permeable Reactive Barrier (PRB) is an in situ method of groundwater treatment in which the reactive membranes are installed across the flow path of the desirable contaminant plume and least permeable to the adjacent aquifer materials. In laboratory studies iron-based materials (elemental iron, zerovalent iron and iron oxyhydroxides), apatites, hydroxyapatites, etc. have been well proven to act as a PRB for uranium removal from groundwater (Simon et al., 2003; Kornilovych et al., 2018). Application of microbial agents (bacteria, fungi, etc.) has been reported to be the reduction of bioavailability of the U ions, by altering the extracellular binding sites and biological solution pH, thus restricting the movement of U ions to the food chain (Kalin et al., 2005; Malekzadeh et al., 2002). Phytoremediation is also another promising approach for the treatment of lower concentrations of U in groundwater, owing to certain advantages namely cost-effectiveness, no generation of secondary source of pollution and a natural way of treatment process (Pratas et al., 2006; Dushenkov et al., 1997).

The adsorption process for the removal of U ions from aqueous matrices has been well explored and established, which governs the mechanism of transport of the contaminant from the aqueous phase to the active sites of the adsorbent surfaces, resulting in the trace concentration of the pollutant in the treated water up to a safe level (Barnett et al., 2000; Tolkou et al., 2020; Tatarchuk et al., 2019; Kausar and Bhatti, 2013; Ma et al., 2020; Tang et al., 2020). The process has certain key advantages namely low cost, process simplicity, availability of a broad range of adsorbents and flexibility of application of the adsorbents in multiple cycles by regenerating the saturated material, which makes it one of the best available technologies for the treatment of contaminated water (Singh et al., 2018; Kyzas and Kostoglou, 2014; Ali, 2012; Crini et al., 2019; Renu and Singh, 2017).

In recent years, nanostructured materials have attracted intensive attention to scientific researchers to exploit the functional adsorbents with exceptional binding capacities (Tatarchuk et al., 2019; Ali, 2012). Graphene is a nanostructured material having a flat layer of sp^2 hybridized carbon atoms, closely packed in a fixed crystal lattice, large surface area (up to 2,600 m^2/g) and possesses π-π conjugation structure, resulting in the material to high physical/thermal and chemical stability (Trivedi et al., 2019). Owing to this exceptional behavior, graphene-based materials have been extensively reported for the removal of environmental contaminants by employing processes like adsorption, photocatalysis, electro catalysis, ion exchange, etc. (Wei et al., 2018; K. Yang et al., 2018). Commercially graphene and its derivatives can be synthesized by following various routes namely oxidative exfoliation-reduction, liquid-phase exfoliation and chemical vapor deposition (Agudosi et al., 2020; Zhu et al., 2012; Singh et al., 2016). Graphene oxide (GO) is the chemical modification form of graphene that contains an elevated amount of oxygen containing surface functional groups, for which it has been reported to be a promising sorbent for the uptake of hazardous uranium species (Tolkou et al., 2020; X. Wang et al., 2016). In the current chapter, GO-based nanostructured materials for U removal from aqueous matrices have been reviewed based on the recent interventions. A summary of reported GO-based adsorbents for U removal from aqueous matrices has been compiled, which is the heart of the chapter, and is presented in Table 8.2.

8.2 GRAPHENE OXIDE (GO) FOR URANIUM UPTAKE IN AQUEOUS MEDIA

Hu et al. studied GO for the removal of UO_2^{2+} from water, in which they reported the maximum adsorption capacity of the material up to 2.3532×10^{-4} mol/g (~56.00 mg/g) at 338 K. They also observed that the sorption process was sensitive toward solution pH, and peak performance of the material was observed at acidic condition (Hu et al., 2017). U removal from the aqueous matrix by single-layered GO was reported by Li et al. The study emphasized the role of the oxygen containing surface groups in the adsorption process. They reported a sorption capacity of 299 mg/g, governing the mechanism of adsorption as the formation of inner-sphere surface complexes (Li et al., 2012). Wang et al. evaluated the GO nanosheets for U removal from a synthetic water sample, which showed an adsorption capacity of 3.969×10^{-4} mol/g (~94.00 mg/g). They observed the negative impact of background electrolyte (NaCl) on the adsorption

Table 8.2 Summary of reported graphene oxide (GO) based adsorbents for the U removal from aqueous matrices

Name of the adsorbent	*BET Surface Area (m^2/g)*	*Adsorption capacity (mg/g)*	*Adsorption model*	*Kinetics model*	*Optimal solution pH*	*Optimal Temp (K)*	*Reference*
GO	—	56.00	Langmuir	Pseudo second order	<7	338	Hu et al. (2017)
Single-layered GO	—	299.00	Langmuir	—	4	333	Li et al. (2012)
Layered GO nanosheets	—	94.00	Langmuir	Pseudo second order	3.45	298	C. L. Wang et al. (2015)
Layered GO nanosheets		97.50	Langmuir		5.0	293	Zhao et al. (2012)
Graphene aerogel	350.80	238.67	Langmuir	Pseudo second order	4.0	298	D. Zhao et al. (2019)
Graphene aerogel capsulated with the degradation products of phytic acid (PAGA)	—	3,550.00	—	Pseudo second order	7	—	Wang et al. (2020a)
Graphene (rGO) hydrogel	—	134.23	Freundlich	Pseudo second order	4.6 to 6.6	—	He et al. (2018)
Metal/metal oxide-incorporated GO composite materials							
Fe-Ni nanoparticle doped GO	—	25.64	Langmuir	Pseudo second order	4	—	Rohith et al. (2020)
Fe-Ni/GO composite	—	384.62	Langmuir	Pseudo second order	5.5	298	Zhang et al. (2019)
Magnetic graphene/iron oxides composite (Fe_3O_4/GO)	—	69.49	Langmuir	—	5.5	293	Zong et al. (2013)
Magnetic graphene/iron oxides composite (Fe_3O_4/GO)	—	83.00	Langmuir	Pseudo second order	—	318	Zhao et al. (2015)
Fe_3O_4/graphene composite (FGC)	42.53	176.47	Langmuir	Pseudo second order	5.5	298	Zhao et al. (2018)
EDTA-functionalized Fe_3O_4/GO	54.32	277.43	Langmuir	Pseudo second order	6.0	318	Zhao et al. (2017)
Manganese dioxide–iron oxide–graphene magnetic nanocomposite (MnO_2–Fe_3O_4–rGO)	58.02	108.70	Langmuir	Pseudo second order	6.0	273	Tan et al. (2015)

(*Continued*)

Table 8.2 (Continued) Summary of reported graphene oxide (GO) based adsorbents for the U removal from aqueous matrices

Name of the adsorbent	*BET Surface Area (m^2/g)*	*Adsorption capacity (mg/g)*	*Adsorption model*	*Kinetics model*	*Optimal solution pH*	*Optimal Temp (K)*	*Reference*
Zerovalent iron-polyaniline-graphene aerogel composite (Fe-PANI-GA)	—	350.47	Langmuir	Pseudo second order	5.5		Chen et al. (2017)
Amidoximated magnetite/GO (AOMGO) composites	—	285.00	Langmuir		5.0	298	Zhao et al. (2013)
Zirconium dioxide containing graphene-like additions	280.00	11.40	Dubinin-Raduchkevich	Pseudo second order	—	—	Perlova et al. (2020)
Zirconium hydrophosphate containing graphene-like additions	190.00	12.50	Dubinin-Raduchkevich	Pseudo second order	—	—	Perlova et al. (2020)
GO-manganese oxide (GOMO) nanocomposite	—	153.85	Langmuir	Pseudo second order	4	333	A. Yang et al. (2018)
Composite adsorbents based on GO and natural/biological materials							
Phosphorylated GO-chitosan composite (GO-CS-P)	—	779.44	Langmuir	—	5.0	293	Cai et al. (2017)
GO nanoribbons/chitosan (GONRs/CTS) composite membrane	—	320.00	Langmuir	Pseudo second order	5	328	Hu et al. (2020)
GO-chitosan aerogel (GO-CS)	—	384.60	Freundlich isotherm (at acidic pH), Langmuir (at alkaline pH)	Pseudo second order	3.0 to 8.0	—	Huang et al. (2017)
Immobilized active yeast cells entrapped in calcium-alginate-PVAGO-cross-linked gel beads	—	34.00	Langmuir and Freundlich	Pseudo first order	3	—	Chen et al. (2020)
GO-immobilized *Saccharomyces cerevisiae* gel beads	—	24.40	Sips	Pseudo second order	3.0 to 9.0	—	Chen and Wang (2016)

(*Continued*)

Table 8.2 (Continued) Summary of reported graphene oxide (GO) based adsorbents for the U removal from aqueous matrices

Name of the adsorbent	*BET Surface Area (m^2/g)*	*Adsorption capacity (mg/g)*	*Adsorption model*	*Kinetics model*	*Optimal solution pH*	*Optimal Temp (K)*	*Reference*
Calcium-alginate GO beads	—	29.40	Langmuir	Pseudo first order	4.0 to 5.0	—	Basu et al. (2018)
Fungus hyphae/GO (FH/GO) composite	—	199.37	Langmuir	Pseudo second order	6.0	293.15	Li et al. (2018)
Fungus hyphae (FH)/Fe_3O_4/GO nanocomposite spheres (FFGS)	—	219.71	Langmuir	Pseudo second order	5.0	293	Zhu et al. (2019)
GO-bovine serum albumin (GO-BSA) composite	—	389.00	Langmuir	Pseudo second order	6	298.15	Yang et al. (2017b)
Konjac glucomannan/GO (KGM/GO) sponges	—	266.97	Langmuir	Pseudo second order	5	313.15	Chen et al. (2018)
Cyclodextrin-modified GO nanosheets (CD/GO)	—	103.90	Langmuir	—	6	—	Song et al. (2014)
GO and tea waste (GO-TW) composite	—	111.61	Langmuir	Pseudo second order	5	333	Yang et al. (2019)
GO-based structural frameworks							
GO-COOH/UiO-66	731.08	998.00	Langmuir	Pseudo second order	6	318.15	Yang et al. (2017a)
Fe_3O_4@HKUST-1/GO		268.82	Langmuir	Pseudo second order	4.0	318	Wang et al. (2020b)
rGO/ZIF-67 aerogel	962.00	1,888.55	Langmuir	Pseudo second order	4.01	313	Zhao et al. (2020)
Chitosan-GO/ZIF (GCZ8A) foam	200.79	361.01	Freundlich	Pseudo second order	8.0	—	Guo et al. (2020)
Multi-level surface functionalization of GO-based materials							
composite hydrogel (PCG)	—	247.00	Langmuir	Pseudo second order	6	308.15	Lu et al. (2020)
Polyamidoxime/polydopamine-decorated GO (GO/PDA/PAO)	—	502.50	Langmuir	Pseudo second order	6.0	318	Dai et al. (2019a)
Polyamidoxime/polyethyleneimine magnetic GO (mGO-PP)	—	606.06	Langmuir	Pseudo second order	6.0	298	Dai et al. (2019b)

(Continued)

Table 8.2 (Continued) Summary of reported graphene oxide (GO) based adsorbents for the U removal from aqueous matrices

Name of the adsorbent	*BET Surface Area (m^2/g)*	*Adsorption capacity (mg/g)*	*Adsorption model*	*Kinetics model*	*Optimal solution pH*	*Optimal Temp (K)*	*Reference*
Copolymer based on amidoxime groups, polypyrrole and GO (GO-APP-AMP)	—	149.57	Langmuir	—	—	—	Qin et al. (2016)
Amidoxime-functionalized GO Nanoribbons (AOGONRs)	—	502.60	Langmuir	Pseudo second order	4.5	298	Y. Wang et al. (2015)
GO/amidoxime hydrogel (AGH)	—	398.40	Langmuir	Pseudo second order	6.0	—	F. Wang et al. (2016)
Sulfonated GO (GOS)	—	309.09	Langmuir	Pseudo second order	6	323.15	Zhang et al. (2016)
Sulfonated GO (SGO)	—	45.05	Langmuir	Pseudo second order	2	—	Sun et al. (2017)
GO and poly (3,4-ethylene dioxythiophene): poly(styrenesulfonate) composite (GO/PEDOT:PSS)	—	384.51	Langmuir	Pseudo second order	4.5	298	Song et al. (2019)
Carboxylated GO (GO-COOH)	—	315.00	Langmuir	Pseudo second order	6.0	298.15	Ma et al. (2019)
Defect GO (dGO)	—	131.00	Langmuir	—	2.2	—	Boulanger et al., 2020
Ammonia-incorporated GO (NH_3-GO)	—	80.13	Langmuir	Pseudo second order	6	313	Verma and Dutta (2015)
Perhydroxy-cucurbit[6]uril decorated GO composite [HO-CB[6]/GO]	—	301.60	Langmuir	—	6.5	328	Shao et al. (2017)
Diethylenetriaminepentaacetic phenylenediamine (DTPAA)-GO composite	—	485.00	Langmuir	Pseudo second order	6.5	298.15	Liu et al. (2018)
Cetyltrimethyl ammonium bromide (C_{16}TAB) and phytic acid (PA)-modified GO (GO-C_{16}TAB-PA)	—	923.00	Langmuir	—	8	298.15	P. Yang et al. (2018)

process (C. L. Wang et al., 2015). Layered graphene nanosheets have been reported for preconcentration of U(VI) ion from a water sample, which showed a higher adsorption capacity as compared to the conventional GO, due to the abundance of oxygen contain surface groups (Zhao et al., 2012). Graphene aerosol prepared through a simple hydrothermal route, having a surface area of 350.8 m^2/g, was investigated for U removal by Zhao et al. They demonstrated the interaction of U(VI) ion with the oxygenated and nitrogenous functional groups based on X-ray photoelectron spectrometry characterization studies (D. Zhao et al., 2019). GA capsulated with the degradation products of phytic acid (PAGA) was synthesized by a novel chemical precipitation strategy and evaluated for emergency removal of uranium. The introduction of the material to the solution results in the release of degradation products from the GA skeleton followed by a combination with uranium and sediment formation. The novel material showed fast kinetics and an exceptional elimination capacity of 3550 mg/g and suffered no interferences from the coexisting ions (Wang et al., 2020a). He et al. reported the synthesis of a 3D porous networked graphene (rGO) hydrogel, which was proven to be an impressive material for fast recovery of uranium with high regeneration capacity (He et al., 2018).

8.3 STRATEGIES TO IMPROVE THE SORPTION EFFICIENCY OF GO-BASED MATERIALS FOR U ADSORPTION

Although many scientific researchers have explored large numbers of GO-based materials with various functional groups for the removal of toxic uranium species from aqueous matrices, their performance is not satisfactory for field applications, and therefore several strategies have been proposed to improve the sorption efficiency. Some of the strategies may be summarized as follows:

- Incorporation of metal/metal oxides onto GO sheet.
- Composition GO with materials of natural/biological origin.
- Development of GO composite structural frameworks (metal organic frameworks and zeolitic imidazolate framework).
- Multi-level surface functionalization of GO with oxygen/nitrogen rich groups (amidoximation, carboxylation, sulfonation, amination, etc. and incorporation of surfactants).

8.3.1 Metal/metal oxide-incorporated GO composite materials

8.3.1.1 GO-iron/iron oxide composites

Incorporation of iron and iron oxides onto the GO sheet has been widely reported for the enhanced sorption of U ions from aqueous solutions. Go sheet doped with Fe-Ni metallic nanoparticles was synthesized by modified Hummer's method and evaluated as U(VI) uptake from water samples, which shows 15% enhancement in the adsorption capacity as compared to the unmodified one (Rohith et al., 2020). The same metallic composition, i.e., Fe-Ni, was evaluated for removal of U(VI) ions from wastewater matrix by Zhang et al., which showed an outstanding adsorption capacity of 384.62 mg/g (Zhang et al., 2019). Iron oxides/graphene composite (Fe_3O_4/GO) has been studied as a promising adsorbent for the preconcentration of U(VI) ion from larger solution volume by Zong et al. (2013). The same type of composite material was also reported by Zhao et al., which could be applied as an efficient and magnetic separable sorbent for U removal (Zhao et al., 2015). Fe_3O_4/graphene composite (FGC) prepared by Zhao et al., by using a single step synthesis route, was proven to be effective for sorption of the U(VI) from a nuclear waste matrix (Zhao et al., 2018). Zhao et al. reported a composite material namely EDTA-mGO, synthesized by fabrication of a chelating agent (EDTA) onto the Fe_3O_4/GO and evaluated for the U(IV) adsorption. The material showed magnetic properties, high stability and good generation efficiency. The adsorption capacity of the material was calculated to be 277.43 mg/g, which was roughly threefold higher as compared to raw mGO (Zhao et al., 2017). Nanocomposite particles based on the fabrication of mixed metal oxides (manganese oxide-iron oxide) on the GO sheet were proved to be a fast and effective adsorbent for U(IV) removal (Tan et al., 2015). Zerovalent iron-polyaniline-graphene aerogel composite (Fe-PANI-GA) was proven to be a cost-effective material for the adsorption of U(VI) from real nuclear waste matrix (Chen et al., 2017). Zhao et al. reported successful application of the amidoximated magnetite/GO (AOMGO) composites for separation and preconcentration of U(VI) from wastewater and seawater matrices. The adsorbent exhibited fast sorption kinetics and better adsorption capacity than the unmodified GO. The material also showed magnetic property, facilitating the easy separation of the materials from the bulk solution by using a magnetic field after application (Zhao et al., 2013).

8.3.1.2 GO-zirconium composite

Zirconium dioxide and zirconium hydrophosphate containing graphene-like addition composites were studied for adsorption of the cationic complexes of UO_2^{2+} compounds from water, which was successfully evaluated as a selective adsorbent for the targeted pollutant in the hard water matrix and also achieved the highest degree of regeneration (Perlova et al., 2020).

8.3.1.3 GO-manganese oxide composite

GO-Manganese Oxide (GOMO) nanocomposite was prepared by using ultrasonic radiation and evaluated for adsorption performance of U ion, which showed roughly 10-fold higher adsorption efficiency as compared to the raw GO (A. Yang et al., 2018).

8.3.2 Composite adsorbents based on GO and natural/biological materials

8.3.2.1 GO-chitosan composites

Chitosan is one of the biomaterials widely reported for application as a composite with GO for U removal. Phosphorylated GO-chitosan composite (GO-CS-P) was demonstrated for the ultrafast kinetics, and the equilibrium was achieved in just 15 min by

Cai et al. Further, the synthesized adsorbent was proved to be selective in the presence of co-ions in the aqueous solution (Cai et al., 2017). Adsorption properties of GO nanoribbons/chitosan (GONRs/CTS) composite membrane for U(VI) removal were investigated by Hu et al., which was proven to be effective for the removal of uranium from wastewater matrix (Hu et al., 2020). GO-chitosan aerogel (GO-CS) was synthesized and evaluated for U(VI) extraction from the synthetic seawater matrix, which retained the high adsorption capacity over a wide range of solution pH (Huang et al., 2017).

8.3.2.2 GO-fungus composites

Chen et al. reported an entrapped active biomass of *Saccharomyces cerevisiae* in calcium-alginate-PVA-GO cross-linked gel bead composite material for the biosorption of U(VI), which showed the applicability of the material in a wide range of pH of from 3 to 7 (Chen et al., 2020). Studies performed by Chen and Wang suggested that GO-immobilized *Saccharomyces cerevisiae* gel beads performed better as compared to the raw yeast cell for the absorption of dissolved U(VI) in aqueous media (Chen and Wang, 2016). Calcium-Alginate-GO composite beads were synthesized by loading GO onto the Ca-alginate matrix, which was proven to be highly effective for uranium removal from the real groundwater samples (Basu et al., 2018). Fungus hyphae/GO (FH/GO) composite material having a net structure was evaluated for U(VI) removal in batch and column modes. The experimental results showed high stability, efficiency and reusability of the material for U(VI) decontamination and for the nuclear wastewater treatment (Li et al., 2018). Zhu et al. prepared a layered core-shell structured fungus hyphae (FH)/Fe_3O_4/GO nanocomposite spheres (FFGS), which was successfully evaluated for the sorption of U(VI) from the wastewater sample (Zhu et al., 2019).

8.3.2.3 Composites based on GO and other natural materials

Yang et al. reported the synthesis of GO coated with bovine serum albumin (BSA), which exhibited higher adsorption efficiency for U(VI) extraction from simulated seawater due to enrichment in amino and carboxyl function groups (Yang et al., 2017b). A natural polymer (Konjac Glucomannan)-GO three dimensional sponge composite was synthesized by the ice template method, which was proved to be highly selective toward U(VI) in the presence of interfering ions (Chen et al., 2018). Cheng et al. reported the synthesis of cyclodextrin-modified GO nanosheets (CD/GO) by an in situ polymerization method and evaluated it for sorption of U(VI) in the presence of humic acid. The sorption capacity of the material was calculated to be 97.328 mg/g for U(VI) and observed to be increased up to 103.935 mg/g with the existence of humic acid, which proves the effectiveness of the material for simultaneous removal of U(VI) and humic acid from the contaminated water sample (Song et al., 2014). GO and tea waste (GO-TW) composite were prepared and studied as a low-cost sorbent for uranium adsorption from aqueous matrices, which exhibited promising adsorption performance (removal rate ~99%). The data obtained from the experiment also suggested the potential application of the adsorbent for remediation of the nuclear industrial effluent (Yang et al., 2019).

8.3.3 GO-based structural frameworks

8.3.3.1 *GO-based metal organic frameworks (MOF)*

Yang et al. designed a MOF composite (GO-COOH/UiO-66) through coordination of the carboxyl groups of GO with the zirconium ion of UiO-66, which showed efficient preconcentration of U(VI) from synthetic seawater matrix (Yang et al., 2017a). Wang et al. reported another type of the MOF composite (Fe_3O_4@HKUST-1/GO), prepared through a green route by mixing magnetic Fe_3O_4 nanoparticles, HKUST-1 nanocrystal and GO, which was successfully evaluated for U(VI) extraction from the contaminated aqueous matrix (Wang et al., 2020b).

8.3.3.2 *GO-based Zeolitic imidazole frameworks (ZIF)*

Zhao et al. reported a fabricated composite of zeolitic imidazole framework-67 (ZIF-67) on reduced GO (rGO) hydrogel, which showed an exceptional adsorption capacity of 1,888.55 mg/g (Zhao et al., 2020). A macro-structured polysaccharide (chitosan)-GO/ZIF (GCZ8A) foam was successfully evaluated for the seawater uranium extraction, which also possessed anti-biofouling and anti-adhesion properties, making it suitable for the field applications (Guo et al., 2020).

8.3.4 Multi-level surface functionalization of GO-based materials

The introduction of surface functional groups onto the GO sheets has been widely reported for enhancement in the adsorption capacity by the functionalized materials as compared to the raw/unmodified sorbents for U(VI) removal in the aqueous matrices.

8.3.4.1 *Amidoximation of GO*

Zhao et al. reported preparation of AOMGO composite for extraction of U(VI) from seawater and wastewater samples, which showed fast sorption kinetics and excellent adsorption capacity compared to the raw GO. The material showed magnetic property, facilitating the easy separation of the materials from the bulk solution after application by using a magnetic field (Zhao et al., 2013). Lu et al. reported a composite hydrogel (PCG) consisting of amidoxime polyacrylonitrile, chitosan and GO, which was successfully evaluated as an efficient material for U(VI) removal (Lu et al., 2020). Dai et al. reported one-pot synthesis of polyamidoxime/polydopamine-decorated GO (GO/PDA/PAO) composite material, which was successfully applied for the reclamation of U(VI) from mine wastewater matrix (Dai et al., 2019a). The same research group also synthesized polyamidoxime/polyethylene mine magnetic GO (mGO-PP) by in situ polymerization of acrylonitrile on magnetic GO (mGO) covalently modified with polyethylenimine (PEI). The experimental data obtained from the study proves the adsorbent as a promising adsorbent for the efficient decontamination of U(VI) in the wastewater sample (Dai et al., 2019b). A group conducting copolymer based on amidoxime groups, polypyrrole and GO (GO-APP-AMP) was synthesized by in situ copolymerization method, which was proven to be effective as a sorption material for the preconcentration and recovery of uranium (Qin et al., 2016). Wang et al. reported the amidoxime-functionalized GO nanoribbons (AOGONRs) composite synthesized by chemical grafting technology, which was observed to be a selective adsorbent for U (Y. Wang et al., 2015). Another composite material namely GO/amidoxime hydrogel (AGH) was proven to be selective for U(VI) in the presence of existing co-ions namely Mg^{2+}, Ca^{2+}, Ba^{2+} and Sr^{2+}. The experimental results signified the potential application of the adsorbent for the treatment of nuclear effluent and uranium extraction from the seawater matrix (F. Wang et al., 2016).

8.3.4.2 Sulfonation of GO

Zhang et al. reported a facile one-pot method for the preparation of sulfonated GO (GOS), which exhibited an excellent adsorption capacity of 309.09 mg/g (Zhang et al., 2016). Sun et al. reported successful evaluation of GOS for the effective enrichment of U(VI) at extreme acidic conditions, which proved to be a promising material for decontamination of nuclear wastewater (Sun et al., 2017). The studies conducted by Song et al. have been focused on various strategies for designing highly efficient adsorbents for U(VI) removal from wastewater samples. They prepared a composite material (GO/PEDOT: PSS), by the direct assembling of GO and poly(3,4-ethylenedioxythiophene): poly(styrenesulfonate) (PEDOT: PSS) through a facile ball milling method, which showed an impressive performance for the U(VI) removal for radioactive wastewater treatment (Song et al., 2019).

8.3.4.3 Carboxylation of GO

Ma et al. prepared carboxylated GO (GO-COOH) via a simple and feasible method and evaluated it as an efficient sorbent for the removal of U(VI). The study reported significant enhancement (65.8%) in the sorption of GO-COOH (315 mg/g) as compared to the unmodified GO (190 mg/g) for U(VI) uptake from an aqueous matrix (Ma et al., 2019). Extremely defective GO (dGO) was prepared through a modified Hummers oxidation procedure by Boulanger et al. The material was abundant in carboxylic groups, as evidenced from the spectroscopic studies, which showed a 15-fold enhancement in the adsorption capacity compared to standard Hummers GO for U(VI) removal (Boulanger et al., 2020).

8.3.4.4 Amidation of GO

Verma et al. reported the synthesis of ammonia-incorporated GO, which was evaluated to be a highly selective material for UO_2^{2+} ion in a multiple component system. The significant improvement in material selectivity was attributed to the amidation of GO, which was marked to be better as compared to the carboxylic acid groups (Verma and Dutta, 2015).

8.3.4.5 *Phosphorylation of GO*

Phosphorylated GO-chitin composite material was rationally designed by Cai et al., which showed an adsorption capacity of 779 mg/g for U(VI) removal under optimal conditions. The material is noted for several advantages, namely high stability, fast sorption kinetics, excellent adsorption capacity, better selectivity and potential reusability, which proves the potential application of the adsorbent for nuclear wastewater decontamination and remediation of the aqueous environmental matrices (Cai et al., 2017).

8.3.4.6 *Hydroxylation of GO*

Shao et al. reported "perhydroxy-cucurbit [6] uril decorated GO composite [HO-CB [6]/GO]" for U(VI) adsorption from contaminated water samples. The composite material exhibited strong dispersion stability, fast sedimentation velocity and an excellent adsorption capacity toward U(VI) removal, which suggests the application of the material as a preferred adsorbent for the wastewater treatment (Shao et al., 2017).

8.3.4.7 *Treatment of GO with chelating agent*

A composite material (GO-DTPAA) was synthesized by the covalent bonding of a chelator, i.e., diethylenetriaminepentaacetic phenylenediamine (DTPAA) with supporting matrix of GO and successfully evaluated as a sorbent for uranium (VI) adsorption in the aqueous solution. The adsorption capacity of GO-DTPAA composite material was calculated to be 485.0 mg/g, which was significantly better than the unmodified GO (97.3 mg/g), performed under the same experimental conditions (Liu et al., 2018).

8.3.4.8 *Phosphatidyl-assisted fabrication of GO*

Yang et al. reported the preparation of a novel kind of GO-based adsorbent by chemical modification with cetyltrimethyl ammonium bromide (C_{16}TAB) and phytic acid (PA) for the removal of U(VI). The functional composite material (GO-C_{16}TAB-PA) showed a sorption capacity of 923 mg/g at optimal conditions, which was 527% higher as compared to unmodified GO under the same experimental conditions. Further, the kinetics of the functionalized materials was enhanced and the equilibrium time was significantly shortened by 50% as compared to that of raw GO. The composite adsorbent possessed superior recyclability and was successfully evaluated for the seawater extraction of U(VI) (P. Yang et al., 2018).

8.4 MODELING THE SORPTION OF U(VI) ON GO AND GO COMPOSITE ADSORBENTS

8.4.1 Adsorption isotherms

The data obtained from the adsorption isotherm study has been analyzed according to the Langmuir, Freundlich, and Temkin isotherm models. Most of the literature reported adsorption isotherms of U(VI) on the pure GO and its modifications can be better explained and fitted to the Langmuir isotherm model, suggesting the monolayer adsorption of U species on the homogenous GO surfaces having identical active sites without any lateral interactions.

8.4.2 Adsorption kinetics

From Table 8.2, it can be noted that the kinetic data can be better fitted by the pseudo second order model as compared to the pseudo first-order model, thereby suggesting that U(VI) adsorption on GO and its composite materials is chemisorption rather than physisorption (Huang et al., 2017).

8.5 EFFECT OF SOLUTION COMPONENTS ON URANIUM SORPTION

8.5.1 Solution pH

It may be observed that the sorption of the U ion is highly dependent on the solution pH. In extreme acidic conditions ($pH<2$) a high concentration of H^+ ions competes with U(VI) ions for the binding sites over the material surface, which results in a decrease in the adsorption capacity (Song et al., 2011). However, as the solution pH tends toward the neutral/near alkaline range, the adsorbent surface is dominated by a negative charge due to deprotonation of the oxygenated surface groups. In that solution pH, uranium prevails as the positive forms, namely UO_2^{2+}, UO_2OH^+, and $(UO_2)_2(OH)_2^{2+}$, making the adsorption process favorable through columbic attraction or electrostatic interaction between the anionic adsorbent surface and cationic uranyl ion (Ai et al., 2018). At the high alkaline solution ($pH>10$), the stable anionic uranium carbonate complexes (i.e., $UO_2(CO_3)^0$, $UO_2(CO_3)_2^{2-}$ and $UO_2(CO_3)_3^{4-}$) or hydroxyl complexes (UO_2OH^+, $(UO_2)_2(OH)_2^{2+}$ and $(UO_2)_3(OH)^{5+}$) dominate in the aqueous solutions (Sutton and Burastero, 2004), as a result of which the electrostatic repulsion between the adsorbate surface and adsorbent species occurs and leads to unfavorable adsorption at the alkaline condition. So, the optimal adsorption of uranyl species on most of the GO-based composite materials was reported to be in the solution pH ranging from 4.5 to 6.0.

8.5.2 Presence of co-ions

Besides the presence of H^+, OH^- and CO_3^{2-} ions, other ions also affect the adsorption process (Zhu et al., 2019; Tatarchuk et al., 2019). Literature information on the effect of ionic strength on the adsorption behavior is rather inconsistent. Wang et al. suggest that at a lower ionic strength of NaCl, the sorption process was not affected. However, an opposite effect was observed at higher ionic strength solutions (C. L. Wang et al., 2015). The adsorption capacity of fungus hyphae/GO (FH/GO) was found to be reduced from 121 to 106 mg/g in the presence of metallic cations (Li et al., 2018). Rohith et al. reported that the presence of SO_4^{2-} and NO_3^- at the concentration range of 0.1–0.5M in the solution appeared to significantly impact the adsorption process (Rohith et al., 2020). Sorption studies of U(VI) onto AOMGO composite as a function of pH in 0.01 M $NaClO_4$, $NaNO_3$, NaCl, Na_2SO_4 and Na_2CO_3 solutions was studied by Zhao et al. It was observed that NO_3^- and CL_2^- have no significant effect on U(VI) removal over a wide range of pH. However, the presence of SO_4^{2-} and CO_3^{2-} has an opposite effect for the same (Zhao et al., 2013).

Most of the studies reported that adsorption of U(VI) is independent of ionic strength, attributed to the formation of inner-sphere surface complexes of metal ions (Zong et al., 2013; A. Yang et al., 2018; Kang et al., 2002; K. Zhao et al., 2019). Perchlorate (ClO_4^-) has been widely studied as a competitive ion for U(VI) adsorption in aqueous solution. However, no substantial differences in the sorption capacity for U(IV) species were observed due to the presence of ClO_4^- ion in the solution, in the cases of GO nanosheet (Li et al., 2012), GO-chitosan composite (Huang et al., 2017), and sulfonated GO (Zhang et al., 2016) materials. The efficient sorption selectivity of many composite materials, namely rGO hydrogel (He et al., 2018), bovine serum albumin-coated GO (GO-BSA) (Yang et al., 2017b), GO-COOH/UiO-66 MOF composite (Yang et al., 2017a), and GO/ZIF-67 aerogel (Zhao et al., 2020), was observed to be independent for U(VI) sorption in the presence of metallic cations in the solution. A study conducted by Chen et al. reported a natural polymer Konjac Glucomannan-mediated assembly of GO for selective adsorption of U(VI) in a multi-ion system containing cations (Cs(I)), Sr(II), Mn(II), Cu(II)) and Fe(III)) and anions (SO_4^{2-}, Cl^-, ClO^-, NO_3^-, Br^-, CO_3^{2-}, F^- and I^-). The experimental results are the following affinity sequence: U(VI) > Sr > Cs > Cu > Mn > Fe, which indicates the preferential selectivity of the adsorbent toward U(VI) in the presence of the competitive co-ions (Chen et al., 2018). Selective adsorption of the U(VI) ion in a multicomponent system has also been reported for other GO-based composite materials, namely composite hydrogel (PCG) (Lu et al., 2020), ethyleneimine polymer GO composite (GO/PEI) (Dai et al., 2019b), amidoximated GO nanoribbons (AOGONRs) (Y. Wang et al., 2015), GO/amidoxime hydrogel (F. Wang et al., 2016), sulfonated GO (Sun et al., 2017; Song et al., 2019), ammonia-modified GO (Verma and Dutta, 2015), phosphorylated GO-chitosan composite (GO-CS-P) (Cai et al., 2017), perhydroxy-CB[6]-decorated GO composite (HO-CB[6]/GO) (Shao et al., 2017) and phosphatidyl-assisted fabricated GO nanosheets (GO-C16TAB-PA) (P. Yang et al., 2018).

The presence of humic acids in the solution showed variable impacts on the UO_2^{2+} adsorption as a function of solution pH. At acidic conditions (pH 2.5–7.0), the adsorption efficiency of the materials was reported to be enhanced for U(VI) sorption in the presence of humic acids, whereas at alkaline conditions (pH 7–10) the sorption phenomenon was observed to be independent (Wang et al., 2017). The same trend was also observed by other studies (Hu et al., 2017; Song et al., 2014). The most probable reason may be the formation of ternary complexes on the adsorbate surface (Tatarchuk et al., 2019; Wang et al., 2019) and the enhancement of the dispersion and stabilization of GO in an aqueous solution, attributed to the prevalence of humic acid in the solution (Hummers and Offeman, 1958).

8.5.3 Temperature

Temperature has a significant impact on the adsorption of UO_2^{2+} over the GO surfaces. Some studies reported a favorable adsorption process at elevated temperature, i.e., the amount of the sorbed uranium was observed to be increasing substantially with the rise in the temperature (Li et al., 2012; Zhao et al., 2012; Zhang et al., 2019; Zhao et al., 2015; Tan et al., 2015; Chen et al., 2017; Zhao et al., 2018; Zhao et al., 2017). In that case, the value of the standard Gibbs free energy ($\Delta G°$) was observed to be decreased with an increase in the temperature. The positive enthalpy change may be due to the high degree of dissolution of U(VI) ions in the aqueous matrix and partial removal of their hydration sheaths prior to sorption onto the adsorbent surface. Here, the endothermicity of the desolvation process exceeds the exothermicity of U(VI) ions attachment to GO surfaces (Zhang et al., 2019; Zhao et al., 2015). The positive values of the standard enthalpy ($\Delta H°$) and the standard entropy ($\Delta S°$) suggest the endothermic nature of the sorption process and the adsorption affinity of GO for the uranium ions.

Some studies contradict the trend, and the amount of adsorption was observed to be decreasing with the increase in temperature (D. Zhao et al., 2019; A. Yang et al., 2018). In this case, the ($\Delta H°$) and ($\Delta S°$) have negative values, which suggests the adsorption process as exothermic and decrease in the affinity of GA surface for the adsorption of U(VI).

8.6 MECHANISM OF THE U ADSORPTION ONTO GO-BASED ADSORBENTS

Interaction between the UO_2^{2+} ion and the surface functional groups over the GO surfaces plays an important role and the major driving force responsible for the adsorption process. The mechanism of the adsorption of uranium species onto the GO surfaces involves ion exchange (Tan et al., 2015; Perlova et al., 2020; Chen and Wang, 2016), inner-sphere surface complexation (Li et al., 2012; C. L. Wang et al., 2015; Zhao et al., 2012; Rohith et al., 2020; Zong et al., 2013; Zhao et al., 2013), electrostatic interactions (Chen et al., 2017; Chen and Wang, 2016; Zhao et al., 2020; Guo et al., 2020) and coordination/chelation (Zhao et al., 2017; Zhao et al., 2013). The incorporation of metal ions on the GO surface promotes the U(VI) reduction and simultaneous oxidation of the metal ions, which results in the enhancement in the adsorption process (Rohith et al., 2020; Zhang et al., 2019).

The process of interaction between the UO_2^{2+} ion and the GO surfaces can be understood by the utilization of spectroscopic methods namely Fourier transform infrared spectroscopy (FTIR), energy X-ray absorption fine structure (EXAFS) and X-ray photoelectron spectroscopy (XPS).

FTIR spectroscopy has been used to explore the presence of the adsorbed uranium species on the GO surface by observing the typical vibrational characteristic of the O = U = O group and red shifting of the surface functional groups. For example, the existence of the adsorbed uranium onto graphene aerogel results in the appearance of a new peak at 918 cm^{-1}, which confirms the presence of the O=U=O group due to the asymmetric vibration frequency caused by UO_2^{2+} (D. Zhao et al., 2019). FTIR spectra of Fe-Ni/GO after the adsorption of U(VI) showed a new peak at 917 cm^{-1}, suggesting the presence of antisymmetric vibration of O = U = O (Zhang et al., 2019). After the adsorption of U(VI) on the Fe-PANI-GA material, two new sharp peaks at 1385 cm^{-1} and at 1033 cm^{-1} were observed, which was attributed to the co-sorbed between NO_3^- group and U(VI) and the vibrational band of linear $[O{=}U{=}O]^{2+}$ respectively (Chen et al., 2017). The sorption of U(VI) by GO-CS-P composite results in the appearance of the characteristic uranyl absorption peak at 927 cm^{-1}. Further, the characteristic P-O vibration at 969 cm^1 disappeared and the absorption peak of P-O at 1,029 cm^{-1} shifts to the higher wavenumber region; vibrations of C-O at 1200 cm^1 and -OH at 3314 cm^1 exhibited blue shifts after the uranyl uptake, suggesting the participation of the phosphonate groups and hydroxyl sites for adsorption of U(VI) (Cai et al., 2017). The FTIR spectra of U(VI) saturated GO-CS material result in the appearance of a new strong absorption band at 906 cm^{-1}, which was assigned to the chemically coordinated uranyl ions. It was also observed that the intensities of –COO- and –OH groups at 1,730 and 1,250 cm^{-1} wavelength region significantly reduce and broaden –NH2 group at 645 cm^{-1} toward the lower wavelength region, suggesting the interaction of these functional groups with U(VI) (Huang et al., 2017). The characteristic appearance of O = U = O group and red shifting of the of aqueous UO_2^{2+} complexes has also been observed in the FTIR spectra of uranium sorbed adsorbents namely GO impregnated calcium-alginate (Cal-Alg-GO) beads (Basu et al., 2018), fungal hyphae/GO bioassembly (Li et al., 2018), bovine serum albumin-coated GO (GO-BSA) composite (Yang et al., 2017b), UiO-66-GO (Yang et al., 2017a), Fe_3O_4@HKUST-1/GO composite (Wang et al., 2020b), chitosan-GO/ZIF foam (Guo et al., 2020), polyamidoxime/polydopamine-decorated GO (GO/PDA/PAO) (Dai et al., 2019a), polyamidoxime/polyethyleneimine magnetic GO (mGO-PP) (Dai et al., 2019b), GO/amidoxime hydrogel (F. Wang et al., 2016), GO and poly(3,4-ethylenedioxythiophene):poly(styrenesulfonate) (PEDOT:PSS) (Song et al., 2019) and phosphatidyl-assisted fabricated GO nanosheets (GO-C16TAB-PA) (P. Yang et al., 2018).

EXAFS spectroscopy is a unique fingerprinting technique which provides information on the local atomic structure of the adsorbate. The mechanism of interaction of the oxygenated groups with U(VI) may be characterized by the formation of typical peaks due to backscattering of O atoms of the (UO_2^{2+}) ion and constriction of the coordination number UO_2^{2+} to 5 in the equatorial plane. Li et al. interpreted the mechanism of adsorption of the UO_2^{2+} over the GO nanosheets based on EXAFS study at different solution pH (3.9, 6.1, and 7.6). From the EXAFS spectra of the U(VI) ladenned GO sheets, two prominent peaks at ~1.3 and ~1.9 Å were observed due to the backscattering from two axial O atoms and equatorial O atoms of UO_2^{2+}, respectively. When the solution pH increased toward the alkaline range, these peaks shifted slightly to higher R values, which suggests the gradual transformation of the U(VI) structure adsorbed on the GO surfaces. The coordination number of UO_2^{2+} in the equatorial plane was calculated to be about 5 at the pH 3.9 spectrum, which also suggests the formation of the inner-sphere complexes of the oxygenated groups with UO_2^{2+} and favorable conditions for high adsorption capacity. At the pH 7.6 spectrum, a new peak appeared at ~3.7, which was assigned to the background scattering of the distal oxygen of a carbonate group, bidentatedly coordinated to the UO_2^{2+} ion. The pH 6.1 spectrum was observed to be the transition state of pH 3.9 and 7.6 spectra, suggesting a superposition of surface complexation of U(VI) with oxygen functional groups and precipitation onto GO (Li et al., 2012). The EXAFS analysis of uranium adsorbed phosphorylated chitosan-

GO composite suggests the coordination of reduced uranium, i.e., U(IV) with phosphate group (Cai et al., 2017). Huang et al. studied the adsorption behavior of GO-chitosan (GO-CS) composite for U(VI) uptake and observed a predominant FT feature at R + ΔR ~1.3 Å, which can be well fitted by two axial oxygen atoms at 1.77–1.79 Å. The total coordination number of U-O/Neq was constrained to 5, suggesting the presence of the UO_2^{2+} ions on the GO-CS surface (Huang et al., 2017). EXAFS spectra of sulfonated GO materials after U(VI) adsorption was observed to be similar to that of UO_2^{2+} species at R < 2 Å. However, a slight modification was observed at R > 2 Å. The FT features at 1.4 and 1.9 Å was assigned to two axial oxygen (O_{ax}) at ~1.82 Å and ~5 equatorial oxygen (Oeq) at 2.545 ± 0.03 Å, respectively. A significant splitting of the $U\text{-}O_{eq}$ shell of sulfonated GO after the U(VI) sorption ($R_{U\text{-}Oeq}$ = 2.577 Å) was observed as compared to UO_2^{2+} ($R_{U\text{-}Oeq}$ = 2.514 Å), suggesting the role of $-OSO_3H$ groups for U(VI) adsorption at acidic condition (pH=2.0). However, at near-neutral condition (pH = 6.0) the FT peak at ~2.3 Å corresponded to U-C shell at 2.9 Å, highlighting the role of –COOH/-OH groups for uranyl adsorption at pH 2.0 (Sun et al., 2017).

XPS is helpful for the understanding of the nature of bonding between the surface functional groups on the adsorbent surface with uranium species. The recorded XPS spectrum of the uranium loaded GO materials could be characterized by the formation of a new free uranyl peak, appearance of the doublets of U 4f peak, and a shifting/lowering in the binding energies of the surface oxygen, nitrogen or phosphorous atom. Wang et al. characterized the GO before and after sorption, in which they observed a significant shift in the binding energies of O1s from 532.79 to 532.63 eV before and after sorption respectively. They interpreted the observation of a shift in the binding energy level as a result of the decrease in the electron density around the O atom, resulting from the electron transfer to U(VI) (C. L. Wang et al., 2015). The XPS spectra recorded after the adsorption of U(VI) on the GA surface showed a new peak at a binding energy of 383 eV, which was regarded as U 4f. The binding energy of O 1s at 532.0 eV was observed to be increased after the U(VI) adsorption, suggesting the complexation of U(VI) with surface oxygenated groups. It was also observed that in the N1s spectrum, intensities of the $-CH_2\text{-}NH_2$ groups were significantly decreased and the position of two peaks shifted to the higher binding energy, suggesting the increase in electron density for the nitrogen atoms due to possible interaction of U(VI) with nitrogen surface groups. The double peak characteristics of U 4f were observed at 392.3 and 381.4 eV, which were assigned to U $4f_{5/2}$ and U $4f_{7/2}$, respectively (D. Zhao et al., 2019). The XPS analysis of U (VI) saturated graphene aerogel material showed the typical strong bimodal peaks U $4f_{5/2}$ and U $4f_{7/2}$ at 393.1 and 382.3 eV, respectively, due to antisymmetric vibration of UO_2^{2+}. The binding energy value of O 1s and P 2p were also observed to be decreased after the complexation with uranium, suggesting the chemical bonding of the adsorbent with O and P functional groups (Wang et al., 2020a). Sorption of U(VI) on Fe-PANI-GA material results in the characteristic peak of U 4f at 382.1 eV. The C 1s and N 1s spectra were observed to be a shift to the lower binding energy after sorption, suggesting the coordination of U(VI) with carbon and amine functional groups. The presence of U(VI) may also be proved from the U 4f spectrum concentrated on 381.4 and 391.7 eV energy levels (Chen et al., 2017). After adsorption of U(VI) onto the magnetic Fe_3O_4/graphene composites, the double peak of U 4f was observed at 381.7 and 392.6 eV, which were corresponded to U $4f_{5/2}$ and U $4f_{7/2}$ respectively. Further, the binding energies of O 1s were slightly shifted to lower binding energies after sorption, which suggests the surface coordination of U(VI) with oxygenated functional groups (Zhao et al., 2018). The appearance of U(VI) characteristic peaks at 381.7 eV suggests the successful adsorption of U(VI) onto EDTA-mGO (Zhao et al., 2017). The evidence for simultaneous reduction of U(VI) and oxidation of the incorporated metal ions onto the GO surface, which facilitate the adsorption process, has been established by XPS spectra (Zhang et al., 2019; Chen et al., 2017). Cai et al. observed the shifting of binding energies of P-O and -OH bonds to lower values and decrease in the relative peak areas of these groups after adsorption of U(VI) on phosphorylated chitosan-GO composite material. The study also revealed that the coordination of U(VI) takes place more selectively with the phosphonate groups as compared to the carboxyl sites (Cai et al., 2017). The XPS spectra of the U(VI) saturated GO-chitosan (GO-CS) composite material showed the characteristic U $4f_{5/2}$ and U $4f_{7/2}$ peaks at 392.3 and 381.5 eV energy levels, respectively. As expected the intensities of O1s peaks were significantly decreased at solution pH of 5.0 and unchanged at solution pH 8.3, suggesting the involvement of $-COO^-$ groups at acidic pH. However, inhibition of the N 1s intensities was observed consistently at all the solution pH, possibly due to effective coordination of $-NH_2$ groups with U(VI) (Huang et al., 2017). The formation of a new free uranyl peak, appearance of doublets of U 4f peaks and lowering in the binding energies of surface oxygen, nitrogen or phosphorous atom were observed in the XPS spectra of many other uranium saturated adsorbents namely fungal hyphae/GO bioassembly (Li et al., 2018), fungal hyphae/Fe_3O_4/GO composite (Zhu et al., 2019), bovine serum albumin-coated GO (GO-BSA) composite (Yang et al., 2017b), Konjac glucomannan/GO (KGM/GO) sponges (Chen et al., 2018), UiO-66-GO (Yang et al., 2017a), Fe_3O_4@HKUST-1/GO composite (Wang et al., 2020b), chitosan-GO/ZIF foam (Guo et al., 2020), polyamidoxime/polydopamine-decorated GO (GO/PDA/PAO) (Dai et al., 2019a), polyamidoxime/polyethyleneimine magnetic GO (mGO-PP) (Dai et al., 2019b), amidoximated GO nanoribbons (AOGONRs) (Y. Wang et al., 2015), GO/amidoxime hydrogel (F. Wang et al., 2016), sulfonated GO (Sun et al., 2017), GO and poly(3,4-ethylenedioxythiophene):poly(styrenesulfonate) (PEDOT:PSS) composite (Song et al., 2019) and phosphatidyl-assisted fabricated GO nanosheets (GO-C16TAB-PA) (P. Yang et al., 2018).

Adsorption of uranyl species over the GO composite materials can also be interpreted from other techniques namely EDS mapping (Zhao et al., 2017; Chen et al., 2020; Basu et al., 2018), XANES (Cai et al., 2017), radial structural functions (Cai et al., 2017), SEM-EDX (Chen and Wang, 2016; Verma and Dutta, 2015) and TEM-EDX (Song et al., 2019).

8.7 REGENERATION AND REUSABILITY OF THE SATURATED MATERIAL

The regeneration and reusability of the saturated adsorbent are the important indices for the calculation of cost benefit analysis and applicability of the adsorbent in the real field. The uranium ladenned GO-based adsorbents can be satisfactorily regenerated by using aqueous solution of reagents namely HCl (Li et al., 2012; He et al., 2018; Zhang et al., 2019; Tan et al., 2015; Zhao et al., 2018; Zhao et al., 2017; Chen and Wang, 2016; Li et al., 2018; Yang et al., 2017b; Chen et al., 2018; Dai et al., 2019a; Dai et al.,

2019b; Y. Wang et al., 2015), HNO_3 (Perlova et al., 2020; A. Yang et al., 2018; Cai et al., 2017; Chen and Wang, 2016; Yang et al., 2017a; Zhao et al., 2020; Verma and Dutta, 2015; P. Yang et al., 2018), KNO_3 (Zong et al., 2013), $NaNO_3$ (Zhao et al., 2015), Na_2CO_3 (Song et al., 2019), $NaHCO_3$ (Lu et al., 2020), NaOH (Chen et al., 2017), $C_2H_7NO_2$ (Cai et al., 2017) and EDTA (Guo et al., 2020).

It has been well reported that the GO-based materials have good mechanical stability and showed excellent reusability for U(VI) adsorption. For example, the amount of adsorbed U(VI) on rGO hydrogel was maintained at 94.76% at the tenth regeneration cycle, as compared to the first cycle (He et al., 2018). The recyclability test of Fe-Ni/GO composite showed only a 7% decrease in the adsorption capacity after the fifth cycle (Zhang et al., 2019). The reclamation and utilization study of Fe_3O_4/GO for the decontamination of U(VI) suggests the effectiveness of the adsorbent for at least six times (Zong et al., 2013). Adsorption efficiency of MnO_2–Fe_3O_4–rGO composite was insignificantly decreased (from 92.58% to 85.46%) after the fourth regeneration cycle (Tan et al., 2015). The result obtained from the adsorption study of U(VI) onto the EDTA-mGO showed a decrease in the sorption efficiency from 91% to 81% after the fifth cycle (Zhao et al., 2017). The GO-BSA adsorbent retained an elution efficiency of 79% after cycling the material five times and its structure was unaltered as compared to the pristine form (Yang et al., 2017b). Several other GO-based adsorbents namely phosphorylated GO-chitosan composite (Cai et al., 2017), GO-immobilized *Saccharomyces cerevisiae* gel beads (Chen and Wang, 2016), GO nanoribbons/chitosan (GONRs/CTS) composite (Hu et al., 2020), fungus hyphae/GO (FH/GO) (Li et al., 2018), magnetic metal organic frameworks (MOFs)/GO (Fe_3O_4@HKUST-1/GO) (Wang et al., 2020b), polyamidoxime/polydopamine-decorated GO (GO/PDA/PAO) (Dai et al., 2019a), Amidoxime-functionalized GO nanoribbons(AOGONRs) (Y. Wang et al., 2015), GO/PEDOT:PSS nanocomposite (Song et al., 2019), ammonia-modified GO (Verma and Dutta, 2015) and cetyltrimethyl ammonium bromide (C_{16}TAB) and phytic acid (PA) modified GO (GO-C_{16}TAB-PA) (P. Yang et al., 2018) also showed efficient adsorption efficiency after multiple sorption/desorption cycles and proved to be potential sorbents for uranium removal for the real field applications.

8.8 CONCLUSION AND RECOMMENDATIONS

The chapter presents a review of the recent application of GO based functionalized adsorbents for the removal of toxic uranium species from various aqueous matrices. A large number of GO-based adsorbents have been explored for uranium adsorption and recent research has been focused on the synthesis of adsorbents having high adsorption capacity and selectively toward uranium for the decontamination of the aqueous matrices. In general, it may be observed that the unmodified/raw GO has less efficiency for uranium removal and the maximum adsorption capacity was reported only up to 56 mg/g. Hence, various strategies have been adopted in the form of physical, chemical and biological modifications of the GO-based sorbents in order to enhance the adsorption capacity of the material and make it feasible for practical applications. Physical modifications in the form of synthesis of single layer GO nanosheets have proven to be drastic enhancement of the adsorption capacity, which has been reported up to 299.00 mg/g. Similarly, the chemical modifications of the GO sheets namely incorporation of metal/metal oxides, multilayer surface functionalization with specific groups, rational designing of structural frameworks (MOF and ZIF), synthesis of aerosols/hydrosols, etc. have been reported to achieve better adsorption capacity. Although several researchers have evaluated metal/metal oxide-incorporated GO-based materials for uranium sorption, their performance is not encouraging for field application owing to insignificant increase in the adsorption capacity after modification. The maximum sorption capacity of these groups has been reported up to 384.62 mg/g, which has been achieved by incorporation of metallic Fe and Ni onto the GO sheets. Surface functionalization of GO materials namely amidoximation, sulfonation, carboxylation, amidation, phosphorylation, hydroxylation, incorporation of surfactants, etc. has shown improved adsorption capacity up to 12 folds as compared to the unmodified one and recorded up to 923.00 mg/g. The structural framework materials namely MOF and ZIF are proven to be candidates of preferential choice for field application considering their exceptional structural stability and sorption capacity (reported up to 1,888.55 mg/g) for treatment of real uranium contaminated water samples. However, the best adsorption capacity among all the GO-based materials was achieved by GO-based hydrosols and aerosols, which has been reported up to 3,550.00 mg/g. The conjugation of GO with materials of biological origins, namely chitin, fungal hyphae, alginate, etc., has shown improved adsorption capacity (up to 779.44 mg/g) and also facilitates easy recovery of the material after the application. So far the adsorption capacity of the GO-based materials are concerned, the modifications may be arranged in the sequence as hydrogels/aerosols > MOF/ZIF > multilayered surface functionalization > conjugation with biological materials > metal/metal oxide incorporation > single layer GO sheet.

GO-based adsorbents possess exceptional thermal, mechanical and chemical stability, which is an important aspect for recovery of the material from the bulk solution after application. The composite material based on GO and materials of biological origin may be explored for simultaneous adsorption and bioremediation of toxic uranium species, which is a challenging aspect for future research.

Although the GO-based adsorbents are proven to possess high adsorption capacity for uranium removal from aqueous matrices, some challenges and research gaps still need to be addressed for further studies, considering the application of these materials for practical feasibility.

Most of the studies have been reported in limited experimental conditions namely low initial uranium concentration, in the presence of fewer interfering ions, at narrow working temperature and solution pH. However, real environmental systems are more complicated and the adsorption process is significantly influenced by multiple factors, which need to be evaluated for application of the graphene-based adsorbents at field conditions. Hence, systematic studies should be conducted for the adsorption studies under relevant environmental conditions for the treatment of real contaminated water samples.

The recent studies have proven that GO-based materials have exceptional adsorption capacity and selectivity for the treatment of uranium contaminated groundwater matrices, where lower concentrations of interfering ions are expected. However, the selectivity of some of the GO-based composites is still not satisfactory in the presence of competitive ions, which is questionable for application of the adsorbents for the treatment of wastewater samples, where elevated concentration of numerous inferring ions is expected to be encountered. On the other hand, some functional groups possess exceptional selectivity for uranium in a multicomponent system. In this context, the synthesis of adsorbents with modification with specific surface functional groups should be emphasized to enhance the selectivity of the material for uranium. Hence, the future research direction should be encouraged for development of graphene-based materials having high selectivity toward uranium for the treatment of real wastewater matrices.

It may be noted that the optimal solution pH for adsorption of uranium on to the GO-based adsorbent is in the narrow range of 4.5–6.0 and the adsorption efficiency observed drastically decreased at alkaline pH range, which poses the practical challenge of extraction of uranium from sea water and treatment of contaminated water samples having elevated pH. Hence, future research should be focused toward the optimization of the protocol for the synthesis of GO-based material for its optimal performance at a wide range of solution pH, covering the alkaline pH range.

Long-term use of the adsorbents leads to another practical problem of biofilm growth on the material surface, which significantly reduces the adsorption efficiency and regeneration ability of the adsorbent. Therefore, it's a great challenge to develop materials with novel anti-biofouling or self-cleaning properties.

The adsorption data has been well described in various adsorption isotherms/kinetics plots and spectroscopic studies to understand the mechanism of uranium adsorption onto GO surfaces. However, there is a need for the depiction of the detailed mechanism of adsorption with the combined study of the experimental data set and theoretical in silico simulation studies, for example, the density function theory.

The current chapter summarizes the GO-based materials for uranium adsorption from aqueous matrices. Various strategies namely incorporation of metal/metal oxide ions, bioassembly GO, rational design of structural frameworks, multi-level surface functionalization of GO-based materials for the enhancement of the adsorption efficiency of these materials have been presented. The effect of the influencing parameters such as solution pH, presence of competitive ions in the solution and temperature has been discussed. The mechanism of the adsorption of the uranium ions onto the GO-based materials has been interpreted by fitting the experiential data by isotherm/kinetics models and observations from spectroscopic studies namely FTIR, EXAFS and XPS. A short note has been presented for regeneration/reusability of the saturated adsorbent, which is an important aspect for consideration of the adsorbents for real field application. Finally, the chapter has been concluded with some broad perspectives with regard to some advantages, key challenges and future directions of the research for the application of the GO-based materials for uranium decontamination from various aqueous matrices.

ACKNOWLEDGMENTS

The author acknowledges the Divisional Head of Instrumentation Laboratory and the Member Secretary of Central Pollution Control Board, Delhi, for the necessary support and granting permission to publish this chapter.

REFERENCES

Agudosi, E.S., Abdullah, E.C., Numan, A., Mubarak, N.M., Khalid, M., Omar, N., 2020. A review of the graphene synthesis routes and its applications in electrochemical energy storage. *Crit. Rev. Solid State Mater. Sci.* 45, 339–377. https://doi.org/10.1080/10408436.2019.1632793

Ai, Y.J., Liu, Y., Lan, W.Y., Jin, J.R., Xing, J.L., Zou, Y.D., Zhao, C.F., Wang, X.K., 2018. The effect of pH on the U(VI) sorption on graphene oxide (GO): A theoretical study. *Chem. Eng. J.* 343, 460–466. https://doi.org/10.1016/j.cej.2018.03.027

Alam, M.S., Cheng, T., 2014. Uranium release from sediment to groundwater: Influence of water chemistry and insights into release mechanisms. *J. Contam. Hydrol.* 164, 72–87. https://doi.org/10.1016/j.jconhyd.2014.06.001

Ali, I., 2012. New generation adsorbents for water treatment. *Chem. Rev.* 112, 5073–5091. https://doi.org/10.1021/cr300133d

Aly, M.M., Hamza, M.F., 2013. A review: Studies on uranium removal using different techniques. Overview. *J. Dispers. Sci. Technol.* 34, 182–213. https://doi.org/10.1080/01932691.2012.657954

Barnett, M.O., Jardine, P.M., Brooks, S.C., Selim, H.M., 2000. Adsorption and transport of uranium (VI) in subsurface Media. *Soil Sci. Soc. Am. J.* 64, 908–917. https://doi.org/10.2136/sssaj2000.643908x

Basu, H., Singhal, R.K., Pimple, M.V., Saha, S., 2018. Graphene oxide encapsulated in alginate beads for enhanced sorption of uranium from different aquatic environments. *J. Environ. Chem. Eng.* 6, 1625–1633. https://doi.org/10.1016/j.jece.2018.01.065

Boulanger, N., Kuzenkova, A.S., Iakunkov, A., Romanchuk, A.Y., Trigub, A.L., Egorov, A. V., Bauters, S., Amidani, L., Retegan, M., Kvashnina, K.O., Kalmykov, S.N., Talyzin, A. V., 2020. Enhanced sorption of radionuclides by defect-rich graphene oxide. *ACS Appl. Mater. Interfaces* 12, 45122–45135. https://doi.org/10.1021/acsami.0c11122

Briffa, J., Sinagra, E., Blundell, R., 2020. Heavy metal pollution in the environment and their toxicological effects on humans. *Heliyon* 6, e04691. https://doi.org/10.1016/j.heliyon.2020.e04691

Cai, Y., Wu, C., Liu, Z., Zhang, L., Chen, L., Wang, J., Wang, X., Yang, S., Wang, S., 2017. Fabrication of a phosphorylated graphene oxide-chitosan composite for highly effective and selective capture of U(VI), *Environmental Science: Nano* 4, 1876–1886. https://doi.org/10.1039/c7en00412e

Central Ground Water Board (CGWB), 2020. Report on Uranium occurrence in shallow aquifers in India, Ministry of Jal Shakti, Government of India. http://cgwb.gov.in/WQ/URANIUM_REPORT_2019-20.pdf

Chen, C., Hu, J., Wang, J., 2020. Uranium biosorption by immobilized active yeast cells entrapped in calcium-alginate-PVA-GO-crosslinked gel beads. *Radiochim. Acta* 108, 273–286. https://doi.org/10.1515/ract–2019–3150

Chen, C., Wang, J., 2016. Uranium removal by novel graphene oxide-immobilized Saccharomyces cerevisiae gel beads. *J. Environ. Radioact.* 162–163, 134–145. https://doi.org/10.1016/j.jenvrad.2016.05.012

Chen, L., Feng, S., Zhao, D., Chen, S., Li, F., Chen, C., 2017. Efficient sorption and reduction of U(VI) on zero-valent iron-polyaniline-graphene aerogel ternary composite. *J. Colloid Interface Sci.* 490, 197–206. https://doi.org/10.1016/j.jcis.2016.11.050

Chen, T., Shi, P., Zhang, J., Li, Y., Duan, T., Dai, L., Wang, L., Yu, X., Zhu, W., 2018. Natural polymer konjac glucomannan mediated assembly of graphene oxide as versatile sponges for water pollution control. *Carbohydr. Polym.* 202, 425–433. https://doi.org/10.1016/j.carbpol.2018.08.133

Crini, G., Lichtfouse, E., Wilson, L.D., Morin-Crini, N., 2019. Conventional and non-conventional adsorbents for wastewater treatment. *Environ. Chem. Lett.* 17, 195–213. https://doi.org/10.1007/s10311-018-0786–8

Dai, Z., Sun, Y., Zhang, H., Ding, D., Li, L., 2019a. Rational synthesis of polyamidoxime/polydopamine-decorated graphene oxide composites for e ffi cient uranium (VI) removal from mine radioactive wastewater. https://doi.org/10.1021/acs.iecr.9b03934

Dai, Z., Sun, Y., Zhang, H., Ding, D., Li, L., 2019b. Highly efficient removal of uranium (VI) from wastewater by polyamidoxime/polyethyleneimine magnetic graphene oxide. *J. Chem. Eng. Data* 64, 5797–5805. https://doi.org/10.1021/acs.jced.9b00759

Dushenkov, S., Vasudev, D., Kapulnik, Y., Gleba, D., Fleisher, D., Ting, K.C., Ensley, B., 1997. Removal of uranium from water using terrestrial plants. *Environ. Sci. Technol.* 31, 3468–3474. https://doi.org/10.1021/es9702201

Guo, X., Yang, H., Liu, Q., Liu, J., Chen, R., 2020. A chitosan-graphene oxide/ZIF foam with anti-biofouling ability for uranium recovery from seawater. *Chem. Eng. J.* 382, 122850. https://doi.org/10.1016/j.cej.2019.122850

Hakonson-Hayes, A.C., Fresquez, P.R., Whicker, F.W., 2002. Assessing potential risks from exposure to natural uranium in well water. *J. Environ. Radioact.* 59, 29–40. https://doi.org/10.1016/S0265-931X(01)00034–0

Hanson, S., Wilson, D., Gunaji, N., Hathaway, S., 1987. Removal of uranium from drinking water by ion exchange and chemical clarification. U.S. EPA Report 600/52-87/076.

He, Y.R., Li, S.C., Li, X.L., Yang, Y., Tang, A.M., Du, L., Tan, Z.Y., Zhang, D., Chen, H.B., 2018. Graphene (rGO) hydrogel: A promising material for facile removal of uranium from aqueous solution. *Chem. Eng. J.* 338, 333–340. https://doi.org/10.1016/j.cej.2018.01.037

Hu, B., Hu, Q., Li, X., Pan, H., Huang, C., Chen, C., 2017. Graphene oxide as a novel adsorbent for highly efficient removal of uo22+ from water. *Desalin. Water Treat.* 79, 178–187. https://doi.org/10.5004/dwt.2017.20796

Hu, X., Wang, Y., Yang, J.O., Li, Y., Wu, P., Zhang, H., Yuan, D., Liu, Y., Wu, Z., Liu, Z., 2020. Synthesis of graphene oxide nanoribbons/chitosan composite membranes for the removal of uranium from aqueous solutions. *Front. Chem. Sci. Eng.* 14, 1029–1038. https://doi.org/10.1007/s11705-019-1898–9

Huang, Z., Li, Z., Zheng, L., Zhou, L., Chai, Z., Wang, X., Shi, W., 2017. Interaction mechanism of uranium(VI) with three-dimensional graphene oxide-chitosan composite: Insights from batch experiments, IR, XPS, and EXAFS spectroscopy. *Chem. Eng. J.* 328, 1066–1074. https://doi.org/10.1016/j.cej.2017.07.067

Hummers, W.S., Offeman, R.E., 1958. Preparation of graphitic oxide. *J. Am. Chem. Soc.* 80, 1339. https://doi.org/10.1021/ja01539a017

Kalin, M., Wheeler, W.N., Meinrath, G., 2005. The removal of uranium from mining waste water using algal/microbial biomass. *J. Environ. Radioact.* 78, 151–177. https://doi.org/10.1016/j.jenvrad.2004.05.002

Kang, M.J., Han, B.E., Hahn, P.S., 2002. Precipitation and adsorption of uranium (VI) under various aqueous conditions. *Environ. Eng. Res.* 7, 149–157

Katsoyiannis, I.A., Zouboulis, A.I., 2013. Removal of uranium from contaminated drinking water: A mini review of available treatment methods. *Desalin. Water Treat.* 51, 2915–2925. https://doi.org/10.1080/19443994.2012.748300

Kausar, A., Bhatti, H.N., 2013. Adsorptive removal of uranium from wastewater: A review. *J. Chem. Soc. Pakistan* 35, 1041–1052.

Kornilovych, B., Wireman, M., Ubaldini, S., Guglietta, D., Koshik, Y., Caruso, B., Kovalchuk, I., 2018. Uranium removal from groundwater by permeable reactive barrier with zero-valent iron and organic carbon mixtures: Laboratory and field studies. *Metals* 8, 1–15. https://doi.org/10.3390/met8060408

Kyzas, G.Z., Kostoglou, M., 2014. Green adsorbents for wastewaters: A critical review. *Materials* 7, 333–364. https://doi.org/10.3390/ma7010333

Lee, S.Y., Bondietti, E.A., 1983. Removing uranium from drinking water by metal hydroxides and anion-exchange resin. *J. Am. Water Work. Assoc.* 75, 536–540. https://doi.org/10.1002/j.1551-8833.1983.tb05216.x

Li, J., Zhang, Y., 2012. Remediation technology for the uranium contaminated environment: A review. *Procedia Environ. Sci.* 13, 1609–1615. https://doi.org/10.1016/j.proenv.2012.01.153

Li, Y., Zou, G., Yang, S., Shi, P., Chen, T., Lian, Y., Duan, T., Zheng, K., Dai, L., Zhu, W., 2018. Bioassembly of fungal hyphae/graphene oxide composite as high performance adsorbents for U(VI) removal. *Appl. Surf. Sci.* 458, 226–235. https://doi.org/10.1016/j.apsusc.2018.07.081

Li, Z., Chen, F., Yuan, L., Liu, Y., Zhao, Y., Chai, Z., Shi, W., 2012. Uranium(VI) adsorption on graphene oxide nanosheets from aqueous solutions. *Chem. Eng. J.* 210, 539–546. https://doi.org/10.1016/j.cej.2012.09.030

Limson Zamora, M., Tracy, B.L., Zielinski, J.M., Meyerhof, D.P., Moss, M.A., 1998. Chronic ingestion of uranium in drinking water: A study of kidney bioeffects in humans. *Toxicol. Sci.* 43, 68–77. https://doi.org/10.1006/toxs.1998.2426

Liu, S., Ouyang, J., Luo, J., Sun, L., Huang, G., Ma, J., 2018. Removal of uranium(VI) from aqueous solution using graphene oxide functionalized with diethylenetriaminepentaacetic phenylenediamine. *J. Nucl. Sci. Technol.* 55, 781–791. https://doi.org/10.1080/00223131.2018.1439415

Lu, W., Dai, Z., Li, L., Liu, J., Wang, S., Yang, H., Cao, C., Liu, L., Chen, T., Zhu, B., Sun, L., Chen, L., Li, H., Zhang, P., 2020. Preparation of composite hydrogel (PCG) and its adsorption performance for uranium (VI). *J. Mol. Liq.* 303, 112604. https://doi.org/10.1016/j.molliq.2020.112604

Ma, F., Gui, Y., Liu, P., Xue, Y., Song, W., 2020. Functional fibrous materials-based adsorbents for uranium adsorption and environmental remediation. *Chem. Eng. J.* 390, 124597. https://doi.org/10.1016/j.cej.2020.124597

Ma, F., Nian, J., Bi, C., Yang, M., Zhang, C., Liu, L., Dong, H., Zhu, M., Dong, B., 2019. Preparation of carboxylated graphene oxide for enhanced adsorption of U(VI). *J. Solid State Chem.* 277, 9–16. https://doi.org/10.1016/j.jssc.2019.05.042

Malekzadeh, F., Farazmand, A., Ghafourian, H., Shahamat, M., Levin, M., Colwell, R.R., 2002. Uranium accumulation by a bacterium isolated from electroplating effluent. *World J. Microbiol. Biotechnol.* 18, 295–302. https://doi.org/10.1023/A:1015215718810

Perlova, O. V., Dzyazko, Y.S., Palchik, A. V., Ivanova, I.S., Perlova, N.O., Danilov, M.O., Rusetskii, I.A., Kolbasov, G.Y., Dzyazko, A.G., 2020. Composites based on zirconium dioxide and zirconium hydrophosphate containing graphene-like additions for removal of U(VI) compounds from water. *Appl. Nanosci.* 10, 4591–4602. https://doi.org/10.1007/s13204-020-01313–1

Pratas, J., Rodrigues, N., Paulo, C., 2006. Uranium accumulator plants from the centre of Portugal - Their potential to phytoremediation. In: Merkel B.J., Hasche-Berger A. (eds) *Uranium in the Environment.* Springer, Berlin, Heidelberg. https://doi.org/10.1007/3-540-28367-6_47

Qin, H., Jin, Y., Gong, T., Cho, Y., Ahn, C., Shin, C., Lee, C., Kim, T., 2016. A conductive copolymer based on graphene oxide and poly (amidoxime-pyrrole) for adsorption of uranium (VI). *Nano* 11, 1–8. https://doi.org/10.1142/S1793292016500454

Renu, M.A., Singh, K., 2017. Heavy metal removal from wastewater using various adsorbents: A review. *J. Water Reuse Desalin.* 7, 387–419. https://doi.org/10.2166/wrd.2016.104

Rohith, S., Ramanan, K.K., Srinivas, N.S., Jegadeesan, G.B., 2020. Fe-Ni-doped graphene oxide for uranium removal—Kinetics and equilibrium studies. *Water. Air. Soil Pollut.* 231, 1–17. https://doi.org/10.1007/s11270-020-04804–7

Rosenberg, E., Pinson, G., Tsosie, R., Tutu, H., Cukrowska, E., 2016. Uranium remediation by ion exchange and sorption methods: A critical review. *Johnson Matthey Technol. Rev.* 60, 59–77. https://doi.org/10.1595/205651316X690178

Seidel, B., Tuchscherer, T., Hennebach, E., Partisch, M., 2005. Uranium in drinking water. *Gas- und Wasserfach. Wasser, Abwasser* 146, 418–421.

Shao, L., Zhong, J., Ren, Y., Tang, H., Wang, X., 2017. Perhydroxy-CB[6] decorated graphene oxide composite for uranium(VI) removal. *J. Radioanal. Nucl. Chem.* 311, 627–635. https://doi.org/10.1007/s10967-016-5067–z

Shen, J., Schäfer, A., 2014. Removal of fluoride and uranium by nanofiltration and reverse osmosis: A review. *Chemosphere* 117, 679–691. https://doi.org/10.1016/j.chemosphere.2014.09.090

Simon, F.-G., Biermann, V., Csovári, M., Csicsák, J., 2003. Uranium removal using elemental iron and hydroxyapatite in permeable reactive barriers. in: *Proceedings of the 8th International Congress on Mine Water & the Environment*, 19–22 October 2003, Johanesburg, pp. 43–55.

Singh, N.B., Nagpal, G., Agrawal, S., Rachna, 2018. Water purification by using adsorbents: A review. *Environ. Technol. Innov.* 11, 187–240. https://doi.org/10.1016/j.eti.2018.05.006

Singh, R.K., Kumar, R., Singh, D.P., 2016. Graphene oxide: Strategies for synthesis, reduction and frontier applications. *RSC Adv.* 6, 64993–65011. https://doi.org/10.1039/c6ra07626b

Song, J., Kong, H., Jang, J., 2011. Adsorption of heavy metal ions from aqueous solution by polyrhodanine-encapsulated magnetic nanoparticles. *J. Colloid Interface Sci.* 359, 505–511. https://doi.org/10.1016/j.jcis.2011.04.034

Song, S., Wang, K., Zhang, Y., Wang, Y., Zhang, C., Wang, Xin, Zhang, R., Chen, J., Wen, T., Wang, X., 2019. Self-assembly of graphene oxide/PEDOT:PSS nanocomposite as a novel adsorbent for uranium immobilization from wastewater. *Environ. Pollut.* 250, 196–205. https://doi.org/10.1016/j.envpol.2019.04.020

Song, W., Shao, D., Lu, S., Wang, X., 2014. Simultaneous removal of uranium and humic acid by cyclodextrin modified graphene oxide nanosheets. *Sci. China Chem.* 57, 1291–1299. https://doi.org/10.1007/s11426-014-5119–6

Sun, Y., Wang, X., Ai, Y., Yu, Z., Huang, W., Chen, C., Hayat, T., Alsaedi, A., Wang, Xiangke, 2017. Interaction of sulfonated graphene oxide with U(VI) studied by spectroscopic analysis and theoretical calculations. *Chem. Eng. J.* 310, 292–299. https://doi.org/10.1016/j.cej.2016.10.122

Sutton, M., Burastero, S.R., 2004. Uranium(VI) solubility and speciation in simulated elemental human biological fluids. *Chem. Res. Toxicol.* 17, 1468–1480. https://doi.org/10.1021/tx049878k

Tan, L., Wang, J., Liu, Q., Sun, Y., Jing, X., Liu, L., Liu, J., Song, D., 2015. The synthesis of a manganese dioxide-iron oxide-graphene magnetic nanocomposite for enhanced uranium(VI) removal. *New J. Chem.* 39, 868–876. https://doi.org/10.1039/c4nj01256a

Tang, N., Liang, J., Niu, C., Wang, H., Luo, Y., Xing, W., Ye, S., Liang, C., Guo, H., Guo, J., Zhang, Y., Zeng, G., 2020. Amidoxime-based materials for uranium recovery and removal. *J. Mater. Chem. A* 8, 7588–7625. https://doi.org/10.1039/c9ta14082d

Tatarchuk, T., Shyichuk, A., Mironyuk, I., Naushad, M., 2019. A review on removal of uranium(VI) ions using titanium dioxide based sorbents. *J. Mol. Liq.* 293, 111563. https://doi.org/10.1016/j.molliq.2019.111563

Tchounwou, P.B., Yedjou, C.G., Patlolla, A.K., Sutton, D.J., 2012. Molecular, clinical and environmental toxicicology Volume 3: Environmental Toxicology. *Mol. Clin. Environ. Toxicol.* 101, 133–164. https://doi.org/10.1007/978–3–7643–8340–4

Tirmarche, M., Baysson, H., Telle-Lamberton, M., 2004. Exposition à l'uranium et risque de cancer : une revue des études épidémiologiques. *Rev. Epidemiol. Sante Publique* 52, 81–90. https://doi.org/10.1016/s0398-7620(04)99024–4

Tolkou, A.K., Katsoyiannis, I.A., Zouboulis, A.I., 2020. Removal of arsenic, chromium and uranium from water sources by novel nanostructured materials including graphene-based modified adsorbents: A mini review of recent developments. *Appl. Sci.* 10, 1–18. https://doi.org/10.3390/app10093241

Trivedi, S., Lobo, K., Ramakrishna Matte, H.S.S., 2019. Synthesis, properties, and applications of graphene. In: Hywel, M., Rout, C.S, Dattatray, J. L. (eds) *Fundamentals and sensing applications of 2D materials.* Elsevier Ltd, Amsterdam. https://doi.org/10.1016/B978-0-08-102577-2.00003-8

United States Environmental Protection Agency, 2009. National Primary Drinking Water Guidelines. Epa 816-F-09-004 1, 7.

Verma, S., Dutta, R.K., 2015. A facile method of synthesizing ammonia modified graphene oxide for efficient removal of uranyl ions from aqueous medium. *RSC Adv.* 5, 77192–77203. https://doi.org/10.1039/c5ra10555b

Wang, C.L., Li, Y., Liu, C.L., 2015. Sorption of uranium from aqueous solutions with graphene oxide. *J. Radioanal. Nucl. Chem.* 304, 1017–1025. https://doi.org/10.1007/s10967-014-3855–x

Wang, F., Li, H., Liu, Q., Li, Z., Li, R., Zhang, H., Liu, L., Emelchenko, G.A., Wang, J., 2016. A graphene oxide/amidoxime hydrogel for enhanced uranium capture. *Sci. Rep.* 6, 1–8. https://doi.org/10.1038/srep19367

Wang, J., He, B., Wei, X., Li, P., Liang, J., Qiang, S., Fan, Q., Wu, W., 2019. Sorption of uranyl ions on TiO_2: Effects of pH, contact time, ionic strength, temperature and HA. *J. Environ. Sci.* 75, 115–123. https://doi.org/10.1016/j.jes.2018.03.010

Wang, X., Yu, S., Jin, J., Wang, H., Alharbi, N.S., Alsaedi, A., Hayat, T., Wang, X., 2016. Application of graphene oxides and graphene oxide-based nanomaterials in radionuclide removal from aqueous solutions. *Sci. Bull.* 61, 1583–1593. https://doi.org/10.1007/s11434-016-1168–x

Wang, Y., Liu, X., Huang, Y., Hayat, T., Alsaedi, A., Li, J., 2017. Interaction mechanisms of U(VI) and graphene oxide from the perspective of particle size distribution. *J. Radioanal. Nucl. Chem.* 311, 209–217. https://doi.org/10.1007/s10967-016-4924–0

Wang, Y., Wang, Z., Ang, R., Yang, J., Liu, N., Liao, J., Yang, Y., Tang, J., 2015. Synthesis of amidoximated graphene oxide nanoribbons from unzipping of multiwalled carbon nanotubes for selective separation of uranium(VI). *RSC Adv.* 5, 89309–89318. https://doi.org/10.1039/c5ra15977f

Wang, Z., Hu, H., Huang, L., Lin, F., Liu, S., Wu, T., Alharbi, N.S., Rabah, S.O., Lu, Y., Wang, X., 2020a. Graphene aerogel capsulated precipitants for high efficiency and rapid elimination of uranium from water. *Chem. Eng. J.* 396, 125272. https://doi.org/10.1016/j.cej.2020.125272

Wang, Z., Zhao, D., Wu, C., Chen, S., Wang, Y., Chen, C., 2020b. Magnetic metal organic frameworks/graphene oxide adsorbent for the removal of U(VI) from aqueous solution. *Appl. Radiat. Isot.* 162, 109160.

Wei, Y., Zhang, Y., Gao, X., Ma, Z., Wang, X., Gao, C., 2018. Multilayered graphene oxide membrane for water treatment: A review. *Carbon* 139, 964–981. https://doi.org/10.1016/j.carbon.2018.07.040

World Health Organization, 2008. *Guidelines for drinking-water quality*. Third edition, Geneva https://www.who.int/water_sanitation_health/publications/gdwq3/en/

Yang, A., Zhu, Y., Huang, C.P., 2018. Facile preparation and adsorption performance of graphene oxide-manganese oxide composite for uranium. *Sci. Rep.* 8, 1–10. https://doi.org/10.1038/s41598-018-27111–y

Yang, A., Zhu, Y., Li, P., Huang, C.P., 2019. Preparation of a magnetic reduced-graphene oxide/tea waste composite for high-efficiency sorption of uranium. *Sci. Rep.* 9, 1–12. https://doi.org/10.1038/s41598-019-42697–7

Yang, K., Wang, J., Chen, X., Zhao, Q., Ghaffar, A., Chen, B., 2018. Application of graphene-based materials in water purification: From the nanoscale to specific devices. *Environ. Sci. Nano* 5, 1264–1297. https://doi.org/10.1039/c8en00194d

Yang, P., Liu, Q., Liu, J., 2017a. Interfacial growth of a metal–organic framework (UiO-66) on functionalized graphene oxide (GO) as a suitable seawater adsorbent for extraction of uranium (VI). *J. Mater. Chem. A* 5, 17933–17942. https://doi.org/10.1039/C6TA10022H

Yang, P., Liu, Q., Liu, J., Zhang, H., Li, Z., Li, R., Liu, L., Wang, J., 2017b. Bovine serum albumin-coated graphene oxide for effective adsorption of uranium(VI) from aqueous solutions. *Ind. Eng. Chem. Res.* 56, 3588–3598. https://doi.org/10.1021/acs.iecr.6b04532

Yang, P., Liu, Q., Zhang, H., Liu, J., Chen, R., Li, R., Wu, D., Bai, X., Wang, J., 2018. Phosphatidyl-assisted fabrication of graphene oxide nanosheets with multiple active sites for uranium(VI) capture. *Environ. Sci. Nano* 5, 1584–1594. https://doi.org/10.1039/c8en00401c

Zhang, Q., Zhao, D., Ding, Y., Chen, Y., Li, F., Alsaedi, A., Hayat, T., Chen, C., 2019. Synthesis of Fe–Ni/graphene oxide composite and its highly efficient removal of uranium(VI) from aqueous solution. *J. Clean. Prod.* 230, 1305–1315. https://doi.org/10.1016/j.jclepro.2019.05.193

Zhang, Z.B., Qiu, Y.F., Dai, Y., Wang, P.F., Gao, B., Dong, Z.M., Cao, X.H., Liu, Y.H., Le, Z.G., 2016. Synthesis and application of sulfonated graphene oxide for the adsorption of uranium(VI) from aqueous solutions. *J. Radioanal. Nucl. Chem.* 310, 547–557. https://doi.org/10.1007/s10967-016-4813–6

Zhao, D., Chen, L., Sun, M., Li, F., 2015. Preparation and application of magnetic graphene oxide composite for the highly efficient immobilization of U(VI) from aqueous solutions. *J. Radioanal. Nucl. Chem.* 306, 221–229. https://doi.org/10.1007/s10967-015-4064–y

Zhao, D., Wang, Y., Zhao, S., Wakeel, M., Wang, Z., Shaikh, R.S., Hayat, T., Chen, C., 2019. A simple method for preparing ultra-light graphene aerogel for rapid removal of U(VI) from aqueous solution. *Environ. Pollut.* 251, 547–554. https://doi.org/10.1016/j.envpol.2019.05.011

Zhao, D., Zhang, Q., Xuan, H., Chen, Y., Zhang, K., Feng, S., Alsaedi, A., Hayat, T., Chen, C., 2017. EDTA functionalized Fe_3O_4/graphene oxide for efficient removal of U(VI) from aqueous solutions. *J. Colloid Interface Sci.* 506, 300–307. https://doi.org/10.1016/j.jcis.2017.07.057

Zhao, D., Zhu, H., Wu, C., Feng, S., Alsaedi, A., Hayat, T., Chen, C., 2018. Facile synthesis of magnetic Fe_3O_4/graphene composites for enhanced U(VI) sorption. *Appl. Surf. Sci.* 444, 691–698. https://doi.org/10.1016/j.apsusc.2018.03.121

Zhao, G., Wen, T., Yang, X., Yang, S., Liao, J., Hu, J., Shao, D., Wang, X., 2012. Preconcentration of U(VI) ions on few-layered graphene oxide nanosheets from aqueous solutions. *Dalt. Trans.* 41, 6182–6188. https://doi.org/10.1039/c2dt00054g

Zhao, K., Chen, C., Cheng, T., Shang, J., 2019. Graphene oxide-facilitated uranium transport and release in saturated medium: Effect of ionic strength and medium structure. *Environ. Pollut.* 247, 668–677. https://doi.org/10.1016/j.envpol.2019.01.037

Zhao, M., Tesfay Reda, A., Zhang, D., 2020. Reduced graphene oxide/ZIF-67 aerogel composite material for uranium adsorption in aqueous solutions. *ACS Omega* 5, 8012–8022. https://doi.org/10.1021/acsomega.0c00089

Zhao, Y., Li, J., Zhang, S., Chen, H., Shao, D., 2013. Efficient enrichment of uranium (VI) on amidoximated magnetite/graphene oxide composites. *RSC Adv.* 3, 18952–18959. https://doi.org/10.1039/c3ra42236d

Zhu, W., Lei, J., Li, Y., Dai, L., Chen, T., Bai, X., zhou, J., Wang, L., Duan, T., 2019. Procedural growth of fungal hyphae/Fe_3O_4/graphene oxide as ordered-structure composites for water purification. *Chem. Eng. J.* 355, 777–783. https://doi.org/10.1016/j.cej.2018.08.215

Zhu, Y., James, D.K., Tour, J.M., 2012. New routes to graphene, graphene oxide and their related applications. *Adv. Mater.* 24, 4924–4955. https://doi.org/10.1002/adma.201202321

Zong, P., Wang, S., Zhao, Y., Wang, H., Pan, H., He, C., 2013. Synthesis and application of magnetic graphene/iron oxides composite for the removal of U(VI) from aqueous solutions. *Chem. Eng. J.* 220, 45–52. https://doi.org/10.1016/j.cej.2013.01.038

Chapter 9

Fluoride contamination and abatement measures

A geoenvironmental perspective

Shiva Shankar Y
Ujjain Engineering College

Devendra Mohan
Indian Institute of Technology, Banaras Hindu University

CONTENTS

DOI: 10.1201/9781003247869-11

9.1 INTRODUCTION

Groundwater is being extensively used for satisfying the irrigational, agricultural and domestic needs in the developing world. Overexploitation of the groundwater combined with intrusion of geogenic pollutants has been a matter of concern necessitating immediate attention. Reasons for pollutant intrusion could be either natural or manmade, but the anthropogenic factors such as excessive utilization, improper handling of industrial effluents and mining activities, etc. have been aggravating the problem. These issues are expected to worsen the future scenarios under the influence of global warming–induced climate change. Researchers have predicted that climate change could predominantly influence the valuable components of groundwater recharge, i.e., hydrologic cycle through alterations in precipitation, runoff and evapotranspiration. These changes will have direct effect on food security and human wellbeing in future (Chakraborti et al., 2011; Kumar et al., 2005; Panwar and Chakrapani, 2013).

9.1.1 Fluorides in natural environment

Fluoride is commonly associated element of the earth's crust, due to its association with different types of rocks in varying concentration ranging between 100 and 2,000 mg/kg. The influence of fluoride in ecosystem and food chain has been gaining importance as numerous researchers have been highlighting the impact of fluoride attributed toxicity on plants and living organisms. Figure 9.1 depicts the overview of the fluoride contamination showing the sources, pathways, drivers and its health effects. In atmosphere, fluorine gas is generated from various natural and manmade sources as particulate matter and gases (shown in Figure 9.1). These released fluorinated compounds into atmosphere and fluorides existing in the earth's crust find their way into food chain through various drivers such as wet and dry deposition for gaseous and particulate pollutants, weathering, precipitation, ion exchange and dissolution for fluoride-bearing minerals, and these deposited pollutants in soil reach the nearby water courses through leaching or surface runoff affecting the ecosystem (Babu et al., 2015; Gupta and Ayoob, 2016; Jagtap et al., 2012).

9.1.2 Mechanisms influencing fluoride attenuation into groundwater

Fluoride ingestion into plants, animals and human beings through groundwater has been a major source. Fluoride attenuation into groundwater depends on the type of rock mineral, soil chemistry, soil characteristics, aquifer characteristics (geological nature of surrounding rocks, physical and chemical characteristics), meteorological condition, water table depth and other external chemical factors. Rock water contact plays key role in the dissolution and decomposition of fluorides into groundwater. The fluoride content in the groundwater could be directly correlated to their concentration in parent rock. Regions containing crystalline igneous and metamorphic rocks formed due to volcanic and hydrothermal activity contribute to excessive fluorides into groundwater. Characteristics of the aquifer were found to be the next prominent

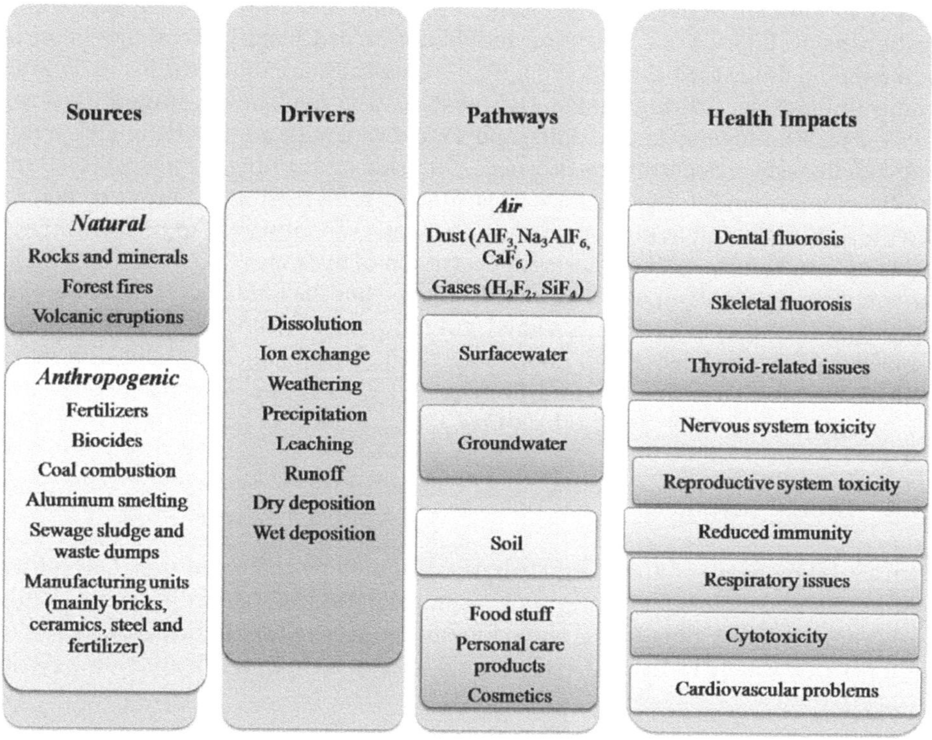

Figure 9.1 Overview of sources, pathways and drivers of fluoride contamination and its health effects.

factor. Elevated levels of fluoride were observed in deeper aquifers compared with shallow aquifers due to higher residence time for rock water interaction. Soil (predominantly clays/silty soils) remains another key factor as fluorides released from weathering of rocks are retained on clays by sorption at low pH. Fluorides are generally stable in abundance of calcium (as CaF_2) in contact with alkaline water (pH > 7.5), and fluoride ions are exchanged with anions resulting in precipitation of calcite. Higher total dissolved solids content favors more such interactions enhancing the fluoride solubility into groundwater either directly from parent rock or from soil. This phenomenon was observed to be predominant in arid climates that experience reduced groundwater recharge, higher evaporation and prolonged rock water interaction (Sivasankar, 2016; Babu et al., 2015).

9.1.3 Effects of fluoride on plants, animals and human beings

Permissible limits for fluoride intake were prescribed by WHO (for water), FAO (water for irrigation) and food and nutritional board of USA (total intake per day). For drinking water at lower concentrations (i.e., 1–1.5 mg/L), fluorides are beneficial to teeth, but at concentrations above 1.5 mg/L, it causes wide range of health affects as shown in Figure 9.1. These effects were commonly associated with animals. Fluorosis (dental and skeletal) has been widely prevalent globally, but the increasing fluoride toxicity in food chain has been resulting in the other health-related impacts. For irrigation water, fluoride concentration should not exceed 1 mg/L. Total daily fluoride intake should not exceed the following limits, quantities are indicated for various age groups in brackets: 0.01 mg/kg/day (0–6 months), 0.05 mg/kg/day (>6 months), 1 mg/day (4–8 years), 4 mg/day (male adults) and 3 mg/day (female adults). Estimates predict that with fluoride concentrations of 1 mg/L, fluoride intake ranges from 0.2 to 1.3 mg/day for children and 3 mg/day for adults. But with increasing pathways of fluoride into the food chain, all living organisms and plants were observed to be insensitive to higher concentrations. Atmospheric concentration of hydrogen fluoride greater than 1 μg/m^3 has shown effect on phenology of the plants, but the influence of the compound entering the plants through air, soil and groundwater varied with species. Majority of plants and insects have shown intolerance to the higher concentrations influencing their biological stages (Johnson and Bretzler, 2015; Food and Agriculture Organization, 2021; Ozsvath, 2009; Zuo et al., 2018).

9.1.4 Global scenario

Figure 9.2 shows map of fluoride distribution adopted from Rasool et al. (2017) highlighting the global fluoride concentrations in groundwater. Chowdhury et al. (2019) have reported that China has largest population affected by fluorosis with 300 (in millions) followed by India (64.64), Tanzania (49.25), Kenya (44.35) and South Africa

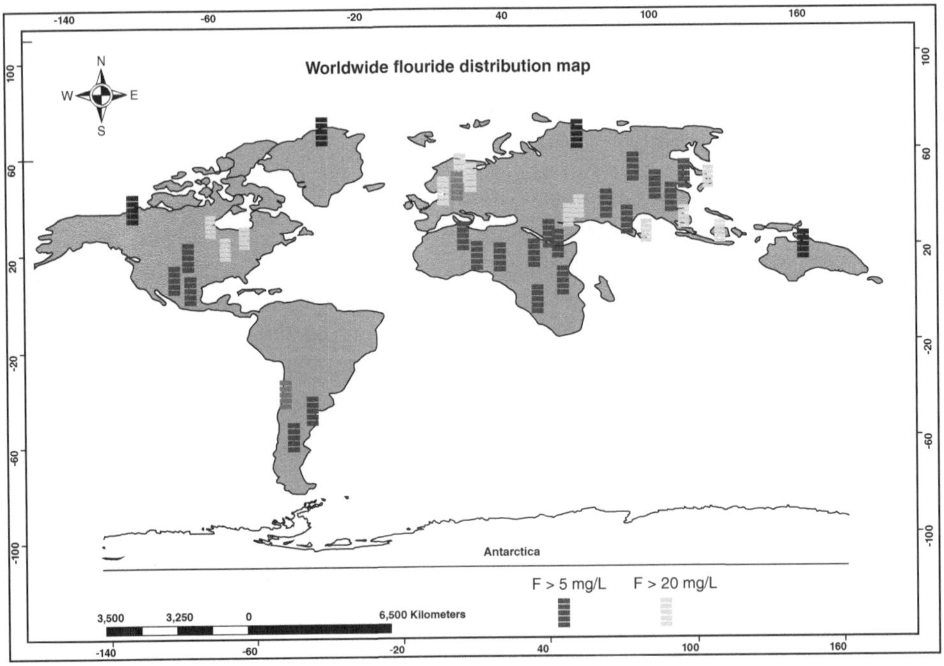

Figure 9.2 Global fluoride contamination in groundwater. (FromRasool et al. 2017.)

(41.9). The highest level of recorded fluoride concentration (i.e., 2800 mg/L) was observed in lake Nakuru, Kenya. Africa and Asia have been worst affected nations in terms of fluorosis impacted population. In Asian countries fluoride-bearing formations in recharge areas are the main source for higher fluoride concentration. Karst aquifers and springs passing through agricultural lands using fluoride-bearing fertilizers, coal mining and volcanic rocks are the minor sources. In European nations such as Poland, Greece and Spain, natural sources combined with phosphate industry and alumina production plants were found to be the major sources. Volcanic activities, geologic formations combined with the meteorological conditions were observed as key factors for higher fluoride concentration in African conditions. Natural factors combined with industrial activity were found to be the major reason in many parts of America (Lacson et al., 2021; Srivastava and Flora, 2020; Vithanage and Bhattacharya, 2015).

9.2 DEFLUORIDATION TECHNIQUES

Table 9.1 highlights the commonly adopted defluoridation techniques with their relative merits, demerits and way ahead for effective application (Ayoob et al., 2008; Jagtap et al., 2012; Vithanage and Bhattacharya, 2015).

9.2.1 Relevance of the work

Based on the aforementioned aspects, the present work is of high relevance on account of the following reasons:

- Many poor countries of Asia and Africa have been worst effected with fluoride contamination. Additional ecological issues such as co-occurrence of other geogenic toxic contaminants and increased dependence on groundwater resources have been exerting for urgent need to employ low-cost locally available materials for defluoridation of groundwater.
- A wide range of defluoridation techniques have been applied in many parts of Asia and Africa in collaboration with global agencies, but these were observed to be much less effective in longer run due to a number of technical and financial constraints.
- This review work discusses a broad range of global literature classified under different sections on the basis of geomaterials type studied over a period of time for effective application.
- The work also sheds light on approaches for regeneration or remediation of exhausted adsorbents for their reuse or safe disposal due to feasibility of such geomaterials as building materials.

9.3 GEOMATERIALS

Natural resources like plants, natural polymers, locally available materials, etc. offer a viable solution for low-cost treatment. This concept has led to exploration of natural materials in treatment using activated carbon/charcoal processed from plants, soils,

Table 9.1 Frequently employed defluoridation techniques: comparison at a glance

Method	*Materials/approaches*	*Merits*	*Demerits*	*Way ahead for application*
Coagulation and precipitation	Alum, lime, magnesium oxide, calcium and phosphate compounds	Wider acceptance, ease in application, cost effective	Higher chemical dosages, secondary issues (excessive aluminum concentration), cumbersome sludge handling	Nalgonda technique has been globally adopted (using alum with lime) and the method has proven to be effective at lower fluoride concentrations. Contact precipitation with Ca and phosphate compounds minimizes limitations of the process (as compared with alum). Pretreatment with contact precipitation followed by other efficient process can optimize the cost
Electro- chemical method	Electro-coagulation	Higher efficiency, lesser secondary issues such as sludge handling and other water quality issues in comparison to conventional coagulation	Excessive power consumption, need for skilled supervision, higher operational costs	Exploration of the non-conventional energy resources has been done for optimizing the process. CSIR-NEERI, India, has successfully developed a solar-based electrolytic defluoridation system as a preferable option (Council of Scientific & Industrial Research-National Environmental Engineering Research Institute, 2017)
Electricity- based methods	Electrodialysis, capacitive deionization, Donnan analysis	Simultaneous defluoridation and desalination, economical in comparison to the membrane-based treatment processes	Membrane scaling, water wastage, higher costs due to power requirement, extent of pretreatment	Capacitive deionization effectively feasible for moderate concentrations has been gaining importance due to benefits such as energy efficiency, cost effectiveness, lower pressure and pumping requirements (Porada et al., 2013)

(Continued)

Table 9.1 (Continued) Frequently employed defluoridation techniques: comparison at a glance

Method	*Materials/approaches*	*Merits*	*Demerits*	*Way ahead for application*
Membrane-based processes	Nanofiltration, ultrafiltration, reverse osmosis, etc.	Highly efficient and reliable processes for defluoridation as well as for simultaneous removal of other ions	Membrane fouling, higher costs, pretreatment and post adjustment, water wastages	These processes can offer the best assurance and the maximum removal under the given conditions. Cost-optimization by combination with other methods enhances the advantages of these processesfurther
Adsorption	Wide range of materials employed as adsorbents	Local availability, economical, ease of operation, effective for much wider application	Relatively lower efficiency, various issues with different adsorbents used for field application, large variations in efficiency with water quality and operational conditions, difficulties in regeneration, handling issues with exhausted adsorbents and the liquid effluent obtained post-regeneration	Numerous researchers have studied wide range of adsorbents classified as natural materials, carbon based, chitosan based, polymers and resins, metal-based and metal organic framework, layered double hydroxides, industrial byproducts, biomaterials, etc. Adsorbents like bone charcoal and activated alumina have been identified as feasible for application. Other adsorbents have also proven effective, but commercial applications with suitable tieups between stakeholders needto be explored further (He et al. 2020).

building materials (bricks and construction wastes), clays and its associated minerals. Clays (or certain soils) and certain minerals have been globally studied for relevant applications in water treatment. The presence of organic matter and certain metal oxides has been observed to be the reason for fluoride uptake. Studies have indicated enhanced potential by modification of its composition through mechanical or chemical means (Ayoob et al., 2008; Jagtap et al., 2012; Tomar and Kumar, 2013).

9.3.1 Soils and clays

Clayey soils and minerals were observed to uptake negatively charged ions at neutral pH due to their positively charged surface. Geomaterials having minerals such as alumino-silicates, oxides of Al and Fe, apatites, calcareous soil exhibited significant potential for fluoride ion uptake. A wide range of processing techniques are available such as calcination of clays at higher temperatures/thermal treatment (Bjorvatn and Bardsen, 1995), chemical modification using salts of magnesium, manganese, hydrogen peroxide, etc. (Ayse and Muge, 2014; Dilip et al., 2010; Gitari et al., 2015; Shengyu et al., 2016), mechano-chemical treatment (using ball mill for grinding the chemicals followed by calcination and chemical treatment) (Guo and Eric, 2012) and coated sands using bone char, goethite and hematite (Mlilo et al., 2010). The results obtained by researchers were observed to be in conformance at lower concentration, higher adsorbent dosage and contact time. However, studies have established the sorption capacity of fluorides variably with time under the influence of change in the characteristics of water, fluoride concentration, dosage of adsorbent and properties of clay. In addition to the above, further investigations are needed in relation to regeneration/disposal of exhausted adsorbent combined with optimum conditions for operation (Vinati et al., 2015).

Chidambaram et al. (2003) studied the efficiency of column constructed with materials such as charcoal, red soil, brick, flyash and serpentine. Materials were added in proportion to their fluoride sorption capacity, the order of their efficiency was found to be as follows red soil>brick> serpentine> flyash> charcoal. The study has established the feasibility of adopting the local soils for potential defluoridation. Chidambaram et al. (2013) have examined the defluoridation capacity of local red soil with sand in ratio 4:1 through column study. The results prove the efficiency of process for rural areas due to effectiveness at neutral pH and performance after regeneration.

Yanxin and Eric (2001) have studied the defluoridation property of weathered soil from Xinzhou, China. Characterization studies point out that the soil is composed of minerals like quartz, feldspar, illite and goethite, with an iron oxide content of 6.75%. The results indicate fluoride uptake in 1.5 h due to the iron oxide coatings on soil particles and lesser influence of other cations. Heat treatment at 400–500°C for 2 h modified the surface characteristics of adsorbent by increasing granularity and permeability. The study has proved that the potential of local soil though the soil sorption capacity (150 μg/g) was lesser (i.e., one fourth) in comparison with commercial activated alumina.

Mekonnen et al. (2014) used locally available materials such as sand, calcined clay, scoria, pumice and bone char for the removal of fluoride and phosphates from drinking water through designed filter. The filters were prepared separately with different materials of varying grain sizes like stone and gravel (0.6–4.75 mm), sand (0.15–0.35 mm) and, calcined clay, pumice, scoria and bone char (0.25–0.5 inch). Filters were tested by water samples containing fluoride concentrations of 6 and 8 mg/L and phosphate concentration of 4 mg/L. The results revealed that bone char and pumice have achieved the highest defluoridation capacities with 89.65% and 82.4% removal, respectively, as compared to calcined clay and sand. Sand was found to be effective for phosphate removal (70%) followed by calcined clay and scoria.

Hamdi and Srasra (2009) studied the defluoridation efficiency of clayey soils collected from different regions in Tunisia, i.e. Sebkhat El Malah-El Mkhacherma (MK), Jebel Haidoudi (H), and Zemlet El Beidha (ZB). The potential of these clays was evaluated for treating industrial waste water with pH 2.75–3.0 and fluoride concentration of 2360 mg/L and NaF stock solutions (50–10,000 mg/L). The results proved that MK soil possessed higher fluoride sorption followed by H and ZB. The results for sodium fluoride solution and acidic wastewater were observed as 71.94, 66.62 and 55.80 mg F/g soil and 93.45, 84.03 and 69.44 mg F/g soil, respectively, for MK, H and ZB soils. The study indicated that the adsorption of fluorides was mainly influenced by structure of clay, carbonates (%) in the sample and pH.

9.3.1.1 Potter clay

Moges et al. (1996) have investigated the fluoride removal using clay pots, bricks and red ash; processed to size of 0.7–1.4 mm. Materials such as processed bricks and red ash have not shown any fluoride uptake, and hence further studies were undertaken using clay pots only. Results show sorption capacity of 200 mg fluoride/kg of adsorbent for treating 1–2 L of water, using 120–240 g of the adsorbent, fluoride content reduced from 5–20 to 1.5 mg/L. Column studies conducted have also shown similar performance. The study suggests though the performance of adsorbent was inferior in comparison to activated alumina, it could effectively serve as an economical option in treating the water supplies for Ethiopia.

Beraki et al. (2014) have examined the potential of crushed burned clay pot as sorbent as household defluoridation system in Eritrea. The study has identified that fluoride levels varied between 1.40 and 3.98 mg/L in majority of areas of the region. Preliminary experimental studies were conducted varying the column depth (15, 20 and 25 cm), flow rate (2.5, 5, 10 and 15 mL/min), and fluoride concentration of 5 mg/L. Based on the results obtained, the parameters for efficient operation was identified as 25 cm bed depth at a flow rate of 2.5 mL/min. Pilot scale unit developed with these parameters suited household requirements (consuming 315 L with 5–7 persons) of defluoridation for 15 days.

9.3.1.2 Bricks

Rafique et al. (2018) have investigated the fluoride sorption characteristics of red brick paving blocks or red pavers in batch mode. The study has shown a maximum removal of 62% at a pH range of 6–7. Yadav et al. (2006) have conducted studies with brick powder as an adsorbent for defluoridation of groundwater in Hussainpur and Thondwal, Haryana, India. The results reveal that under optimum conditions for maximum removal, fluoride removal of synthetic samples with fluoride concentration of 5 mg/L was observed as 54.4% and 80.4% for brick powder and activated charcoal, respectively, in 120 min. The results suggest that brick powder could be advantageous due to the wider availability, efficiency at pH 6–8 and lesser influence of other anions, whereas the commercial activated charcoal has shown higher efficiency but effective only at acidic pH.

Padmasiri (1997) has investigated the performance of defluoridators prepared with locally available brick pieces and brick and cement for treating water containing higher fluoride concentrations. Designed household setup was operated in an upflow mode and was kept in contact with fluoride-rich water for 12 h before the operation. The results obtained have shown removal efficiencies of 85% and 25% at the beginning and end of the life, the after operation periods for filter was observed between 90 and 250 days; adsorbent needs to be replaced after this operation period.

9.3.2 Bentonite

Yuxin et al. (2011) have examined the defluoridation capacity of granular acid-treated bentonite. The results obtained in batch mode have shown maximum fluoride removal in duration of 40 min at a pH of 4.95. Fluoride sorption was higher in column studies as compared to batch mode. The results of column studies suggested that better efficiency could be achieved by reducing flow rate and raising the initial fluoride concentration. The study has shown good sorption capacity, after the regeneration using alkali/alum treatment.

Dilip et al. (2010) have studied the performance of magnesium oxide modified bentonite clay for fluoride removal. The results proved the efficiency of modified bentonite at wider pH (3–10) and with a maximum adsorption capacity of 2.26 mg/g with an initial fluoride concentration of 5.52 mg/L. Bicarbonate ions influenced the process by occupying available active sites. Regeneration with sodium hydroxide has reduced efficiency from 95.47% to 73%.

Gitari et al. (2015) have conducted studies using raw bentonite modified by Fe^{3+} aqueous solution for fluoride removal from the groundwater in Sub-Saharan Africa. The results proved that modified bentonite (100%) was found to be efficient in comparison to raw bentonite (95%) at fluoride concentration of 10 mg/L. Fe^{3+}-modified bentonite exhibited performance at wider pH (2–10).

9.3.3 Montmorillonite

Tor (2006) has pointed out the potential of montmorillonite with sorption capacity of 0.263 mg/g and maximum removal at pH of 6, with the system reaching equilibrium after 180 minutes. Adsorption of fluoride was observed to be due to interaction between metal oxides and fluoride ions at surface of the mineral, and desorption (or regeneration) could be done by washing the adsorbent with solution having a pH of 12.

Meetu et al. (2002) proved the enhancement in fluoride sorption (up to 99%) using acidified montmorillonite as compared with montmorillonite and kaolinite in the pH range of 5–7. The study outlines the release of Al^{3+}, Fe^{3+} and silica from acidified montmorillonite lattice as main reason for efficiency that minimized the formation of hydroxy silicates.

Karthikeyan et al. (2005) have examined the performance of montmorillonite for understanding the influence of particle size (75, 150 and 300 μm) and temperature (30°C, 40°C, 50°C and 60°C) at neutral pH. The results revealed that removal efficiency increased with temperature, reaching optimum level at 50th min with particle size of 75 μm.

9.3.4 Kaolinite

Gogoi and Ruby (2008) have studied the fluoride removal of kaolinite obtained from local traditional potter of Majuli river island, Assam. The study has attempted using raw kaolinite and activated kaolinite clay with conc. H_2SO_4. The results suggest the feasibility of using both the materials for defluoridation, but raw kaolinite has shown lower performance as compared to activated adsorbent.

9.3.5 Limestone

Gogoi et al. (2015) have examined the efficiency of phosphoric acid enhanced limestone. Phosphoric acid was observed to enhance the fluoride removal capacity of limestone from 0.39 to 1.10 mg/g. Fluoride removal varied with changes in phosphoric acid concentration and pH. The predominant mechanism for removal was observed to be the sorption of fluoride by calcium phosphates produced in the reactor combined with precipitation of CaF_2 and sorption by limestone. The study has suggested that regeneration could be done easily by scrubbing and rinsing, with lime or NaOH.

Jain and Jayaram (2009) have examined capacity of lime stone and aluminum hydroxide-impregnated limestone. The results proved that at pH 8 aluminum hydroxide-impregnated lime stone has shown much higher sorption (84.03 mg/g) as compared to limestone (43.1 mg/g).

9.3.6 Bauxite

Sajidu et al. (2008) have examined the efficiency of locally available bauxite of Southern Malawi. The results indicate that defluoridation capacity of the raw bauxite as 93.8% at a dose of 2.5 g/200 mL of 8 mg/L F^- solution. High defluoridation capacity was achieved in the study due to the presence of gibbsite ($Al(OH)_3$) and kaolinite ($Al_2Si_2O_5(OH)_4$), precipitation of AlF_3 and exchange of hydroxide in the gibbsite layers of the kaolinite.

Samson et al. (2012) have also worked on defluoridation using local raw bauxite for treatment of groundwater containing 6.17 mg/L of F^- in Maching a district of Southern Malawi. Raw bauxite was used for treatment in combination with sand and charcoal for clarification in order to optimize the adsorbent dosage and contact time. The results obtained show that fluoride removal was efficient at optimum bauxite dosage of 0.150 kg/L within a contact time of 15 min, and optimum combined dosage of sand and charcoal was observed to be 0.72 kg/L.

Thole (2011) has examined the efficiency of composite filter prepared with local materials such as calcined bauxite, gypsum and magnesite. Materials were processed to sizes of 1.2–1.4mm diameter followed by heat treatment at 200°C for 2h. Composite filter was developed after the initial studies with material in batch mode and then the arrangement in filter was done in the ratio of their loading capacities. Mean loading capacities for bauxite, gypsum and magnesite were 5.6, 3.4 and 1.7mg F^-/g, respectively. Loading capacities of materials reduced at higher dosages due to the increase in sorption capacity. The results have shown variation in pH with different materials, bauxite and gypsum have reduced the pH, while magnesite increased water pH. Higher breakthrough times (120, 210, 255 and 360min) were achieved for higher doses (45, 75, 120 and 150g). Water quality parameters such as pH, apparent color, hardness and residual concentrations of SO_4^{2-}, Cl^-, Fe^{2+} and Al^{3+} were observed to be within permissible limit.

Thole et al. (2013) have studied the performance of composite filter prepared with bauxite, gypsum and magnesite, for fluoride removal in Tanzania. Materials considered in the study were locally procured within Tanzania, characterization studies reveal the presence of the following composition for individual materials; bauxite: Al_2O_3, SiO_2 and Fe_2O_3; gypsum: CaO, SO_3 and SiO_2; and magnesite: MgO and SiO_2. The materials were calcined at 350°C, 400°C, 450°C and 500°C, and the activated materials were then mixed in mass ratios of 1:2:3, 1:3:2, 2:1:3, 2:3:1, 3:1:2 and 3:2:1 (bauxite: gypsum: magnesite). Batch studies were conducted with fluoride concentration of 12.62 mg/L. The results suggest a maximum defluoridation capacity of 11.89mg F^-/g with the 3:2:1 processed at 500°C. The composite filter was observed to restrict Cl^-, Al^{3+}, Ca^{2+}, Mg^{2+}, Fe^{2+}within permissible limits as per WHO standards, but less effective for treating the sulfates and iron.

9.3.7 Laterites

Sarkar et al. (2006) indicated the potential of laterite soil collected from Bankura District, West Bengal, India, for defluoridation at temperature range 293–313 K for 10 mg/dm^3 using 500 μm particle size for achieving maximum removal in the range of 83.3%–74.4%. Maiti et al. (2011) have studied defluoridation potential of the acid-base treated raw laterite collected from Kharagpur. Adsorption capacities of treated laterite were 36.3, 37.9 and 39.1 mg/g at 288, 305 and 315 K, respectively. The removal of fluoride was found to be optimal in the pH range of 3–5 within equilibrium time of 3h.

9.3.8 Hematite

Teutli-Sequeira et al. (2012) have examined the performance of hematite modified with aluminum oxide. The results have shown that modified hematite was more effective compared to unmodified adsorbent. Maximum sorption of modified hematite was observed as 116.75 mg/g at an optimum pH range of 2.34–6.26.

Patel et al. (2014) have studied the defluoridation using natural hematite iron ore and synthetic magnesite for Chukru (Jharkhand state) and Karlakot (Orissa state), India. Maximum fluoride concentration of 10.3 mg/L was observed in Chukru and 4.6 mg/L in the case of Karlakot. Maximum efficiency (i.e., 84% removal) was achieved in the case of synthetic magnetite after 2 min with an F concentration of 6 mg/L, an adsorbent dose of 10 g/L at an initial pH of 7. In the case of natural hematite iron ore, the defluoridation efficiency was less (i.e., 66% removal) compared to synthetic magnetite at a higher contact time with an initial pH of 3.2, a fluoride concentration of 3 mg/L and an adsorbent dose of 7 g/L.

9.3.9 Calcite

Ghada and Suzanne (2010) have studied the feasibility of using raw marine sediment collected from Egyptian Mediterranean coast, composed of calcite, magnesium-calcite and aragonite. The results suggest suitability up to 100% removal at pH values of 5 and 6.2, within contact time of 60 min, using 0.1 g of raw marine sediment, with fluoride concentrations of 10 and 15 mg/L.

9.3.10 Siderite

Shan and Guo (2013) have examined the performance of natural siderite modified by aluminum sulfate and aluminum solution. Defluoridation studies were undertaken by natural siderite powders, $Al_2(SO_4)_3$ and AlOOH mixed with different ratios, granulated and calcined at different temperatures for varying time intervals. The optimum modified natural siderite was obtained with the mass ratio of 50:0.3:10 g (natural siderite powders: $Al_2(SO_4)_3$: AlOOH), calcined at 450°C for 3 h. The results proved that process was efficient at wider pH (3.5–10.0), and adsorption was affected with bicarbonates and phosphates. Column studies suggested that the performance of the system varied with increasing concentration and flow rate. The fluoride uptake by the optimum modified siderite was observed mainly due to the coexistence of hematite and γ-Al_2O_3 from aluminum chemicals in the material and favorable characteristics of adsorbent such as large specific surface area, larger pore volume and mean pore diameter.

9.3.11 Lignite

Pekař (2009) has proved the feasibility of utilizing the lignite from South Moravia (Czec Republic) for adsorption of fluoride ion ranging between 5×10^{-5} and 8×10^{-4} mol/L in smaller contact time. The study found that lignite had the capacity to bind majority of fluorides releasing minor part (13%) in pure water. Sivasamy et al. (2001) have investigated the feasibility of coal-based sorbents, i.e., lignite, bituminous coal and fine coke as potential defluoridation materials. The results have shown that the fluoride sorption capacity varied with pH, at acidic pH, fine coke and bituminous coal have shown better performance, and lignite has shown greater efficiency at basic pH.

9.3.12 Hydroxyapatite

Ramdani et al. (2015) have examined the performance of activated alumina, sodium clay and hydroxyapatite for enhancing the groundwater quality of Saharan region,

Algeria. The results proved that maximum removal could be achieved with activated alumina (90%) followed by hydroxyapatite (83.4%) and then sodium clay (73.95%), but the performance varied with anions such as sulfate, bicarbonate and carbonate in defluoridation.

Ismaiel and Wole (2013) have studied the performance of the clay-hydroxyapatite composites (C- HA) for defluoridation. The studywas conducted by varying temperature (500°C, 600°C, 700°C, 800°C and 900°C) and mixture ratio (clay and hydroxyapatite as 100:0, 90:10, 80:20, 75:25, 0:100), C-HA composites were prepared from hydroxyapatite, redart clay and fine wood chips (sawdust) in well-controlled proportions. The results indicate that the sorption capacity of adsorbent containing only hydroxyapatite decreased with increasing temperature, but C-HA adsorbents have shown the increasing trend. The studyhas pointed out that composites could be effectively applied for defluoridation as well as for removal of microbial pathogens.

9.3.13 Other geomaterials

Fekadu et al. (2013) have investigated the potential of termite mold containing silicon, aluminum, iron and titanium oxides for defluoridation. The results suggest that higher percentage (i.e., 90%) was achieved in pH rangeof 3–8 and equilibrium was attained within 10 min of agitation. Anions such as chloride, nitrate and sulfate had a favorable influence on the performance but carbonate and phosphates limited its efficiency. Termite mold proved to be successful in reducing the fluoride content of groundwater from7.56 mg/L to less than 1.5 mg/L. The studyhas indicated possible regenerationof the adsorbent using calcined eggshell or NaOH solution.

Yin et al. (2015) have proved the application of calcined attapulgite at 700°C. The results have also confirmed better efficiency at higher pH (7–10) and lesser influence of anions. Characterization studies confirmed that the sorption characteristics of F^- at 700°C was due to formation of calcium fluoride precipitates and the complexation of fluoride with the –OH group.

Biswas et al. (2016) have conducted studies on thermally activated (at 550°C) shale collected from coal mines of West Bengal, India. The study proves better efficiency at lower pH and treated water also contained lower pH with a removal efficiency of 88%, at an initial fluorideconcentration of 10 mg/L, pH 3, and an adsorbent dose of 70 g/L.

Beekam et al. (2016) have investigated the processed iron ore for fluoride removalfrom high concentration water of Rift valley, Ethiopia. The results revealed its suitability as sorption capacity of iron ore was 1.72 mg/g at a pH of 6, and the equilibrium was attained after 120 min. The results were verified under the natural conditions using the original groundwater samples, and the fluoride level was observed to decreasefrom 14.22 to 1.17 mg/L.

Kim et al. (2013) have also examined the use of pyrophyllite, and the results have shown sorption capacity of 0.737 mg/g, at wider pH (4–9), but the presence of carbonates and sulfates has influenced its performance. The studyhas also observed that adsorbent treated at 400°C had the better sorption capacity as compared to untreated adsorbent.

Malakootian et al. (2011) have examined the adsorption capacity of pumice obtained from Anar mine, Iraq. The results proved effectiveness at a pH of 7 in 180 min.

The performance reduced on increasing the fluoride concentration. At optimum conditions, removal of 74.64%could be achieved with Kuhbonan water having a fluoride content of 2.8 mg/L.

9.4 REMEDIATION OF FLUORIDE CONTAMINATED SOILS

The following are the different approaches for remediation of fluoride contaminated soils:

- Phytoremediation is a promising clean technology for removal of fluorides from air, water and soil. This method serves as suitable alternative for exhausted fluoride contaminated soil for insitu decontamination. Plant species such as *Raphanus sativus*, *Oryza sativa*, and *Lupinus luteus* have shown the uptake of fluoride from contaminated soil (Weerasooriyagedara et al., 2020).
- Bioremediation using algal biomass (spirogyra IO1, spirogyra IO2, etc.), bacteria (*Pseudomonas aeruginosa*, *Acinetobacter* sp. RH5 and *Providencia vermicola*) and fungal species (*Aspergillus penicilloides*, *Mucor racemosus*, etc.) is another green technology for defluoridation (Kabir et al., 2019).
- Soil washing using acids such as hydrochloric acid (HCl), nitric acid (HNO_3), sulfuric acid (H_2SO_4), tartaric acid ($C_4H_6O_6$) and bases such as sodium hydroxide (NaOH) has been proven for remediation with proper consideration for its corrosive behavior in handling and release of hydrogen fluoride fumes (Deok et al., 2015).
- Electrokinetic remediation technique adopting low-intensity direct current through soil is being successfully adopted for soil remediation. Certain studies have also highlighted the combination of this technique with permeable reactive barrier filled for maximum efficiency (Ming et al., 2016).

9.5 CONCLUSIONS

Defluoridation of groundwater with special emphasis on adsorption using geomaterials has been discussed in the present study. The work has discussed various studies adopting locally available geomaterials or waste byproducts with apposite processing for low-cost fluoride removal. Itprovides a thorough understanding of geological, hydrological, economic and ecological scenarios for the choice of suitable technique(s). The studies can be planned with synergistic tieups among stakeholders in the following stages:

- **Identification:** Involves exploration regarding geological, hydrological, industrial (brick, ceramic, flyash or other byproducts) and soil profiles for understanding feasible adsorbents for defluoridation. Hydrological profile may provide guidance for understand the scope for developing alternate water resources and other realistic management options.
- **Analysis:** Includes characterization studies of identified materials, aiming at understanding of physical, chemical and biological aspects of the geomaterials and water. The results of the analysis followed by initial trials (batch mode) in diverse

scenarios may provide crucial information regarding enhancement achievable in defluoridation capacity through suitable processing technique(s) to be applied effectively for identified adsorbent(s) and/or for pretreatment of raw water.

- **Field trials:** Initially conducted batch-mode trials need to be followed by studies on columns designed for ensuring daily water treatment objectives on field scale. These combined trials can provide important operational parameters for continued processing of raw water.
- Regeneration process for exhausted adsorbents as well as the related management strategy needs to be properly established for beneficial reuse adsorbents and to ensure compliance with safe disposal guidelines ultimately.

REFERENCES

Ayoob, S., Gupta, A.K., Bhat, V.T., 2008. A conceptual overview on sustainable technologies for the defluoridation of drinking water. *Crit. Rev. Environ. Sci. Technol.* 38(6), 401–470. https://doi.org/10.1080/10643380701413310

Ayse, D.A., Muge, O.S., 2014. Adsorption of fluoride on the raw and modified cement clay. *Clean Soil Air Water* 42 (4), 415–420. https://doi.org/10.1002/clen.201300074

Babu, S.S., Kumar, S., Roychowdhury, T., Vidyadharan, V., Roychowdhury, N., Samanta, J., Bhowmick, S., 2015. Occurrence and impacts of fluoride in drinking water. *Indian Groundwater* 5, 40–54.

Beekam, K., Abebe, B., Fekadu, F., Moa, M., Michael, B., 2016. Experimental evaluation of sorptive removal of fluoride from drinking water using iron ore. *Appl. Water Sci.* 6(1), 57–65. https://doi.org/10.1007/s13201-014-0210–x

Beraki, B.M., Alfred, O.M., Beatrice, K.K., 2014. Development of household defluoridation unit based on crushed burnt clay pot as sorbent medium: A case of Keren community. *Environ. Nat. Resour. J.* 4(3), 67–82.

Biswas, G., Dutta, M., Dutta, S., Adhikari, K., 2016. A comparative study of removal of fluoride from contaminated water using shale collected from different coal mines in India. *Environ. Sci. Pollut. Res.* 23(10), 9418–9431. https://doi.org/10.1007/s11356-015-5815–6

Bjorvatn, K., Bardsen, A., 1995. Use of activated clay for defluoridation of water. In: Dahi, E., Bregnhoj, H. (eds) *Proceedings of the First International Workshop on Fluorosis and Defluoridation of Water*, pp. 49–55. International Society for Fluoride Research, Auckland.

Council of Scientific & Industrial Research- National Environmental Engineering Research Institute. 2017. Electrolytic defluoridation technique. http://www.neeri.res.in/content/electrolytic-defluoridation-edf-technique

Chidambaram, S., Manikandan, S., Ramanathan, A.L., Prasanna, M.V., Thivya, C., Karmegam, U., Thilagavathi, R., Rajkumar, K., 2013. A study on the defluoridation in water by using natural soil. *Appl. Water Sci.* 3, 741–751. https://doi.org/10.1007/s13201-013-0122–1

Chidambaram, S., Ramanathan, A.L., Vasudevan, S., 2003. Fluoride removal studies in water using natural materials. *Water SA*, 29(3), 339–344. https://doi.org/10.4314/wsa.v29i3.4936

Chakraborti, D., Das, B., Murill, T.M., 2011. Examining India's groundwater quality management. *Environ. Sci. Technol.* 45, 27–33. https://doi.org/10.1021/es101695d

Chowdhury, A., Adak, M.K., Mukherjee, A., Dhak, P. Khatun, J., Dhak, D., 2019. A critical review on geochemical and geological aspects of fluoride belts, fluorosis and natural materials and other sources for alternatives to fluoride exposure. *J. Hydrol.* 574, 333–359. https://doi.org/10.1016/j.jhydrol.2019.04.033.

Deok, H.M., Raehyun, J., Agamemnon, K., Kyung, H.C., Jeong Hun, P., 2015. Soil washing of fluorine contaminated soil using various washing solutions. *Bull Environ Contam Toxicol.* 94(3), 334–339. https://doi.org/10.1007/s00128-014-1449–5

Dilip, T., Sadhana, R., Raju, K., Siddharth, M., Subrt, J., Labhsetwar, N., 2010. Magnesium incorporated bentonite clay for defluoridation of drinking water. *J. Hazard Mater.* 180, 122–130. https://doi.org/10.1016/j.jhazmat.2010.04.001

Fekadu, F., Esayas, A., Bernd, L., 2013. Defluoridation of groundwater using termite mound. *Water Air Soil Pollut.* 224, 1552–1567. https://doi.org/10.1007/s11270-013-1552–y

Food and Agriculture Organization, 2021. Details available at http://www.fao.org/3/t0234e/t0234e06.htm

Ghada, F.E., Suzanne, E.O.D., 2010. Physicochemical and geochemical characteristics of raw marine sediment used in fluoride removal. *J. Environ. Sci. Health A* 45(12), 1601–1615. https://doi.org/10.1080/10934529.2010.506117

Gitari, W.M., Ngulube, T., Masindi, V., Gumbo, J.R., 2015. Defluoridation of groundwater using Fe^{3+}-modified bentonite clay: optimization of adsorption conditions. *Desalin. Water Treat.* 53(6), 1578–1590. https://doi.org/10.1080/19443994.2013.855669

Gogoi, S., Nath, S.K., Bordoloi, S., Dutta, R.K., 2015. Fluoride removal from groundwater by limestone treatment in presence of phosphoric acid. *J. Environ. Manage.* 152, 132–139. https://doi.org/10.1016/j.jenvman.2015.01.031

Gogoi, P.K., Ruby, B., 2008. Fluoride removal from water by adsorption on acid activated kaolinite clay. *Indian J. Chem. Technol.* 15(5), 500–503.

Guo, Q., Eric, J.R., 2012. Fluoride removal from water by meixnerite and its calcination product. *Appl. Clay Sci.* 56, 7–15. https://doi.org/10.1016/j.clay.2011.11.013

Gupta, A.K., Ayoob, S., 2016. *Fluoride in drinking water: Status, issues and solutions.* CRC Press, Florida, USA.

Hamdi, N., Srasra, E., 2009. Retention of fluoride from industrial acidic wastewater and NaF solution by three Tunisian clayey soils. *Fluoride* 42(1), 39–45.

He, J., Yang, Y., Wu, Z., Xie, C., Zhang, K., Kong, L., Liu, J., 2020. Review of fluoride removal from water environment by adsorption. *J. Environ. Chem. Eng.* 8, 104516. https://doi.org/10.1016/j.jece.2020.104516

Ismaiel, Y., Wole, S., 2013. Adsorption of fluoride from water using sintered clay-hydroxyapatite composites. *J. Environ. Eng.* 139(7), 995–1003.

Jagtap, S., Yenkie, M.K., Labhsetwar, N., Rayalu, S., 2012. Fluoride in drinking water and defluoridation of water. *Chem. Rev.* 112(4), 2454–2466. https://doi.org/10.1021/cr2002855

Jain, S., Jayaram, R.V., 2009. Removal of fluoride from contaminated drinking water using unmodified and aluminium hydroxide impregnated blue lime stone waste. *Sep. Sci. Technol.* 44(6), 1436–1451. https://doi.org/10.1080/01496390902766074

Johnson, C.A., Bretzler, A., (eds), 2015. *Geogenic contamination handbook- addressing arsenic and fluoride in drinking water.* Swiss Federal Institute of Aquatic Science and Technology (Eawag), Dubendorf, Switzerland.

Kabir, H., Gupta, A.K., Tripathy, S., 2019. Fluoride and human health: Systematic appraisal of sources, exposures, metabolism, and toxicity. *Crit. Rev. Environ. Sci. Technol.* https://doi.org/10.1080/10643389.2019.1647028

Karthikeyan, G., Anitha, P., Alagumuthu, G., 2005. Fluoride adsorption studies of montmorillonite clay. *Indian J. Chem. Technol.* 12, 263–272.

Kim, J., Lee, C., Park, J., Kang, J., Choi, N., Kim, S., 2013. Use of pyrophyllite clay for fluoride removal from aqueous solution. *Desalin. Water Treat.* 51, 3408–3416. https://doi.org/10.1080/19443994.2012.749198

Kumar, R., Singh, R.D., Sharma, K.D., 2005. Water resources of India. *Curr. Sci.* 89(5), 794–811.

Lacson, C.F.Z., Lu, M., Huang, Y., 2021. Fluoride-containing water: A global perspective and a pursuit to sustainable water defluoridation management -An overview. *J. Clean. Prod.* 280, 124236. https://doi.org/10.1016/j.jclepro.2020.124236

Maiti, A., Basu, J.K., De, S., 2011. Chemical treated laterite as promising fluoride adsorbent for aqueous system and kinetic modeling. *Desalination* 265, 28–36. https://doi.org/10.1016/j.desal.2010.07.026

Malakootian, M., Moosazadeh, M., Yousefi, N., Fatehizadeh, A., 2011. Fluoride removal from aqueous solution by pumice: case study on Kuhbonan water. *Afr. J. Environ. Sci. Technol.* 5(4), 299–306.

Meetu, A., Kavita, R., Shrivastav, R., Sahab, D., 2002. A study on fluoride sorption by montmorillonite & kaolinite. *Water Air Soil Pollut.* 141(1), 247–261. https://doi.org/10.1023/A:1021328525215

Mekonnen, B., Alebel, A., Esayas, A., Embialle, M., 2014. Efficiency of locally available filter media on fluoride and phosphate removal for household water treatment system. *Chinese J. Popul. Resour. Environ.* 12(2), 110–115. https://doi.org/10.1080/10042857.2014.910873

Ming, Z., Shufa, Z., Yaoyao, Y., Tingting, Z., 2016. An electrokinetic/activated alumina permeable reactive barrier-system for the treatment of fluorine-contaminated soil. *Clean Technol. Environ. Policy* 18(8), 2691–2699. https://doi.org/10.1007/s10098-016-1156–5

Mlilo, T.B., Brunson, L.R., Sabatini, D.A., 2010. Arsenic and fluoride removal using simple materials. *J. Environ. Eng.* 136(4), 391–398. https://doi.org/10.1061/(ASCE)EE.1943–7870.0000154

Moges, G., Zewge, F., Socher, M., 1996. Preliminary investigations on the defluoridation of water using fired clay chips. *J. Afr. Earth Sci.* 21(4), 479–482. https://doi.org/10.1016/0899–5362(96)00030–9

Ozsvath, D.L., 2009. Fluoride and environmental health: A review. *Rev. Environ. Sci. Biotechnol.* 8(1), 59–79. https://doi.org/10.1007/s11157-008-9136–9

Padmasiri, J.P., 1997. Low cost domestic defluoridation. In: *Pre-prints of the 2nd International workshop on fluorosis and defluoridation: Nazreth, Ethiopia*, pp. 146–150, International Society of Fluoride Research, Dunedin.

Panwar, S., Chakrapani, G.J., 2013. Climate change and its influence on groundwater resources. *Curr. Sci.* 105(1), 37–46.

Patel, S.C., Khalkho, R., Patel, S.K., Sheikh, J.M., Behera, D., Chaudhari, S., Prabhakar, N., 2014. Fluoride contamination of groundwater in parts of eastern India and a preliminary experimental study of fluoride adsorption by natural hematite iron ore and synthetic magnetite. *Environ. Earth Sci.* 72(6), 2033–2049. https://doi.org/10.1007/s12665-014-3112–1

Pekař, M., 2009. Fluoride anion binding by natural lignite (South Moravian Deposit of Vienna Basin). *Water Air Soil Pollut.* 197, 303–312. https://doi.org/10.1007/s11270-008-9812–y

Porada, S., Zhao, R., van der Wal, A., Presser, V., Biesheuvel, P.M., 2013. Review on the science and technology of water desalination by capacitive deionization. *Prog. Mater. Sci.* 58, 1388–1442. https://doi.org/10.1016/j.pmatsci.2013.03.005

Rafique, T., Anasa, M., Chadhara, K.M., Soomrob, F., Alvia, S.K., 2018. Defluoridation of water by wasted red brick paving blocks. *Desalin. Water Treat.* 108, 207–215.

Ramdani, A., Taleb, S., Benghalem, A., Deratani, A., Ghaffour, N., 2015. Enhancement of Saharan groundwater quality by reducing its fluoride concentration using different materials. *Desalin. Water Treat.* 54(12), 3444–3453. https://doi.org/10.1080/19443994.2014.910838

Rasool, A., Farooqi, A., Xiao, T., Ali, W., Noor, S., Abiola, O., Ali, S., Nasim, W., 2017. A review of global outlook on fluoride contamination in groundwater with prominence on the Pakistan current situation. *Environ Geochem Health* 40, 1265–1281. https://doi.org/10.1007/s10653-017-0054–z

Sajidu, S.M.I., Masamba, W.R.L., Thole, B., Mwatseteza, J.F., 2008. Groundwater fluoride levels in villages of southern Malawi and removal studies using bauxite. *Int. J. Phys. Sci.* 3(1), 1–11. https://doi.org/10.5897/IJPS.9000284

Samson, S., Chikumbusho, K., Wellington, M., Jonas, M., 2012. Defluoridation of groundwater using raw bauxite: Rural domestic defluoridation technology. *Environ. Nat. Resour. Res.* 2(3), 1–9. https://doi.org/10.5539/enrr.v2n3p1

Sarkar, M., Banerjee, A., Pramanick, P.P., 2006. Kinetics and mechanism of fluoride removal using laterite. *Ind. Eng. Chem. Res.* 45(17), 5920–5927. https://doi.org/10.1021/ie060016j

Shan, Y., Guo, H., 2013. Fluoride adsorption on modified natural siderite: Optimization and performance. *Chem. Eng. J.* 223, 183–191. https://doi.org/10.1016/j.cej.2013.03.023

Shengyu, Z., Ying, L., Xiaosi, S., Yuya, B., Bowen, Y., Yuling, Z., 2016. Removal of fluoride ion from groundwater by adsorption on lanthanum and aluminum loaded clay adsorbent. *Environ. Earth Sci.* 75, 401–410. https://doi.org/10.1007/s12665-015-5205–x

Sivasankar, V., 2016. *Surface modified carbons as scavengers for fluoride from water,* Springer International Publishing, Switzerland, https://doi.org/10.1007/978–3–319–40686–2

Sivasamy, A., Singh, K.P., Mohan, D., Maruthamuthu, M., 2001. Studies on defluoridation of water by coal- based sorbents. *J. Chem. Technol. Biotechnol.* 76(7), 717–722. https://doi.org/10.1002/jctb.440

Srivastava, S., Flora, S., 2020. Fluoride in drinking water and skeletal fluorosis: A review of the global impact. *Curr. Environ. Health Rep.* 7, 140–146. https://doi.org/10.1007/s40572-020-00270–9

Teutli-Sequeira, A., Solache-Ríos, M., Balderas-Hernández, P., 2012. Modification effects of hematite with aluminum hydroxide on the removal of fluoride ions from water. *Water Air Soil Pollut.* 22, 319–327. https://doi.org/10.1007/s11270-011-0860–3

Thole, B., 2011. Defluoridation kinetics of 200°C calcined bauxite, gypsum, and magnesite and breakthrough characteristics of their composite filter. *J. Fluor. Chem.* 132(8), 529–535. https://doi.org/10.1016/j.jfluchem.2011.05.016

Thole, B., Masamba, W.R.L., Mtalo, F.W., 2013. Water defluoridation by bauxite-gypsum-magnesite (B-G-Mc) based filters calcined at 350–500°C. *Int. J. Phys. Sci.* 8, 956–962. https://doi.org/10.5897/IJPS10.668

Tomar, V., Kumar, D., 2013. A critical study on efficiency of different materials for fluoride removal from aqueous media. *Chem. Cent. J.* 7, 51–66. https://doi.org/10.1186/1752-153X–7–51

Tor, A., 2006. Removal of fluoride from an aqueous solution by using montmorillonite. *Desalination* 201, 267–276. https://doi.org/10.1016/j.desal.2006.06.003

Vinati, A., Mahanty, B., Behera, S.K., 2015. Clay and clay minerals for fluoride removal from water: A state-of-the-art review. *Appl. Clay Sci.* 114, 340–348. https://doi.org/10.1016/j.clay.2015.06.013

Vithanage, M., Bhattacharya, P., 2015. Fluoride in the environment: Sources, distribution and defluoridation. *Environ. Chem. Lett.* 13(2), 131–147. https://doi.org/10.1007/s10311-015-0496–4

Weerasooriyagedara, M., Ashiq, A., Rajapaksha, A.U., Wanigathunge, R.P., Agarwal, T., Magana-Arachchi, D., Vithanage, M., 2020. Phytoremediation of fluoride from the environmental matrices: A review on its application strategies. *Groundw. Sustain. Dev.* 10, 100349. https://doi.org/10.1016/j.gsd.2020.100349

Yadav, A.K., Kaushik, C.P., Haritash, A.K., Kansal, A., Rani, N., 2006. Defluoridation of groundwater using brick powder as an adsorbent. *J. Hazard Mater.* 128, 289–293. https://doi.org/10.1016/j.jhazmat.2005.08.006

Yanxin, W.A., Eric, J.R., 2001. Activation and regeneration of a soil sorbent for defluoridation of drinking water. *Appl. Geochem.* 16(5), 531–539. https://doi.org/10.1016/S0883–2927(00)00050–0

Yin, H., Kong, M., Tang, W., 2015. Removal of fluoride from contaminated water using natural calcium-rich attapulgite as a low-cost adsorbent. *Water AirSoil Pollut.* 226, 425–436. https://doi.org/10.1007/s11270-015-2691–0

Yuxin, M., Fengmei, S., Xilai, Z., Jun, M., Congjie, G., 2011. Removal of fluoride from aqueous solution using granular acid-treated bentonite (GHB): Batch and column studies. *J. Hazard Mater.* 185, 1073–1080. https://doi.org/10.1016/j.jhazmat.2010.10.016

Zuo, H., Chen, L., Kong, M., Qiu, L., Lü, P., Wu, P., Yang, Y., Chen, K., 2018. Toxic effects of fluoride on organisms. *Life Sci.* 198, 18–24. https://doi.org/10.1016/j.lfs.2018.02.001

Chapter 10

Physicochemical and biological methods for treatment of municipal solid waste incineration ash to reduce its potential adverse impacts on groundwater

Basanta Kumar Biswal and Umesh U. Jadhav
Nanyang Technological University

Deeksha Patil
Savitribai Phule Pune University

En-Hua Yang
Nanyang Technological University

CONTENTS

DOI: 10.1201/9781003247869-12

10.1 INTRODUCTION

Solid waste generation has significantly increased in many countries due to rapid growth of population, modernization and spread of industrialization. According to the World Bank report, in 2016, nearly 2.01 billon t (0.74 kg per person per day) of solid waste is generated from world cities, and it is expected that the generation rate will be increased by 70% (3.40 billion t) in 2050 (World Bank, 2016). Management of such a huge amount of solid wastes is a serious concern worldwide. Although disposal at engineered landfill facilities is a common practice, this option is not feasible because of excessive operation costs, reduction of land space, stricter environmental laws and strong public resistance to construct new landfills (Veli et al., 2008). Incineration (also called waste-to-energy process) has been considered as an attractive technology for waste management since it significantly reduces the volume (up to 90%) and the mass (up to 70%) of solid wastes (Lam, 2010). Combustion of solid wastes in incineration plants, however, produces two types of solid by-products termed as incineration bottom ash (IBA) and incineration fly ash (IFA). Among the two types of ashes, IBA presents the main fraction (80% by weight) and the remaining 20% is IFA (Chimenos et al., 1999).

The modern waste management systems are aiming toward zero waste and zero landfill. Thus, efforts have been made to minimize the volume of waste disposal by three important routes namely reduce, reuse and recycle (commonly known as 3 Rs). There are many advantages for reutilization of ash such as reduction of disposal costs, saving landfill space, protection of natural resources and decrease in environmental contamination. A viable solution could be to reutilize these waste materials for various public works. Many European countries (e.g., France, Denmark, Germany, the Netherlands, etc.) use up to 50% of IBA for construction of buildings and road pavements (Chimenos et al., 1999).

The leaching of pollutants including heavy metals, trace elements, oxyanions and salts from these ash materials is a significant environmental concern. It is reported that more than 80% of the heavy metals namely Fe, Cu, Zn, Cr, Pb, As, Ni and Ti in municipal solid waste were found to remain in IBA after incineration. The high amount of heavy metals in IBA shows environmental concerns when it will be disposed at landfilling or reutilization (Yao et al., 2014). Previous researchers have mentioned that most of the heavy metals in IBA may pass the toxicity characteristic leaching procedures (TCLP) test (Feng et al., 2007); however, in longer-term scenario, there may be a chance of leaching of metals which would be risk to the surrounding environments (Lin et al., 2007). Municipal solid waste incineration (MSWI) fly ash is generally considered as highly hazardous materials due to enrichment of alkaline metals, calcium, chlorine, sulfur, heavy metals and persistent organic pollutants, e.g., dioxins (Ahmaruzzaman, 2010). Therefore, many studies have evaluated ecotoxicity and effects on groundwater to evaluate the degree of environmental consequences of incineration ash residues prior considering their reutilization or disposal.

Several studies have reported about contamination of water systems (surface water and ground water) due to leaching of various pollutants specifically heavy metals from ash (Kopsick and Angino, 1981; Prasad and Mondal, 2008). The key parameters usually measured for evaluation of ground water quality include pH, total hardness, total dissolved solids, and metal contents including heavy metals (Prasad and Mondal, 2008). Additionally, a few studies have performed germination bioassay using wheat seed (determination of germination index) and showed that leachate from incineration ash causes phytotoxicity and genotoxicity (Feng et al., 2007; Phoungthong et al., 2016). In total, since both IBA and IFA contain toxic constituents, it is required to treat the ash residues to lower its hazardous level prior landfill dumping or reutilization in civil engineering works.

The primary objective of this chapter is to update the scientific communities on the recent developments in literature on physiochemical, thermal and biological methods for treatment of MSWI bottom ash and fly ash to remove/reduce hazardous contaminants (mostly heavy metals) from ash materials. Moreover, the effect of various environmental factors on treatment processes is reviewed. A brief discussion on the potential impacts of leachate generated from MSWI ash on groundwater qualities is presented. The possible mechanisms for metal mobilization/immobilization due to interactions of ash residues with microorganisms (bacteria and fungus) and the key parameters that influence the fate of the microbe-mediated metal transformation process are reported. For economic advantage, potential reutilization of MSWI ash for civil engineering and geotechnical/geo-environmental applications are described. Finally, this chapter highlights the key knowledge gaps that should be considered in future works which would help for development of environmentally friendly and sustainable technologies for management and treatment of MSWI ash.

10.2 PHYSICOCHEMICAL CHARACTERISTICS OF IBA AND IFA

The properties of MSWI ash can be divided into two categories namely physical and chemical characteristics. The potential reutilization of ash mainly depends on its physical and chemical qualities. The key physical characteristics are moisture quantity, particle size distribution, bulk density, porosity, permeability and compressive strength (Lam, 2010) (Table 10.1). The chemical properties include loss on ignition, chemical composition, heavy metals and leachability, organic components and chloride concentration (Lam, 2010). In general, IFA is gray, alkaline, abrasive and refractory in nature (Ahmaruzzaman, 2010). The IBA is grayish with various particle size distributions and alkaline in nature; the surface morphology shows that it is porous and irregular particle with rough shape (Figure 10.1). The structure of IFA looks very irregularly amorphous, fluffy and flocculent appearance (Wang et al., 2006) (Figure 10.1).

Determination of particle size distribution of MSWI ash is important when the material is selected to use as an alternative for secondary building materials. Moreover, particle characterization of ash residues is also essential for the recovery/extraction of metal contents. IBA generally consists of coarse particles of size 0.1–100 mm, whereas IFA comprises of small diameter particles of size 1–500 μm (Al-Rawas et al., 2005). Furthermore, IBA can be segregated into six different fractions using standard meshes after drying at 105°C. The mean particle size of IFA is 26 μm, whereas the mean size of IBA is 3 mm (Xu et al., 2014).

The composition of IBA and IFA depends on types of waste materials incinerated and the applied combustion method. In general, IBA contains non-combustible materials including stone, glass, ferrous and non-ferrous metals and metal oxides, heavy metals and trace elements, and trace amount of organic matter (Qiao et al., 2008). In IBA, among various oxides, silicon (SiO_2) and calcium oxides (CaO) are prevalent

Table 10.1 Physical characteristics of municipal solid waste IBA and IFA

Parameter	*Unit*	*IBA*	*IFA*
References	-	Traina et al. (2007); Jurič et al. (2006)	Xu et al. (2014); Xu and Ting (2004); Zhang et al. (2016)
pH	-	12.8	11.7
Color	-	Gray	Gray
Electrical conductivity	mS/cm	8.29	-
Hydraulic permeability	cm/s	2.5×10^{-5}	10^{-4}–10^{-6}
LOI[b]	%	2.4	2.0
Density	g/cm^3	1.9	0.4
Water content	%	25	0.35–0.7
Specific gravity	g/cm^3	2.1	3.1
Mean particle size	μm	3,000	26
CEC[a]	meq 100/g	6.7	-
Solubility in water	%	10	23
Gravel	%	54.4	-
Sand	%	42.5	-
Silt/Clay	%	3	-

[a]CEC: cation exchange capacity.
[b]LOI: loss on ignition.

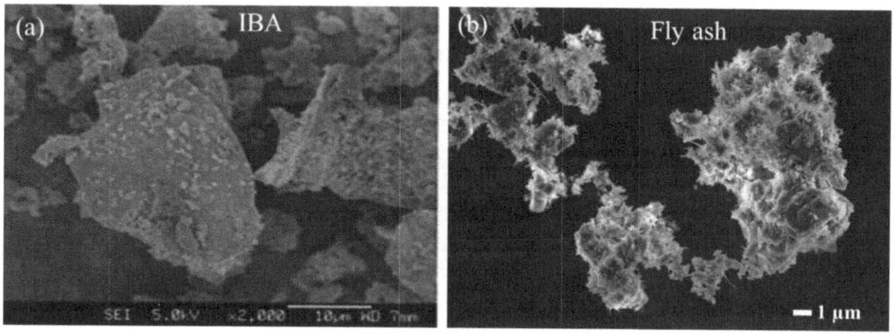

Figure 10.1 SEM images of municipal solid waste incineration bottom ash (IBA) (a) (adopted from Chen et al., 2016) and IFA particles (b) (Adopted from Li et al., 2017.)

Table 10.2 Comparison of chemical composition between municipal solid waste IBA, IFA and Portland cement

Key mineral phase	*Formula*	*IBA (wt. %)[a]*	*IFA (wt. %)[a]*	*Ordinary Portland cement (wt. %)[b]*
References	-	Chen et al. (2016)	Li et al. (2017)	Li et al. (2017)
Lime	CaO	29.06	47.4	62.2
Quartz	SiO_2	32.75	2.2	24.3
Aluminum oxide	Al_2O_3	8.57	1.1	4.6
Hematite	Fe_2O_3	10.02	0.8	4.0
Magnesium oxide	MgO	1.75	1.6	3.3
Potassium oxide	K_2O	1.24	6.7	0.6
Sodium oxide	Na_2O	2.87	4.9	0.2
Phosphorus pentoxide	P_2O_5	4.77	0.7	0.2
Sulfur trioxide	SO_3	3.01	5.3	-
Chloride	Cl	-	25	-

[a]Main mineral phase detected in IBA and IFA collected from an incineration plant in Singapore.
[b]Ordinary Portland cement (OPC) compliant to the Singapore standards SS 26.

(Table 10.2). In addition to inorganic matter, IBA also contains unburnt organic matter ranging between 0.5% and 5% of the total organic carbon (TOC) (Bayard et al., 2009). Various aliphatic compounds (formic, acetic, lactic, oxalic and maleic acids) and aromatic acids (phthalic, terephthalic and benzoic acids) are detected in leachates from IBA, but these contributed to around 5%–13% of dissolved organic carbon (DOC) (van Zomeren and Comans, 2004). The properties of IBA varied with size of IBA particles. IBA with particle size between 4 and 25 mm accounts nearly 50% of the IBA weight, and it consists of mainly glass, synthetic ceramics and minerals (Chimenos et al., 1999). However, magnetic metals are found in particle size of 1–6 mm, and heavy metals are mainly concentrated in smaller particles (<1 mm). To the end, physicochemical data (Tables 10.1–10.3) suggest that characteristics of IBA and IFA can be widely varied based on the source of incineration residues, composition of input waste materials and the operating process of incineration plants (Chang and Wey, 2006).

Table 10.3 A comparison of metal contents between municipal solid waste IBA, IFA and Portland cement

Element	*IBA (mg/kg)*[a]		*IFA (mg/kg)*[a]		*Portland cement (mg/kg)*[b]
References	Biswal et al. (2019)	*Element*	Wu and Ting (2006)	*Element*	Dorileo et al. (2014)
Zn	2,102	Ca	258,000	Ca	316,380
Cu	1,378	K	23,020	Fe	740
Mn	430	Al	19,210	Bi	240
Ba	361	Mg	10,820	Zn	130
Pb	161	Zn	6,288	Mn	55
Sn	113	Fe	5,240	Pb	30
Sr	105	Pb	2,000	Ni	30
Cr	82	Cu	570	Cr	25
Ni	62	Ba	539	Cu	5
B	39	Mn	309	As	3.7
Co	8.8	Sr	151	Cd	<0.002
V	8.5	Cd	95	-	-
As	7	Cr	72	-	-
Cd	1.8	Ni	22	-	-
Ag	0.65	Co	14	-	-

[a]The data for IBA and IFA refer to municipal solid waste incineration plants in Singapore.
[b]Portland cement refers to structural white Portland cement.

10.3 REGULATORY REQUIREMENT FOR DISPOSAL/REUSE OF MSWI ASH RESIDUES

Leaching test has been carried out to monitor hazardous level of waste materials prior their landfill disposal. TCLP is the common method approved by USEPA to classify waste into hazardous or non-hazardous materials. For TCLP, eight metals (As, Ag, Cr, Ba, Cd, Hg, Pb and Se) are regulated by USEPA, and the permissible limit for each metal is presented in Table 10.4. In the TCLP test, if it is found that concentration of any regulated metal exceeds the regulatory level, then the waste is considered as "hazardous." The cost of disposal of the waste is increased if it is declared as "hazardous waste" by TCLP test. Thus, the waste requires pre-treatment to bring the concentrations of metals below the regulatory limit. Based on the results reported by various researchers (Table 10.4), it is found that leaching concentration of majority heavy metals (except lead) from IBA and cadmium in IFA falls below the USEPA regulatory level. The leaching concentrations of lead (19.06 mg/L) in IBA exceeded the regulatory limit (5 mg/L) (Feng et al., 2007). In IFA, the leaching concentration of cadmium (2.23 mg/L) was above the standard (1.0 mg/L). Another study noticed that the concentration of lead in IFA (37.6 mg/L) was higher than the regulatory level (5 mg/L) (Wu and Ting, 2006). Although most of the heavy metal concentrations in IBA and IFA fall below the TCLP limit (i.e., TCLP is a short-term study in which IBA/IFA interacts with leaching medium for a short period), there could be environmental concern on long-term behavior in landfilling and reutilization because of interactions of IBA/IFA with environmental microorganisms and/or influence of surrounding soil physicochemical characteristics. Thus, it requires further treatment to reduce its potential environmental problems prior to disposal or reutilization.

Table 10.4 TCLP analysis of municipal solid waste IBA and IFA

Contaminants	*Contaminant concentration (mg/L)*			
	IBA leachate	*IFA leachate*	*TCLP limit*	*Drinking water limit*
References	Feng et al. (2007)	Yang et al. (2008)	USEPA (2004)	World Health Organization (2017)
Arsenic (As)	0.12	NA[a]	5.0	0.01
Barium (Ba)	0.41	NA	100	1.3
Cadmium (Cd)	<0.01	**2.23**[b]	1.0	0.003
Chromium (Cr)	0.13	0.19	5.0	0.05
Lead (Pb)	**19.06**[b]	3.10	5.0	0.01
Mercury (Hg)	<0.01	NA	0.2	0.006
Selenium (Se)	0.24	NA	1.0	0.04
Silver (Ag)	NA	NA	5.0	-
Copper (Cu)	0.1	2.37	-	2.0
Zinc (Zn)	1.71	56.02	-	-
Manganese (Mn)	<0.01	1.43	-	0.1

[a]NA: data not available.
[b]The values shown in bold are exceeded the TCLP limit.

10.4 POTENTIAL ADVERSE IMPACTS OF MSWI ASH (IBA AND IFA) LEACHATE ON GROUNDWATER

One of the burning challenges in many countries is the safe disposal of MSWI bottom ash and fly ash because they have been seen as significant organizational limitations and environmental pollution. Recently, soil and water contamination around the IFA disposal region has been the topic of great worldwide concern. Trace metals present in ash particles leach out and contaminate the sources of groundwater and surface water. IFA contains abundant heavy metals that can pose risks to human health if human beings inhale or consume it in higher volume; it may lead to certain disorders like respiratory disorder and gastrointestinal diseases as well as it may also act as a carcinogen (Pandey et al., 2020). Many researchers have analyzed groundwater pollution due to landfills using the groundwater contamination index to map the degree of contamination (Singh et al., 2014). Literature shows considerable works have been performed on the impacts of ash generated from thermal plants (e.g., coal fly ash) on groundwater qualities (Kopsick and Angino, 1981; Pandey et al., 2020; Prasad and Mondal, 2008). However, limited information is available on the effects of MSWI ash (IBA and IFA) on groundwater characteristics (Phoungthong et al., 2018).

When municipal solid wastes IBA and IFA are used in construction of road, there are high chances that hazardous components present in these materials may leach out due to runoff. Thus, runoff containing toxic contaminants when meets surface or groundwater may pose negative impacts on ecosystems (Phoungthong et al., 2018). Baun et al. (2004) found that in one of the studied landfill sites, leachates contained non-volatile organic compounds and they can be transferred to genotoxic compounds by liver enzymes. Pandey et al. (2020) noticed that ionic concentration in the water increased due to leaching of elements like bicarbonate, calcium, sodium and sulfate ions and these ions increases the overall conductivity of water.

Areas near the landfills can have substantial risk toward local groundwater users or natural environment since the leachate from landfill may reach to the groundwater. Nagarajan et al. (2012) studied the effect of leachate percolation on groundwater characteristic from a landfill site at Vendipalayam, Tamil Nadu, India. The authors have attempted to find out the possible correlation between the distance of landfill site/depth of sampling well and the groundwater contamination level. It is seen that the pollutant concentration was higher in the groundwater samples collected near to the landfill site. For example, the concentrations of total dissolved solids, chloride, sulfate, nitrate, sodium and iron were higher in groundwater sampled close to the landfill area. Notably, the contaminants concentrations steadily decreased with increase of the distance of sampling location from the landfill site as well as increase of depth of the sampling well which suggest percolation rate of leachate also drops with rise of distance between sampling point/depth and landfill site. Birgisdóttir et al. (2006) assessed environmental impacts of a road constructed with municipal solid waste IBA using a life cycle assessment model (ROAD-RES) and found that high quantity of salt present in the IBA (50%) can be leached to the groundwater; thus, use of IBA in road construction may not be safe from environmental perspective due to high potential for groundwater pollution.

Prasad and Mondal (2008) analyzed groundwater samples collected near an open cast mine filled with thermal power plant coal fly ash, and they found that fluoride (F: 0.251–4.78 mg/L) and manganese (Mn: 0.1102–6.1175 mg/L) concentrations were significantly higher in groundwater sampled near to the mine site than uncontaminated site (F: 0.168–0.431 mg/L; Mn: 0.0908–0.168 mg/L). The concentration of fluoride in groundwater at the mine site exceeds the drinking water standard (1.0–1.5 mg/L). Another study by Dandautiya et al. (2018) reported that the concentration of iron (Fe: 0.301–1.545 mg/L) and selenium (Se: 0–0.077 mg/L) in groundwater collected from a coal thermal power plant was significantly higher than the WHO drinking water standard (Fe: 0.3 mg/L, Se: 0.04 mg/L). Singh et al. (2014) noticed that acidic pH accelerates leaching of metals from coal fly ash; however, the leaching rate is lower in alkaline pH condition. Moreover, the authors have observed that the concentration of a few metals (Cd, Cr, Cu and Fe) in groundwater collected near to an ash pond was higher than the prescribed drinking water standard. Overall, results of these studies suggest that reuse or disposal of untreated MSWI ash could pose serious threats to environments including groundwater pollution and damage of surrounding ecosystems.

10.5 PHYSICOCHEMICAL METHODS FOR TREATMENT OF IBA AND IFA

10.5.1 Physical treatment: screening and magnetic separation

Screening is a process in which MSWI ash (mainly IBA) are separated based on the size of particles. Chen and Yang (2017) separated IBA into five categories based on the size of IBA particles: >6.3 mm (biggest), 1.18–6.3 mm, 0.3–1.18 mm, 0.075–0.3 mm and <0.075 mm (smallest). The largest fraction (>6.3 mm) mainly consists of glass. The authors have noticed that the quantity of silicon oxide (SiO_2) was more in the larger IBA particles (>6.3 mm, and 1.18–6.3 mm), whereas calcium oxide (CaO) was mostly concentrated in the smallest size particles (46 wt.%, <0.075 mm size). Magnetic separation can be used for recovery of ferrous material, and ferrous oxide is usually rich in the relatively bigger size IBA particles (0.3–6.3 mm) (Chen and Yang, 2017). Removal of ferrous components is advantageous for downstream treatment specifically metal dissolution improved by bacterial (iron- and sulfur-oxidizing bacteria-based) bioleaching for IBA without ferrous materials (Panda, 2020). Non-ferrous materials can be recovered by using eddy current separation (ECS).

10.5.2 Chemical treatment

10.5.2.1 Aging

Aging is the stockpiling of large amount of ash in the open air which allows some draining and evaporation. It is seen that aging process reduces leachability of heavy metals and TOC, changes mineral phases, as well as decreases pH level from alkaline to neutral values. There are various methodologies used for weathering namely accelerated carbonation, air treatment and treatment with inert gas (N_2 gas) (Polettini and Pomi, 2004).

It is well known that solubility depends on pH of the aqueous medium. Thus, pH adjustment of the MSWI ash can be effective to reduce the solubility of various metals. Normally, adjustment of pH is done by adding lime to the ash which causes rising of the ash pH into a range where metals are less soluble. Polettini and Pomi (2004) investigated metal leaching behavior of IBA by accelerated aging (e.g., carbonation, treatment with air and N_2) and found that the leaching rate of heavy metals was a function of both the types of aging methods and duration of treatment.

10.5.2.2 Solidification/stabilization (S/S) method

In the S/S process, the hazardous contents present in MSWI ash are immobilized by chemical and/or physical methods. The IFA is encapsulated using binders like cement. This minimizes the solubility and leachability of contaminates (Lam, 2010). Hence, the cement-based solidification technology is very popular and quite effective on reduction of pollution due to MSWI fly ash. In the cement-based S/S technology, cement is used as a binder, and the potential mechanisms for immobilization of hazardous contaminants (e.g., heavy metals) include precipitation, surface complexation, sorption and encapsulation (Chen et al., 2009). The reaction between cement and water is exothermic and the mixture turns into alkaline. The main chemical product in the hydrated cement is calcium silicate hydrogel (C-S-H), and it is primarily responsible for immobilization of heavy metals due to its network-like structure. A study on S/S of MSWI fly ash using Portland cement demonstrated that several factors such as cement-to-IFA (C/F) ratio and curing duration had considerable impacts on the leaching behaviors of heavy metals in the stabilized IFA (Tang et al., 2016). They have noticed that with the increase of C/F ratio from 4:6 to 8:2, the extent of heavy metal leaching (Cd, Pb, Ni, Cr, Cu, etc.) reduced by 2.33%–85.23%. Similarly, with enhancement of curing period from 3 to 56 days, the degree of leaching of metals significantly declined (i.e., by

16.5% to 88.7%). The authors have stated that the reduction of leaching is possibly due to adsorption or migration of heavy metals into the minerals formed (e.g., ettringite) by reaction of IFA and cement. To avoid the heavy metal release from solidified products, pre-treatment like water washing of IFA is recommended. Wang et al. (2015) used four different types of chemical reagents (also called solidifying agents such as sixthio guanidine acid (SGA), tetrathio bicarbamic acid (TBA), sodium dimethyldithiocarbamate (SDD) and Na_2S) to test degree of S/S of MSWI fly ash. Among the four reagents, only SGA and TBA showed higher S/S effectiveness, which possibly link to their chemical structure, i.e., presence of multiple hydrosulfide groups which bind with heavy metals very strongly, thus reducing their leachability. The extent of heavy metal leaching in IBA can be reduced by addition of stabilizing reagents namely phosphate, sulfide, carbonate and other commercial products. Overall, S/S is a well-accepted method for immobilization of heavy metals in waste materials using various approaches such as use of cement or other binders, geopolymer formation and application of chemically synthesized solidifying agents.

10.5.2.3 Chemical extraction (chemical leaching)

Chemical leaching is a process to extract heavy metals from MSWI ash residues. The process involves the detachment of soluble constituents from a solid phase by a solvent (Wong et al., 2020). In the leaching phase of MSWI ash, several distinct chemical reagents have been used to remove and recover heavy metals from ash. The leaching behavior is highly influenced by the type of chemical reagents used and other environmental factors. Generally mineral acids like HCl, HNO_3, and H_2SO_4, alkaline solutions namely NaOH, aqueous NH_3 and other chelating agents such as ethylenediaminetetraacetic acid (EDTA), ammonium nitrate and ammonium chloride are used for leaching of heavy metals (Fedje et al., 2010). It has been found that hydrochloric acid can be used to strip Cr, Cu, Pb and Zn from IFA. Fedje and co-workers drew a conclusion that the common organic acids like formic acid, acetic acid, lactic acid and oxalic acid were not as effective as mineral acids for metal leaching (Fedje et al., 2010), even if it was true that these organic acids form soluble metal complexes (Tang and Steenari, 2016). Sulfuric acid and oxalic acid are found to be inefficient in the leaching of Pb due to the low solubility of $PbSO_4$ and PbC_2O_4. Except for Ca and Pb, H_2SO_4 dissolves many metals (Wu and Ting, 2006).

Zhang and Itoh (2006) performed metal extraction from MSWI fly ash using water washing pre-treatment plus acid-assisted hydrothermal treatment. Three types of inorganic acids (HCl, HNO_3, and H_2SO_4) and two types of organic acid (oxalic and citric acid) were used to examine their leaching efficiency. The water washing helps removal of high quantity of light elements like Na (67%), K (76%) and Ca (48%). Among the five types of acids tested, HCl was the best for extraction of metals (Cr, Cd and Pb) from ash; however, application of hydrothermal conditions (150°C for 5h) further accelerated the metal dissolution efficiency (Zhang and Itoh, 2006). Nagib and Inoue (2000) examined removal rate of Pb and Zn from MSWI fly ash using acids, namely H_2SO_4, HCl and acetic acid as well as alkaline solution (NaOH). The study found that H_2SO_4 is effective for extraction of Zn, whereas HCl or acetic acid shows promising performance for solubilization of both Pb and Zn. The alkaline treatment was not effective specifically for Zn (low solubility). In total, organic acid, inorganic acid, alkaline solution and chelating agents are used for extraction of heavy metals from MSWI ash materials in chemical leaching method, and the removal efficiency depends on the selection of chemical reagents and the target metals.

10.5.3 Factors affecting chemical leaching process

10.5.3.1 Characteristics of MSWI ash

The key parameters that influence properties of MSWI ash include operational conditions like incineration temperature, incineration technology, composition of the wastes, and its bulk content, particle size distribution. The major factors that impact for selection of an appropriate physicochemical treatment process include ash composition and its size, hazardous level, solubility of elements present in ash, hydraulic conductivity, specific gravity, and pH (Rendek et al., 2007). Mobility of ash during chemical leaching also depends on its physical form namely shape and structure (Dandautiya et al., 2018).

10.5.3.2 Liquid-to-solid ratio (L/S)

L/S plays a keen role in leachability of heavy metals in chemical leaching process. Its influence has been studied using sequential and column tests as well as using mathematical models (Grathwohl and Susset, 2009; Guyonnet, 2010; Quina et al., 2011). Heavy metal leaching is fundamentally controlled by solubility as L/S critically affects the leaching of heavy metals. In some cases, increase of the L/S ratio, the leaching efficiency of heavy metals, enhanced. However, the impact of L/S is element specific like with slow release initially and higher mobility with enhancing cumulative L/S ratios.

10.5.3.3 pH

Leaching of organic and inorganic elements from MSWI ash is also controlled by pH, and it differs in different metal species. A recent study which evaluated effect of acid washing of IBA on leaching of heavy metals showed that the amount of leaching was increased for most of the metal with rise of acid concentration (Sun and Yi, 2021). Another study noticed that the release of Cr, Cu, Pb and Zn was high in alkaline as well as acid conditions (Silva et al., 2019). Zhang et al. (2016) evaluated effects of pH ($pH = 2–14$) on leaching behavior of several elements including heavy metals from MSWI fly ash, and they reported that leaching characteristics of alkaline elements were not considerably changed with change of pH; however, leaching profile of heavy metals significantly impacted by pH and the highest metal extraction (e.g., Al and As) observed at acidic environment ($pH < 2.0$). The leaching of Se and Ba was optimum at alkaline condition ($pH > 11$). A geotechnical modeling study shows that leaching of metals (except As, Se, Ag and Cr) was controlled by various mechanisms such as precipitation or dissolution of their carbonate, oxides, hydroxides or sulfate solids (Zhang et al., 2016).

10.5.3.4 Temperature

A few studies have reported positive effects of temperature on leaching of heavy metals from MSWI ash, which means the reaction may be endothermic (Agcasulu and Akcil, 2017; Nagib and Inoue, 2000). Agcasulu and Akcil (2017) conducted acid leaching of IBA using 2M HCl and found that the recovery of Cu, Fe, Mo, Ni and V increased (3%–14%) with rise of temperature from 25°C to 80°C. Nagib and Inoue (2000) performed alkaline leaching tests of Pb and Zn in IFA using 0.75 N NaOH and observed that Pb removal increased from 20% to 53% by elevation of temperature from 30°C to 90°C. But removal of Zn was not dependent on temperature since only 8% of Zn was recovered with the rise of temperature. Similar findings were reached by Tang and Steenari (2016) where the authors carried out acid leaching of IFA using HCl and HNO_3 (pH = 2.0) with temperature ranging between 20°C and 60°C. They stated that the leaching of Fe enhanced with rise of temperature, whereas temperature has negative impacts on Zn leaching under both types of acid solutions.

10.5.3.5 Weathering/aging

A few mineralogical and chemical properties of MSWI ash could undergo drastic changes during weathering, along with oxidation of certain metals (e.g., Al, Fe and Cu), solubilization and precipitation of the main cation hydroxides and salts, carbonation, pH neutralization and neo-formation of glass-like clay minerals (Chimenos et al., 2003). The pH and natural weathering reactions namely carbonation and metallic hydroxide production depend on the composition of the bulk and size of the ash particles. The macroscopic properties that can be altered due to weathering/aging are pH, redox potential, acid neutralization capacity, ion-exchange and sorption capacities (Polettini and Pomi, 2004). Weathering has long been recognized to have huge influence on the leaching of trace heavy metals, while at the same time immobilizing trace elements by newly formed reactive sorptive solid substrates and clay-like minerals (Chimenos et al., 2003). IBA is normally alkaline in nature, and thus prior to reuse for any construction-related works, the material must be subjected to a weathering step in which the carbonation process helps to decrease its pH to nearly environmentally safe neutral level (Silva et al., 2019).

10.5.3.6 Presence of organic compounds

Organic compounds like polychlorinated biphenyl (PCB), chlorinated compounds (chlorobenzenes, chlorophenols, halogenated organic acids, etc.), phthalates, aldehydes, ketones, organic acids, alkanes and alkenes are generally present at varying proportions in MSWI ash which may affect the ash treatment process (e.g., chemical leaching) (Lisk, 1988). Among those organic compounds like humic substances, namely humic acid and fulvic acid, are the major compounds detected in the ash leachate. These substances increase the solubility of hydrophobic substances by complexation mechanism as well as they may also change the biotoxicity and bioavailability of hazardous components, thereby influencing their biodegradation or bio-uptake behavior (Fan et al., 2006).

10.5.4 Thermal treatments of IBA and IFA

10.5.4.1 Vitrification

Vitrification is the process in which ash treatment is carried out at relatively high temperature which converts ash into a chemically stable dense glassy material (Kuo et al., 2012). The treatment is usually carried out between 1,000°C and 1,500°C (Nowak et al., 2010; Park and Heo, 2002). During vitrification, the ash produces three types of products: vitreous product, metallic product and matte product. The vitreous material can be used as aggregates for civil engineering works including concrete production and asphalt, while the metallic and matte products can be recycled for other purposes. The vitrification technology has three major advantages (Kuo et al., 2012). First, the process can immobilize heavy metals and convert the toxic metals into stabilized slag. Second, it can disintegrate various organic compounds including polyaromatic hydrocarbons (PAHs). Third, it can transform certain valuable metals to recoverable forms. Nowak et al. (2010) conducted heavy metal removal from MSWI fly ash by two steps: first chlorination and then thermal treatment. In this work, first a mixture was formed using IFA, calcium chloride ($CaCl_2$) and water, and then the mixture was taken for thermal treatment using rotary reactor at 1,000°C. It was found that various metals including Cd, Cu, Pb and Zn were efficiently (60%–90%) stabilized from IFA. Park and Heo (2002) vitrified MSWI fly ash at 1,500°C for 30 min, and the produced glassy material possessed impressive mechanical properties (e.g., Vickers hardness of 4,000–5,000 MPa) as well as it showed excellent resistance to heavy metal leaching, i.e., Pb <0.2 mg/L, Cd and Cu <0.04 mg/L and Cr <0.02 mg/L. Together, vitrification is an effective method for stabilization of hazardous components of waste materials (specifically waste having high chlorine content).

10.5.4.2 Melting

Melting is another thermal stability process by which the metallic components are immobilized in the molecular structure of the waste material, thus hindering them from mobilization and diffusing into the immediate environment (Sakai and Hiraoka, 2000). In the melting process, the ash is usually melted at an elevated temperature (1,100°C–1,400°C). During the process, the ash residues transform into both physical and chemical state, and the melted product is solidified by bringing down the temperature (called cooling). In this way, the mass and volume of the ash materials are significantly decreased, forming a high-density melted material. The melted slag product shows high stability and is often considered as a resource. The solidified slag can be employed for civil engineering application such as road constructions and land reclamation. Sakai and Hiraoka (2000) reported that during melting of MSWI fly ash at 1,400°C, certain dioxins namely polychlorinated dibenzo-p-dioxins (PCDDs) and polychlorinated dibenzofurans (PCDFs) are decomposed as well as volatile heavy metals are concentrated in the IFA. Collectively, the melting technology converts MSWI ash into stable and non-hazardous product which can be recycled for use in construction industry.

10.5.4.3 Sintering

Sintering is one of the thermal treatment approaches to handle MSWI ash (Wey et al., 2006). The technique of sintering is considered promising as it combines the possibility to detoxify and recycle the ash as building materials (Wang et al., 2001). In sintering, the particles are bonded coherently further enhancing the strength and other characteristics of the compact particle (Wang et al., 1998). For treatment of IFA by sintering, the IFA is bonded together after firing it at high temperatures. The finished product is sintered matrices that have a good amount of strength and low chances of heavy metal leaching, and it depends on the particle size of IFA (Wang et al., 2001). Using high temperatures below the melting point of waste between 700°C and 1,100°C, the organic residues are decomposed, and inorganic components and heavy metals are stabilized. The stabilization takes place because of incorporation of oxides formed by elemental constituents into a material that has texture and appearance like ceramic (Wey et al., 2006). The leaching characteristics of heavy metal in the sintered IFA are generally based on the speciation of heavy metals produced during the sintering process. In addition, the speciation variance is influenced by the sintering atmosphere and the sintering temperature as well. It is irrefutable that the degree of compaction of powders, the temperature and duration of sintering along with retention time and the chemical composition of IFA may also have a significant impact on various properties: the form and magnitude of porosity and, consequently, on the specific gravity, mechanical strength and leaching tendency of heavy metals of the sintered products as well as on their chemical stability in aqueous medium (Wey et al., 2006). In IFA, there are several minor compounds such as alkali metal, alkaline-earth metal, chloride and sulfate; the partitions of different compounds affect the sintering process temperature and the quality of products (Wey et al., 2006).

10.6 BIOLOGICAL METHOD (BACTERIA- AND FUNGAL-BASED BIOLEACHING) FOR TREATMENT OF IBA AND IFA

Conventional pyro- or hydro-metallurgical methods for treatment of MSWI ash and heavy metal recovery provide a fast treatment and nearly total detoxification of toxic compounds. But these techniques suffer major drawbacks such as high energy consumption and possible release of secondary hazardous by-products to the environment during treatment. Since most of the conventional methods (including pyro- or hydro-metallurgical) are highly expensive, energy intensive and not environmentally friendly, researchers have explored biohydrometallurgical method (also called bioleaching and considered as a green technology) as an alternative for removal of heavy metals from MSWI ash residues (Mishra and Rhee, 2014; Xu et al., 2014). Biological leaching not only helps in resource (i.e., metal) recovery from ash residues but also detoxifies the wastes which can be disposed or reused safely. Generally, bioleaching method is broadly classified into two types (i.e., one-step or two-step) based on the leaching organism's (bacteria or fungus) growth pattern (presence or absence of ash). The one-step approach is like a conventional growth system in which the leaching organism is incubated with the metal-containing waste material. This method could be suitable for waste containing low amounts of toxic components since the growth of the leaching organisms may be impacted by the presence of toxic metals in waste. High liquid-to-solid ratio could be an option to compensate/reduce potential toxic effects to microorganisms. However, in two-step approach, the leaching organism is initially inoculated into the growth media and incubated for a certain time to reach the complete growth, and then leaching is initiated by addition of metal-containing waste to the pre-growth culture media (Xu et al., 2014). This method could be suitable for waste having complex toxic elements that could inhibit growth of the leaching organisms. In the above two types of leaching methods, physical contact occurs between leaching organism and waste material. However, a few studies have carried out leaching of waste materials using cell-free solution (called spent medium) (Wu and Ting, 2006). In this method, the cells are initially cultivated in the growth medium, and then leaching was attempted by introducing waste materials into the cell-free spent medium (removed cells by centrifugation-filtration or other techniques). This process can be applied for wastes containing highly toxic elements. The biological leaching is mainly driven by metabolic products (e.g., bioacids) produced by the leaching organisms during growth. The quality and quantity of bioacids depend on many factors including nature of leaching organisms, the substrate used for growth, nature of waste materials, L/S ratio, etc. (Wu and Ting, 2006) (Table 10.5).

Table 10.5 Production of organic acids (bioacids) during fungal bioleaching of municipal solid waste IFA

MSWI ash	*Pulp density, % (w/v)*	*Type of bioacid*	*Bioacid concentration (mM) in Aspergillus niger bioleaching*			*References*
			Pure culture	*One-step leaching*	*Two-step leaching*	
IFA	1%–6%	Citrate	~50	NA[a]	30–36	Xu and Ting (2009)
IFA	5%	Citrate	110	ND[b]	NA	Bosshard et al. (1996)
		Gluconate	ND	283	NA	
IFA	1.0%	Citrate	51.6	38.6	33	Xu et al. (2014)
		Oxalate	4.62	1.93	1.27	
		Gluconate	16	171.16	163.7	
IFA	1%–2%	Citrate	70	10.5–13.6	14.3–29.3	Wu and Ting (2006)
		Oxalate	3	1.3–1.9	1.8–2.9	
		Gluconate	ND	171.2–208.1	143.4–172.1	
IFA	1%–4%	Citrate	3.95	2.23	0.09–3.90	Wang et al. (2009)
		Oxalate	ND	0.66	0.36–0.54	
		Gluconate	24.31	0.42–60.77	0.36–30.10	
IFA[c]	5%–7%	Citrate	7.23–9.16	NA	8.44–63.71	Yang et al. (2009)
		Oxalate	2.08–3.25	NA	0.97–1.82	
		Gluconate	0–0.31	NA	121.62–217.84	
IFA	1%–4%	Citrate	0.1	0.1–5.3	7.5–11.1	Yang et al. (2008)
		Oxalate	6.8	0.5–7.2	0.2–16.8	
		Gluconate	13	28.4–36.6	23.1–43.5	

[a]NA: not available.
[b]ND: not detected.
[c]This study used two types of *Aspergillus niger* strains namely *A. niger* AS 3.879 and *A. niger* AS 3.40.

The first batch-scale bioleaching study was carried out by Bosshard et al. (1996) on MSWI fly ash under one-step and two-step processes using a fungal species, *Aspergillus niger*. In the one-step method with gluconic acid as the leaching agent, the proportion of heavy metal (Cd, Pb, Zn, Mn, Cu and Al) removal varied between 30% and 57%. However, the heavy metal removal efficiency increased in two-step process with citrate as the dominant leaching agent and the metal removal increased (i.e., from 27% to 81%). Wu and Ting (2006) performed bioleaching of IFA using *A. niger* following both one-step and two-step techniques. They have noticed that the metal leaching efficiency was similar in both processes with gluconic acid as the lixiviant in bioleaching agent. The proportion of metal dissolution varied based on the nature of metals, i.e., removal of Al, Mn and Zn was 80%–100%, Cu and Pb was 60%–70% and only 30% for Fe. Yang et al. (2009) reported that presence of Fe and Al in MSWI ash reduced the growth of two *A. niger* strains (named as *A. niger* AS 3.879 and AS 3.40). Here, the authors have explored to adopt the two *A. niger* strains to the higher metal concentration (Fe = 700 mg/L and Al = 3500 mg/L), and then conducted two-step bioleaching of heavy metals from MSWI fly ash. The metal adopted fungal strains were effective for the metal extraction than the control strain (without adaptation). For example, with inoculation of un-adopted *A. niger* (AS 3.879 with 50 g/L IFA), the extraction yield of Cd, Cr, Cu and Zn was 32%, 9.48%, 16.69% and 19.26%, respectively, while the extraction yield significantly increased to 98.01%, 26.82%, 73.78% and 61.30%, respectively, when the leaching test was conducted using aluminum-adopted *A. niger* AS 3.879 strain. Since MSWI ash contains several hazardous elements including alkali salts, to reduce their negative effect on bioleaching, Wang et al. (2009) pre-treated MSWI fly ash using simple water washing, and then carried out bioleaching experiments using *A. niger* (AS 3.40). It is seen that the pre-treatment removed several metals such as K (50.6%), Na (41.1%), Ca (5.2%) and Cr (1%). Moreover, pre-treatment improves dissolution of IFA particles due to removal of chloride-containing alkali salts. Notably, *A. niger* growth lag phase and bioleaching duration were declined by 45% and 30%, respectively, which is possibly due to removal of toxic components by water washing.

In addition to experimental works, a few studies have studied modeling of fungal growth and bioleaching rate kinetics. Xu and Ting (2009) investigated growth kinetics of *A. niger*, bioacid production and kinetics for dissolution of metals at different pulp densities of IFA (1%–6%). The growth pattern of *A. niger* can be well explained and best fit to the modified Gompertz model. The specific growth rate of *A. niger* in the two-step process significantly decreased with increase of pulp density, i.e., reduced from 0.100 day^{-1} (1% pulp density) to 0.033 day^{-1} (6% pulp density). Xu and Ting (2004) applied central composite design (CCD) to find an optimum condition for fungal bioleaching of IFA received from an MSWI plant. They have tested four parameters namely concentration of caron source (sucrose), inoculum size (fungal spore concentration), pulp density (IFA quantity), and the time at which IFA was added into the fungal-based bioleaching system. Using empirical models, authors have found that two factors namely substrate concentration (sucrose) and pulp density are most critical than others for bioleaching study.

The following mechanisms have been suggested on metal removal in fungal biological leaching due to production of citric acid as a metabolite in sucrose ($C_6H_{12}O_6$) medium (Hocheng et al., 2014).

$$C_6H_{12}O_6\,(\text{sucrose}) + 1.5O_2 \rightarrow C_6H_8O_7\,(\text{Citric acid}) + 2H_2O \tag{10.1}$$

$$C_6H_8O_7 \leftrightarrow (C_6H_7O_7)^- + H^+ \tag{10.2}$$

$$(C_6H_7O_7)^- + M^+\,(\text{Metal}) \leftrightarrow \text{Metal Citrate complex} \tag{10.3}$$

In addition to fungus, many reports on bioleaching of MSWI ash have applied inorganic acid producing bacteria, namely iron-oxidizing bacteria (IOB) and/or sulfur-oxidizing bacteria (SOB) (Brombacher, 1998; Ishigaki et al., 2005). Like fungal bioleaching which is mediated by biogenic acid (gluconic, citric or oxalic acid), SOB produce sulfuric acid, while IOB oxidize reduced metal compounds. In a lab-scale leaching plant (LSLP) study with 50 g/L of IFA pulp density and using a mixed bacteria of *Thiobacillus thiooxidans* and *Thiobacillus ferrooxidans* (Brombacher, 1998), high removal efficiency of various metals was observed: Cd (100%), Zn (81%), Cu (89%), Ni (64%), Cr (12%) and Al (52%). The authors have suggested that the removal of Cu was due to metabolic activity of *T. ferrooxidans*, whereas release of other metals was due to biologically produced H_2SO_4. In another study, researchers have used pure and mixed culture of SOB and IOB for bioleaching study as well as proposed the mechanism of heavy metal removal from MSWI fly ash (Ishigaki et al., 2005). Mixed culture system shows better performance than pure culture system, and the removal efficiency for various metals is as follows: Cr (100%), Cd (100%), Zn (78%) and Cu (67%). Ferric iron (Fe^{3+}) was detected in the mixed culture, and higher metal recovery was observed due to redox mechanisms coupling with the dissolution by sulfate. Among the two bacteria, IOB was better for metal leaching as well as tolerance to high concentration of heavy metals, but SOB showed tolerance to higher dose of IFA.

The microbial-mediated solubilization of metals from waste materials can be due to two possible mechanisms namely direct and indirect mechanism (Chen and Lin, 2000; Ishigaki et al., 2005). The direct mechanism is mainly suitable for sulfide-containing ores, whereas the indirect mechanism is applicable for non-sulfide-containing waste materials. The following equations represent the metal removal by two types of mechanisms (Chen and Lin, 2000).

i. **Direct mechanism:**

$$MS + 2O_2 \xrightarrow{\textit{thiobacilli}} MSO_4 \tag{10.4}$$

ii. **Indirect mechanism:**

$$S^0 + H_2O + \frac{3}{2}O_2 \rightarrow H_2SO_4 \quad (\text{Production of } H_2SO_4 \text{ by } \textit{thiobacilli}) \tag{10.5}$$

$$H_2SO_4 + M \rightarrow MSO_4 + 2H^+\,(M = \text{Metal}) \tag{10.6}$$

In *Acidithiobacillus ferrooxidans* bioleaching process, the metal solubilization is performed based on the oxidation mechanism. *A. ferrooxidans* uses ferrous iron (Fe^{2+}) as an energy source and forms ferric ion (Fe^{3+}). The ferric ion is a strong oxidizing agent, and it reacts with metals and solubilizes them (Hocheng et al. 2014). The reaction pathways are described below.

$$2Fe^{2+} + 2H^+ + 0.5O_2 \rightarrow 2Fe^{3+} + 2H_2O \quad (10.7)$$

$$Fe^{3+} + e^- \rightarrow Fe^{2+} \quad (10.8)$$

$$M_{(s)} - e^- \rightarrow M^+{}_{(aq)}\,(M = \text{Metal}) \quad (10.9)$$

In addition to the wild application of autotrophic chemolithotrophic bacteria (*A. thiooxidans* and *A. ferrooxidans* which use inorganic substrate such as ferrous iron and reduced sulfur as electron donor and energy source), a few studies have reported bioleaching of heavy metals from IBA using heterotrophic bacteria namely *Pseudomonas aeruginosa* PAO1 that can use glucose as a C-source (Biswal et al., 2020, 2019). The study by Biswal et al. (2020) showed that the heavy metal (As, Ba, Co, Cr, Cu, Pb and Zn) removal varied between 20.2% and 95.1%. Gluconic acid is one of the key metabolic products due to glucose metabolism by *P. aeruginosa* PAO1, and the potential metal removal mechanism is presented below (Biswal et al., 2020; 2019). A comparison of metal removal efficiency from MSWI ash in fungal and bacteria bioleaching and chemical leaching performed with equimolar concentration of biogenic acids is summarized in Table 10.6.

$$C_6H_{12}O_6\ (\text{Glucose}) + 0.5\ O_2 \rightarrow C_6H_{12}O_7\ (\text{Gluconic acid}) \quad (10.10)$$

$$C_6H_{12}O_7\ (\text{Gluconic acid}) \leftrightarrow (C_6H_{11}O_7)^-\ (\text{Gluconate}) + H^+ \quad (10.11)$$

$$(C_6H_{11}O_7)^- + M^+\,(\text{Metal}) \rightarrow [\text{M-Gluconate}]\ \text{complex} \quad (10.12)$$

10.7 METAL MOBILIZATION AND IMMOBILIZATION IN BIOLOGICAL TREATMENT PROCESSES

10.7.1 Microbes-metal interactions: metal mobilization

The ubiquitous characteristics of microorganisms have made them use for treatment of various environmental pollutants including toxic metals. Many microorganisms (e.g., heterotrophic and autotrophic microbes) can facilitate mobilization of heavy metals from the solid materials by transferring the metals from insolubilized to solubilized form. Moreover, microbes can change the bioavailability and possible toxicity of metal by altering its valence state through oxidation or reduction reactions. The metal mobilization can be due to the following biological processes (Brandl and Faramarzi, 2006) (Figure 10.2).

Table 10.6 Comparison of metal removal yield from municipal solid waste incineration ash by abiotic (chemical) and biotic (fungal and bacteria) leaching

Mode of study	*Ash type*	*Metal*	*Abiotic leaching (chemical) (%)*	*Biotic leaching (%)*	*Additional information*	*References*
IBA						
Column	IBA	Zn	-	42.9	Chemical: no data available	Mäkinen et al. (2019)
		Cu	-	5.8	Biotic: mixed bacteria culture[a]	
		Fe	-	18.1		
		Al	-	30.5		
Shaking flask	IBA	Zn	-	80	Chemical: no data available	Funari et al. (2019)
		Cu	-	100	Biotic: mixed bacteria culture[b]	
		Pb	-	20		
Shaking flask	IBA	Cu	-	20.2	Chemical: no data available	Biswal et al. (2019)
		As	-	24.7	Biotic: bacteria (*P. aeruginosa* PAO1)	
		Pb	-	90.4		
		Cr	-	93.0		
		Ba	-	95.1		
		Zn	-	87.4		
		Co	-	91.3		
		Ni	-	94.4		
Shaking flask	IBA	Cu	-	73.98	Chemical: no data available	Panda (2020)
		Ni	-	98.68	Biotic: mixed bacteria culture[c]	
		Zn	-	59.09		
		Al	-	58.84		
		Ti	-	92.85		
IFA						

(*Continued*)

Table 10.6 (Continued) Comparison of metal removal yield from municipal solid waste incineration ash by abiotic (chemical) and biotic (fungal and bacteria) leaching

Mode of study	*Ash type*	*Metal*	*Abiotic leaching (chemical) (%)*	*Biotic leaching (%)*	*Additional information*	*References*
STR[d]	IFA	Zn	85.2	88.7	Chemical: H_2SO_4 + Fe(III) (pH: 1.8)	Li et al. (2020)
		Cd	83.8	93.2	Biotic: mixed bacteria culture[e]	
		Mn	86.8	99.9		
		Pb	11.4	13.8		
Glass reactor	IFA	Zn	-	91	Chemical: H_2SO_4 (pH: 2.0)	Funari et al. (2017)
		Cu	95	74	Biotic: mixed bacteria culture[f]	
		Pb	45	59		
		Cr	70	63		
		Co	28	55		
		As	13	18		
		Ni	93	66		
Shaking flask	IFA	Cd	100	81	Chemical: 110 mM citric acid (pH: 2.3)	Bosshard et al. (1996)
		Zn	~70	66	Biotic: fungus (*A. niger*)	
		Cu	~62	57		
		Pb	~55	52		
		Mn	~34	32		
		Al	~30	27		
Shaking flask	IFA	Al, Mn Zn	80–100 (Al, Cu); 48–80 (Mn, Zn); <60 (Pb, Fe)	80–100	Chemical: mixed acids[g]	Wu and Ting (2006)
		Cu, Pb		60–70	Biotic: fungus (*A. niger*)	
		Fe	-	30		
LSLP[h]	IFA	Zn	100	81	Chemical: 5N H_2SO_4 (pH: 2)	
		Al	64	52	Biotic: mixed bacteria culture[i]	Brombacher et al. (1998)

(Continued)

Table 10.6 (Continued) Comparison of metal removal yield from municipal solid waste incineration ash by abiotic (chemical) and biotic (fungal and bacteria) leaching

Mode of study	*Ash type*	*Metal*	*Abiotic leaching (chemical) (%)*	*Biotic leaching (%)*	*Additional information*	*References*
		Cd	89	100		
		Cu	88	89		
		Ni	22	64		
		Cr	11	12		
Shaking flask	IFA	Mn	~29–58	98.7	Chemical: mixed organic acids[j] Biotic: fungus (*A. niger*)	Yang et al. (2008)
		Cd	80–85	87.6		
		Cu	78–82	69.7		
		Zn	50–57	68.5		
		Cr	20–30	42.1		
		Fe	~17–25	31.7		
		Pb	58	36.5		
Shaking flask	IFA	Zn	-	90	Chemical: no data available Biotic: mixed bacteria culture[b]	Funari et al. (2019)
		Cu	-	90		
		Pb	-	10		

a Mixed culture of iron and sulfur-oxidizing microorganisms: *Marinobacter* sp., *Acidithiobacillus* (namely *Acidithiobacillus ferrooxidans, Acidithiobacillus thiooxidans, Acidithiobacillus albertensis, Acidithiobacillus ferrivorans*), *Leptospirillum* (*L. ferrooxidans*), *Cuniculiplasma, Nitrosotenius* and *Ferroplasma*.
b Mixed bacteria culture: Mixed culture of *Acidothiobacillus thiooxidans* and *Acidothiobacillus ferrooxidans*.
c Mixed bacteria culture: *Acidithiobacillus ferrooxidans, Leptospirillum ferrooxidans, Leptospirillum ferriphilum*, and *Acidithiobacillus thiooxidans*.
d STR: Stirred tank reactors.
e Mixed bacteria culture: Mixed species of iron-oxidizing acidophiles namely *Leptospirillum ferriphilum, Ferroplasma thermophilum* and *Sulfobacillus thermosulfidoxidans*.
f Mixed acidophilic bacteria culture: *Acidithiobacillus ferrooxidans, Acidithiobacillus thiooxidans, Acidithiobacillus caldus, Leptospirillum ferrooxidans, Sulfobacillus thermosulfidooxidans, Sulfobacillus thermotolerans* and some members of *Alicyclobacillus* genus.
g Mixed acids: Mixture of organic (citric, oxalic and gluconic acids) (0.1 M) and inorganic acids (sulfuric and nitric acid) (0.5 M).
h LSLP: Laboratory-scale leaching plant.
i Mixed bacteria culture: *Thiobacillus thiooxidans* and *Thiobacillus ferrooxidans*.
j Mixed organic acids: 50 mM citric and gluconic acids.

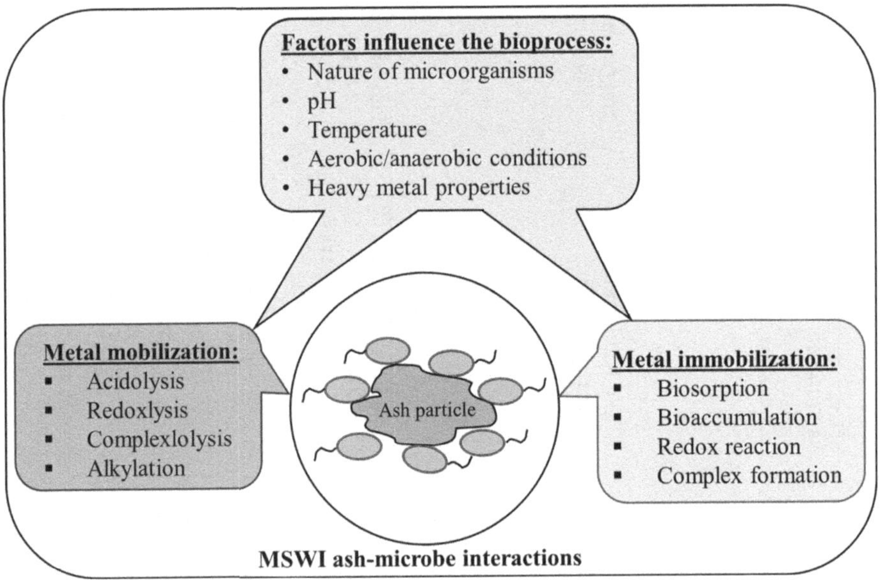

Figure 10.2 Metal mobilization and/or immobilization due to interactions between MSWI ash particles and microbes as well as the key environmental factors that impact the bioprocesses.

10.7.1.1 Acidolysis

The inorganic acid (sulfuric acid) or organic acids (e.g., citric acid and oxalic acid) produced from the bacteria and fungus by metabolism of inorganic/organic substrates contribute to acidolysis-based metals solubilization. The process is also called as proton (e.g., H^+) induced metal dissolution in which bioleaching agents (bacteria and fungus) secrets protons that enhance metal mobility (Brandl, 2002). Moreover, in this mechanism, protons are bound to the solid material surface causing the weakening of chemical bonds, and then displacement of metal ions from the surface of solid material.

10.7.1.2 Redoxolysis

Redoxolysis is a process of microbe-mediated oxidation-reduction of metals (Brandl, 2002). Metal mobilization enhances depending on the nature of metals, metal oxidation state and its mineral phase (Mishra and Rhee, 2014). The metal solubilization can be through "direct (contact)" or "indirect" leaching. In direct mechanism, physical contact between minerals and bacterial cells are needed, and microorganisms oxidize minerals (e.g., iron sulfides) by acquiring electrons from the reduced minerals (Brandl, 2008). In indirect mechanism, oxidation of reduced metals in minerals is mediated by microbially produced ferrous iron (Fe^{2+}). The ferrous iron present in the iron-containing minerals is oxidized (regeneration) by IOB (e.g., *Acidithiobacillus ferrooxidans* and *Thiobacillus ferrooxidans*) (Rawlings, 2004). In this process, iron acts as an electron carrier, and no direct physical contact is required for the oxidation of iron. The direct and indirect leaching are presented in the following equations (Bosecker 1997; Brandl, 2008; Rawlings, 2004).

i. **Direct mechanism:**

$$FeS_2 + \frac{7}{2}O_2 + H_2O \xrightarrow{\textit{thiobacilli}} FeSO_4 + H_2SO_4 \tag{10.13}$$

ii. **Indirect mechanism:**

$$2FeSO_4 + \frac{1}{2}O_2 + H_2SO_4 \xrightarrow{\textit{T. ferrooxidans}} Fe_2(SO_4)_3 + H_2O \tag{10.14}$$

$$FeS_2 + Fe_2(SO_4)_3 \xrightarrow{\text{Chemical oxidation}} 3FeSO_4 + 2S \tag{10.15}$$

$$2S + 3O_2 + 2H_2O \xrightarrow{\textit{T. thiooxidans}} 2H_2SO_4 \tag{10.16}$$

10.7.1.3 Complexolysis

The complexolysis is also called ligand-induced metal dissolution (Brandl, 2002). In this mechanism, microorganisms excrete complexing or chelating compounds that enhance metal mobilization. Complex compounds are produced on surface of metals by ligand exchange polarizing critical bonds and enable the removal of metals from the surface (Brandl, 2002). The organic acid anions can form complexes with positively charged metals, and thus enhance mobility (Gadd, 2001). Some microbes (e.g., some members of *Shewanella* species) are capable of formation of low molecular weight iron-chelating compounds (called siderophores) which mobilize mainly ferric (Fe^{3+}) ion (Gadd, 2001).

10.7.1.4 Alkylation

Biological alkylation (e.g., methylation) of metal yields volatile derivatives, and the process is also called "metal volatilization" because alkylated metals exhibit higher volatility than to their elemental or ionic forms (Brandl, 2008; Gadd, 2001). In this process, enzymes released from organisms such as bacteria, algae and fungi facilitate covalent attachment of alkyl groups to the metal. The alkylation process is often used for decontamination of heavy metal (As, Se, etc.) from polluted soil and water (Gadd, 2001). Fungal species namely *Penicillium* sp., *Candida humicola* and *Gliocladium roseum* can transfer monomethylarsenic acid to trimethylarsine as well as some fungus can convert selenite and selenate to methyl products namely dimethylselenide and dimethyldiselenide (Gadd, 2001).

10.7.2 Microbes-metal interactions: metal immobilization

Besides solubilization, microbe-metal interactions could facilitate metal immobilization, and biological immobilization can be classified as direct mineralization through enzymatic actions or indirect/passive mode by microbial induced mineralization (MIM) (Jain et al., 2010). The following mechanisms have been described for biological immobilization of metals (Brandl and Faramarzi, 2006).

10.7.2.1 Biosorption

Biosorption is a process of uptake of soluble and insoluble metals on the surface of living or dead cells, and the physicochemical process is called adsorption (Gadd, 2001). In this mechanism, the negatively charged functional groups namely carboxyl, hydroxyl, phosphate, sulfonate and amino present in the microbial cell can attract and adsorb a variety of dissolve metals on to its surface (Brandl and Faramarzi, 2006; Jain et al., 2010). Biosorption can be due to the following pathways: physicochemical interactions (electrostatic forces, and ion or proton transposition), complexation or chelation (Diep et al., 2018). The degree of metal sorption depends on chemical nature of microbial cells and environmental parameters such as salinity and pH (Brandl and Faramarzi, 2006). For example, marine algae namely *Sargassum* and *Ascophyllum* can sequester lead and cadmium up to 30% of biomass dry weight, while mycelia of fungi such as *Absidia* and *Rhizopus* can accumulate diverse metals (copper, zinc, lead, cadmium and uranium) up to 25% of biomass dry weight (Volesky, 1994).

10.7.2.2 Bioaccumulation

Bioaccumulation is a metabolically active process, and the dissolved metals are imported into the microbial cell through import system (e.g., cell wall or membrane) (Brandl and Faramarzi, 2006; Diep et al., 2018). Once inside the cell, cellular materials like proteins and peptide legends can uptake and sequester heavy metals, and they can be used for several cellular processes including catalysis for enzymes, charge stabilization of biomolecules and signaling (Diep et al., 2018). Bioaccumulation capacity depends on the several parameters including nature of microorganisms and physicochemical parameters such as pH, temperature, contact time between biomass and metals, biomass density and age as well as toxicity of heavy metals to living biomass (Zabochnicka-Świątek and Krzywonos, 2014). For example, As(III) bioaccumulation on *Escherichia coli* BLR biomass was 0.17 mg/g (dry weight basis), while its accumulation slightly increased to 0.22 mg/g when a yeast species, *Saccharomyces cerevisiae* BY4742, was used (Diep et al., 2018).

10.7.2.3 Redox reaction

In redox reaction, metal converts to lower redox states by microorganisms, and this leads to reduction of metal mobility and toxicity, and it depends on the oxidation state of metals (Brandl and Faramarzi, 2006). This process is often accompanied by indirect reductive precipitation of metal, for example reduction of Cr(VI) due to indirect reduction of ferrous iron (Fe (II)) and sulfide by sulfate-reducing bacteria (SRB) (Gadd, 2001). Several metal reducing bacteria can use oxidized from of metals such as Cr(VI), Fe(III) and Mn(IV) as terminal electron acceptor. For example, *Shewanella putrefaciens* and *Shewanella oneidensis* MR-1 can reduce Fe(III) and Mn(IV) to Fe(II) and Mn(II), respectively (Bretschger et al., 2007), while *Desulfovibrio desulfuricans* can reduce Pd(II) to Pd(0) (Gadd, 2001).

10.7.2.4 Complex formation

In this mechanism, immobilization of dissolved metal occurs by formation of complex compounds (as precipitated salts) due to reaction of metal with biologically produced complexing agents (e.g., sulfide), and SRB mainly drive this bioprocess (Gadd, 2001). It is known that solubility products of several metal sulfides are very low, i.e., in between 4.65×10^{-14} (Mn) and 6.44×10^{-54} (Hg), and thus metal precipitation can take place at lower sulfide concentration. Moreover, microbial metabolisms can change the pH of local environment which can lead to precipitation of metals (e.g., metal hydroxides) (Brandl and Faramarzi, 2006).

10.8 FACTORS AFFECTING MICROBIAL MOBILIZATION AND IMMOBILIZATION OF HEAVY METALS

The performance of biological leaching of heavy metals from MSWI ash residues (IBA and IFA) primarily depends on types of leaching organisms, physicochemical characteristics of ash materials (e.g., heavy metal characteristics) and environmental conditions (e.g., pH, temperature, aerobic/anaerobic conditions) (Bosecker, 1997; Brandl, 2008) (Figure 10.2). The leaching conditions that support growth and activities of leaching organisms accelerate metal removal efficiency as well.

10.8.1 Nature of microorganisms

Microorganisms present in environment play a crucial role on existence of heavy metal in soluble or insoluble phase. In general, autotrophic and heterotrophic microorganisms are responsible for mobilization/immobilization of heavy metals from the ash residues (Gadd, 2001). In autotrophic process, chemolithotroph and acidophilic microorganisms use CO_2 as electron donor and inorganic compounds such as Fe(II) and reduced sulfur species (sulfate, thiosulfate, etc.) as electron acceptor which oxidize to Fe(III) and H_2SO_4, respectively, and the oxidation reaction provides energy to the autotrophic bacteria. SOB (e.g., *Thiobacillus thiooxidans*) and IOB (e.g., *Leptospirillum ferrooxidans*) are the key microorganisms that drive the autotrophic-mediated metal leaching (Gadd, 2001). The heterotrophic organisms generally use organic carbon (sucrose, $C_{12}H_{22}O_{11}$, is the most common C-source) as a source of carbon and energy, and several organic acids (e.g., citrate, oxalate, gluconate, etc.) are produced depending on the enzymes involved in the sucrose metabolism (Biswal et al., 2020). Fungal species namely *Aspergillus niger* (Xu and Ting, 2004) and bacteria namely *Pseudomonas aeruginosa* PAO1 (Biswal et al., 2020, 2019) have been used for leaching of heavy metals from various waste materials including IBA and IFA.

10.8.2 pH

The pH of the leaching medium should be appropriately maintained to attain the maximum growth and activity of leaching organisms which facilitate metal mobilization/immobilization. For metal solubilization, pH ranging between 2.0 and 3.0 is suitable for optimum growth of inorganic acid producing bacteria namely *Acidithiobacillus ferrooxidans* and *Acidithiobacillus thiooxidans* (Bosecker, 1997; Xu et al. 2014). However, under strong acidic conditions, i.e., pH less than 1.4, biological activities of these bacteria are significantly declined. The fungal species can grow under both acidic and alkaline conditions, i.e., a broad pH, ranging from 1.5 to 9.8 (Xu et al. 2014). The acidic metabolites produced due to metabolism of substrates by leaching organisms could change the pH of the leaching environment.

10.8.3 Temperature

The growth of the leaching organisms (bacteria and fungus) is strongly influenced by the temperature, and thus the growth of leaching organisms is considered as the rate-limiting step for metal dissolution from ash residues. Most of the IOB and SOB as well as fungal species can grow between 25°C and 30°C (Bosecker, 1997). In a recent study, bioleaching efficiency of Zn from a zinc sulfide mineral (sphalerite) using *Leptospirillum ferriphilum* increased from 80% to 97% with rise of temperature from 25°C to 40°C (Sundramurthy et al., 2020). However, further increase of temperature to 45°C caused considerable reduction of Zn bioleaching efficiency (84%). Another study that used fungal strain of *A. niger* showed higher iron removal from kaolin at lower temperature (47.7% at 25°C), but reduction of removal with rise of temperature (33.8% at 30°C) (Hajihoseini and Fakharpour, 2019).

10.8.4 Aerobic and anaerobic condition

Most of the leaching microorganisms grow in aerobic environment, and thus a sufficient amount of air/oxygen should be supplied to achieve optimum biological activities and high metal extraction efficiency (Bosecker, 1997). These aerobic organisms get energy by oxidation of substrates (e.g., oxidation of Fe(II) to Fe(III) by *Acidithiobacillus ferrooxidans*) and dissolved O_2 acts as a terminal electron acceptor (Liang et al., 2016). Various methods such as stirring, shaking or aeration are used to enhance dissolved oxygen (DO) level in the leaching medium (Bosecker, 1997). A study on recovery of copper from waste printed circuit boards using *Acidithiobacillus ferrooxidans* shows that with increase of DO level from 10% to 20%, the Cu leaching efficiency remarkably rose from 62.3% to 90.6% (Liang et al., 2016). However, with further increase of DO level to 25%, the Cu extraction reduced to nearly 80%. Thus, appropriate DO level should be maintained to obtain maximum metal recovery efficiency, and the DO requirement may vary according to type of leaching organisms.

Metal immobilization by reduction mechanism (specifically dissimilatory reduction process) is mostly carried out by anaerobic bacteria (e.g., SRB). Organic carbon (e.g., lactate) and hydrogen (H_2) are commonly used as a terminal electron donor (Lloyd et al., 2001). The dissimilatory metal reduction is usually conducted in anaerobic environments (e.g., sediments and ground water), and the process is influenced by organic and inorganic geochemistry of waste materials (Lovley, 1993).

10.8.5 Heavy metal characteristics

The quantity and quality (e.g., toxicity) of heavy metals present in MSWI ash can influence the activities of leaching organisms, which could then change metal leaching behavior. Thus, while designing bioleaching study, ash quantity (also called pulp density) should be appropriately selected in order to reduce potential toxic effects of heavy metals on leaching microorganisms. According to a study by Xu and Ting (2009), with increase of pulp density (incineration ash) from 0% to 5% (w/v), the specific growth of *A. niger* decreased from 0.134 to 0.003 day^{-1}. For ash containing toxic metals, bacteria having capacity to tolerate high metal concentration should be considered to achieve better leaching efficiency. Among the two commonly used bacteria, IOB and SOB, SOB shows greater tolerance to heavy metals than IOB (Ishigaki et al., 2005), and tolerance characteristics vary according to the nature of metals. For example, *Thiobacillus* species can tolerate heavy metals namely Ni, Cu and Zn up to 50, 55 and 112 g/L, respectively (Bosecker, 1997). For better leaching performance, specifically for highly toxic metals (e.g., As), mixed culture system could be more effective than single culture system (Ishigaki et al., 2005).

10.9 POTENTIAL REUTILIZATION OF IBA AND IFA FOR CIVIL ENGINEERING APPLICATIONS

Instead of considering MSWI ash as waste materials, it can be considered as a resource like natural materials and can be recycled for utilization in various civil engineering and geotechnical works. Several studies have been conducted on potential reutilization of MSWI ash for various construction-related works specifically for production of concrete and geopolymer (Biswal et al., 2020; 2019). MSWI ash contains significant quantities of oxides of silicon (SiO_2) and calcium (CaO) (Table 10.2), and thus the feasibility of its application specifically in the production of low-strength and lightweight concrete by partially substituting natural resources like sand was explored previously (Al-Rawas et al., 2005; Jurič et al., 2006; Qiao et al., 2008). The chemical composition of MSWI ash is also similar to the Portland cement (Tables 10.2 and 10.3), and thus ash can be considered as cementitious materials. Jurič et al. (2006) found that satisfactory quality of concrete with low compressive strength requirement (up to 40MPa) can be achieved by replacing 15 wt.% cement by IBA. Qiao et al. (2008) produced lightweight concrete of compressive strength higher than 10 MPa (density less than 1.8 g/cm^3) using IBA.

MSWI ash can be used as a sub-base material for construction of road pavement (Lam, 2010). In road construction, natural aggregates are usually used as sub-base materials. As availability of natural aggregates is limited, replacement by MSWI ash can reduce the demand on natural resources and also the associated construction cost. Many countries including Sweden and France have made road using IBA. In Sweden study, it was seen that replacing gravel in the road base with the IBA did not affect the release of metals such as Ca, Co, Fe, Mn, Ni and Pb and other inorganic contaminants such as NO_3-N (nitrate-nitrogen) to the environment (Ore et al., 2007).

The type of geotechnical applications of MSWI ash depends on its geotechnical and chemical properties which could vary depending on the nature of feedstock materials and the adopted incineration technology. Muhunthan et al. (2004) studied geotechnical engineering properties of MSWI ash and found that the permeability of IBA is very close to the sand, while permeability of IFA falls within the range of silts and clays. They also noticed that IBA subjected to 100% compaction at the highest water content showed lower density compared to other filling materials. Thus, IBA can be utilized as a filling or embankment material.

Since MSWI ash contains several hazardous components which could contaminate environmental ecosystems and pose negative impacts on human health, a few studies have suggested for pre-treatments (such as washing with water, treatment with low concentrated acid, thermal treatment, chemical stabilization, etc.) to upgrade MSWI ash quality prior reutilization (Xuan et al., 2018). Collivignarelli and Sorlini (2002) pre-treated MSWI fly ash by water washing followed by cement-lime stabilization, and then used the stabilized ash for concrete production. The produced concrete acquired very good mechanical and other properties, e.g., minimum compressive strength of 15 MPa. Overall, the potential use of MSWI ash for civil engineering applications is a path toward a circular economy, and consideration of MSWI ash for construction and geotechnical works depends on its physical, chemical and geotechnical characteristics. A summary of physical, chemical, thermal and biological methods that have been employed for treatment of MSWI ash based on the target applications is presented in Figure 10.3.

10.10 CONCLUSIONS AND FUTURE PROSPECTS

This chapter reported the recent developments on treatments of IBA and IFA generated from MSWI plants. MSWI ash contains several hazardous components including heavy metals which hinder their possible reutilization opportunities. Landfill disposal or reutilization of untreated MSWI ash could pose adverse impacts on environment, specifically groundwater contamination due to percolation of leachate containing heavy metals and soluble salts. Physicochemical, thermal and/or biological methods are employed for treatment of MSWI ash to convert it to relatively cleaner material and reduce the chances of environmental pollution. Among the three methods, physicochemical and thermal treatments are commonly adopted. However, these approaches are often associated with high cost and high energy requirement and result in high environmental impacts. Biological method could be an attractive alternative due to several advantages such as low cost, less energy demand, minimal potential for release of secondary pollutants and most importantly the use of non-chemical agents (i.e., environmental microorganisms). Bacteria namely IOB and SOB, and *Aspergillus niger* fungal species are the main organisms used for bioleaching. Interactions of microorganisms with MSWI ash can mobilize and/or immobilize heavy metals depending on

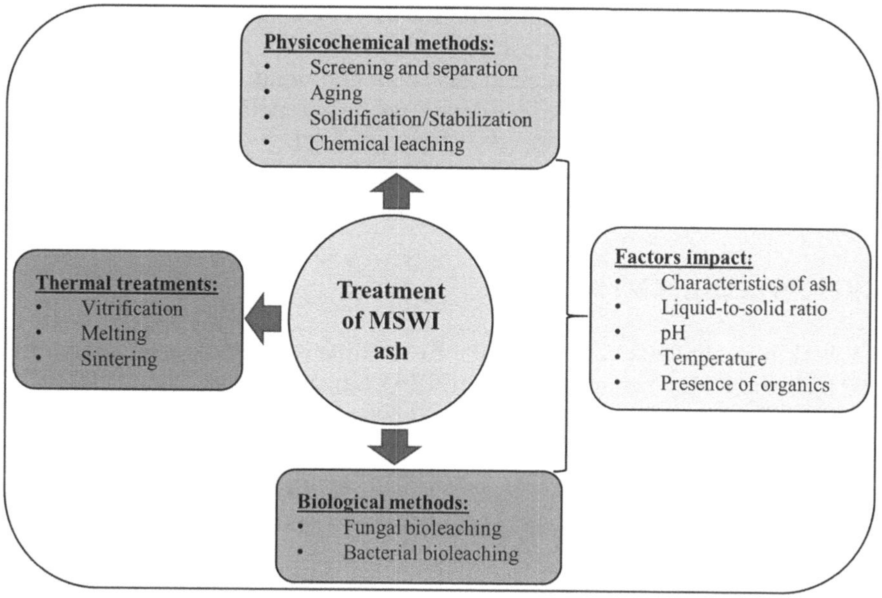

Figure 10.3 Summary of key physicochemical, thermal and biological methods applied for treatment of MSWI ash (IBA and IFA) as well as the important factors which influence the treatment performance (specifically chemical leaching and bioleaching).

the nature of microorganisms, ash properties and environmental conditions. Overall, treatment of MSWI ash with an appropriate method, and then recycling of ash materials for civil engineering and geotechnical applications could be a sustainable management strategy since it would save landfill space, disposal cost, consumption of natural resources, reduction of environmental pollution and a path toward a circular economy.

The future perspectives of the research topic are highlighted here which should be considered in future studies. Literature search indicates that a significant amount of work has been done on IBA for detoxification and recycling as construction materials, but little attention is given on IFA for civil engineering and geotechnical applications since studies on IFA are mainly on treatment for landfill disposal. Thus, future research should explore for feasibility of reutilization of IFA to a maximum extent for construction-related works. Most of the leaching studies on MSWI ash are carried out in lab-scale mode; however, field-scale experiments are required to evaluate the leaching behavior of heavy metals when the waste material is exposed to the natural environment having diverse microorganisms and complex environmental conditions for longer time. This would help to better understand the leaching behavior and mechanisms of heavy metals. To assess groundwater pollution, most of the past studies are conducted using coal fly ash from thermal plants, and thus future research should focus on MSWI ash to quantify the magnitude of pollution. Soil is the immediate environment of contact for concrete produced using MSWI ash. The toxic components in MSWI ash could impact the soil ecosystem; thus, control (e.g., microcosm/

mesocosm-based) studies are required for better assessment the effects of toxic elements in ash on dynamics of soil microbes and the associated enzyme activities. This would help for consideration of MSWI ash for soil stabilization or other geo-environmental applications. From the environmental and economic standpoint, an effective biological treatment technology should be developed for heavy metal solubilization, immobilization or stabilization to upgrade MSWI ash qualities as well as recovery of valuable and precious metals from MSWI ash through chemical and/or microbially mediated methods, and then recycle the metals for industrial applications based on their purity.

ACKNOWLEDGMENTS

This work is partially supported by the Environment Technology Research Program (ETRP), National Environment Agency (NEA), Singapore (project no. ETRP 1301 104).

CONFLICT OF INTEREST

The authors declare no conflict of interest.

REFERENCES

Agcasulu, I. and Akcil, A. 2017. Metal recovery from bottom ash of an incineration plant: Laboratory reactor tests. *Mineral Processing and Extractive Metallurgy Review* 38, 199–206.

Ahmaruzzaman, M. 2010. A review on the utilization of fly ash. *Progress in Energy and Combustion Science* 36(3), 327–363.

Al-Rawas, A.A., Hago, A.W., Taha, R. and Al-Kharousi, K. 2005. Use of incinerator ash as a replacement for cement and sand in cement mortars. *Building and Environment* 40(9), 1261–1266.

Baun, A., Ledin, A., Reitzel, L.A., Bjerg, P.L. and Christensen, T.H. 2004. Xenobiotic organic compounds in leachates from ten Danish MSW landfills—Chemical analysis and toxicity tests. *Water Research* 38, 3845–3858.

Bayard, R., Pestre, C. and Gourdon, R. 2009. Aerobic microbial activity in fresh and aged bottom ashes from municipal solid waste incineration (MSWI). *International Biodeterioration & Biodegradation* 63(6), 739–746.

Birgisdóttir, H., Pihl, K.A., Bhander, G., Hauschild, M.Z. and Christensen, T.H. 2006. Environmental assessment of roads constructed with and without bottom ash from municipal solid waste incineration. *Transportation Research Part D: Transport and Environment* 11, 358–368.

Biswal, B.K., Chen, Z. and Yang, E.-H. 2019. Hydrothermal process reduced *Pseudomonas aeruginosa* PAO1-driven bioleaching of heavy metals in a novel aerated concrete synthesized using municipal solid waste incineration bottom ash. *Chemical Engineering Journal* 360, 1082–1091.

Biswal, B.K., Zhu, W. and Yang, E.-H. 2020. Investigation on *Pseudomonas aeruginosa* PAO1-driven bioleaching behavior of heavy metals in a novel geopolymer synthesized from municipal solid waste incineration bottom ash. *Construction and Building Materials* 241, 118005.

Bosecker, K. 1997. Bioleaching: Metal solubilization by microorganisms. *FEMS Microbiology Reviews* 20(3–4), 591–604.

Bosshard, P.P., Bachofen, R. and Brandl, H. 1996. Metal leaching of fly ash from municipal waste incineration by *Aspergillus niger*. *Environmental Science & Technology* 30(10), 3066–3070.

Brandl, H. 2002. Metal-microbe-interactions and their biotechnological applications for mineral waste treatment. *Recent Research Developments in Microbiology* 6, 571–584.

Brandl, H. 2001. Microbial leaching of metals. In: Rehm, H.J., Reed, G. (eds), *Biotechnology*, pp. 191–224, Wiley-VCH Verlag GmbH, Weinheim.

Brandl, H. and Faramarzi, M.A. 2006. Microbe-metal-interactions for the biotechnological treatment of metal-containing solid waste. *China Particuology* 4(2), 93–97.

Bretschger, O., Obraztsova, A., Sturm, C.A., Chang, I.S., Gorby, Y.A., Reed, S.B., Culley, D.E., Reardon, C.L., Barua, S., Romine, M.F., Zhou, J., Beliaev, A.S., Bouhenni, R., Saffarini, D., Mansfeld, F., Kim, B.-H., Fredrickson, J.K. and Nealson, K.H. 2007. Current production and metal oxide reduction by *Shewanella oneidensis* MR-1 wild type and mutants. *Applied and Environmental Microbiology* 73, 7003–7012.

Brombacher, C. 1998. Development of a laboratory-scale leaching plant for metal extraction from fly ash by *Thiobacillus* strains. *Applied and Environmental Microbiology* 64(4), 1237.

Chang, F.-Y. and Wey, M.-Y. 2006. Comparison of the characteristics of bottom and fly ashes generated from various incineration processes. *Journal of Hazardous Materials* 138(3), 594–603.

Chen, S.Y. and Lin, J.G. 2000. Influence of solid content on bioleaching of heavy metals from contaminated sediment by *Thiobacillus* spp. *Journal of Chemical Technology and Biotechnology* 75(8), 649–656.

Chen, Q.Y., Tyrer, M., Hills, C.D., Yang, X.M. and Carey, P. 2009. Immobilisation of heavy metal in cement-based solidification/stabilisation: A review. *Waste Management* 29, 390–403.

Chen, Z., Liu, Y., Zhu, W. and Yang, E.-H. 2016. Incinerator bottom ash (IBA) aerated geopolymer. *Construction and Building Materials* 112, 1025–1031.

Chen, Z. and Yang, E.-H. 2017. Early age hydration of blended cement with different size fractions of municipal solid waste incineration bottom ash. *Construction and Building Materials* 156, 880–890.

Chimenos, J.., Fernández, A., Miralles, L., Segarra, M. and Espiell, F. 2003. Short-term natural weathering of MSWI bottom ash as a function of particle size. *Waste Management* 23(10), 887–895.

Chimenos, J.M., Segarra, M., Fernández, M.A. and Espiell, F. 1999. Characterization of the bottom ash in municipal solid waste incinerator. *Journal of Hazardous Materials* 64(3), 211–222.

Collivignarelli, C. and Sorlini, S. 2002. Reuse of municipal solid wastes incineration fly ashes in concrete mixtures. *Waste Management* 22, 909–912.

Dandautiya, R., Singh, A.P. and Kundu, S. 2018. Impact assessment of fly ash on ground water quality: An experimental study using batch leaching tests. *Waste Management & Research* 36, 624–634.

Diep, P., Mahadevan, R. and Yakunin, A.F. 2018. Heavy metal removal by bioaccumulation using genetically engineered microorganisms. *Frontiers in Bioengineering and Biotechnology* 6, 157.

Dorileo, M.C.G.O., Bandeca, M.C., Pedro, F.L.M., Volpato, L.E.R., Guedes, O.A., Dalla Villa, R., Tonetto, M.R. and Borges, A.H. 2014. Analysis of metal contents in Portland Type V and MTA-based cements. *ScientificWorld Journal*. 2014, 983728.

Fan, H., Shu, H.-Y., Yang, H.-S. and Chen, W.-C. 2006. Characteristics of landfill leachates in central Taiwan. *Science of the Total Environment* 361(1–3), 25–37.

Fedje, K.K., Ekberg, C., Skarnemark, G. and Steenari, B.-M. 2010. Removal of hazardous metals from MSW fly ash—An evaluation of ash leaching methods. *Journal of Hazardous Materials* 173(1–3), 310–317.

Feng, S., Wang, X., Wei, G., Peng, P., Yang, Y. and Cao, Z. 2007. Leachates of municipal solid waste incineration bottom ash from Macao: Heavy metal concentrations and genotoxicity. *Chemosphere* 67(6), 1133–1137.

Funari, V., Gomes, H.I., Cappelletti, M., Fedi, S., Dinelli, E., Rogerson, M., Mayes, W.M. and Rovere, M. 2019. Optimization routes for the bioleaching of MSWI fly and bottom ashes using microorganisms collected from a natural system. *Waste and Biomass Valorization* 10, 3833–3842.

Funari, V., Mäkinen, J., Salminen, J., Braga, R., Dinelli, E. and Revitzer, H. 2017. Metal removal from municipal solid waste incineration fly ash: A comparison between chemical leaching and bioleaching. *Waste Management* 60, 397–406.

Gadd G.M. 2001. Metal transformations. In: Gadd G.M. (ed), *Fungi in bioremediation*, pp. 359–382. Cambridge University Press, Cambridge.

Grathwohl, P. and Susset, B. 2009. Comparison of percolation to batch and sequential leaching tests: Theory and data. *Waste Management* 29(10), 2681–2688.

Guyonnet, D. 2010. Comparison of percolation to batch and sequential leaching tests: Theory and data. *Waste Management* 30(8–9), 1746–1747.

Hajihoseini, J. and Fakharpour, M. 2019. Effect of temperature on bioleaching of iron impurities from kaolin by *Aspergillus niger* fungal. *Journal of Asian Ceramic Societies* 7, 82–89.

Hocheng, H., Su, C. and Jadhav, U.U. 2014. Bioleaching of metals from steel slag by *Acidithiobacillus thiooxidans* culture supernatant. *Chemosphere* 117(1), 652–657.

Ishigaki, T., Nakanishi, A., Tateda, M., Ike, M. and Fujita, M. 2005. Bioleaching of metal from municipal waste incineration fly ash using a mixed culture of sulfur-oxidizing and iron-oxidizing bacteria. *Chemosphere* 60(8), 1087–1094.

Jain, S.K., Khan, A.A. and Rai, M. 2010. *Geomicrobiology*, Enfield, NH: Science Publishers; Boca Raton: Distributed by CRC Press, c2010.

Jurič, B., Hanžič, L., Ilić, R. and Samec, N. 2006. Utilization of municipal solid waste bottom ash and recycled aggregate in concrete. *Waste Management* 26(12), 1436–1442.

Kopsick, D.A. and Angino, E.E. 1981. Effect of leachate solutions from fly and bottom ash on groundwater quality. *Journal of Hydrology* 54, 341–356.

Kuo, Y.M., Huang, K.L. and Lin, C. 2012. Metal behavior during vitrification of municipal solid waste incinerator fly ash. *Aerosol and Air Quality Research* 12(6), 1379–1385.

Lam, C.H.K. 2010. Use of incineration MSW ash: A review. *Sustainability* 2(7), 1943.

Li, J., Dong, Z. and Yang, E.-H. 2017. Strain hardening cementitious composites incorporating high volumes of municipal solid waste incineration fly ash. *Construction and Building Materials* 146, 183–191.

Li, S., Tian, Z., Liu, R., Zhou, W., Cheng, H., Sun, J., Zhao, K., Wang, Y. and Zhou, H. 2020. Effective multi-metal removal from plant incineration ash via the combination of bioleaching and brine leaching. *RSC Advances* 10, 1388–1399.

Liang, G., Li, P., Liu, W. and Wang, B. 2016. Enhanced bioleaching efficiency of copper from waste printed circuit boards (PCBs) by dissolved oxygen-shifted strategy in *Acidithiobacillus ferrooxidans*. *Journal of Material Cycles and Waste Management* 18, 742–751.

Lin, C.-F., Wu, C.-H. and Liu, Y.-C. 2007. Long-term leaching test of incinerator bottom ash: Evaluation of Cu partition. *Waste Management* 27(7), 954–960.

Lisk, D. J. 1988. Environmental implications of incineration of municipal solid waste and ash disposal. *Science of the Total Environment* 74, 39–66.

Lloyd, J.R., Mabbett, A.N., Williams, D.R. and Macaskie, L.E. 2001. Metal reduction by sulphate-reducing bacteria: Physiological diversity and metal specificity. *Hydrometallurgy* 59, 327–337.

Lovley, D.R. 1993. Dissimilatory metal reduction. *Annual Review of Microbiology* 47, 263.

Mäkinen, J., Salo, M., Soini, J. and Kinnunen, P. 2019. Laboratory scale investigations on heap (bio) leaching of municipal solid waste incineration bottom ash. *Minerals* 9(5), 290.

Mishra, D. and Rhee, Y. 2014. Microbial leaching of metals from solid industrial wastes. *Journal of Microbiology* 52(1), 1–7.

Muhunthan, B., Taha, R. and Said, J. 2004. Geotechnical engineering properties of incinerator ash mixes. *Journal of the Air & Waste Management Association* 54, 985–991.

Nagarajan, R., Thirumalaisamy, S. and Lakshumanan, E. 2012. Impact of leachate on groundwater pollution due to non-engineered municipal solid waste landfill sites of erode city, Tamil Nadu, India. *Iranian Journal of Environmental Health Science & Engineering* 9, 35.
Nagib, S. and Inoue, K. 2000. Recovery of lead and zinc from fly ash generated from municipal incineration plants by means of acid and/or alkaline leaching. *Hydrometallurgy* 56(3), 269–292.
Nowak, B., Pessl, A., Aschenbrenner, P., Szentannai, P., Mattenberger, H., Rechberger, H., Hermann, L. and Winter, F. 2010. Heavy metal removal from municipal solid waste fly ash by chlorination and thermal treatment. *Journal of Hazardous Materials* 179, 323–331.
Ore, S., Todorovic, J., Ecke, H., Grennberg, K., Lidelöw, S. and Lagerkvist, A. 2007. Toxicity of leachate from bottom ash in a road construction. *Waste Management* 27(11), 1626–1637.
Panda, S. 2020. Magnetic separation of ferrous fractions linked to improved bioleaching of metals from waste-to-energy incinerator bottom ash (IBA): A green approach. *Environmental Science and Pollution Research* 27, 9475–9489.
Pandey, V., Ray, M. and Kumar, V. 2020. Assessment of water-quality parameters of groundwater contaminated by fly ash leachate near Koradi Thermal Power Plant, Nagpur. *Environmental Science and Pollution Research* 27, 27422–27434.
Park, Y.J. and Heo, J. 2002. Vitrification of fly ash from municipal solid waste incinerator. *Journal of Hazardous Materials* 91, 83–93.
Phoungthong, K., Shao, L.-M., He, P.-J. and Zhang, H. 2018. Phytotoxicity and groundwater impacts of leaching from thermal treatment residues in roadways. *Journal of Environmental Sciences* 63, 58–67.
Phoungthong, K., Zhang, H., Shao, L.-M. and He, P.-J. 2016. Variation of the phytotoxicity of municipal solid waste incinerator bottom ash on wheat (*Triticum aestivum* L.) seed germination with leaching conditions. *Chemosphere* 146, 547–554.
Polettini, A. and Pomi, R. 2004. The leaching behavior of incinerator bottom ash as affected by accelerated ageing. *Journal of Hazardous Materials* 113, 209–215.
Prasad, B. and Mondal, K.K. 2008. The impact of filling an abandoned open cast mine with fly ash on ground water quality: A case study. *Mine Water and the Environment* 27, 40–45.
Qiao, X.C., Ng, B.R., Tyrer, M., Poon, C.S. and Cheeseman, C.R. 2008. Production of lightweight concrete using incinerator bottom ash. *Construction and Building Materials* 22(4), 473–480.
Quina, M. J., Bordado, J. C. M. and Quinta-Ferreira, R. M. 2011. Percolation and batch leaching tests to assess release of inorganic pollutants from municipal solid waste incinerator residues. *Waste Management*, 31(2), 236–245.
Rawlings, D. 2004. Microbially-assisted dissolution of minerals and its use in the mining industry. *Pure and Applied Chemistry* 76, 847–859.
Rendek, E., Ducom, G. and Germain, P. 2007. Influence of waste input and combustion technology on MSWI bottom ash quality. *Waste Management* 27(10), 1403–1407.
Sakai, S.-I. and Hiraoka, M. 2000. Municipal solid waste incinerator residue recycling by thermal processes. *Waste Management* 20(2–3), 249–258.
Silva, R. V., de Brito, J., Lynn, C. J. and Dhir, R. K. 2019. Environmental impacts of the use of bottom ashes from municipal solid waste incineration: A review. *Resources, Conservation and Recycling* 140, 23–35.
Singh, R.K., Gupta, N.C. and Guha, B. 2014. pH dependence leaching characteristics of selected metals from coal fly ash and its impact on ground water quality. *International Journal of Chemical and Environmental Engineering* 5, 218–222.
Sun, X. and Yi, Y. 2021. Acid washing of incineration bottom ash of municipal solid waste: Effects of pH on removal and leaching of heavy metals. *Waste Management* 120, 183–192.
Sundramurthy, V.P., Rajoo, B., Srinivasan, N.R. and Kavitha, R. 2020. Bioleaching of Zn from sphalerite using *Leptospirillum ferriphilum* isolate: Effect of temperature and kinetic aspects. *Applied Biological Chemistry* 63, 44.

Tang, J. and Steenari, B.-M. 2016. Leaching optimization of municipal solid waste incineration ash for resource recovery: A case study sman.2015.1of Cu, Zn, Pb and Cd. *Waste Management* 48, 315–322.

Tang, Q., Liu, Y., Gu, F. and Zhou, T. 2016. Solidification/stabilization of fly ash from a municipal solid waste incineration facility using portland cement. *Advances in Materials Science and Engineering* 2016, 7101243.

Traina, G., Morselli, L. and Adorno, G.P. 2007. Electrokinetic remediation of bottom ash from municipal solid waste incinerator. *Electrochimica Acta* 52(10), 3380–3385.

USEPA, 2004. Test methods for evaluating solid waste, physical/chemical methods, chapter seven: Characteristics introduction and regulatory definitions. https://www.epa.gov/sites/production/files/2015-10/documents/chap7_0.pdf

van Zomeren, A. and Comans, R.N.J. 2004. Contribution of natural organic matter to copper leaching from municipal solid waste incinerator bottom ash. *Environmental Science & Technology* 38(14), 3927–3932.

Veli, S., Kirli, L., Alyuz, B. and Durmusoglu, E. 2008. Characterization of bottom ash, fly ash, and filter cake produced from hazardous waste incineration. *Polish Journal of Environmental Studies* 17(1), 139–145.

Volesky, B. 1994. Advances in biosorption of metals: Selection of biomass types. *FEMS Microbiology Reviews* 14, 291–302.

Wang, F.-H., Zhang, F., Chen, Y.-J., Gao, J. and Zhao, B. 2015. A comparative study on the heavy metal solidification/stabilization performance of four chemical solidifying agents in municipal solid waste incineration fly ash. *Journal of Hazardous Materials*, 300, 451–458.

Wang, J., Ban, H., Teng, X., Wang, H. and Ladwig, K. 2006. Impacts of pH and ammonia on the leaching of Cu(II) and Cd(II) from coal fly ash. *Chemosphere* 64(11), 1892–1898.

Wang, K., Chiang, K., Perng, J. and Sun, C. 1998. The characteristics study on sintering of municipal solid waste incinerator ashes. *Journal of Hazardous Materials* 59(2–3), 201–210.

Wang, K.-S., Sun, C.-J. and Liu, C.-Y. 2001. Effects of the type of sintering atmosphere on the chromium leachability of thermal-treated municipal solid waste incinerator fly ash. *Waste Management*, 21(1), 85–91.

Wang, Q., Yang, J., Wang, Q. and Wu, T. 2009. Effects of water-washing pretreatment on bioleaching of heavy metals from municipal solid waste incinerator fly ash. *Journal of Hazardous Materials* 162(2–3), 812–818.

Wey, M.-Y., Liu, K.-Y., Tsai, T.-H. and Chou, J.-T. 2006. Thermal treatment of the fly ash from municipal solid waste incinerator with rotary kiln. *Journal of Hazardous Materials* 137(2), 981–989.

Wong, S., Mah, A.X.Y., Nordin, A.H., Nyakuma, B.B., Ngadi, N., Mat, R., Amin, N.A.S., Ho, W.S. and Lee, T.H. 2020. Emerging trends in municipal solid waste incineration ashes research: A bibliometric analysis from 1994 to 2018. *Environmental Science and Pollution Research* 27, 7757–7784.

World Bank, 2016. Solid waste management. https://www.worldbank.org/en/topic/urbandevelopment/brief/solid-waste-management.

World Health Organization, 2017. Guidelines for drinking-water quality: First addendum to the fourth edition. https://www.who.int/publications/i/item/9789241549950

Wu, H.-Y. and Ting, Y.-P. 2006. Metal extraction from municipal solid waste (MSW) incinerator fly ash—Chemical leaching and fungal bioleaching. *Enzyme and Microbial Technology* 38(6), 839–847.

Xu, T.-J., Ramanathan, T. and Ting, Y.-P. 2014. Bioleaching of incineration fly ash by *Aspergillus niger*—Precipitation of metallic salt crystals and morphological alteration of the fungus. *Biotechnology Reports* 3(0), 8–14.

Xu, T.-J. and Ting, Y.-P. 2009. Fungal bioleaching of incineration fly ash: Metal extraction and modeling growth kinetics. E*nzyme and Microbial Technology* 44(5), 323–328.

Xu, T.-J. and Ting, Y.-P. 2004. Optimisation on bioleaching of incinerator fly ash by *Aspergillus niger*—Use of central composite design. *Enzyme and Microbial Technology*. 35, 444–454.
Xuan, D., Tang, P. and Poon, C.S. 2018. Limitations and quality upgrading techniques for utilization of MSW incineration bottom ash in engineering applications—A review. *Construction and Building Materials* 190, 1091–1102.
Yang, J., Wang, Q., Luo, Q., Wang, Q. and Wu, T. 2009. Biosorption behavior of heavy metals in bioleaching process of MSWI fly ash by *Aspergillus niger*. *Biochemical Engineering Journal* 46(3), 294–299.
Yang, J., Wang, Q., Wang, Q. and Wu, T. 2008. Comparisons of one-step and two-step bioleaching for heavy metals removed from municipal solid waste incineration fly ash. *Environmental Engineering Science* 25, 783–789.
Yao, Q., Samad, N.B., Keller, B., Seah, X.S., Huang, L. and Lau, R. 2014. Mobility of heavy metals and rare earth elements in incineration bottom ash through particle size reduction. *Chemical Engineering Science* 118, 214–220.
Zabochnicka-Świątek, M. and Krzywonos, M. 2014. Potentials of biosorption and bioaccumulation processes for heavy metal removal. *Polish Journal of Environmental Studies* 23, 551–561.
Zhang, F.-S. and Itoh, H. 2006. Extraction of metals from municipal solid waste incinerator fly ash by hydrothermal process. *Journal of Hazardous Materials* 136(3), 663–670.
Zhang, Y., Cetin, B., Likos, W.J. and Edil, T.B. 2016. Impacts of pH on leaching potential of elements from MSW incineration fly ash. *Fuel* 184, 815–825.

Chapter 11

Concern for heavy metal ion water pollution

Their strategic detection and removal opportunities

Sasireka Velusamy and Anurag Roy
University of Exeter

Senthilarasu Sundaram
Edinburgh Napier University

Tapas K. Mallick
University of Exeter

CONTENTS

DOI: 10.1201/9781003247869-13

11.1 INTRODUCTION

Water is a crucial element for human consumption, agriculture purposes, industries, and other energy generation processes. About 70% of the Earth is covered by water; however, only 30% of the water on the planet is freshwater which is suitable for human and animal consumption. Water is an excellent solvent that can dissolve various substances than any other liquid. From that point of view, water serves as a carrier medium that supplies nutrients to plants and animals. At the same time, unfortunately toxic substances and pollutants are also transferred through water. Due to increase in the human population and industrial growth throughout the world, demand of water has increased tremendously. At present, available natural water resources are contaminated drastically through activities of industrialisation. Figure 11.1 represents the schematic representation of anthropogenic activities securing for water pollution. The manmade activities that lead to the pollution of water; for example, textile industries release wastewater containing dyes, specific chemicals, and finishing chemicals into the waterways which leads to degradation of water quality and spread of toxicity to the ecosystem (Velusamy et al., 2021). Many sectors have their treatment plants to treat wastewater before its discharge; however, few other sectors are discharging wastewater without any treatment, which leads to deterioration of aquatic environment. Population growth due to urbanisation has contributed significantly to the contamination of accessible water resources through discharge of untreated industrial effluents without treatment, run-off from agricultural fields, and reckless utilisation of natural resources (Vardhan et al., 2019; Kumar et al, 2021). Industrial activities have caused a diverse range of pollutions such as solid waste pollution, water pollution, and air pollution. There are more than 700 organic, inorganic, and microbial pollutants that are potent to cause water pollution when discharged without treatment (Dubey et al.,

Figure 11.1 Schematic representation of anthropogenic activities securing for water pollution.

2021; Mohapatra et al., 2021; Pantha et al., 2021). These diverse range of noxious wastes discharged into the water may contain mutagenic, carcinogenic, and genotoxic pollutants (Menon et al., 2020; Mohapatra et al., 2020). Deterioration of water quality is dangerous to aquatic flora and fauna. Therefore, it is essential to remove contaminants from industrial wastewater to prevent contamination of surface and groundwater. According to the World Water Council report, the water scarcity area would expand to about 3.9 billion peoples by 2030 (Xu et al., 2018), which may make wastewater treatment essential for its reuse in domestic and industrial purposes. Pollutants such as pesticides, dyes, surfactants, heavy metal ions, and pharmaceutical and chemical wastes are hazardous to environmental substances (Grace Pavithra et al., 2019; Villarín and Merel, 2020). Heavy metal contamination is becoming a significant global issue due to the risk of its spreading, toxicity, and destructive effects on our water environment. Industrial wastewater discharges increased the concentration of heavy metal ions in water resources. Heavy metals are mostly non-biodegradable; hence, their bioaccumulation through consumption of water over a significant period increases their concentration in living organisms (Xu et al., 2017; Kumar et al., 2020). The quality of water is estimated by few parameters such as dissolved oxygen (DO), alkalinity, pH, hardness, conductivity, turbidity and salinity, which are associated with the ecological and geological characteristics of the water body (Sousa et al., 2018). The inadequate accessibility of drinking water demands a pure water supply. Before moving into heavy metal ion pollution, the types of pollutants and their sources will be briefly discussed.

11.1.1 Sources of water pollution

Inorganic wastewater pollution is caused mainly through agricultural run-offs, various industrial effluents, and municipal sewage releases, and its makes water unfit for consumption. Agricultural run-offs contain pollutants like pesticides, herbicides, and other associated chemicals, which generate unfavourable effects on water resources and human health. Sewage water discharged from many scattered areas to waterways without proper treatment and effluents from industries contain a broad range of chemical pollutants. In general, water pollution can be divided into two categories: groundwater contamination and surface water contamination. Groundwater is one of the most critical sources for irrigation process. It is susceptible to contamination of many toxic substances such as pesticides, fertilisers, road salt (used in cold weather), toxic substances from various industries, and used motor oil. Surface water contamination has two subdivisions: point source pollutants (PS) and non-point source pollutants (NPS). Point source pollution includes sewage treatment plants, factories, mines, electric power plants, and here the origin of contaminants could be identified through the exact point of entry.

In contrast, sources of non-point source water pollution includes toxic chemicals, herbicides, and fertilisers from agricultural fields. Furthermore, hazardous substances from run-off of urban residential areas, sediments from improperly managed construction, irrigation chemicals, and faulty septic systems are usually much diluted when discharged from their point source; therefore, it is very difficult to identify and resolve these non-point source discharges. Water pollution is also associated with manufacturing industries, namely, tanneries, industrial estates, product-manufacturing industries, paper industries, textile industries, chemical-manufacturing industries, industrial dumpsites, lead smelting, food industries, mining industries, lead-acid battery industry, etc. These industries release water contained with toxic dye pollutants. Another crucial source of pollutants is spillage of oil into water resources. The breakage or leakage of underground pipelines of oil industries kills marine organisms such as seaweeds and fishes, thus leading to intake of contaminated seafood.

11.1.2 Impact of industrial wastewater

Water pollution has become a universal concern that causes serious problems worldwide, and it causes the death of 14,000 people per day (Pink, 2006). Industries release effluents containing pollutants that eventually reach the water bodies and cause harmful effects to the aquatic environment. Various chemicals from different industries cause various health consequences and affect the geochemical cycles of our environment. Industrial effluents have a higher range of biological oxygen demand (BOD), chemical oxygen demand (COD), dissolved solids (DS), acids, metals, alkali, and other harmful pollutants, which degrade the water quality and strongly influence the photosynthesis activity in plants. High pollution levels of water affect light penetration and oxygen consumption. As a result of these, the microorganisms existing in the water resources are unable to continue their purification processes of the environment (Nosheen et al., 2000). Another critical factor is an increase in the temperature of the water body due to the release of effluents from different industries, which interrupts the aquatic environment's metabolic process. Various factors such as inconsistent pH range, increased turbidity, and inadequate oxygen level diminish the marine creatures' liveliness and cause distinct health complications in animals that consume those including human beings. For example, arsenic in drinking water sources causes skin cancer, keratosis, and hyperpigmentation.

11.1.3 Types of water pollutants

Pollutants are generally categorised into three major types: inorganic (chemical contaminants), organic (natural pollutants), and biological pollutants.

Water pollutants can be divided mainly into the following categories:

i. **Disease-causing agents:** Consumption of water containing microorganisms causes several diseases, namely, cholera, typhoid, dysentery, and hepatitis.
ii. **Oxygen-consuming wastes:** Dissolved oxygen in water resources is depleted if organic materials are released into the water. Industrial waste from paper mills, tanneries, and food-processing industries is included in the oxygen-consuming wastes. When organic wastes are overloaded in the water resources, it decreases the dissolved oxygen level. If the dissolved oxygen level falls below 5 ppm (parts per million), it severely affects the aquatic species specifically.
iii. **Plant nutrients:** Agriculture run-off containing fertilisers causes water pollution. Plants require few mineral nutrients from the soil for their growth. Mineral nutrients are split into three groups: primary nutrients (potassium, nitrogen, and phosphorous), secondary nutrients (calcium, magnesium, and sulphur), and

micronutrients (iron, copper, chlorine, manganese, molybdenum, and zinc). Whenever the permissible limit exceeds, then the toxicity seeps through the soil and contaminate the groundwater.

iv. **Suspended solids and sediments:** Anthropogenic activities such as housing developments, cutting timbers, mining, and other construction-related works increase deposits. The increasing quantity of residues in water resources causes various environmental problems; suspended particles increase the turbidity of water, reducing the sunlight penetration and eventually affecting the photosynthesis activity of plants. When released into the turbid water, toxic substances such as metals and pesticides stick to suspended particles and slowly dilute to insignificant concentrations.

v. **Dissolved solids:** Freshwater always contains some DS, the ions in water are essential for the growth of most life forms, but when their limit exceeds, they become toxic and wipe out living aquatic organisms.

vi. **Thermal pollution:** Water pollution caused by the heat factor is called thermal pollution. For example, electric power industries consume an enormous amount of water from water resources for cooling purposes. The industrial process for generating electric power produces immense heat, which needs heat removal to stabilise the plant. As a result, the temperature of circulation water of the process increases from 10°C to 20°C. Discharge of this warm water affects aquatic organisms by increasing their body temperatures. Together with these pollutants, few other more pollutant types are given in Table 11.1.

According to a report, by 2025, the entire world will face water stress due to inadequate accessibility of water, which would increase the need for reusing water the

Table 11.1 Different types of water pollutants and their sources

Category	*Type of pollutants*	*Sources*	*Health issues*
Disease-causing pathogens	Bacteria, viruses, and protozoa	Raw sewage, human, and animal waste	Typhoid, cholera, dysentery, and infectious hepatitis
Oxygen-demanding wastes	Biodegradable animal waste and plant waste	Sewage, animal waste, and paper mill waste	Oxygen demand kills aquatic environment organisms
Plant nutrients	Nitrates (NO_3^-) and phosphates (PO_4^{3-})	Inorganic fertilisers, sewage, and animal wastes	Excessive algae growth depletes the oxygen level and kills fish
Organic chemicals	Oil, gasoline, plastics, pesticides, fertilisers, and cleaning chemical solvents	Mining sites, industrial waste, household waste, and run-off from roads	Poisonous and further make sick those who drink it and also toxic to aquatic species

(Continued)

Table 11.1 (Continued) Different types of water pollutants and their sources

Category	*Type of pollutants*	*Sources*	*Health issues*
Inorganic chemicals	Acids, bases, salts, and metal compounds	Industrial wastes, mining site wastes, household wastes, and run-off from roads	Cancers, toxic to aquatic species, disorders of nervous systems and reproductive systems
Sediments	Soil and sludge from wastes	From construction sites, mining sites, and land erosion from farms	Block rivers, lakes, and reservoirs and reduce the photosynthesis activity of aquatic plants and disrupt the food production chain
Heavy metals	Lead, mercury, arsenic, copper, chromium, cadmium, nickel, and zinc	Various industrial wastewater, mining wastes, domestic chemicals, landfills, and agricultural run-off	Causes health issues to humans and aquatic species
Radioactive and toxic substances	Toxic chemicals	Bioaccumulation in fish and other organisms	Cancers, congenital disabilities, and genetic mutations
Oil pollutants	Oil spills, offshore oils, mining, and oil refineries	Accidental discharge of petroleum into the water resources	Kills a large quantity of marine species
Thermal pollutants	Heat	Industrial plants and electric power	Thermal discharge controls the plant growth and aquatic species development
Medical wastes	Medicines and antibiotics	Dumping medicinal wastes into the water resources and wastewater treatment	Infertility in aquatic organisms and other unknown symptoms
Acids	Releases the trace metals from soils, rocks, and metal surface such as water pipes	Atmospheric deposition, mine effluents, and degrading plant materials	Decreases the availability of fishes and other aquatic species

pollutants (Chowdhary et al., 2020). Due to these concerns, the World Health Organization (WHO) and the United States Environmental Protection Agency (USEPA) have set the permissible discharge limit for heavy metals to reduce environmental pollution. Therefore, this chapter primarily overviews sources and consequences of heavy metal pollution by investigating the monitoring and detection processes, particularly biosensor detection operation and wastewater treatment methods to remediate these harmful substances primarily, focusing on the photocatalytic together with adsorption techniques.

11.2 HEAVY METAL POLLUTION

Water resources are being desperately contaminated with various heavy metals and other hazardous chemicals due to anthropogenic activities. The extensive use of heavy metals in agricultural, domestic, and industrial applications expedited their diffusion into water bodies across the environment. The noxiously harmful heavy metal contamination generates a detrimental effect on human health, which has raised the distress for removing heavy metals contaminating water from various sources. The expeditious development of industrialisation has resulted in the demand many metals due to manufacturing of various products like batteries, pigments, photographic films, automobile fuels, steel making, and coating materials. These activities generate an immense range of environmental pollution impacts because of their potential harmfulness to the environment and human health. Heavy metal ions, such as lead and cadmium, are broadly used in metallurgy, photographic materials, and the electronic industry, which generates a substantial unavoidable quantity of wastewater. The toxicity due to heavy metal contamination is spreading across the food chain. Certain heavy metals are essential for the health and growth of human beings, such as zinc, copper, manganese, and nickel, but they cause toxicity when their limits exceed. Heavy metals released from various industries can have adverse effects on human health, thus causing several diseases. Heavy metals are groups of metals and metalloids with high densities exceeding 5 mg/cc and atomic weights ranging between 63.5 and 200.6, and they are non-biodegradable due to their chemical stability (Fu and Wang, 2011). The industries producing heavy metals include textile, paper, pharmaceuticals, petroleum, rubber, tannery, battery, mining, wood and metal-finishing industries. Release of toxic heavy metal-containing wastages, mainly mercury, chromium, lead, arsenic, cadmium, silver, antimony, nickel, zinc, tin, vanadium, and titanium, from diverse industries generates a massive impact on our health and the environment. Hence, before discharging, the effluents need to undergo treatment processes to protect our ecosystem. Heavy metals are inherently carcinogenic; their consumption through drinking water causes severe health problems like cancers, nervous system diseases, pulmonary diseases, and organ damages, and their long-term exposure in extreme conditions may lead to death. Hazardous heavy metal ions such as lead, cadmium, copper, arsenic, nickel, chromium, zinc, and mercury are highly soluble in water; moreover, these metals are highly toxic. Accumulation of toxic substances in living species creates various health-related issues because of their carcinogenicity. Discharge of effluents containing heavy metals seeps through the soil surface in due course. They enter into the groundwater and other available natural water bodies, then are adsorbed by plants from contaminated soil and accordingly penetrating through the food supply chain to generate incompatible environmental effects.

11.2.1 Types of heavy metal ions

Industrial sectors release several metal ions, such as cadmium, arsenic, copper, nickel, chromium, nickel, zinc, mercury, and lead-containing wastewater mentioned below, to familiarise the sources and adverse effects to the living organisms.

11.2.1.1 Cadmium (Cd)

Cd is highly carcinogenic; even at low concentrations, most commonly, cadmium ion noxiously toxicity spreading oxidation states are +2 and +1 in the environment (Khan et al., 2017). Cadmium (Cd) broadly utilised in electroplating industries, textile industries, batteries, pigments, petroleum products, metallurgical industries, photography plastics and synthetic chemicals. The composition of wastes from these industries is highly variable. Heavy metal contamination of water bodies disrupts the marine environment by decreasing the reproduction rate of aquatic organisms, resulting in the miserable inhabitant's percentage rate (Hamilton et al., 2016). Cd heavy metal ion toxicity brings significant health effects on the lung, kidney, liver and respiratory organs, and reproductive organs. According to the USEPA, the permissible limit for the cadmium ion is 0.005 mg/L (Gautam et al., 2016), and followingly the maximum contaminant (MCL) limit is 0.01 mg/L[1], once the limit exceeds, the carcinogenic symptoms will develop (Bhanjana et al., 2017). Accordingly, cadmium containing effluent discharges should be below 0.01 mg/L[1] per EU environmental quality standards. Followingly, the USA international agency for cancer research, together with the national toxicology programme, acknowledged cadmium as one of the carcinogenic metal ions examined toxic harshly at 15.0 mg/L[1] concentration. Cadmium containing water provokes an unfavourable effect on plant growth and the lipid content developmental factor. A decrease in cadmium utilisation containing products prevents living organisms and our atmospheric environment conditions.

11.2.1.2 Arsenic (As)

Arsenic is a naturally occurring metalloid that exists in organic and inorganic forms in water sources. Thermodynamically, arsenic (As) has four stable oxidation states: +5 (arsenate), +3 (arsenite), 0 (arsenic), and –3 (arsine). As compounds from industrial effluents can easily dissolve in water and contaminate water resources like lakes, rivers, and groundwater by dissolving in rain and snowfall. Therefore, arsenic contamination of groundwater is a severe health threat worldwide. According to USEPA, the permissible limit of the arsenic ion is 0.05 mg/L (Gautam et al., 2016). The WHO has also set the guidelines to maintain all the metal ions, particularly arsenic. The WHO decreased their recommendation from 0.05 to 0.01 mg/L to protect our health from arsenic contamination. The groundwater can contain high arsenic concentration that should be tested to carefully identify the arsenic contamination level. It has to be monitored to avoid the perplexity of arsenic-containing water consumption. In general, the arsenic concentration in the Earth's coating ranges from 1.5 to 5 mg/kg; higher concentrations found in sedimentary rock, mainly in iron and manganese ores. Arsenic released from these ores enters the soil, surface water, groundwater, and other possible sites.

Further, consumption of this polluted water spreads toxicity to humans and produces negative effects such as hyperpigmentation, keratosis, and various cancer types (Gibb et al., 2011). In addition, long-term arsenic exposure can cause cancers in different organs such as lungs, skin, kidneys, nasal passages, liver, and prostate.

Additionally, intake of arsenic in small concentrations through dust or soil causes adverse effects in young children. Arsenic can stick to small particles in the air, and can stay in the air for several days and travel long distances through the air, which cannot be destroyed unfortunately. Thus, arsenic pollution has been considered a significant environmental issue in many countries worldwide. Potential causes of arsenic contamination are not entirely determined, and few known sources are drinking water, worldwide food distribution, industrial manufacturing products, wood preservatives, smoking cosmetics products, and natural environmental sources.

11.2.1.3 Copper (Cu)

For human health, copper is an essential element and necessary in the formation of haemoglobin in the red platelets. Cu^+ generally found in enzymes, fit appropriate for carrying oxygen similar to haemoglobin. On the other hand, the divalent copper (Cu^{2+}) is toxic and carcinogenic when consumed in excess through ingestion; it also causes mutagenesis in humans (Shawabkeh et al., 2004). Copper is used in various industries such as paint and pigments, mining fertilisers, and textile industry. If consumption of copper exceeds the limit, liver damage and lung cancer are developed. According to the USEPA and the WHO, the permissible limits of Cu^{2+} in discharge of industrial effluents are 1.3 and 1.5 mg/L respectively. (Kalavathy et al., 2005).

11.2.1.4 Nickel (Ni)

The extensive utilisation of Ni-containing products leads to environmental pollution by Ni and its by-products. Nickel oxidation states are from −1 to +4, but the +2 oxidation state is crucial and highly harmful (Mobasherpour et al., 2012). According to USEPA, the permissible limit of nickel ion is 0.1 mg/L (Gautam et al., 2016). Various sources of Ni and their relative health effects are discussed in short in Table 11.2.

11.2.1.5 Chromium (Cr)

Chromium (Cr) is another toxic metal present in industrial effluents that has adverse effects on living organisms. Its oxidation states range from +2 to +6. Interestingly, the trivalent Cr (III) and hexavalent Cr (VI) forms are considered more toxic, non-biodegradable, carcinogenic, teratogenic, and mutagenic because of their oxidising properties (Liu et al.,2016). Cr (VI) is dangerous to health when its limit (0.05 mg/L) exceeds in potable water (Shahid et al.,2017). European Union requires that Cr (VI) discharge to

Table 11.2 Heavy metal sources and their health complication

Heavy metal	*WHO limit (in mg/L)*	*Main sources*	*Impacts on human health*
Copper (Cu)	1.0	Mining and metallurgy, plating, electrical and electronic waste, mining and metallurgy, pesticides, fertilisers, paints and pigments, textile industry and explosive	Risk of liver cancer, carcinogenic and mutagenic, malignant disease and cellular injury, for example, DNA damage
Mercury (Hg)	0.001	Chemical production, pesticides, electrical switches, fluorescent bulbs, mercury arc lamps, released from gold and antimony mining, medical waste, production of mercury products (batteries, thermometers, mercury amalgam), explosive, rubber, and plastics making industries	Diarrhoea, fever, vomiting, nausea, pink disease, and paralgesia
Cadmium (Cd)	0.005	Plastic, pigments and paints, mining batteries, electroplating, metal plating, coating operations, welding, electronic components additive, phosphorus fertilisers, textile manufacturing	Pneumonitis, lung cancer, hypertension, proteinuria, osteocalcin, diseases, renal loss, kidney damage, renal failure, carcinogenic, respiratory disease, gastrointestinal, and kidney stone formation
Arsenic (As)	0.01	Phosphate fertilisers, mining, paints, textile, medicinal, pharmaceutical wastewater, iron and steel, smelting, and glass industries	Breathing problems, lung and skin cancers, nausea, vomiting, and diarrhoea
Lead (Pb)	0.05	Paint, pesticides, insecticides, herbicides, textile, burning coal and mining, paints and pigments, storage batteries, manufacturing of lead-acid batteries, mining, and metallurgy	Kidney failure, cerebral and renal disorders, affect the sense organs, damage reproductive, cardiovascular and gastrointestinal systems, causes anaemia, neurotoxic, damage the developing of infant brain, impaired intellectual ability, and behavioural problems children

(*Continued*)

Table 11.2 (Continued) Heavy metal sources and their health complication

Heavy metal	*WHO limit (in mg/L)*	*Main sources*	*Impacts on human health*
Zinc (Zn)	5.0	Alloy production, electrical devices, brass plating, wood pulp production, newsprint paper production, steelworks with galvanising lines, paints, pesticides, zinc and brass metal works, refineries, rubber and tyre industries	Kidney and liver failure, haematuria, seizures, depression, prostate cancer and vomiting, skin irritations, cramps, vomiting, anaemia, neurological signs, increased thirst, and dizziness.
Chromium (Cr)	0.1	Tanning, leather industry, chemical industry, fertilisers, textile industry, paints and pigments, metal plating, mining, coating operation, photography, steel industries, electroplating, glass industries, catalysts, fungicides, chrome alloy, and metallurgic industries	Carcinogenic and genotoxic, acute gastrointestinal pain, liver and kidney damage, diarrhoea, nausea, lung cancer, asthma attacks, rhinitis, pulmonary fibrosis, prostate cancer.
Iron (Fe)	0.1	Construction materials, paint and pigment industries, coagulants, water pipes, and sanitary industries	Genetic disorder
Nickel (Ni)	0.1	Electroplating, production of iron and steel, mining and metallurgy, battery, chemical industries, and food-processing industries	Nausea, vomiting, diarrhoea, diabetes, cancer, and hyperpigmentation
Antimony (Sb)	0.02	Coal combustion, smelting, and mining processes	Cardiovascular diseases, nasal ulceration, liver diseases, cancer, dermatitis, and conjunctivitis

surface water be below 0.05 mg/L to control its toxicity (Pagana et al., 2017). According to USEPA, the permissible limit of chromium ion is 0.05 mg/L (Gautam et al., 2016). Cr (III) is an essential element for human health at a low non-toxic concentration of 0.20 mg per day. However, Cr (VI) is considered a carcinogenic solid and destructive to living entities.

11.2.1.6 Zinc (Zn)

Zn is a mineral present in various forms, namely, Zn O, ZnS, $ZnCO_3$, Zn_2SiO_4, etc. Main sources of Zn contamination are brass metal works, steel processing, mining, coal combustion, paper, and electroplating industries. Furthermore, Zn is an essential element in the building and housing industries. Zn is also utilised in residential plumbing systems as a substitute for copper content. Zn contamination causes health issues such as liver failure, gastric and abdominal cramps, epigastric pain, nausea, diarrhoea, lower levels of high-density lipoprotein cholesterol, anaemia, stomach cramps (Jakobik Kolon et al., 2017). According to the Indian Standards Bureau (BIS), the permissible zinc level in the drinking water is 5 mg/L, and according to the USEPA, the allowable limit of zinc ion is 0.5 mg/L (Gautam et al., 2016) to control the pollutants load from industrial discharges. These produced pollutants negatively impact human health and the environment, requiring effective treatments for removing zinc from the wastewater effluents.

11.2.1.7 Lead (Pb)

The WHO set the permissible limit of Pb for lead drinkable water as 0.01 mg/L. Environmental Protection Agency (EPA) established the maximum permitted discharge limit in both the consumable water and wastewater as 6.0×10^{-3} mg/L (Helena et al., 2000; Arbabi et al., 2015). According to the USEPA, the permissible limit of lead ion is 0.05 mg/L (Gautam et al., 2016). Majorly textile and agricultural industries are responsible for Pb water pollution. The impact of Pb hazards is summarised in Table 11.2.

11.2.1.8 Mercury (Hg)

Mercury exists in different forms like elemental mercury (Hg^0), mercurous ion (Hg_2^{2+}), and mercuric ion (Hg^{2+}) (Kumari and Chauhan, 2014). It creates various environmental problems; generally, mercury waste occurs in the refinery, pharmaceuticals, and fertiliser industries (Parham et al., 2012). Hg causes possible damages to the brain, reproductive system, kidney, and respiratory system. According to the USEPA, the permissible limit of mercury ion is 0.002 mg/L (Gautam et al., 2016). Hg is dangerous even in low concentrations for aquatic organisms, and the other significant factor is its non-biodegradable characteristic.

11.2.2 Detection methods

Different scattered harmful materials present in wastewater are unfavourably produced and discharged due to domestic, industrial and agricultural activities. The structure and concentration of chemical substances in wastewater mainly come under three categories: microorganisms, inorganic chemicals, and organic chemicals, which all cause adverse effects on the environment, namely, disruption of hormone systems, cytotoxicity, genotoxicity, and carcinogenic effects (Korostynska et al., 2013). Multiple indicators are used to measure the characteristics of discharged wastewater. Further, a variety of analytical methods have been used to monitor the quantity of heavy metal and biological contaminants present in effluents. Detection methods of heavy metals includes inductively coupled plasma mass spectrometry (ICP-MS), atomic fluorescence spectrometry (AFS), atomic absorption spectroscopy (AAS), atomic emission spectroscopy (AES), and voltammetric and potentiometric techniques. These methods have high detection sensitivity and great accuracy for the detection of heavy metals. However, these instruments are expensive and need high maintenance cost compared with conventional methods. Monitoring and detecting heavy metal water pollution is significant to protect environment and human health. Light portable sensors can also be used to measure heavy metal concentration at the point of discharge. Specifically, portable instruments are advantageous for reporting live measurement and continuous monitoring analysis and controlling the heavy metal ions in released wastewater. Sensors have been developed to determine the wide range of analytes in the environmental, which helps to monitor and screening the accurate toxicity level instantly that tends to minimises the toxicity risk. From that perspective, biosensors play a significant role in the detection of pollutants.

11.2.3 Biosensors

For monitoring the pollutants, various tools are available, yet they are inconvenient and complex to use and need trained professionals in a laboratory setting. In the past decades, attention is paid to use of biosensors and chemical sensors to detect biological pollutants in the environment (Nag et al. 2021). Sensors have been increasingly used to monitor water quality, for example, salinity, pH, DS, and dissolved organic substances. Biosensors have also been developed for disposable applications, for example, single-use glucose biosensors designed for identifying diabetes (Yoo and Lee, 2010). In the past years, biosensors has received the most incredible attention in detection of environmental pollutants in wastewaters; the main benefit of these biosensors is detection of pollutants even at low concentrations. Until the 20th century, biosensors were used in laboratory applications, but soon their development is exceedingly fast with three generations of evolution. In the first generation, the action process depends on the electrical response; in the second generation of biosensors, intermediaries were developed to improve the link between the reaction and the transducer to get better results. In the third generation of biosensors, the response is received without an intermediary medium (Mandal et al., 2020). A biosensor is a biological sensor device made up of transducers and a biological component that measures biological or chemical reactions by calculating an analyte proportioning to the signal response. Figure 11.2 represents the general working principle of biosensors. Biosensors have different applications in disease monitoring, drug discovery, detection of pollutants, identification of disease-causing microorganisms, and markers of disease indicators in bodily fluids (blood, urine, saliva, and sweat). Biomolecules are responsible for precise analyte

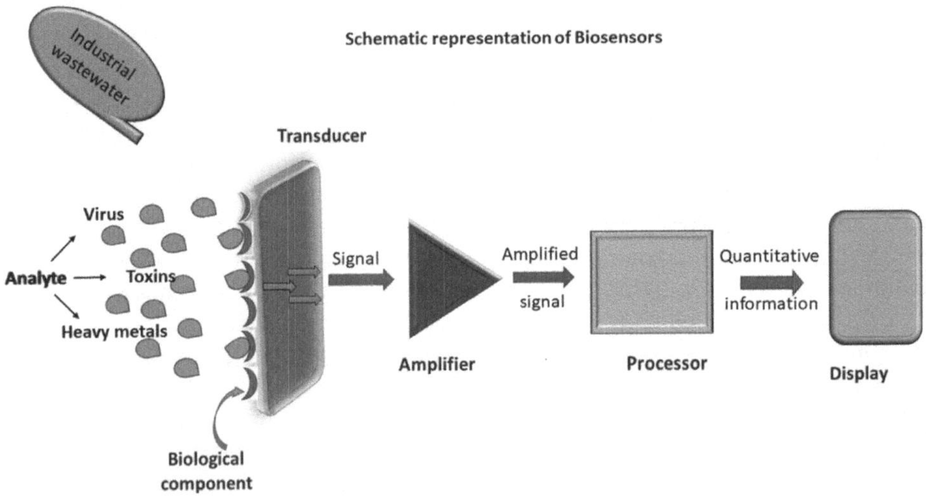

Figure 11.2 General working principle of a biosensor.

recognition, while physicochemical converter provides an electric output signal expanded by an electronic unit (Scheller and Schubert, 1992). Biosensors are valuable for various applications such as food quality control, agriculture, drug detection, and pollutant detection (Amine et al., 2006). A biosensor is a simple, cost-effective, and speedy outcome-providing detection method for monitoring water pollution caused by versatile heavy metals like zinc, cobalt, cadmium, and lead. Biosensors can be used in *in situ* applications, which prevents long-term measurement processes in laboratories without additional sample preparation, and the other main benefit is a speedy outcome that is essential for environmental control and monitoring.

The classic combination of nanomaterials and the most desired component for next-generation technologies such as metal ion (bio-wastes) moves from the sphere of dreams to the panorama of modern science and technology. It opens the door to novel bio- and nanotechnology applications. Apart from the technological aspect of nano-biotechnology, a critical understanding of how nanomaterials interact with metal ions present in industrial wastes of different compositions and how it impacts metal ion sensing has recently attracted increasing attention of the research community. Following the developments in this field, it has become apparent that nanomaterials' exploration and diversity in metal ion sensing are closely associated with nanomaterials' physicochemical properties and structural aspects of the metal ion. Several novel biotechnologies based on fundamental elements of nanomaterial-metal ion binding and their interactions have been developed in various fields, such as designing new types of sensing material, nanodevices, diagnostic application, and food industry bio-electronics, bio-computation, and forensic sciences in recent years. In addition to the knowledge of chemical and structural properties (biocompatibility, water solubility, and biodegradability) of nanomaterials, more efforts have been directed towards identifying and designing nanomaterials for site-specific and sequence-specific metal ion binding. Only one-to-one interactions between a single metal ion-molecule and nanoparticles can achieve high sensitivity through a specific or non-specific molecular binding. A significant part of the study of nanomaterial-metal ion interaction deals with the functionalised gold and silver nanoparticles (NPs) conjugation with thiolated oligonucleotides for specific metal ion hybridisation and identification of complementary sequences of interest.

Over the years, fluorescent property having an "on-off-on" technique of various nanomaterials (carbon dots, Au, Ag) has been used to detect different metal ions, viz. Cu^{2+}, Fe^{3+}, Hg^{2+}, Cd^{2+}, and Au^{3+} with below permeable concentration in water. In most cases, these metal ions act as a fluorescent quencher. Numerous efforts are being explored to develop functionalised fluorescent nanomaterials for intercellular metal ion sensing and imaging.

11.2.4 Classification of biosensors

The biosensor can be divided into two elements: transducing elements and biocomponent elements. Figure 11.3 represents the categories of biosensor-based detection methods. Various approaches involving biosensors for detection of heavy metals are summarised in Table 11.3. Transducers include electrochemical transducers, optical transducers, and mass-based transducers. When the analyte gets in contact with the biocomponent, a chemical reaction is induced that produces or consumes ions and changes the analyte's electrical properties, such as electrical current. Transducers detect these changes by creating an electrochemical signal associated with the analyte amount of aquatic solution. The electrochemicals are further divided into four

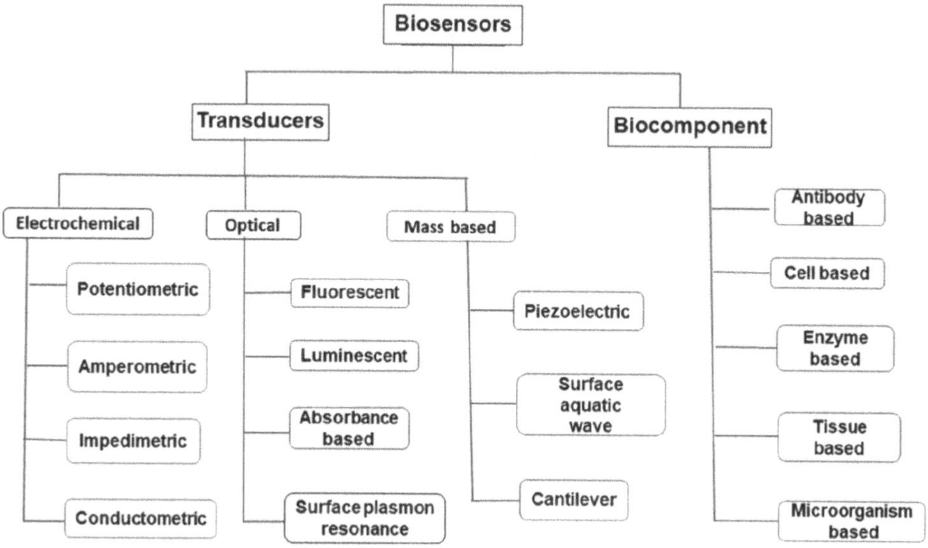

Figure 11.3 Categories of biosensor-based detection methods.

Table 11.3 Different types of biosensors for the detection of heavy metals

Biosensor type	*Analysed heavy metal*	*Detection range (μM)*	*Reference*
Enzyme-based	Hg (II), Cd (II), Pb (II), and Cr (VI)	Hg (II) = 0.0025–0.1 Cd (II) = 0.01–0.1 Pb (II) = 0.01–0.12 Cr (VI) = 0.0025–0.06	Silva et al. (2020)
Enzyme-based	Cu (II), Hg (II), Cd (II), and Pb (II)	Cu (II) = 0.079–16 Hg (II) = 0.025–5 Pb (II) = 0.10–15 Cd (II) = 0.04–62	Ayenimo and Adeloju (2015)
Enzyme-based	Cu (II), Pb (II), and Cd (II)	0.01–0.001	Syshchyk et al. (2015)
Enzyme-based	Pb (II)	0.001–0.1	Tang et al. (2019)
Cell-based	Cd (II), Hg (II), and Pb (II)	Cd = 6.25–7.99 Hg = 8.80–9.82 Pb = 44.18–60.69	Zhu et al. (2017)
DNA-based biosensor	Hg (II)	0.0001–0.01	Cai et al. (2017)

subdivisions: potentiometric, amperometric, impedimetric, and conductometric. The optical transducer component includes fluorescent, luminescent, absorbance-based, and surface-oriented resonance. Piezoelectric, surface aquatic wave, and cantilever come under mass-based transducers. The biocomponent category consists of antibodies, cells, enzymes, tissues, and microorganism-based biocomponents. Among these classifications, an electrochemical biosensor is an advantageous method for detecting heavy metal pollutants from effluents. The working principle of an electrochemical biosensor is computed by the changes in microbial activity inside the biosensor, that determined by the response to the toxicant, which will be defined by the oxygen consumption (Agostino, 2020). For example, the microbial fuel cell (MFC)-based biosensor is a cost-effective device that runs for a more extended period and gives detection results in *in situ* applications.

11.3 HEAVY METAL REMOVAL TREATMENT TECHNIQUE USING ADSORPTION AND PHOTOCATALYTIC REDUCTION

Pollutants such as heavy metals, persistent organic pollutants, metalloid, and other chemical substances have extensively contaminated the aquatic ecosystem worldwide. Elimination of these poisonous and hazardous substances from water is a significant environmental challenge. From this perspective, various approaches have evolved to determine the pollutants monitor their load and treat their toxicity using conventional treatment methods. Different physicochemical methods are available to treat the metal-containing effluents, namely, adsorption, reverse osmosis, evaporation techniques, chemical precipitation, filtration, chemical oxidisation, electrochemical process, ion exchange, and photocatalytic process. However, these methods have their limitations, such as membrane fouling, high energy requirement, high cost, high maintenance cost, and incomplete removal of metals with high sludge production which has to undergo secondary treatment. Several industries generate a considerable amount of toxic dye-containing wastewater, which are carcinogenic and pose severe threats to humans and other living organisms. Many researchers have used the methods mentioned above to eliminate heavy metal ions; most of these techniques provided outstanding results. Among the conventional methods, photocatalytic and adsorption techniques are more convenient and advantageous comparatively. While removing heavy metals from industrial wastewater, coagulation generates secondary pollutants that are harmful to the ecosystem and need secondary treatments to ensure complete removal of metals. Moreover, it produces an enormous amount of sludge due to extensive utilisation of chemical substances. The resin-based ion-exchange process is also considered an essential process for removal of metal ions that substitutes the ions for high removal efficiency. However, this process suffers from regeneration of secondary pollutants; therefore, it cannot be utilised for industrial-scale applications due to the cost of resin. The membrane filtration method consists of various separations through ultrafiltration, reverse osmosis, microfiltration, and electrodialysis. This process is very facile due to its easy and reliable operation, low cost, and less maintenance. However, the selection of pollutant, specificity, cost, and membrane fouling are considered as significant disadvantages. High membrane fouling and improper selectivity results in loss of accuracy, and high financial maintenance is unsuitable for industrial-scale applications. On the other hand, being affordable and comfortable, the chemical precipitation method is used to treat wastewater effluents. Both hydroxide precipitation and sulphide precipitation have limitations, such as separation issues in sulphide, filtration complications, and sedimentation processes. In contrast, the hydroxide method requires numerous chemicals to decrease the metals to an adequate level for release into the environment. This chemical precipitation method requires a large amount of capital investment, more power supply, and a considerable volume of chemicals, which limits its industrial applications. Electrochemical techniques include electrocoagulation, electrodeposition, and electro flotation techniques, all of which have been analysed to remove heavy metal ions from wastewater. Significant advantages of using these methods are recovery of metals in the purest form, less operating cost, and no complication in the disposal of sludges. Bioremediation utilises microorganisms for removal of industrial wastewater effluents. This microbe-assisted method is widely used to remove toxic heavy metal pollutants, which shows moderate removal efficiency; besides, the traditional wastewater treatment is associated with high production code. Bioremediation produces unfavourable secondary pollutants, has low efficiency in the aquatic environment, and leaves unwanted wastes in waterways. The adsorption process is also a well-defined and well-established method for heavy metal ion screening. Adsorption happens when a liquid solute (adsorbate) gets adsorbed on some solid (adsorbent) surface. Figure 11.4 represents the schematic diagram of the adsorption process. Adsorption efficiency is related to the characteristics of the adsorbent. There are several low-cost adsorbents available to treat water contaminants. Specifically, carbon-based materials like activating carbon, agricultural waste, etc., play a predominant role in adsorption. Two types of adsorption happen on the surface of the adsorbent: physical adsorption and chemical adsorption. Desorption of its reversible techniques can regenerate adsorbents, and the regenerated materials can be utilised many more times further, which can be degradable. Regeneration of

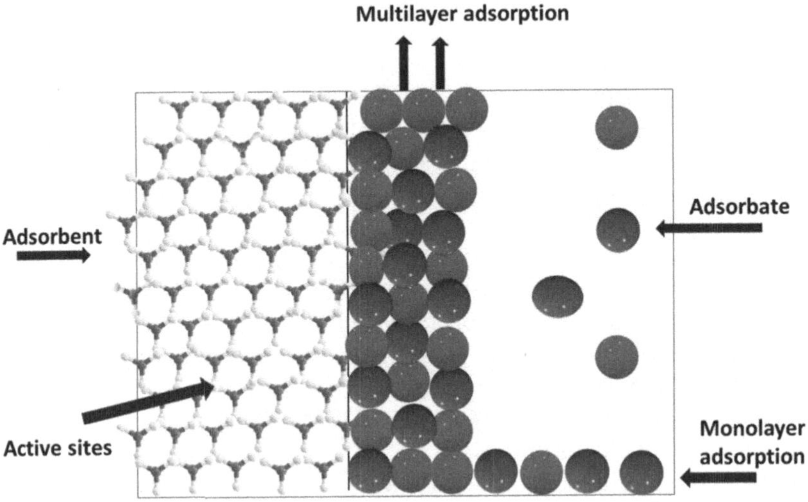

Figure 11.4 Schematic diagram of the adsorption process.

adsorbents can be achieved through the thermal generation pressure swing method and electrochemical regeneration, confirming the adsorption process in an environmentally friendly manner. Environment supportive adsorption process is facile and flexible while applying, and there is no harmful waste production. Moreover, this process has become one of the leading industrial wastewater effluent treatment methods. Premier absorbent properties depend on their surface area, pore size, pore volume, surface functionality, polarity, and surface treatment. This leads to enhanced removal efficiency of the heavy metal ion pollutants and controls the removal kinetics. Effluent pollutants' efficiency depends on adsorbent properties, temperature, contact time, mixing speed, ionic strength, and pH. As a result, nanomaterials are considered a promising candidate that can exhibit various easily tunable properties, as mentioned above, for successful metal pollutant removal. Researchers widely use the adsorption method to treat industrial wastewater effluents. Table 11.4 summarises the functional properties of different adsorbents originated for metal ion removal.

11.3.1 Photocatalysis

Photocatalysis of any semiconducting material is related primarily to its size and morphology. The essential criteria for an efficient photocatalyst are (a) suitable optical bandgap (E_g) with high light-harvesting capacity, (b) effective charge separation and migration to the surface rather than charge recombination, (c) proper conduction band (CB) and valance band (VB) edge potential to conduct the redox reaction for pollutant mineralisation or fuel production, and (d) photochemical stability of the photocatalyst. The semiconducting nanomaterials efficiently satisfy this candidature by providing a high surface-to-volume ratio, tunable morphology with high dimensionality and/or porosity, abundant surface-active sites for the photo-redox reactions, efficient

Table 11.4 Different adsorbents for the adsorption of the industrial heavy metal ions

Adsorbents	*Metal ion*	*Adsorption capacity (mg/g)*	*Reference*
DES modified MWCNT (multi-wall carbon nanotube functionalised by glycerol-based deep eutectic solvent)	Ni (II)	115.8, t = 68 min, pH = 7.21, T= 55°C	Rahmati et al. (2020)
Nanoflower-like CdS	Cr (VI)	396.47, pH= 2, T = 120 min at room temperature	Roy et al. (2020)
Phosphoric acid-activated *Chrysopogon zizanioides* roots-activated carbon	Cu (II) and Ni (II)	176.92 and 167.92, pH 7, T = 2 min	Thanarasu et al. (2021)
Flower globular magnesium hydroxide	Ni (II)	287.11, pH = 6.07–7.71 at room temperature	Jiang et al. (2019)
Graphene oxide/alginate hydrogel membranes	Pb (II) Cr (III)	327.9 118.6, pH = 5–6 at room temp.	Bai et al., (2019)
Electroplating sludge (raw) Electroplating sludge (calcined)	Ni (II)	Raw –210.9 Calcined –163.6	Peng et al. (2020)
Mesoporous carbon nitride CTAB-modified geopolymer	Cu (II) Cr (VI)	108.2 95.3 T = 303.15 K	Yu et al. (2020)
Activated carbon	Ag (I) Cr (VI)	104.0 96.3 pH = 2.5 at room temperature	Wu et al. (2020)
GOCS (carbon disulphide functionalisation on graphene oxide)	Pb (II)	280.2 pH= 5.7 333K	Lian et al. (2020)
Sulphur-functionalised graphene oxide nanosheet	Pb (II) Cd (II) Ni (II) Zn (II)	312 384 344 285 at room temperature	Pirveysian and Ghiaci (2018)
MGO@TNs (magnetic graphene oxide-titanate)	Pb (II)	322.7 pH = 5	Yang et al. (2020a)
GO-SF (graphene oxide/silk fibroin aerogel)	Zn (II) Ni (II) Cu (II)	72.1 13.9 83.4	Wang et al. (2018)

light-harvesting capacity, and prolonged charge carrier transfer with reduced recombination as compared to the bulk material. Besides, the presence of oxygen-related surface defects in the nanostructured materials is an advantageous factor to retard the charge recombination. There are three significant steps in photocatalysis on light irradiation wavelength $(\lambda) \geq E_g$ energy of the photocatalyst: (a) the photogenerated electrons jump to the CB, leaving the holes in the valence band (VB); (b) charge separation and migration of these excitons to the photocatalyst surface rather than their recombination; and finally, (c) redox reaction with the adsorbed moieties. In principle, the CB electrons initiate the reduction reaction, whereas the holes in the VB involve the oxidation reaction. Therefore, the reduction or oxidation capability strictly depends on the band edge positions of CB and VB, respectively. Semiconducting metal oxides such as TiO_2, ZnO, and SnO_2 are widely used as photocatalysts. Figure 11.5 represents the schematic illustration of the photocatalytic reaction process.

Advanced oxidation processes (AOPs), such as photocatalysis, are widely considered to suppress organic pollutants of contaminants of emerging concern (CECs). The photocatalytic technique has gained significant research attention for removal of heavy metal ions from industrial wastewater. Photocatalysis is also called artificial photosynthesis activity. Changes in the reaction rate caused by the light in photocatalyst adsorb the light for converting photonic energy to chemical energy. The photocatalysis mechanism comprises hydroxyl radical (■OH) and oxide anion (■O_2^-); among them, OH■ exhibits high oxidation potential, which acts as an influential oxidising agent even though both acts as an oxidant to degrade organic compounds. An increase in irradiation time and photocatalyst dosage had a positive impact on photocatalytic reduction. Figure 11.5 illustrates a photocatalysis reaction. For example, oxide-based semiconductors have a filled VB and empty CB with excellent light adsorption properties for electrons/holes generation on the surfaces; when the light energy is applied on the oxide semiconductors, electron/holes pairs are formed in the CB and VB, respectively. During photocatalytic reduction, electrons are transferred to the CB to adsorb the heavy metal ions, and the holes in the VB are supposed to oxidise water. The photocatalytic method is an environmentally friendly, low-cost, high energy-saving, flexible catalyst loading and highly efficient method for the degradation of organic pollutants. Electrons and positive holes are generated under light irradiation (photons) in

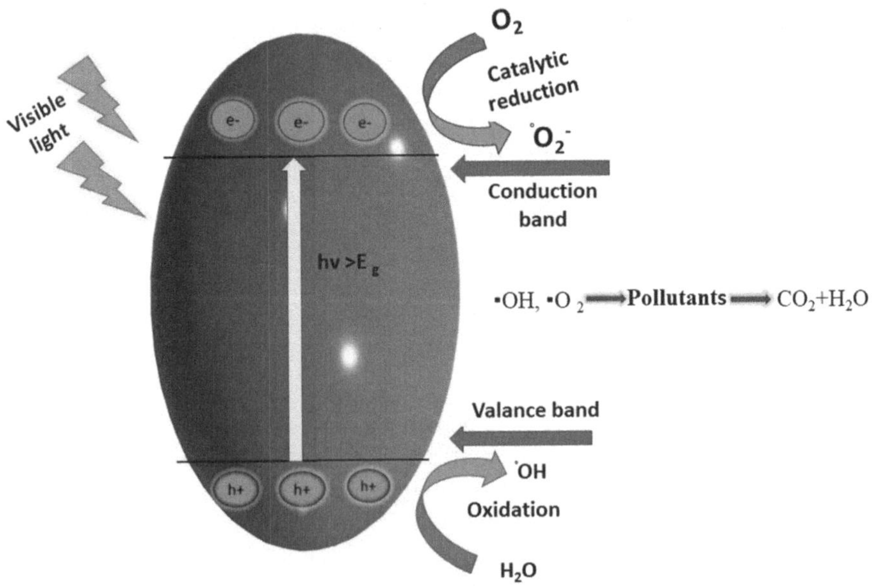

Figure 11.5 Schematic illustration of the photocatalytic reaction process.

the conduction and VBs of photocatalysts such as CdS, ZnS, ZnO, WO_3, and TiO_2. According to the mobility, two types of photocatalysts are available, namely, mobilised (powder form) and immobilised (thin film) photocatalysts (Pedanekar et al., 2020). A thin film is a promising photocatalyst because of its cost-effectiveness and recycling capabilities after washing all the pollutant loads, which is not possible in the powdered form. In the powdered photocatalyst form, recycling is impossible due to a large power load of aggregation. Due to the chemical stability, efficiency, and non-toxicity in the photocatalysis method, TiO_2, CdS, ZnS, ZnO, WO_3, SnO_2, and Fe_3O_3 are the most commonly used photocatalysts and have been well studied for environmental protection. The photocatalytic reduction efficiency of heavy metals depends on the photocatalyst's physical properties and morphology, for example, narrow E_g, ample surface, pore volume, and mesopore structure. For instance, CdS semiconductor as a photocatalyst has a comprehensive E_g (~2.5 eV), which permits light absorption.

11.3.2 Metal oxides usage in photocatalytic application

Semiconducting metal oxides exhibit an optimum E_g (~3eV) suitable for photocatalytic application. During the photocatalytic process, the metal oxide adsorbed the photons and further escalated for the catalytic process, resulting in either degradation or transformation of a less toxic oxidation state of the pollutant metal ion. TiO_2 is widely used in photocatalytic degradation because of its cost-effectiveness, chemical stability, highly active material, non-toxicity, and high resistance against photochemical deterioration (Hitam and Jalil, 2020). However, due to unfavourable properties: rapid charge recombination of the photogenerated electron and pair hole, unsatisfactory sensitivity of TiO_2 under solar light, aggregation of TiO_2 nanoparticle which all preventing the efficiency of the photo catalysis reactions. To improve the reaction efficiency, modification of TiO_2 needs to focus for better performance. ZnO is a low-cost, chemically stable, non-toxic, environmentally friendly material used in photo catalysis, and it provides exceptionally good results in nanomaterials when compared with TiO_2 material. Its photocatalytic activity is highly increased when coupled with other semiconductors such as TiO_2, WO_3, and SnO_2 (Mugunthan et al., 2019). WO_3 is a low-cost and low-toxicity material with stable physicochemical properties and ability to oxidising the holes in the VB, which shows promising materials in photocatalytic performance, including dye-sensitised solar cells (DSSCs), sensors, and photo electrochemical water splitting. When united with graphene, graphene oxide, and other carbon-based composite materials, the photocatalytic activity of WO_3 improved in various energy-related applications and environmental remediation (Sajjad et al., 2019). Copper oxides (CuO and Cu_2O) also play a significant role in photocatalytic degradation; copper oxide combined with carbon-based composites produces a promising performance in photocatalysis. However, the removal efficiency of metal oxide-based composite is highly influenced by the morphology, size, and microstructure of particles. For example, TiO_2 and ZnO semiconductors with nearly identical E_g (~3.2 and 3.37 eV) are commonly used as catalysts because of their cost-effectiveness, non-toxicity nature, regeneration capabilities, and effectiveness in photodegradation processes (Padmanabhan et al., 2021). Another novel mesoporous CuO/ZnO heterojunction material shows a synergistic effect for enhancing photocatalytic reduction/removal of Hg (II) ion with nearly 100%

Table 11.5 Photocatalytic techniques using different materials for degradation of metal ions

Materials	*Concentration amount of metal ion (mg/L)*	*Metal ion photocatalytic in percentage (%)*	*Time*	*Reference*
CdS	10	Cr (VI)-99.1%	140 min	Peng et al. (2019)
$CoFe_2O_4/ZrO_2$	20	Cr (VI)-83%	180 min	Emadian et al. (2020)
WO_3/TiO_2	0.01–4	As (III) -99%	60 min	Nanarrete-Magana et al. (2021)
CdO/ZnO (3%)	100	Hg (II)-100%	60 min	Mohamad et al. (2021b)
Fe_3O_4-ZnAl-LDH-TiO_2	20	Cr (VI)-98.4%	8 h	Yang et al. (2020b)
Wurtzite/natural sphalerite heterostructure	0.5	Cr (VI) -97.11%	30 min	Hu et al. (2021)
Mesoporous $CoFe_2O_4$	100	Hg (II) -100%	60 min	Kadi et al. (2020a)
Mesoporous α-Fe_2O_3/g-C_3N_4	100	Hg (II) -90%	60 min	Kadi et al. (2020b)

degradation within 60 min (Mohamed et al., 2021a). Metal oxide photocatalysts have been modified by chemical and structural changes to improve the catalytic performance, which increases the removal of heavy metals (Danish et al., 2021). Table 11.5 describes the photocatalytic reaction using a different material for the degradation of a metal ion.

11.4 FUTURE SCOPE

In this chapter, different sources of heavy metal contamination are explored and explained briefly. Traditional analytical detection techniques no longer satisfy the need for continuous monitoring because of their high price, longer analysis time, and requirement of well-trained analysts. From that perspective, biosensors play a predominant role in detecting heavy metals and overcoming those limitations of traditional methods. Comparatively, biosensors have attracted significant attention in the detection of wastewater environmental pollutants in the past decades. The advantage of biosensors is facilitated to detect pollutants even at low concentrations. Still, there is a need for upgrading biosensors with suitable materials, which will enhance the detection of heavy metal ions from wastewater. Revolutionary equipment in chemical, optical, and biomonitoring ensures water quality information across various water bodies. There is a scope to establish further mechanisms of interaction between nanomaterials and biomolecules on the surface of electrodes or nanofilms to generate a facile biofilm. Employing nanomaterials for sensor fabrication and utilising their E_g, thermal, magnetic, and optical properties could be projected as a futuristic work in this field.

Nevertheless, nanomaterial-based biosensors exhibit promising attractive prospects and are suitable for food analysis, clinical diagnosis, process control, and environmental monitoring. The environmental economy encompasses activities and products that serve environmental protection and resource management, which even control part of a nation's gross domestic product (GDP). Industrial wastewater pollution is a particular threat to water resources in India, home to some 100 million people and a growing economy. An estimated 38354 million litres per day (MLD) of sewage is generated in major cities of India, but the sewage treatment capacity is only 11786 MLD. Similarly, only 60% of industrial wastewater, primarily from large-scale industries, is treated (Benson et al., 2018). The environmental technology sector in India is expanding rapidly, with evident business opportunities for pollution abatement technology innovations. The United Nations (2015: 6.3) has specifically targeted improvements to "water quality by reducing pollution, eliminating dumping and minimising release of hazardous chemicals and materials, having the proportion of untreated wastewater and substantially increasing recycling and safe reuse globally" by 2030. The research addresses other sustainable development goals (SDGs), including, among other things, SDG3 (good health), SDG11 (sustainable communities), and SDG12 (responsible consumption). It is estimated that the total cost for establishing a treatment system for the entire domestic wastewater is ~90k USD. The Government of India's financial budget allocated ~96 k USD in 2019–2020 compared to ~70 k USD in 2018–2019 for different water treatment and cleaning processes. The programme – a component of the National Rural Drinking Water Mission – saw a massive increase of 69% in allocation, up from 53 k USD in 2018–2019 as mentioned in the Union Budget of Finance Department, Government of India (PRS Legislative Research, 2021).

As India faces severe wastewater treatment problems leading to human health risks through consumption of contaminated drinking water, water scarcity, and climate change, a sustainable clean solution is highly desired. The textile industries play a vital role in increasing India's GDP value. The Indian textiles industry, currently estimated at around US$108 billion, is expected to reach US$223 billion by 2021. The industry is the second largest employer after agriculture, employing over 45 million people directly and 60 million people indirectly. The Indian textile industry contributes approximately 5% to India's GDP and 14% to the overall Index of Industrial Production (IIP). The growth implies domestic sales to rise to US$315 billion from currently US$68 billion. At the same time, exports are suggested to increase to US$185 billion from approximately US$41 billion now (Central Water Commission, 2018). However, the use of chemicals and dyes the manufacture of textiles generates enormous waste as sludge, fibres, and chemically polluted waters. Since the textile industry is highly water-intensive and India had been identified as a highly water-scarce region, the long-term viability of the Indian textile industry hinges heavily on sustainable water management in India.

11.5 CONCLUSION

This chapter briefly examined the possibilities of water pollution from different scattered sources. It mainly investigated the industrial wastewater heavy metal pollution resulting from artificial activities to develop the nation's economy. Heavy metal pollution's hazardous impacts on the environment, human life, and aquatic environment and related issues have discussed concisely. Consequently, global urbanisation causes contaminants and industrial development producing contaminants from unknown point sources have been released directly and indirectly to the environment. From that perspective, industrial heavy metal pollution becomes the topmost concern worldwide. Here, the industrial origins of metal ion contamination and its causes to living organisms briefly examined.

Further reviewed the general information about detection methods, particularly biosensor working principles and their categories have evaluated to detect a heavy metal. Biosensors showed the highest sensitivity and potentially a promising technology for the detection of heavy metal ions. In particular, electrochemical methods have designed to detect heavy metals like mercury, cadmium, lead, silver, and copper. Additionally, sensors should detect more metals even in low concentrations in water bodies, thus preventing adverse effects caused by their contamination. The conventional treatment methods to treat these industrial effluents are discussed; among them, adsorption and photocatalytic techniques are explained more specifically. Adsorption and photocatalysis are highly proficient and low-cost alternative techniques in wastewater treatment. Adsorbent's adsorption efficiency primarily depends on the physicochemical properties of the adsorbent. The high surface area containing adsorbents enhance the metal ion removal due to the availability of more functional groups on the surface. Further regeneration of adsorbents plays a predominant role in adsorption that allows reusing the adsorbents many times to prevent the investment. Alternatively, photocatalysis is another promising and environmentally friendly for photodegradation of various environmental pollutants. From that perspective, awareness related to the reuse or recycle of wastewater is significantly necessary due to the current water scarcity situation anywhere in the world.

ACKNOWLEDGEMENTS

S.V. acknowledges the College of Engineering, Mathematics, and Physical Sciences, the University of Exeter, for the PhD UKRI studentship (EP/R 513210/1).

CONFLICT OF INTEREST

The authors declare no competing financial interest.

REFERENCES

Agostino V., 2020. Environmental electroactive consortia as reusable biosensing element for freshwater toxicity monitoring. *New Biotechnol* 55, 36–45.

Amine A., Mohammadi H., Bourais I., Palleschi G., 2006. Enzyme inhibition- based biosensors for food safety and environmental monitoring. *Biosens Bioelectron* 21, 1405–1423.

Arbabi M., Hemati S., Amiri M., 2015. Removal of lead ions from industrial wastewater: A review of removal methods. *Int J Epidemiol Res* 2(2), 105–109.

Ayenimo JG., Adeloju SB., 2015. Talanta inhibitive potentiometric detection of trace metals with ultrathin polypyrrole glucose oxidase biosensor. *Talanta* 137, 62–70.

Bai C., Wang L., Zhu Z., 2019. Adsorption of Cr(III) and Pb(II) by graphene oxide/alginate hydrogel membrane: Characterization, adsorption kinetics, isotherm and thermodynamics studies. *Int J Biol Macromol* 147, 898–910.

Benson, D., Parker, B., Mishra, V., Muthukumar, K., Sundaram, S., 2018. *India-UK Workshop on knowledge transfer on the sustainability of innovative wastewater treatment technologies to India: Final Report*. Penryn: Environment and Sustainability Institute.

Bhanjana G., Dilbaghi N., Kim K., Kumar S., 2017. Carbon nanotubes as sorbent material for removal of cadmium. *J Mol Liq* 242, 966–970.

Cai, W., Xie, S., Zhang, J., Tang, D., Tang, Y., 2017. An electrochemical impedance biosensor for Hg^{2+} detection based on DNA hydrogel by coupling with DNAzyme-assisted target recycling and hybridization chain reaction. *Biosens Bioelectron* 98, 466−472.

Central Water Commission, 2018. *Status of Trace and Toxic Heavy Metals in Indian Rivers*. New Delhi: CWC. Central Water Commission, Department of Water Resources, River Development and Ganga Rejuvenation, Government of India

Chowdhary P., Bharagava R.N., Mishra S., Khan N., 2020. Role of industries in water scarcity and its adverse effects on environment and human health. In: Shukla V., Kumar N. (eds.) *Environmental Concerns and Sustainable Development*, pp. 235–256. Singapore: Springer.

Danish M.S.S., Estrella L.L., Alemaida, I.M.A., Lisin, A., Moiseev, N., Ahmadi, M., Nazari, M., Wali, M., Zaheb, H., Senjyu, T., 2021. Photocatalytic applications of metal oxides for sustainable environmental remediation. *Metals* 11(1), 1–25.

Dubey M., Mohapatra S., Tyagi V. K., Suthar S., Kazmi, A. A., 2021. Occurrence, fate, and persistence of emerging micropollutants in sewage sludge treatment. *Environ Pollut 273*, 116515.

Emadian S.S., Ghorbani M., Bakeri G., 2020. Magnetically separable $CoFe_2O_4/ZrO_2$ nanocomposite for the photocatalytic reduction of hexavalent chromium under visible light irradiation. *Synth Met* 267, 116470.

Fu F., Wang Q., 2011. Removal of heavy metal ions from wastewaters: A review. *J Environ Manage* 92, 407–418.

Gautam P.K., Gautam R.K., Banerjee S., Chattopadhyaya M.C., Pandey J.D., 2016. Heavy metals in the environment: Fate, transport, toxicity and remediation technologies. In: Pathania D. (ed.) *Heavy Metals: Sources, Toxicity and Remediation Technique*, pp. 101–130, Hauppauge, NY: Nova Science Publishers, Inc.

Gibb H., Haver C., Gaylor D., Ramasamy S., Lee J.S., Lobdell D., 2011. Utility of recent studies to assess the National Research Council 2001 estimates of cancer risk from ingested arsenic. *Environ Health Perspect* 119(3), 284–290.

Grace Pavithra K., Jaikumar V., Kumar P.S., SundarRajan P., 2019. A review on cleaner strategies for chromium industrial wastewater: Present research and future perspective. *J Clean Prod* 228, 580–593.

Hamilton P.B., Cowx I.G., Oleksiak M.F., Griffiths A.M., Grahn M., Stevens J.R., Carvalho G.R., Nicol E., Tyler C.R., 2016. Population-level consequences for wild fish exposed to sublethal concentrations of chemicals—A critical review. *Mar Biol Ecol* 17(3), 545–566.

Helena B., Pardo R., Vega M., Barrado E., Fernandez JM., Fernandez L., 2000.Temporal evolution of groundwater composition in an alluvial aquifer (Pisuerga River, Spain) by principal component analysis. *Water Res* 34(3), 807–816.

Hitam C.N.C., Jalil A.A., 2020. A review on exploration of Fe_2O_3 photocatalyst towards degradation of dyes and organic contaminants. *J Environ Manag* 258, 110050.

Hu W., Dong F., Liu M., Yang D., Luo N., Chen C., 2021. Applied surface science photocatalytic reduction of Cr (VI) using a wurtzite/natural sphalerite heterostructure: Synergistic effects of exposed active facets, vacancies and a heterophase junction. *Appl Surf Sci* 550, 149267.

Jakobik- Kolon A., Bok-Badura J., Karon K., Mitko K., Milewski A., 2017. Hybrid pectin-based bio sorbents for zinc ions removal. *Carbohydr Polym* 169, 213–219.

Jiang D., Yang Y., Huang C., Huang M., Chen J., Rao T., 2019. Removal of the heavy metal ion nickel (II) via an adsorption method using flower globular magnesium hydroxide. *J Hazard Mater* 373, 131–140.

Kadi M.W., Mohamed R.M., Ismail A.A., Bahnemann D.W., 2020a. Decoration of g-C_3N_4 nanosheets by mesoporous $CoFe_2O_4$ nanoparticles for promoting visible-light photocatalytic Hg (II) reduction. *Colloids Surf A Physicochem Eng Asp* 603, 125206.

Kadi M.W., Mohamed R.M., Ismail A.A., Bahnemann D.W., 2020b. Performance of nctionmesoporous α-Fe_2O_3/g-C_3N_4 heteroju for photoreduction of Hg (II) under visible light illumination. *Ceram Int* 14, 23098–23106.

Kalavathy M.H., Karthikeyan T., Rajgopal S., Miranda L.R., 2005. Kinetic and isotherm studies of Cu (II) adsorption onto H_3PO_4-activated rubber wood sawdust. *J Coll Interface Sci* 292, 354–362.

Khan M.A., Khan S., Khan A., Alam M., 2017. Soil contamination with cadmium, consequences and remediation using organic amendments. *Sci Total Environ* 601, 1591–1605.

Korostynska O., Mason A., Al-Shamma'a A., 2013. Monitoring pollutants in wastewater: Traditional lab based versus modern real-time approaches. In: Mukhopadhyay S., Mason A. (eds) *Smart Sensors for Real-Time Water Quality Monitoring*, pp. 1–24. Berlin, Heidelberg: Springer.

Kumar A., Cabral-Pinto M., Kumar A., Kumar M., Dinis P.A., 2020. Estimation of risk to the eco-environment and human health of using heavy metals in the Uttarakhand Himalaya, India. *Appl Sci* 10, 1–18.

Kumar M., Upadhyay R., Mohapatra S., Dhangar K., 2021. Transport of black carbon in the presence of perchlorate in aquatic environment: A perspective of cogenerated species in the Gangetic plain of India. *Case Stud Chem Environ Eng* 3, 100065.

Kumari S., Chauhan G.S., 2014. New cellulose-lysine Schiff-base-based sensor-adsorbent for mercury ions. *ACS Appl Mater Interfaces* 6, 5908–5917.

Lian Q., Ahmad U.Z., Gang D.D., Zappi M.E., Fortela, D.L.B., Hernandez, R., 2020. The effects of carbon disulfide driven functionalization on graphene oxide for enhanced Pb(II) adsorption: Investigation of adsorption mechanism. *Chemosphere* 248, 126078.

Liu J., Huang K., Xie K., Yang Y., Liu H., 2016. An ecological new approach for treating Cr(VI)-containing industrial wastewater: Photochemical reduction. *Water Res* 93,187–194.

Mandal N., Adhikary S., Rakshit R., 2020. Nanobiosensors: Recent developments in soil health assessment. In: Rakshit A., Ghosh S., Chakraborty S., Philip V., Datta A. (eds) *Soil Analysis: Recent Trends and Applications*. Singapore: Springer, pp. 285–304.

Menon N.G., Mohapatra S., Padhye L.P., Tatiparti S.S.V., Mukherji S. (2020) Review on occurrence and toxicity of pharmaceutical contamination in Southeast Asia. In: Kumar M., Snow D., Honda R. (eds) *Emerging Issues in the Water Environment During Anthropocene. Springer Transactions in Civil and Environmental Engineering*. Singapore: Springer. https://doi.org/10.1007/978-981-32-9771-5_4

Mobasherpour I., Salahi E., Ebrahimi M., 2012. Removal of divalent nickel cations from aqueous solution by multi-walled carbon nano tubes: Equilibrium and kinetic processes. *Res Chem Intermed* 38, 2205–2222.

Mohamed R.M., Ismail A.A., 2021a. Doping cadmium oxide into mesoporous zinc oxide matrix for enhanced removal/reduction of Hg (II) ions. *Ceram Int* 47(7PA), 9896–906.

Mohamed, R.M., Ismail, A.A., 2021b. Photocatalytic reduction and removal of mercury ions over mesoporous CuO/ZnO S-scheme heterojunction photocatalyst. *Ceram Int* 47(7), 9659–9667.

Mohapatra S., Menon N.G., Mohapatra G., Pisharody L., Pattnaik A., Menon N.G., Bhukya P.L., Srivastava M., Singh M., Barman M.K., Gin K.Y., 2020. The novel SARS-CoV-2 pandemic: Possible environmental transmission, detection, persistence and fate during wastewater and water treatment. *Sci Total Environ* 765, 142746.

Mohapatra S., Sharma N., Mohapatra G., Padhye L.P., Mukherji S., 2021. Seasonal variation in fluorescence characteristics of dissolved organic matter in wastewater and identification of proteins through HRLC-MS/MS. *J Hazard Mater* 413, 125453.

Mugunthan E., Saidutta M.B., Jagadeeshbabu P.E., 2019. Photocatalytic activity of ZnO-WO_3 for diclofenac degradation under visible light irradiation. *J Photochem Photobiol A Chem* 383, 111993.

Nag P., Sadani K., Mohapatra S., Mukherji S., Mukherji, S., 2021. Evanescent wave optical fiber sensors using enzymatic hydrolysis on nanostructured polyaniline for detection of β-lactam antibiotics in food and environment. *Anal Chem 93*(4), 2299–2308.

Navarrete-Magaña M., Estrella-González A., May-Ix L., Cipagauta-Díaz S., Gómez, R., 2021. Improved photocatalytic oxidation of arsenic (III) with WO_3/TiO_2 nanomaterials synthesized by the sol-gel method. *J Environ Manage* 282, 111602.

Nosheen S., Nawaz H., Rehman, K.U., 2000. Physicochemical characterization of effluents of local textile industries of Faisalabad–Pakistan. *Inter J Agric Biol* 2(3), 232–233.

Padmanabhan N.T., Thomas N., Louis J., Mathew D.T., Ganguly P., John H., Pillai S.C., 2021. Graphene coupled TiO_2 photocatalysts for environmental applications: A review. *Chemosphere* 271, 129506.

Pagana A.E., Sklari S.D., Kikkinides E.S., Zaspalis V.T., 2017. Combined adsorption permeation membrane process for the removal of chromium (III) ions from contaminated water. *J Memb Sci* 367(1–2), 319–324.

Pantha K., Acharya K., Mohapatra S., Khanal S., Amatya N., Ospina-Betancourth C., Butte G., Shrestha S.D., Rajbhandari P., Werner D., 2021. Faecal pollution source tracking in the holy Bagmati River by portable 16S rRNA gene sequencing. *npj Clean Water* 4(1), 1–10.

Parham H., Zargar B., Shiralipour R., 2012. Fast and efficient removal of mercury from water samples using magnetic iron oxide nanoparticles modified with 2-mercapto benzo thiazole. *J Hazard Mater* 205–206, 94–100.

Pedanekar R.S., Shaikh S.K., Rajpure K.Y., 2020. Thin film photocatalysis for environmental remediation: A status review. *Current Appl Phys* 20, 931–952.

Peng G., Deng S., Liu F., Li T., Yu G., 2019. Superhigh adsorption of nickel from electroplating wastewater by raw and calcined electroplating sludge waste. *J Clean Prod* 246, 118948.

Peng G., Deng S., Liu F., Li T., Yu G., 2020. Superhigh adsorption of nickel from electroplating wastewater by raw and calcined electroplating sludge waste. *J Clean Prod* 246, 118948.

Pink D.H., 2006. Investing in tomorrow's liquid gold. *World J Anal Chem* 2, 42–46.

Pirveysian M., Ghiaci M., 2018. Synthesis and characterization of sulfur functionalized graphene oxide nanosheets as efficient sorbent for removal of Pb^{2+}, Cd^{2+}, Ni^{2+} and Zn^{2+} ions from aqueous solution: A combined thermodynamic and kinetic studies. *Appl Surf Sci* 428, 98–109.

PRS Legislative Research, 2021. Legislative Research (prsindia.org) accessed on 15 April 2021. https://prsindia.org/

Rahmati N., Rahimnejad M., Pourali M., Muallah SK., 2020. Effective removal of nickel ions from aqueous solution using multi-wall carbon nanotube functionalized by glycerol-based-deep eutectic solvent. *Colloid Interface Sci Commun* 40, 100347.

Roy A., Velusamy S., Mallick TK., Sundaram S., 2020. Synergistic effect of nanoflower-like CdS for removal of highly toxic aqueous Cr(VI). *Mater Lett* 270, 127734.

Sajjad S., Khan M., Leghari S., Ryma N., Farooqi S., 2019. Potential visible WO_3/GO composite photocatalyst. *Int J Appl Ceram Technol* 16, 1218–1227.

Scheller F., Schubert F., 1992. *Biosensors*. New York: Elsevier Science Publishers.

Shahid M., Shamshad S., Ra M., Khalid S., Bibi I., Khan N., 2017. Chromium speciation, bioavailability, uptake, toxicity and detoxi fi cation in soil-plant system: A review. *Chemosphere* 2(178), 513–533.

Shawabkeh R., Al-Harahsheh A, Al-Otoom A., 2004. Copper and zinc sorption by treated oil shale ash. *Sep Purif Technol* 40, 251–257.

Silva W., Ghica M.E., Brett C.M.A., 2020. Biotoxic trace metal ion detection by enzymatic inhibition of a glucose biosensor based on a poly (brilliant green)–deep eutectic solvent/carbon nanotube modified electrode. *Talanta* 208, 120427.

Sousa J.C., Ribeiro A.R., Barbosa M.O., Pereira M.F.R., Silva A.M.T., 2018. A review on environment monitoring of water organic pollutants identified by EU guidelines. *J Hazard Mater* 344,146–162.

Syshchyk O., Skryshevsky V.A., Soldatkin O.O., Soldatkin A.P., 2015. Enzyme biosensor systems based on porous silicon photoluminescence for detection of glucose, urea and heavy metals. *Biosens Bioelectron* 66, 89–94.

Tang D., Xia B., Tang Y., Zhang J., Zhou Q., 2019. Metal-ion-induced DNAzyme on magnetic beads for detection of lead (II) by using rolling circle amplification, glucose oxidase, and readout of pH changes. *Micro Chimica Acta* 186, 318.

Thanarasu A., Periyasamy K., Manickam P., Devaraj T., 2021. Proceedings comparative studies on adsorption of dye and heavy metal ions from effluents using eco-friendly adsorbent. *Mater Today Proc* 36, 775–81.

Vardhan K.H., Kumar, P.S., Panda, R.C., 2019. A review on heavy metal pollution, toxicity and remedial measures: Current trends and future perspectives. *J Mol Liq* 290, 111197.

Velusamy S., Roy A., Sundaram S., Mallick T.K., 2021. A review on heavy metal ions and containing dyes removal through graphene oxide-based adsorption strategies for textile wastewater treatment. *Chem Rec* 21, 1570–1610.

Villarín M.C., Merel, S., 2020. Assessment of current challenges and paradigm shifts in wastewater management. *J Hazard Mater* 390, 122–139.

Wang S., Ning H., Hu N., Huang K., Weng S., Wu X., 2018. Preparation and characterization of graphene oxide/silk fibroin hybrid aerogel for dye and heavy metal adsorption. *Compos Part B Eng* 163, 716–22.

Wu F., Zhao T., Yao Y., Jiang T., Wang B., Wang M., 2020. Recycling supercapacitor activated carbons for adsorption of silver (I) and chromium (VI) ions from aqueous solutions. *Chemosphere* 238, 124638.

Xu J., Cao Z., Zhang Y., Yuan Z., Lou, Z., Xu X., Wang X., 2018. A review of functionalized carbon nanotubes and graphene for heavy metal adsorption from water: Preparation, application, and mechanism. *Chemosphere* 195, 351–364.

Xu Y., Zhang C., Zhao M., Rong H., Zhang K., Chen Q., 2017. Comparison of bioleaching and electrokinetic remediation processes for removal of heavy metals from wastewater treatment sludge. *Chemosphere* 168, 1152–1157.

Yang X., Guo, N., Yu, Y., Li, H., Xia, H., Yu, H., 2020a. Synthesis of magnetic graphene oxide-titanate composites for efficient removal of Pb(II) from wastewater: Performance and mechanism. *J Environ Manage* 256, 109943.

Yang Y., Li J., Yan T., Zhu R., Yan L., Pei Z., 2020b. Journal of colloid and interface science adsorption and photocatalytic reduction of aqueous Cr (VI) by Fe_3O_4-ZnAl-layered double hydroxide/TiO_2 composites. *J Colloid Interface Sci* 562, 493–501.

Yoo E.H., Lee S.Y., 2010. Glucose biosensors: An overview of use in clinical practice. *Sensors* 10(5), 4558–4576.

Yu Z., Song W., Li, J., Li, Q., 2020. Improved simultaneous adsorption of Cu(II) and Cr(VI) of organic modified metakaolin-based geopolymer. *Arab J Chem* 13, 4811–4823.

Zhu X., Wu G., Lu N., Yuan X., Li B., 2017. A miniaturized electrochemical toxicity biosensor based on graphene oxide quantum dots/carboxylated carbon nanotubes for assessment of priority pollutants. *J Hazard Mater* 324, 272–280.

Section 3

Environmental assessment pathways, and socio-ecosystem framework

Chapter 12

Environmental fate assessments to understand surface water pollution from metaldehyde-based molluscicide

Nathan Keighley
FERA Science Ltd.
Newcastle University

CONTENTS

12.1 AN OVERVIEW OF METALDEHYDE POLLUTION IN THE UK

12.1.1 Contamination of surface water from metaldehyde

The widespread use of molluscicide has resulted in the active ingredient metaldehyde being frequently detected in drinking water above acceptable levels. The recent concern of metaldehyde as an emerging agricultural pollutant has arisen as water companies have frequently observed concentrations in drinking water in exceedance of the

DOI: 10.1201/9781003247869-15

current statutory limit of 0.1 μg/L (Castle et al., 2018; UKWIR, 2015). Current methods to remove metaldehyde from finished water are expensive. This has necessitated modelling of metaldehyde concentrations in surface waters to predict peak concentrations and enable management of abstraction timing. These models rely on input parameters derived from laboratory experiments. The levels of metaldehyde observed in drinking water are not considered a risk to human health; however, exceedance of the regulation concentration level means that metaldehyde is a cause of concern for water companies, who may be subjected to financial penalties for non-compliance.

Protection of agricultural crops is of crucial importance for society, as populations continue to grow. It is also important to ensure sustainable farming, with a consciousness for the environmental impact of farming practices. Slug damage to crops can cause significant losses in agricultural production. While slug damage can be mitigated by particular agricultural practices such as soil cultivation by ploughing and deeper seed drilling, these are often insufficient and chemical control is required. Oilseed rape and wheat are among the most susceptible crops to slugs, with up to 59% and 22% of the total crop area affected, respectively (Nicholls 2014). If all molluscicides were withdrawn, the calculated annual tonnage lost equates to around a value of £42 million for these crops alone (Nicholls, 2014). It has been estimated (Dewar, 2016) that the cost to UK crop production if chemical control methods were unavailable could be £100 million per year.

While molluscicides are crucial for the protection of agricultural crops and are needed for efficient food production, the active ingredient, metaldehyde, has recently become prominent as a source of pollution in finished drinking waters, with concentrations frequently in exceedance of the current statutory limit of 0.1 μg/L (EC 1107/2009). This has necessitated modelling of metaldehyde concentrations in surface waters to predict peak concentrations to help manage abstraction times (e.g., Asfaw et al., 2018; Kay and Grayson, 2014; Lu et al., 2017; Castle et al., 2017, Singh et al., 2020, Patel et al., 2020). Current methods to remove metaldehyde from finished water, for example by using granular activated carbon, are too expensive to be commercially viable. This means that water companies are reliant on the predictive capability of models to inform when peak concentrations in surface water are expected and extensive monitoring programmes, which are expensive.

Initiative schemes were put forward for the responsible use of metaldehyde by the metaldehyde stewardship group (MSG) which put in place guidelines for the application of metaldehyde to agricultural fields. It suggests using the minimum amount of active ingredient, with a maximum application rate of 210 g (ai)/ha; advisors may recommend rates lowered to 160 g (ai)/ha near water sources. The maximum total dose rate is 700 g (ai)/ha/year. No pellets are to be spread within 6m of a water course, and growers are advised not to apply when heavy rain is forecast, or when drains are flowing. The majority of failures due to pesticides have been the result of metaldehyde pollution, for example, in 2013, 110 of 2,269 raw water abstraction points were confirmed to have actual risk of metaldehyde contamination (Colbourne, 2013). However, these initiatives did not go far enough to mitigate the pollution of surface water by metaldehyde, with levels in surface water continuing to exceed the statutory limit (UKWIR, 2015). Consequently, novel research is required to understand the dynamics of metaldehyde degradation in the environment, so that catchment management initiatives could be designed that take a targeted approach to metaldehyde pollution. This could serve to protect water quality, but perhaps also allow for the continued use of the product, which is an effective means of protecting agricultural crops from slug damage.

Metaldehyde enters water sources primarily as a diffuse pollutant from agricultural fields. Metaldehyde readily leaches into water courses because it is a mobile compound and has a relatively low binding affinity with soil (EFSA, 2010). In water and low-oxygen environments, it remains very stable; therefore, in reservoirs, metaldehyde concentrations do not readily decline. The only significant route by which metaldehyde is removed from the environment is through biodegradation in soil. Models used to generate environmental fate assessments for metaldehyde rely on biodegradation rates in soil determined during batch laboratory experiments, where the decline is metaldehyde concentration measured over time. However, the degradation measured in batch laboratory studies in the literature, such as EFSA (2010), occurs readily, which is not expected for a compound that presents a pollution concern.

12.1.2 Metaldehyde degradation in soil

There are a small number of studies detailing the degradation of the pesticide metaldehyde. Work by Zhang et al. (2011) looked at metaldehyde degradation in cabbage and soil in field studies at three locations and the half-life (DT_{50}) in soil was determined and ranged between 0.75 and 1.02 days when applied at two times the recommended high dose. A comparable study by Dong et al. (2017) looked at the dissipation of metaldehyde in pakchoi and soil and found the half-life in soil to be 2.3–2.4 days, but the half-life was only assessed for soils at two locations. Metaldehyde degradation was found to occur quickly and was not affected by weather. The study did not find any links between degradation and soil properties. In another study (Calumpang et al., 1995), residue trials to determine the fate of metaldehyde in a rice paddy, using 6% metaldehyde pellets applied at 4 kg/ha, were conducted. In water, dissolution of the pellets produced residue levels that remained fairly constant between 1.58 and 1.47 mg/L for 1–3 days after application. Residues could only be detected up to 9 days after application; degradation followed pseudo first-order kinetics, with a half-life of 0.27 days. In soil, metaldehyde deposition from pellets was only apparent 3 days after application. Residue levels ranged from 0.053 to 0.127 mg/kg from day 3, peaking at day 14 before undergoing rapid degradation. It was concluded that metaldehyde residues were not persistent in rice paddy ecosystems.

Metaldehyde demonstrated rapid dissipation in these studies, suggesting that it should not be persistent. However, analytical capability has improved since the publication of this work, and this could affect the detection of metaldehyde in environmental media. The detection procedures use Gas Chromatography-Mass Spectrometry (GCMS) and Liquid Chromatography-Mass Spectrometry (LCMS), which is standard procedure; however, the sensitivity of these instruments has improved since the publication of these studies, so contemporary studies will be more accurate.

Degradation experiments were conducted more recently by the European Food Standards Authority (EFSA, 2010) for the draft assessment report on metaldehyde and the Ministry for Agriculture Fisheries and Food, for the Food and Environment Protection Act evaluation on metaldehyde (Ministry of Agriculture Fisheries and Food, 1996) to produce reports for regulatory purposes. These studies were conducted in line with standard regulatory guidelines, according to OECD 307. However, these laboratory conditions are environmentally unrealistic. One of the main discrepancies is the use of laboratory-grade metaldehyde, when the active compound, in reality, is contained in a formulated pellet. To investigate the influence of pellet formulation versus standard test paradigms, Bond (2018) conducted an OECD 307 degradation experiment to investigate whether degradation will be different for a pelleted metaldehyde rather than laboratory-grade test compound. Bond's results revealed that degradation of metaldehyde contained within pellets took slightly longer, compared to laboratory-grade metaldehyde, with derived DT_{50} values of 3.43 and 3.98 days for wet and dry processed pellets, respectively, compared to 1.19 days for laboratory-grade metaldehyde. Bond also examined the leaching potential of metaldehyde from pellets in intact soil columns. Other parameters, such as soil moisture content and temperature were not accounted for in these studies. These parameters could have a greater effect on degradation and hence were included in this investigation.

The potential for microbial degradation of organic pollutants depends on the nature of the microbial community, which in turn depends on the nature of the soil. Soil particle size fractions, i.e. different soil types, provide different surface properties and micro-environments (Hemkemeyer et al., 2015). Metaldehyde-degrading bacteria have been extracted from soil (Thomas et al., 2017; Castro-Gutiérrez et al., 2020, Singh et al., 2020) and their relative abundance in the microbial population will be governed by the soil environment. There are many properties of soil to consider that will have an influence on biodegradation. Favourable conditions for bacterial communities include higher organic matter content (Bausenwein et al, 2008), and larger soil particle size fractions mean greater porosity, therefore more space for water storage and more aeration in soil (although smaller soil particle size provides a larger surface area for microbes to live on) (Hemkemeyer et al., 2015). Differences in soil moisture content will have an effect on the actions of soil microorganisms. The standard moisture conditions of 40%–60% water holding capacity (OECD 307, 2002) used in regulatory studies is likely to yield a different result for biodegradation rate than if soil moisture content was higher, representative of UK spring/autumn conditions when slug pellets are applied. Furthermore, during these seasons, temperatures will be much lower than 20°C, suggested by OECD 307. The temperature dependency of degradation therefore must be estimated using the Arrhenius equation. However, the parameters for this model are based on averages for a range of plant protection products (Matthies and Beulke, 2017), and this might not be representative for metaldehyde. Hence the temperature dependency and effects of soil moisture content on metaldehyde degradation were examined in this study.

Leaching into lower soil horizons that are less biologically active may reduce metaldehyde degradation, followed by access to field drains which would enable the compound to enter water sources more quickly. Similarly, cracks in clay create preferential flow pathways, resulting in quicker access to field drains and reduce the opportunity for biodegradation (Bond, 2018, Kumar et al., 2019). Metaldehyde residues that bind to soil may become biologically unavailable; however, metaldehyde has a relatively low K_{OC} value, so this may be less significant, but it does mean that there would be more leaching occurring. As such, the amount of metaldehyde retained and degraded in soil may be less than some models predict. Therefore, understanding metaldehyde degradation and leaching in representative soils, such as lysimeters made of intact soil collected from agricultural fields, would be important for accurately assessing the environmental fate of metaldehyde.

Concerns over wildlife poisoning recently led to the decision to ban outdoor use of metaldehyde in the UK (Department for Environment, Food & Rural Affairs, Health and Safety Executive, and The Rt Hon Michael Gove MP, 2020). Metaldehyde can still be used elsewhere in Europe. Despite the ban on the outdoor use of metaldehyde, there is still a risk of water quality failures, as metaldehyde is persistent in water and has been observed in water sources from catchment that have not had recent applications (UKWIR, 2015). With the withdrawal of metaldehyde, growers probably have to switch to using ferric phosphate-based molluscicide. Ferric phosphate has low solubility in water, and the environmental fate of the product is not thought to significantly raise the natural concentration of iron and phosphate ions in water, when compared to artificial fertiliser usage. However, it is unclear how the product, formulated with chelate molecules, interacts with the environment, and there are concerns over toxicity to soil invertebrates, such as earthworms (Edwards, 2009).

12.1.3 Objectives of the study

The discrepancy between low water pollution risks predicted by the testing regime used for regulatory approval of metaldehyde and the real-world occurrence of metaldehyde in drinking water poses fundamental questions about the robustness of current testing procedures, and the need to improve them. Metaldehyde presents a useful case-study to exemplify this limitation of standard-procedure environmental fate assessments. This study sought to examine whether chemical-specific environmental fate assessments might be required for certain problem compounds, such as metaldehyde, where detailed studies can be implemented to predict environmental concentrations more accurately. Standard guidelines for examining pesticide degradation in soil recommend soil moisture contents and incubation temperature that are not always representative of the environmental conditions in the UK when slug pellets are applied. This investigation tested metaldehyde degradation in three soils over a range of temperatures and soil moisture contents in order to improve the model prediction for metaldehyde degradation in soil.

It was hypothesised that the temperature dependency of metaldehyde degradation would be greater than the pre-defined model prediction. This would serve as a partial explanation for why the compound persists in the environment; with the presumption that metaldehyde degradation at average autumn temperatures for the UK is slower in reality than the calculation would suggest. No other study in the literature, to the author's knowledge, considered this aspect of the risk assessment, and deriving new predictions would serve to improve the environmental fate assessments for metaldehyde. The hypothesis was synthesised based upon the consideration of metaldehyde having a simplistic chemical structure, and therefore its degradation would deviate from the standard predictions, which are based on the average degradation of a range of plant protection products (Matthies and Beulke, 2017), which typically have complex chemical formulas.

It was hypothesised that soil moisture content would significantly affect metaldehyde degradation and therefore perturb model predictions. It is generally considered that higher soil moisture contents will benefit the microbial community in the soil, as more water is available for their metabolism. However, it was hypothesised that the degradation of metaldehyde, which is shown to be an aerobic process (EFSA, 2010, Kumar et al., 2020) would be retarded in high soil moisture conditions, which are more representative of condition during autumn in the UK, as less air is available in the soil for the respiration of these microorganisms that use metaldehyde for fuel. The extent of this effect could be measured and used to improve upon model predictions.

To investigate leaching of metaldehyde, intact soil core columns collected from the field were used to construct lysimeters, replicating the field scenario. It was hypothesised that metaldehyde would be more persistent in intact soil, representative of the field scenario, where there is less diffusion of air through the soil. Leachate was collected and analysed in order to determine the legacy of metaldehyde in the field. This experiment aimed to understand how much metaldehyde would be leached from soil after application and determine how much might remain in the soil and contribute towards long-term pollution, as well as assess the compound's susceptibility to degradation in representative soils. The information collected form the study was invaluable for improving environmental fate assessments for this problem compound. The results were compared to the revised model prediction generated from the degradation results obtained from the incubation experiments done earlier in this study.

12.2 BIODEGRADATION OF METALDEHYDE IN UK SCENARIOS IS OVERESTIMATED BY CURRENT RISK ASSESSMENT MODELS

12.2.1 Materials and laboratory methods

In the first part of this study, a series of batch degradation experiments were conducted to identify whether varied experimental conditions would significantly affect the degradation of metaldehyde in soil. Batch degradation studies were performed with three field-fresh soils: a clay loam soil collected from a permanent pasture field site, an arable clay loam and an arable sandy loam soil. The study was designed to test the effect of temperature and soil moisture content on degradation rate. The experimental methods to test each parameter were kept the same, other than for the use of a different soil type. The experimental method consisted of extracting metaldehyde, which was dosed into soil (20 g dry weight) as an aqueous solution (2 mg/L), in accordance with OECD 307 guidelines, using methanol in 50 mL vessels over increasing time points of 0, 1, 3, 7, 14, 21, 28 and 60 days. The samples were left to equilibrate to the laboratory conditions for 2 weeks prior to dosing. Additional replicates were included in the experiment, which were not dosed with metaldehyde to serve as a negative control to prove that there was no metaldehyde already present in the soil. A control experiment using sterile soil was also set up concurrently, to show that metaldehyde disappearance was due to degradation, rather than other factors such as sorption. The decline in concentration with time in the supernatant was then measured using LCMS. This data was then analysed using a software called Computer Assisted Kinetic Evaluation (CAKE) to calculated degradation rates.

12.2.2 Metaldehyde detection and quantification in soil extracts

Extracts from the successive time points were stored in a refrigerator at 4°C prior to analysis by LCMS. Samples for LCMS were prepared by diluting 1/10 with water using a pipette and filtered with a 0.2 μm pore nylon filter. The sequential extracts were analysed by LCMS separately. Repeated analysis of selected samples over 3 months provided stability data that proved that storage of these samples over this period would not lead to dissipation in metaldehyde concentrations while being stored under these conditions. Calibration standards were matrix-matched, produced from the same metaldehyde stock solution used for spiking the batch experiments; made up with blank soil extract. Metaldehyde could be measured in the extracts with an LOQ 0.1 μg/L.

The LCMS method was developed from the one described by Thomas (2016) and optimised for metaldehyde in soil extracts. Changing the C18 column used on the SCIEX instruments at FERA science ltd. to a BEH phenyl column on the Waters instruments helped to notably reduce matrix suppression of the metaldehyde signal, which caused a loss of response with low metaldehyde concentrations. The method used a BEHphenyl (Acquity UPLC) 1.7 μm particles 2.1 × 100 mm column, with a flow rate of 0.4 mL/min, using the same mobile phase as previously; 1 mM ammonium acetate, prepared with ultrapure water, and methanol as the organic component, with a gradient flow. For mass spectrometry, the original parameters used with the SCIEX instruments were selected, with the following amendment: opened the acquisition window earlier (between 0.5 and 5.00 min) with a 0.6 s solvent delay. Samples were stored in a refrigerator at 4°C prior to injection into the instrument. Mass spectrometry involved an optimised protocol with greater response and improved precision that uses 0.5 mL/min flow rate, 5.25 kV capillary voltage and 325°C solvation temperature. To observe product ions, multiple reaction monitoring with a dwell time of 160 ms was used.

12.2.3 Soil incubation set-up and operation

Samples for the batch incubation experiments were prepared in the same manner, following the methods described in Section 11.2.2. To test the effects of temperature on metaldehyde degradation, replicate experiments were performed concurrently in separate incubators set to temperatures of 20°C, 16°C, 12°C, 8°C and 4°C so that the data could be used to construct an Arrhenius plot and thereby quantify the temperature dependency of the degradation in the pasture clay loam soil. The reason that this soil was selected is that it was guaranteed to have not had previous exposure to metaldehyde, which might have perturbed the results, as the microbial community was not conditioned to utilising metaldehyde, so represented a worst-case scenario for persistence.

To examine the effects of soil moisture content on metaldehyde degradation, replicated batch experiments were set up and the moisture content adjusted by the addition of the appropriate amount of deionised water. Experiments were set up using the arable clay loam soil amended to soil moisture contents of 40%, 60%, 80% and 100% of Maximum Water Holding Capacity (MWHC) and the pasture clay loam and sandy arable loam included experiments at moderate (60% MWHC) and high (100% MWHC) soil moisture contents.

12.2.4 The effects of varied temperature and soil moisture content on metaldehyde degradation

Following this experimental procedure, metaldehyde degradation was measured in these three soils at different temperatures and soil moisture contents. The aim of this investigation was to understand whether either of these parameters would perturb the biodegradation of metaldehyde and thereby improve models to assess the environmental fate. The measured degradation half-lives for metaldehyde were greater for the permanent pasture clay loam soil, while in the arable soils, metaldehyde was shown to degrade quickly, in line with findings from the literature. It was found that in general, the tested parameters did reduce the rate of metaldehyde degradation compared to the reference experiment, but not to the extent that it would be classed as persistent.

For each of the three soils used, the results of varied temperature and soil moisture content were compared to a reference experiment conducted in accordance with OECD 307 guidelines, conducted at an incubation temperature of 20°C and soil moisture of 60% MWHC. The spread of dissipation half-lives measured for metaldehyde over a range of soil moisture contents and two temperatures (20°C and 12°C) are shown in Figure 12.1. The most prominent result across the entire suite of batch degradation experiments was that of the pasture clay loam soil under high moisture (100% MWHC) conditions. Metaldehyde became persistent under highly moistened conditions at 100% MWHC, with a half-life on 4,150 days: far exceeding measured degradation times for any other experiment. This result is omitted from Figure 12.1 for better resolution of the other results. Reduced degradation rate under high soil moisture conditions relative to the reference experiment was also observed with the arable clay loam soil, but not to the same extent. Metaldehyde continued to be readily biodegradable, having a

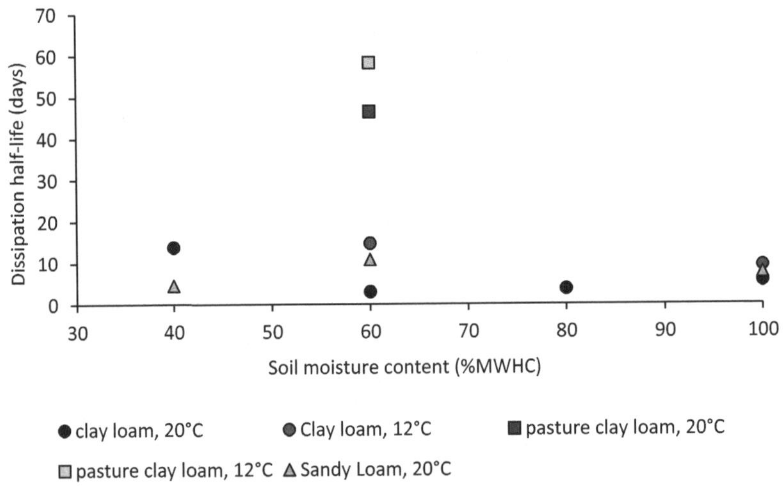

Figure 12.1 Distribution of DT_{50} values for metaldehyde batch degradation experiments conducted over a range of soil moisture contents and incubation temperatures. The results for the pasture clay loam, 100% MWHC experiment, are omitted for improved clarity.

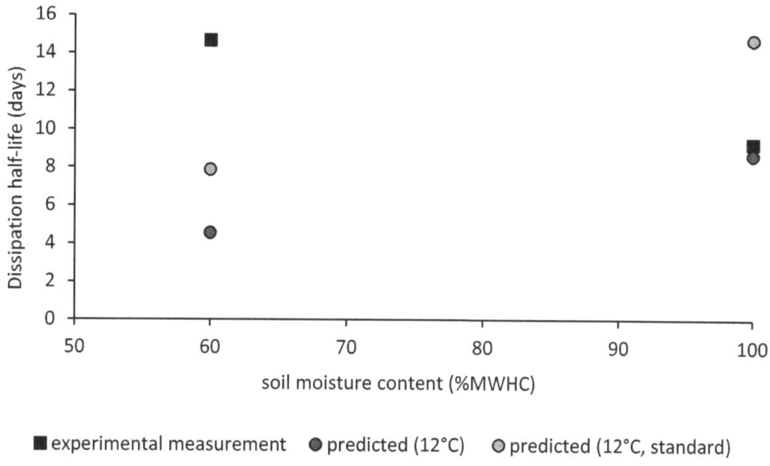

Figure 12.2 Plot of experimentally measured DT_{50} values for metaldehyde overlaid with the values predicted from the Arrhenius equation.

half-life of 5.6 days, despite high (100% MWHC) soil moisture, compared to a DT_{50} of 2.97 days for the reference experiment. Additionally, rapid metaldehyde degradation was observed during experiments using the sandy loam soil, which was also collected field-fresh under arable management. Metaldehyde degradation at high soil moisture content, again 100% MWHC, proceeded with a half-life of 7.7 days. In this case, degradation rate was not reduced compared to the reference experiment.

The effect of varied temperature on the rate of metaldehyde degradation rates was investigated to elucidate whether predicted rates would deviate significantly from the experimental outcomes. The Arrhenius equation was used to predict rate constants at 12°C at two soil moisture contents: 60% MWHC and 100% MWHC. A value for the activation energy for metaldehyde degradation was determined in the pasture clay loam soil over a range of temperatures from 4°C to 20°C. The predictions, shown in Figure 12.2, generally deviated from the experimentally determined values by an appreciable amount for both the experimentally derived activation energy and standard activation energy used in regulatory modelling. The experimentally derived activation energy in this investigation was not substantially more accurate than the standard value, with only the high soil moisture experiment fitting the model well. It was therefore concluded that the activation energy derived for the permanent pasture soil was not applicable to the arable soil of the same texture class. It was presumed that soil specific activation energy values would need to be derived for more accurate environmental fate predictions. However, generating data in the laboratory to derive activation energies involves significant resources and cost. Therefore, it was decided that the standard value for activation energy is a good approximation for environmental fate modelling.

The suite of batch degradation experiments demonstrated a varied set of degradation rates. These could not always be linked to the parameter under investigation, and the greatest variability was observed between soil types as opposed to the tested parameters. The conclusions made from the batch degradation study speculated that other factors pertaining to the soil's physical-chemical and biological nature are prominent in governing the rate of metaldehyde degraded in agricultural soil. Factors such as soil aeration, organic carbon content, related to nutrient availability for active microbes in the soil with the capacity to degrade metaldehyde is likely to influence the abundance of these microorganisms and therefore would be expected to influence the degradation rate of metaldehyde. This might better account for the variation in degradation rate between soil types being greater than the variation in degradation rates observed for varied soil moisture content and temperature.

In the case of the permanent pasture soil, for example, the microbial community of this soil is likely to be conditioned for more anaerobic conditions, created as a result of livestock activity compacting the soil to create a pan, and this soil is not disturbed during cultivation; a process which aerates the soil. It might be expected that, at high moisture content, there is little opportunity for aerobic microorganisms, such as those with the capacity to degrade metaldehyde, to flourish. By contrast, the loose, cultivated arable soils could be expected to have better aeration, which would encourage the aerobic biodegradation of metaldehyde. This might explain the observed faster degradation rates during laboratory incubations.

It is most probable that the nature of the soil microbial population is the influencing feature of agricultural soils that affects the degradation of pesticides, including metaldehyde. The characteristics of the soil microbial community will be dependent on soil properties such as texture class, where the particle size distribution determines the nature of the solid surfaces for particular kinds of microbes to live on, and soil pH and organic carbon content, which affects the types of nutrients available to them. Soil properties will be influenced by the soil management practices used throughout the agricultural regime. Cultivation practices, cropping and the application of manure and artificial fertiliser, as well as pest management chemicals, are all key features of the soil preparation for agricultural use that will influence the composition of the soil microbial community. Additionally, there is the effect of climatic conditions, such as seasonal changes in rainfall and temperature which will also affect the composition of the soil microbiome.

There are many factors of consideration needed for a holistic appraisal of the environmental scenario under which microbes are exposed to plant protection products, such as metaldehyde. Because temperature and soil moisture content will vary throughout the year, seasonality may be significant in determining the soil microbial population's capacity for pesticide degradation. All of these elements that influence biochemical processes in the soil, including in relation to pesticide degradation, need to be accounted for in order to make accurate environmental fate predictions. This sets limitations on the accuracy of batch degradation studies because these conditions cannot wholly be replicated in the laboratory. The conditions when field-fresh soil is collected for these studies cannot be accurately reproduced, and therefore the microbial population will be differentiated for individual batch incubations. This is likely to have implications for mineralisation processes occurring with the test compound under scrutiny.

Since the prevailing environmental conditions, such as rainfall and its relation to soil moisture content, and the monthly average temperature when the soil used for batch degradation experiments was collected are expected to have an impact on the composition of the microbial population, it is important to assess these factors when interpreting batch degradation data for field-fresh soil samples. The seasonal dynamics of soil microbial communities, as they become habituated to certain environmental conditions might have a greater impact on pesticide degradation than the short-term condition fluctuations in moisture content and temperature amended for the degradation studies. If the soil microbial population slowly acclimatises over time, the soil moisture and temperature conditions attributable to the seasonal climate may have a greater effect than the amendments made to these properties during the 60–120-day experiments.

Based on the observed degradation rates measured in this study, the environmental conditions observed when the soil was collected could relate to the measured degradation rates. The results vary a great deal despite maintaining the same laboratory protocols. The general trend is that metaldehyde degradation proceeds fastest in the arable soils, and less quickly in the clay loam permanent pasture soil. The permanent pasture clay loam soil was collected in July, during mild weather and dry ground conditions. Perhaps, if the soil microbial community was conditioned to warmer, dryer conditions, it was less well adapted to metabolise metaldehyde when the conditions in the batch experiments were amended to lower temperature, with incubations conducted at 12°C, 8°C and 4°C, or high soil moisture content (100% MWHC) during which metaldehyde was observed to be highly persistent. The arable clay soil was collected in November and the arable sandy loam in January, when ground conditions were much wetter, and the ambient temperature was cooler. If the soil microbial community was already acclimatised to these conditions, then this could explain why metaldehyde degradation proceeded faster in these soils. This also supports the conclusion that environmental influences prior to collection of the soil will have a greater effect on the capacity for metaldehyde degradation than the experimentally investigated parameters. The implications of this conclusion with respect to environmental fate modelling would be that for accurate biodegradation assessments representative of the field, batch experiments would need to be supported by field dissipation, or lysimeter, trials, as well as characterisation of the degrading microorganisms and quantification of their abundance in the soil.

12.3 RAPID LEACHING TO LOWER SOIL HORIZONS WILL INCREASE THE RISK OF METALDEHYDE POLLUTION

12.3.1 Lysimeter design and operation

To test the effect of ploughed cultivation on the leaching of metaldehyde to lower soil horizons, lysimeters were set up by extracting intact soil columns from the field. Parallel samples were collected in triplicate for undisturbed soil and recently cultivated soil. The cultivation was simulated by digging the surface soil to a depth of 20 cm using a spade, and then breaking the clods to a fine till with a rake. Two contrasting soils were used; the clay loam arable soil and the sandy loam arable soil that were used for the batch degradation experiments. Both fields had previously been used for growing cereals and had not received any metaldehyde application for a number of years.

Soil columns were collected by forcing PVC plastic pipes (35 cm long × 16 cm diameter) into the ground, which were subsequently dug out to extract the column from the ground, with the soil inside intact. These columns were then taken back to the laboratory to be assembled into the lysimeter experiment. The columns were contained in wooden boxes filled with sand to modulate diurnal variations in temperature. Pipes were connected from the bottoms of the columns which fed to collection bottles to collect leachate. Over a period of 120 days, the columns were irrigated artificially in accordance with March rainfall amounts, and the leachate was collected. The columns were set up and equilibrated for two weeks, and then metaldehyde was applied on day 10, day 13 and day 36 of the experiment. At the end of the experiment, the lysimeters were deconstructed and the soil from each column was destructively sampled; being separated into five depth fractions.

12.3.2 Discussion of experiment objectives

The objectives of the lysimeter experiments were to assess the amount of metaldehyde that leaches through intact soil columns, representative of the field and to determine the persistence of metaldehyde in the soil by measuring concentrations at different depths of the soil profile. The residue levels measured in leachate and soil extracts after the 120-day period were compared to the expected residue levels determined from the rate of degradation determined during the batch degradation experiments. This served to assess the accuracy of the laboratory incubations in relation to the environmental scenario. Additionally, the lysimeter experiment aimed to explore the effects of cultivation on the leachability of mobile pesticides, such as metaldehyde.

It was speculated that metaldehyde applied as pellets would result in high local soil concentrations leading to a reduced capacity for biodegradation, as the substrate is in gross excess. This heterogeneity in metaldehyde concentration in agricultural soils leading to a reduced capacity for mineralisation of the chemical relative to the time taken to leach to lower soil horizons might increase the risk of pollution. Varied soil moistures in the field in intact soils, which are less aerated, could also lead to reduced degradation, further increasing the risk of pollution from metaldehyde. These considerations were acknowledged during the lysimeter study. This study above all aimed to assess how representative can batch degradation test be in relation to real environmental scenarios.

At greater soil depth, air can less easily diffuse into the pore spaces, so the environment becomes increasingly anaerobic with greater depth. This will inhibit the actions of aerobic metaldehyde-degrading bacteria, so the rate of degradation would be expected to be decreasing with increasing soil depth. Furthermore, the availability of utilisable organic matter is less at lower soil horizons. No other study, to the author's knowledge, has enumerated concentrations of metaldehyde as a function of soil depth, which would relate to the capacity for degradation, and this may be a key contributor to why metaldehyde is present in drinking water at levels above what might be expected from previous risk assessments. Considering the actions of microbes at lower soil depth will be paramount for understanding the degradation of metaldehyde in the environmental scenario. As previously mentioned, the physical-chemical properties of the soil will govern the composition of the microbial population and therefore the capacity for different chemicals to be degraded. The biodegradation of pesticides in soil will also depend on a number of other factors including: the bioavailability of the pesticide, the preference that the microorganisms have for utilising the pesticide as a substrate to generate metabolic energy over other substrates, and the aeration of the soil for aerobic process to occur.

Reduced bioavailability of metaldehyde due to sorption will have a less significant impact on degradation rate than the increased microbial activity with higher OC, given that metaldehyde has a low binding affinity with soil, indicated by the K_{OC} values reported in the literature. Arable fields with a regular manuring regime may be less disposed to metaldehyde losses, due to greater degradation potential. Methods of seedbed preparation, such as ploughing which aerates the soil and promotes bacterial activity on the organic matter, were considered for the field samples taken for the lysimeter experiment. Farming practice could be an important aspect of consideration for the management of recurring chemical contaminants, such as metaldehyde.

12.3.3 Results and discussion of metaldehyde leaching

The clay loam columns were the freest draining and leachate could be collected from the start of the 120-day experiment. There was no significant difference in drainage rate between cultivated and uncultivated soil. For the sandy loam soils, this was different; the uncultivated soil did not drain well due to soil compaction. Consequently, no leachate was collected until day 49. The cultivated samples were better draining and did provide leachate earlier in the experiment, refer to Figure 12.3.

The timing and pattern of metaldehyde break-through in the leachate was found to be influenced by the drainage of the columns. For the clay loam columns, metaldehyde break-through occurred as three peaks, consistent with the application regime. The peaks occurred on days 21, 45 and 56 for both cultivated and uncultivated columns. The uncultivated columns demonstrated the highest concentrations of metaldehyde

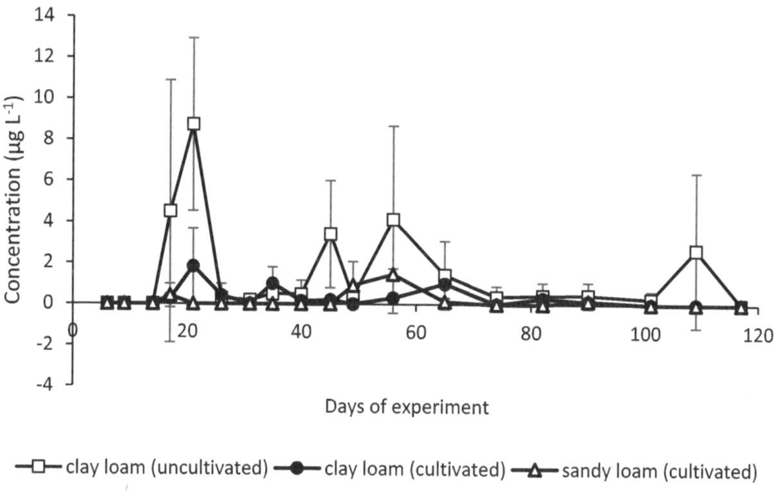

Figure 12.3 Break-through curves for the soil columns used during the lysimeter experiment.

in the leachate, with the greatest concentration of 8.7 μg/L observed on the first peak. This suggests a lag phase of 11 days, between first application and peak concentration in drain water. However, it is worth noting that the disintegration of slug pellets would be likely to reduce metaldehyde leaching. These columns therefore represent a worst-case scenario.

There was a greater degree of variability in metaldehyde leaching observed for the sandy loam columns, likely because of variable drainage velocities. In general, the uncultivated columns, where the soil was compacted, did not drain as readily as the columns which contained soil disturbed by the cultivation process. The implications regarding the leaching of metaldehyde were that, in the uncultivated columns, with the slower drainage rate, metaldehyde did not leach through until later in the experiment, with peak concentration of 0.17 μg/L reached by day 90. This is shown in Figure 12.3; metaldehyde concentrations in leachate are lower than those measured for the clay loam columns. The cultivated sandy soil drained more freely, although it was not until later in the experiment, by day 40, that drainage occurred with comparable velocity to the clay loam columns, and leachate was collected. Metaldehyde leaching demonstrated the characteristic three peaks in metaldehyde concentration over the course of the experiment. However, the concentrations of metaldehyde measured in the leachate were much higher than the results from the other columns, presumably because leaching occurred later in the experiments, allowing metaldehyde to first accumulate before being washed out. The data for the sandy loam uncultivated soils is not included in Figure 12.3 as the peak concentrations of 50.1, 49.6 and 82.3 μg/L (measured on days 56, 90, and 101) are above the scale of the other replicates, showing that in poorly draining soils, metaldehyde may be released in a short period at high concentration when drain flow eventually does occur.

Analogy can be made between the lysimeters and the field scenario. The lysimeter experiment was set up to be representative of field conditions, using intact soil from fields under arable management. Based on the findings of this experiment, it could be postulated that, in well drained fields, drain flow might be expected to contain lower concentrations of metaldehyde for a longer period. This effect might result from the applied metaldehyde being gradually washed out from the soil, if applied during the drain flow period. For fields where rainwater less readily drains, or if metaldehyde application occurs outside of the drain flow period, a higher concentration of metaldehyde might be expected in drain flow waters for a shorter period. As the drains begin to run, accumulated residues of metaldehyde retained in the soil are washed out.

12.3.4 Metaldehyde persistence in the soil profile

Additional information that might help to assess the amount of metaldehyde retained in soil at risk of being washed out into drain flow waters would be to ascertain where in the soil profile metaldehyde residues accumulate. If legacy pools accumulate in proximity to field drains, surface waters could be at risk from further surges in metaldehyde concentrations during successive rainfall events. Therefore, a concentration profile of metaldehyde residues with soil depth was constructed (Figure 12.4) for the lysimeters.

The results showed that the highest concentrations of metaldehyde retained in soil were in the surface cultivated layer. This might represent an adsorbed fraction of

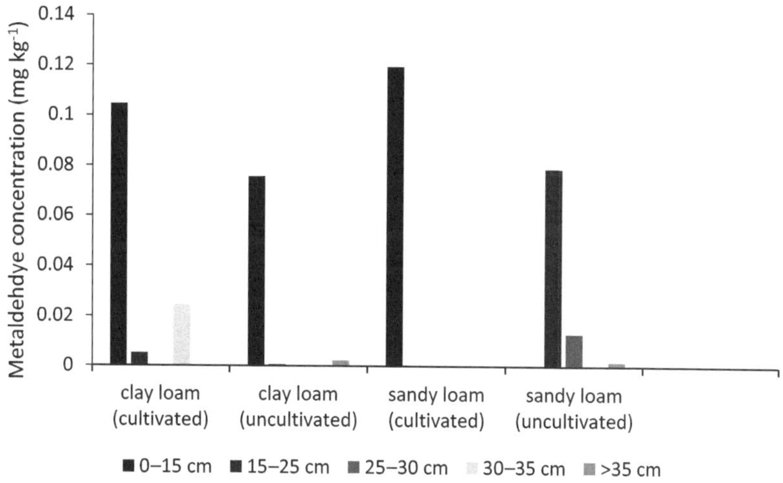

Figure 12.4 Metaldehyde concentration profile with soil depth, determined from destructive sampling of soil columns.

Table 12.1 Mass balance of metaldehyde as a percentage of the initial applied dose

	Clay loam (cultivated)	*Clay loam (uncultivated)*	*Sandy loam (cultivated)*	*Sandy loam (uncultivated)*
Recovery (% applied)	29.23403	9.564039	16.71367	11.29261
RSD	14.47901	10.18386	31.79827	7.359314

metaldehyde that becomes bound to soil shortly after application, and then does not readily move to lower soil depths. Concentrations of metaldehyde above 0.07 mg/kg soil were measured. For the cultivated clay loam columns, 0.024 mg/kg was able to percolate to the subsoil. The cultivated sandy loam columns had the greatest concentration of residual metaldehyde, all of which was present in the surface horizon. In the uncultivated sandy loam columns however, residual metaldehyde was present at a slightly greater depth from 15 to 30 cm, suggesting that the compacted soil provided less numerous or suitable surfaces for metaldehyde binding. Overall, these data suggest that significant concentrations of metaldehyde can remain in soil after one drain flow period. This could, over time, result in accumulation of metaldehyde residues in soil with successive slug pellets applications and therefore increase the risk of pollution of surface water.

By combining the metaldehyde residues extracted from the soil columns with the quantity measured in leachate over the course of the leaching experiment, it is possible to calculate how much metaldehyde has dissipated relative the amount that was originally applied. This is shown in Table 12.1. The results show that significant proportions of metaldehyde (between 9% and 30%) remain after the course of the experiment, which lasted 120 days. This is not in agreement with the batch degradation data, which suggests that metaldehyde would degrade below detectable levels in 60 days in these same soils. Therefore, using degradation half-lives derived in the laboratory setting, during batch incubation experiments, may not be representative of processes in the field. When used as input parameters for modelling and risk assessment, it could result in an underestimation of metaldehyde pollution.

12.3.5 Comparison to model predictions

Model predictions were generated in a drain flow model called MACRO, using the experiment results from the batch incubation experiments. From these results, which are shown graphically in Figure 12.5, it can be seen that a range of metaldehyde concentrations could be expected in drain flow waters for different scenarios. This applied to the simulations and the leachate collected form the lysimeter experiment. The highest predicted concentrations, above 200 μg/L, for the simulations stem from the model input with the worst-case DT_{50} of 4,150 days. This trend would be expected, as less degradation occurs, so more metaldehyde remains over time, which may be washed into drain flow. Interestingly, some of the column drain flow results have comparable metaldehyde concentrations in the leachate, suggesting that degradation rates under more environmentally realistic conditions might be slower than those measured during most of the batch incubation experiments. This would support the hypothesis that metaldehyde degradation rates are overestimated from the regulatory process.

Not all the simulations that use the input parameter $DT_{50}=4{,}150$ days demonstrate high metaldehyde concentrations in drain flow. Simulations S12, S16 and S21 are among the lowest predicted concentrations at 10–4 μg/L, despite the incredibly long dissipation time. These simulations have a lower rainfall, and the scenarios involve heavy clay/silty clay soils. In these scenarios, water movement through soil, and

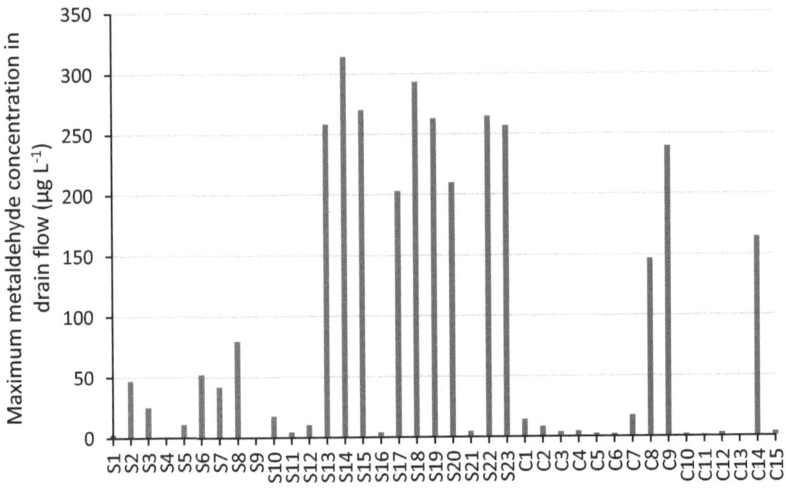

Figure 12.5 Chart showing the maximum concentration of metaldehyde in drain flow predicted from MACRO simulations (S1–23) and measured values from soil columns (C1–15) in the lysimeter experiment.

therefore leaching of chemicals, would be slow. In the lysimeter experiments, columns which did not drain freely tended to give the highest peak concentrations, e.g., C8, C9 and C14. However, these measurements were taken after maximum flow velocity ensued, and prior to this minimal concentration of metaldehyde could be detected. This idiosyncrasy fits with the model predictions and supports the conclusions that drainage rates in the field will have a significant effect on the concentrations of metaldehyde leaving the field via drain flow.

For the simulations that used the moderate reference point $DT_{50}=46$ days, peak concentrations of metaldehyde predicted in drain flow were lower (<80 μg/L), as expected. The same trend of greater rainfall and lighter soil produced the highest concentrations. This reinforces the conclusion that a seasonal "flush" is a governing factor in peak metaldehyde concentrations in surface water. One disparity between the simulations and the lysimeter experiment is that, in the simulations, metaldehyde is applied before the drain flow event, whereas in the lysimeter experiment, the columns are already draining. This allows the metaldehyde to be washed out gradually, as demonstrated in columns C1–C6, which were free draining due to abundant preferential flow pathways. In contrast, columns which did not flow readily initially allowed for accumulation of metaldehyde residues, which washed out in high concentrations once drain flow ensued, which is more representative of the model.

As discussed, drain flow is presumed to a determining factor for metaldehyde wash out and resultant concentration in surface water over a given time frame. However, for a complete assessment of pollution risk, it is necessary to consider residues of metaldehyde that remain in soil as "legacy pools". Figure 12.6 shows the proportion of metaldehyde lost to drains relative to the amount that is dissipated. The trend between the amounts of metaldehyde lost to drains and the amount dissipated reflects the data outlined in Figure 12.5, as expected. For the simulations where less metaldehyde is lost

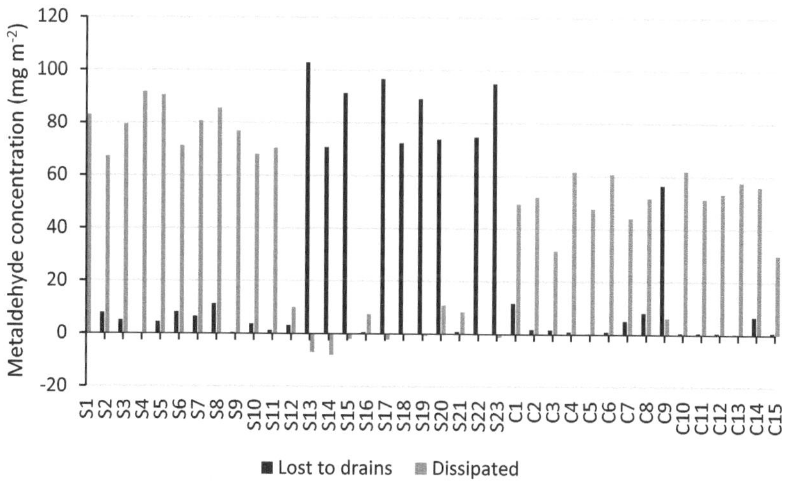

Figure 12.6 Chart showing the proportion of metaldehyde per square metre lost to drains or dissipated, flow predicted from MACRO simulations (S1–23) and measured values from soil columns (C1–15) in the lysimeter experiment.

to drains, greater amounts are dissipated. This is logical, as more degradation leaves less metaldehyde available to wash out in drain flow. Indeed, some of the simulations suggest that metaldehyde residues may accumulate over time, indicated by a negative value for the amount dissipated, e.g., S13, S14, S15 and S17. These scenarios demonstrate high levels of metaldehyde lost to drain flow. The amount of metaldehyde lost in leachate from the soil columns is broadly in agreement with the model predictions.

12.3.6 Final discussion and future research

The nature of metaldehyde prevalence in drain flow will have implications for surface water monitoring initiatives. For surface waters that are intended for abstraction of drinking water, it is vital that concentrations of metaldehyde are below the regulatory limits, otherwise the water company will be liable to penalties. Hence, monitoring programmes are needed. By understanding drainage discharge rates for different parts of a catchment area and being able to ascertain what concentrations of metaldehyde might be expected and over what time frame levels above the regulatory limit might be detected, monitoring programmes can be made more targeted, making them more time-efficient and cost-efficient. This information could also help to improve the reliability of model predictions.

The key to understanding the processes involved in the biodegradation of pesticides is the nature of the degrading microbes. The degradation of metaldehyde in the environment is heavily dependent on the action of soil microbes. Comparing trends in the microbiology will facilitate understanding of the process involved in metaldehyde degradation in soil. Presumably, in samples where microbial activity is higher, the rate of metaldehyde degradation will be greater. The amount of microbial activity is likely to be linked with soil organic content, but may also be affected by other soil properties, which may impact this correlation. However, only certain bacteria can degrade metaldehyde, so for an accurate assessment of the role of microbes in metaldehyde degradation, techniques, such as qPCR and minION sequencing, need to be employed to determine the quantity of these types of microorganisms in the soil. This comprised later work during this project. This produces a holistic appraisal of the processes in the environment that determines the rate of degradation of metaldehyde. The surface soil and deeper soil depth analysed during microbial studies directly compares the two different scenarios where metaldehyde degradation could take place. If increasing soil depth has significant differences in microbial communities to affect metaldehyde degradation, this could explain the discrepancy between measured degradation half-life and persistence in the environment.

What this study has shown is that batch laboratory degradation assessments, following standard guidelines, can have limitations for accurately predicting environmental concentrations of pesticides when used in models. Metaldehyde represents a highly relevant and appropriate case-study because under the standard test paradigm it is not assumed to be a risk for pollution, yet it is one of the most troublesome compounds with regard to UK water quality. The suite of batch degradation experiments conducted during this study exemplified how, for one compound, degradation rates can be highly variable. These experiments also demonstrated that metaldehyde does have the potential to become persistent and that legacy in soil might well depend on soil conditions. These derived half-lives can be input into standard regulatory models to reveal whether more accurately predicted environmental concentrations can be derived. The lysimeter experiments revealed that metaldehyde residues can remain in soil longer than the standard batch incubations would predict and that metaldehyde is at high risk for leaching into drain flow waters.

12.4 CONCLUSIONS AND RECOMMENDATION DRAWN FROM THE STUDY

This study aimed to improve the accuracy of environmental fate assessments for metaldehyde. The standard approach to environmental fate assessment fails to accurately predict the contamination of surface water with metaldehyde. It was hypothesised that lower incubation temperatures and higher soil moisture contents would elucidate how metaldehyde persists in soil. However, the results did not entirely clarify this, with notable persistence only occurring in one experiment. The study did illustrate, however, that metaldehyde can persist in intact soil longer than would be expected, based on the incubation results. In future work, molecular microbiology aims to improve understanding of the mechanisms of the degradation to help explain the observations from the incubation experiments and lysimeters. No previously reported work examines the effects of different experiment conditions on metaldehyde degradation to this level of detail, combined with the microbiology, gives this study its state-of-the art perspective on metaldehyde degradation in soil and therefore presents a significant contribution to understanding the legacy of this compound in the environment. This understanding will be important with regard to implementing new management strategies around the responsible use of metaldehyde elsewhere in the world.

REFERENCES

Asfaw, A., K. Maher, and J. D. Shucksmith. "Modelling of Metaldehyde Concentrations in Surface Waters: A Travel Time Based Approach." *Journal of Hydrology* 562 (2018): 397–410.

Bausenwein, U., A. Gattinger, U. Langer, A. Embacher, H-P. Hartmann, M. Sommer, J. C. Munch, and M. Schloter. "Exploring Soil Microbial Communities and Soil Organic Matter: Variability and Interactions in Arable Soils under Minimum Tillage Practice." *Applied Soil Ecology* 40 (2008): 67–77.

Busquets, Rosa, Oleksandr P. Kozynchenko, Raymond L. D. Whitby, Stephen R. Tennison, and Andrew B. Cundy. "Phenolic Carbon Tailored for the Removal of Polar Organic Contaminants from Water: A Solution to the Metaldehyde Problem?" *Water Research* 61 (2014): 46–56.

Bond, Stephanie Grace. "Assessing the Fate of Metaldehyde Applied to Arable Soils." (2018) MSc thesis (7709390).

Castle, Glenn D., Graham A. Mills, Adil Bakir, Anthony Gravell, Melanie Schumacher, Kate Snowd, and Gary R. Fones. "Measuring Metaldehyde in Surface Waters in the UK Using Two Monitoring Approaches." *Environmental Science: Processes & Impacts* 20, no. 1180 (2018): 1180–90.

Castle, G. D., G. A. Mills, Anthony Gravell, L. Jones, Ian Townsend, D. G. Cameron, and G. R. Fones. "Review of the Molluscicide Metaldehyde in the Environment." *Environmental Science: Water Research & Technology* 3, no. 415 (2017): 415–28.

Calumpang, S. M. F., M. J. B. Medina, A. W. Tejada, and J. R. Medina. "Environmental Impact of Two Molluscicides: Niclosamide and Metaldehyde in a Rice Paddy Ecosystem." *Bulletin of Environmental Contamination and Toxicology* 55 (1995): 494––501.

Castro-Gutiérrez, Víctor, Edward Fuller, John C. Thomas, Chris J. Sinclair, Steven Johnson, Thorunn Helgason, and James W. B. Moir. "Genomic Basis for Pesticide Degradation Revealed by Selection, Isolation and Characterisation of a Library of Metaldehyde-Degrading Strains from Soil." *Soil Biology and Biochemistry* 142 (2020): 107702.

Colbourne, J. (2013). "Letter to the Parliamentary Under-Secretary for Natural Environment and Fisheries." Available at: http://www.dwi.gov.uk/about/annual-report/2012/letter-english.pdf

Department for Environment, Food & Rural Affairs, Health and Safety Executive, and The Rt Hon Michael Gove MP "Restrictions on the Use of Metaldehyde to Protect Wildlife" News Release (2020).

Dewar, Alan M., Andrew Ferguson, Judith K. Pell, Caroline Nicholls, and Jenna Watts. "A Review of Pest Management in Cereals and Oilseed Rape in the UK." In *AHDB Cereals and Oilseeds* (2016). https://ahdb.org.uk/a-review-of-pest-management-in-cereals-and-oilseed-rape-in-the-uk

Dong, Bizhang, Xingjiao Shao, Hongfang Lin, and Jiye Hu. "Dissipation, Residues and Risk Assessment of Metaldehyde and Niclosamide Ethanolamine in Pakchoi after Field Application." *Food Chemistry* 229 (2017): 604–09.

Edwards, Clive A., Norman Q. Arancon, Marcus Vasko-Bennett, Brandon Little, and Ahmed Askar. "The Relative Toxicity of Metaldehyde and Iron Phosphate-Based Molluscicides to Earthworms." *Crop Protection* 28 (2009): 289–94.

EFSA. "Conclusion on the Peer Review of the Pesticide Risk Assessment of the Active Substance Metaldehyde." *EFSA Journal* 8, no. 10 (2010): 1856.

Hemkemeyer, Michael, Bent T. Christensen, Rainer Martens, and Christoph C. Tebbe. "Soil Particle Size Fractions Harbour Distinct Microbial Communities and Differ in Potential for Microbial Mineralisation of Organic Pollutants." *Soil Biology & Biochemistry* 90 (2015): 255–65.

Kay, Paul, and Richard Grayson. "Using Water Industry Data to Assess the Metaldehyde Pollution Problem." *Water and Environment Journal* 28 (2014): 410–17.

Kumar, Manish, Daniel D. Snow, Yusong Li, and Patrick J. Shea. "Perchlorate Behavior in the Context of Black Carbon and Metal Cogeneration Following Fireworks Emission at Oak Lake, Lincoln, Nebraska, USA." *Environmental Pollution* 253 (2019): 930–938.

Kumar, Manish, Tushara Chaminda, Arbind K. Patel, Himaya Sewwandi, Payal Mazumder, Madhvi Joshi, and Ryo Honda "Prevalence of antibiotic resistance in the tropical rivers of Sri Lanka and India." *Environmental Research*, 188 (2020): 109765.

Lu, Q., P. G. Whitehead, G. Bussi, M. N. Futter, and L. Nizzetto. "Modelling Metaldehyde in Catchments: A River Thames Case-Study." *Environmental Science: Processes & Impacts* 19, no. 586 (2017): 586–95.

Ministry of Agriculture Fisheries and Food. "Food and Environment Protection Act." (1996).

Matthies, Michael, and Sabine Beulke. "Considerations of Temperature in the Context of the Persistence Classification in the Eu." *Environmental Sciences Europe* 29, no. 15 (2017): 1–11.

Nicholls, Caroline J. "Implications of Not Controlling Slugs in Oilseed Rape and Wheat in the UK." *Research Review* (2014).

Patel, Arbind Kumar, Chandrasekhar Bhagat, Kaling Taki, and Manish Kumar. "Microplastic Vulnerability in the Sediments of the Sabarmati River of India." In Resilience, Response, and Risk in Water Systems, pp. 127–38. Springer, Singapore (2020).

Salvestrini, S., P. Vanore, A. Bogush, S. Mayadevi, and L. C. Campos. "Sorption of Metaldehyde Using Granular Activated Carbon." *Journal of Water Reuse and Desalination* 7, no. 3 (2017): 280–87.

Singh, Ashwin, Arbind Kumar Patel, Jyoti Prakash Deka, and Manish Kumar. "Natural Recharge Transcends Anthropogenic Forcing that Influences Arsenic Vulnerability of the Quaternary Alluviums of the Mid-Gangetic Plain." *NPJ Clean Water* 3, no. 1 (2020): 1–12.

Singh, Ashwin, Arbind Kumar Patel, Algappan Ramanathan, and Manish Kumar. "Climatic Influences on Arsenic Health Risk in the Metamorphic Precambrian Deposits of Sri Lanka: A Re-Analysis-Based Critical Review." *Journal of Climate Change* 6, no. 1 (2020): 15–24.

Thomas, John C., Thorunn Helgason, Chris J. Sinclair, and James W. B. Moir. "Isolation and Characterization of Metaldehyde-Degrading Bacteria from Domestic Soils." *Microbial Biotechnology* 10, no. 6 (2017): 1824–29.

UKWIR. "Pesticide Risk Mapping and Catchment Interventions." (2015).

Zhang, H. Y., C. Wang, P. J. Xu, and Y. Q. Ma. "Analysis of Molluscicide Metaldehyde in Vegetables by Dispersive Solid-Phase Extraction and Liquid Chromatography-Tandem Mass Spectrometry." *Food Additives & Contaminants* 28, no. 8 (2011): 1034–40.

Chapter 13

Elucidation of vulnerability of groundwater quality to agriculture and surface runoff

A comprehensive review under the backdrop of future scenario of climate change

Koushik Singha Roy
Shell Technology Centre

Johnson Christian and Dharmik Vyas
MEFGI

CONTENTS

13.1 INTRODUCTION

Groundwater is considered as the important source of drinking water for billions of people (Sampat, 2000). Dependency of groundwater is very high especially in geographic regions including China, Eastern part of Europe, North India, and the USA (Hall et al., 2008). The groundwater has been severely deteriorating owing to its large-scale exploitation, contamination through various man-made activities, lack of policies, and lacunae in conservation management (Gogu and Dassargues, 2000). The

DOI: 10.1201/9781003247869-16

level of contamination has been increased significantly to a higher degree due to ongoing trend of changed climatic condition which poses a significant threat to the quality, storage, and recharge of groundwater resource. To address this issue a suitable model-based approach is highly required. In current decades, a plethora of methods have been developed for simulating and mapping of groundwater vulnerability. A plethora of models characteristically comprise the investigation of the interconnections between key hydrogeological attributes. Contaminants in groundwater are usually categorized as (1) natural and anthropocene origin or (2) physical, chemical, and biological. However, sources of groundwater contamination can be classified into various types such as (1) natural and anthropogenic, (2) surface and subsurface, (3) point and non-point, and (4) rural and urban. Some uncertainties are always there in identifying contaminants, in particular emerging contaminants (Menon et al., 2020; Mohapatra et al., 2018, 2021; Priyanka and Mohapatra, 2020; Priyanka et al., 2021). This chapter puts some approaches to document some scenarios of groundwater contamination from surface runoff especially from agriculture and urban sectors in particular under changed climatic condition.

13.2 GROUNDWATER POLLUTION BY AGRICULTURE AND URBAN SURFACE RUNOFF

Land use activity through agricultural practices is considered as one of the important factors which mostly affect the quality of both the surface water and groundwater (Fohrer et al., 2001). The intensification of agricultural activity results in soil erosion and increases sediment load. This further aggravates the leaching of nutrients into the groundwater and surface water bodies including lakes, streams, and rivers (Foley et al., 2005). At present the expansion in cultivation practices puts constraints on natural resources and poses a great threat to the pollution of aquifers (Verburg et al., 2004). An aquifer becomes inoperative for decades or long time as and when it becomes contaminated with harmful contaminants or chemicals (Freitas et al., 2015). Apart from these, end result of toxicity in groundwater does not finish the loss of potable water supply but may lead to health disorders (Conant et al., 2004).

13.2.1 Irrigated agriculture: means of contamination in groundwater

Agriculture is the backbone of nation's economy. However, agriculture requires a significant quantity of fresh water and poses serious constraint on groundwater utilization (Moeck et al., 2018). In India, states like Haryana, Punjab, and West Bengal heavily depend on groundwater resource, and non-judicious utilization of groundwater leads to depletion of groundwater level and results in salinity in soil. Further application of nitrogenous fertilizers and pesticides results in leaching of harmful chemicals into groundwater. The different case studies on nitrate contamination have been documented in Table 13.1

Table 13.1 Groundwater contamination in some select studies in Indian context

Study location	*Nature of groundwater contaminants*	*Remarks/outcome*	*Reference*
Guna district, Madhya Pradesh	Salinity hazard	The observation showed that 28% of groundwater system suitable for irrigation. There was 48% groundwater system useful for salt tolerant crop cultivation and 12% water source unsuitable for irrigation.	Srivastava (2019)
Apulia, Southern Italy	Microbial contamination (*Legionella* sp.)	The groundwater samples were found as the potent source of *Legionella* sp. which leads to legionellosis.	De Giglio et al. (2019)
Chtouka-Massa basin, Morocco	Salt intrusion, nitrate contamination	Nitrate pollution and high salinity were found in heavily exploited aquifer under arid climate conditions.	Malki et al. (2017)
Singanallur, Coimbatore, Tamil Nadu, India	High salinity and low to medium alkalinity	High salinity results in soil infertility and adverse impact on plant growth.	Selvakumar et al. (2017)

13.2.2 Groundwater contamination by nitrate

The NO_3^- concentration has been aggravated in groundwater in various regions in the world including India (Rajeswari et al., 2016). There are numerous processes including application of nitrogenous fertilizer, livestock feeding, manure application, leaching from septic tanks, and industrial effluents that lead to the contamination of groundwater (Curt et al., 2004). In Hissar and Mahendragarh district of Haryana, India, the groundwater has become highly contaminated with NO_3^- and found that the NO_3^- concentration exceeds 1500 mg/L (Govindan et al., 2015). We have documented the status of NO_3^- level in groundwater in recent time (Table 13.2). The elevated level of NO_3^- may impose several health disorders to the susceptible subjects including blue baby syndrome (Methemoglobinemia) in infants, thyroid disorders, lymphatic cancer, goiter, stomach cancer, and gastric cancer (Van Drecht et al., 2001). Therefore, sound strategic management of NO_3^- removal from groundwater must be implemented. There are different physical as well chemical processes available for the removal of NO_3^- from groundwater including photocatalytic exchange, electrodialysis, reverse osmosis, ion exchange, etc. However, all these processes suffer from different limitations from application point of view. Hence, strong research studies are needed for complete removal of NO_3^- from groundwater.

Table 13.2 Status of NO_3^- level in groundwater system in different geological regions

Location	*Lithology/geology*	*Nitrate level (mg/L)*	*Remarks*	*Reference*
San Joaquin valley, California, USA	Tertiary and quaternary alluvial sediments at shallow depth (≤150 m)	46% wells were found to be beyond the maximum concentration level of 10 mg/L	The area was predominantly having agriculture landscape of diverse crop types and the NO_3^- leached because of the nitrogenous fertilizer applications.	Lockhart et al. (2013)
Haean Basin, Gangwon province, North–east Korea	The Jurassic igneous rocks intruding into composite metamorphic rocks. The aquifer developed by weathering of granite rocks with alluvial deposits at its upper side.	The NO_3^--N concentration in the groundwater was found to be the 2.9–15.2 mg/L, and in the rice paddy the average concentration of NO_3^- was 10.7 mg/L.	The contamination took place in the groundwater due to the land use practices and surface erosion.	Kim et al. (2015)
Gharbar, Birsinhipur, Kalipur village in Dhanbad, Jharkhand, India	Unconfined and confined aquifer of Gondwanas and Archaeans rock types	Fifty samples were tested having NO_3^- concentration in the range of 34.1–319.1 mg/L	The NO_3^- concentration was positively correlated with TDS, Mg^{+2}, Ca^{+2}, Cl^-, SO_4^{-2} ions.	Thapa et al. (2019)
Yerraguntla, Kadapa district, Andhra Pradesh, India	Lime stone geology	Forty groundwater samples were collected and tested. The NO_3^- concentration found was 2.5–760.12 mg/L with average concentration of 86.13 mg/L	Hazard quotient varied from 0.05 to 18.53 among infants, children, male, and female.	Suvarna et al. (2020)

(*Continued*)

Table 13.2 (Continued) Status of NO_3^- level in groundwater system in different geological regions

Location	*Lithology/geology*	*Nitrate level (mg/L)*	*Remarks*	*Reference*
Nagpur, Maharashtra, India	Precambrian crystalline rocks comprising gneisses, schists, and granites, sandstones of kamathi formation. Aquifers were unconfined, semi-confined, and confined in nature having the thickness of 1–10 m of weathered rocks	Approximately 50% of the groundwater samples were found to be exceeded of NO_3^- concentration (1.6–432 mg/L)	The origin of nitrate contamination was due to the leaching from municipal wastes, animal wastes, septic tanks, urban garbage, and nitrogenous fertilizer plants.	Marghade (2020)
A Limia (Ourense, North-Western Spain)	The area is having old tertiary depression filled intermittently with quaternary sediments.	The NO_3^- concentration was found in the range of 10–25 mg/L.	The contaminant got its entry into the river and lagoons through livestock wastes, organic loading through soil runoff.	Garzon-Vidueira et al. (2020)
Yadani-Bhongi district, Telengana, India	The geology of the region having granite & gneiss. The Archaean rock not having primary porosity forms water-bearing formation with secondary porosity due to weathering process.	The NO_3^- concentration was found in the range of 14–82 mg/L. It was also observed that almost 43% of the groundwater samples beyond the safe level of 45 mg/L.	The oral exposure was found much higher as compared to the dermal contact and non-carcinogenic risk varied from 0.3 to 2.6 mg/L among male, female and children.	Adimalla et al. (2021)

13.2.3 Fluoride and cadmium contamination in groundwater: surface runoff in cities

In groundwater fluoride (F^-) pollution has become an important issue owing to its natural source. A myriad of researchers have concluded that the geogenic sources are the prime origin of F^- entry into the aquifer as well groundwater system (Sharma et al., 2015; Mukherjee and Singh, 2018). The south and southeast Asia is now considered as the hotspot of F^- toxicity in recent times (World Health Organization, 2011). The F^- content has been found in the range from 20 to 79 mg/L in groundwater in various locations in Pakistan, Republic of Korea, Malaysia, and Afghanistan (Farooqi et al., 2007; Chae et al., 2007; Shamsuddin et al., 2016; Hayat and Baba, 2017). These high F^- contents primarily originated through geological processes, weathering of F^- containing rocks, and occurrence of rock minerals like apatite in groundwater (Msonda et al., 2007). A group of researchers (Sharma et al., 2012; Yadav et al., 2016; Reddy et al., 2016; Satyanarayana et al., 2017; Kumar et al., 2016; Samantara et al., 2017; Thivya et al., 2017; Tirkey et al., 2017; Yadav et al., 2018) have studied the status of F^- concentration in groundwater in different states in India and their observations have been cited in Figure 13.1. The results suggest that a strategic and sound management policy is need of the hour to combat fluoride pollution in groundwater.

Cadmium, one of the toxic metals in groundwater, gets its route in human body through contaminated food and water and shows bioaccumulation in several organs in body (Hajeb et al., 2014). The Cd routed in groundwater is mainly through geogenic origin like weathering of rocks and surface runoff (Nies, 2003). The rocks contain varying concentration of Cd (Figures 13.2–13.4), and upon weathering and disintegration it enters into groundwater. The Cd concentrations in igneous rocks were varied in the order of 0.02–0.25 mg/kg (Figure 13.2). The Cd concentrations were varied significantly in different orders in sedimentary rocks. The maximum concentration of the Cd was found in the Marlstone (Figure 13.3). The metamorphic rocks like Gneisses and Schists were found at the order of 0.04 and 0.02 mg/kg, respectively (Figure 13.4). The human-induced Cd inputs into groundwater are different sources, including landfills, sewage discharge, emissions from combustion process, mining activities, battery industry, and pesticides application (Sprynskyy et al., 2011; Bigalke et al., 2017; Khan

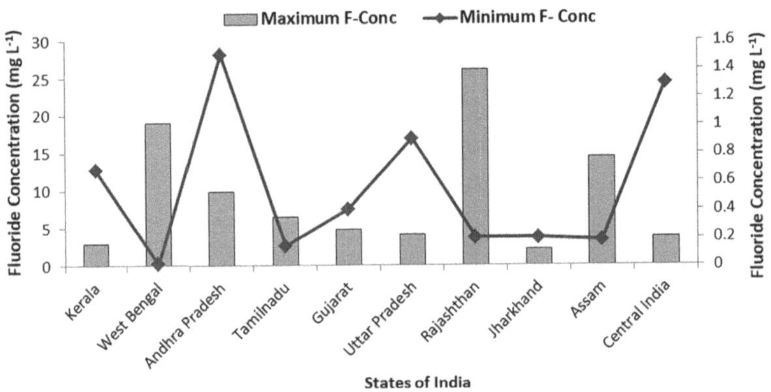

Figure 13.1 Fluoride concentration in groundwater in different states in India.

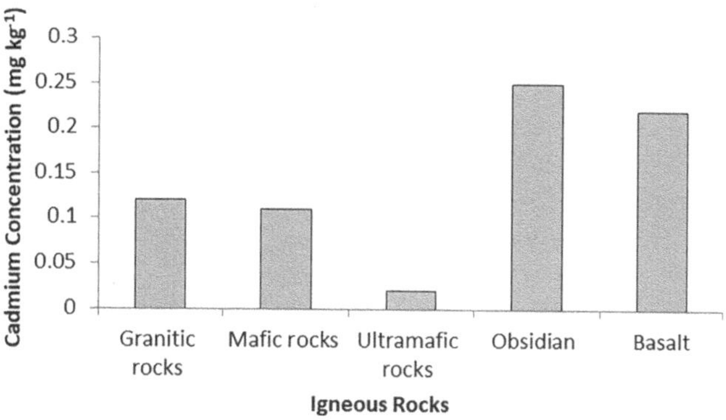

Figure 13.2 Cadmium concentration in various igneous rock types.

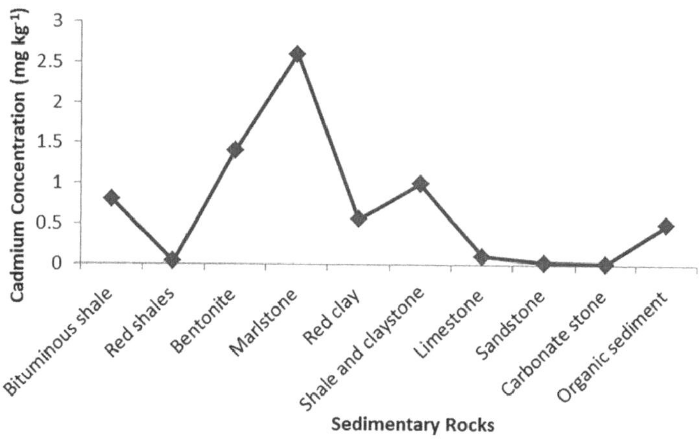

Figure 13.3 Cadmium concentration in different sedimentary rock types.

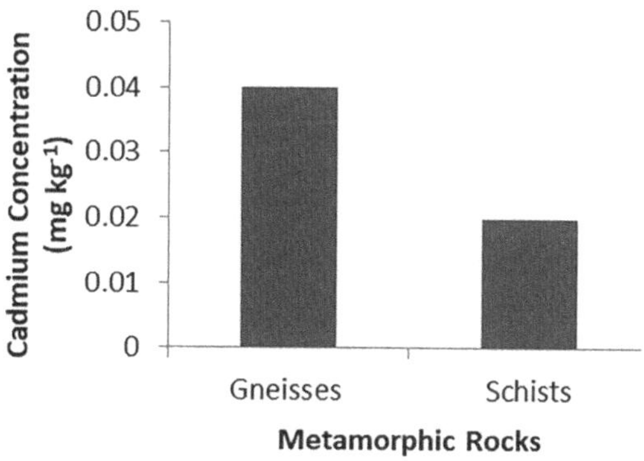

Figure 13.4 Cadmium concentration in different metamorphic rock types.

et al., 2017). In Europe, the Cd concentration in municipal solid wastes and leachate was found in the order of 0.3–12 and 0.5–3.4 mg/kg, respectively. On the other hand, in rice grains, the Cd concentrations were found to be 2.5 times higher as compared to that of rice samples grown in non-polluted areas (Aoshima, 2016). In northwest of Germany, the Cd concentration exceeded the permissible limit of 0.5 μL/L. The evaluation of an extensive data set on hydro-geochemical data set made it convenient to observe various facets of Cd transport in the context of groundwater chemistry and land use activity (Kubier and Pichler, 2019).

13.3 CLIMATE CHANGE AND MOBILIZATION OF CONTAMINANTS IN GROUNDWATER

Surface water along with groundwater is significantly facing high pressure owing to the ongoing trend of climate warming as imposed by the climate change scenario (Kløve et al., 2014). The direct influence is mainly manifested by recharge of groundwater, and on the other hand the indirect process includes the various usage of it (Taylor et al., 2013). In their study, the significance of human activity and land use changes impact on the groundwater is highlighted. However, to check the vulnerability status of groundwater the main focus would be to monitor the parameters that result in water resource management and free from pollution/or contamination (Green et al., 2011). Therefore, the specialty of aquifer (confined, unconfined, Aquitard, etc.), available precipitation, edaphic and terrain characteristics, land cover, and land use pattern are the important parameters that need to be monitored in order to correlate with groundwater quality vulnerability with anticipated climate change condition. All these said parameters are subjected to be influenced by climate change; hence, it will directly or indirectly influence the groundwater availability as well to retain its quality. In this context, various types of simulation model come into play to elucidate the status of groundwater and its quality. The models are useful to predict the availability of water resources, level of groundwater table, and sea level also. There are different land use practices including agricultural activities, minerals extraction, urbanization, and human settlement which affect groundwater quality immensely. The science is that particular land use activity contributes to the atmospheric exchange through the evapo-transpiration process and serves as the important parameter for the determination of water balance, surface runoff, and aquifer recharge. Hamann et al. (2013) and Dezsi et al. (2018) carried out studies based on climate data projections using different models from various sources and concluded that the water availability was in a declining trend during 1990–2020 and could be aggravated by 2050 in the European continent. Hence proper monitoring and management should be there to cope up with the situation.

13.3.1 Climate change and associated factors

The Earth surface receives most of its energy from the sun as electromagnetic radiation, in the visible spectrum along with small amount of infrared (IR) and ultraviolet (UV) radiation. The Earth's surface absorbs the incoming short-wave solar energy and releases it back to the atmosphere as long-wave terrestrial radiation. This long-wave radiation emitted by the Earth surface again is absorbed by some atmospheric trace gases causing a warming known as greenhouse effect. Greenhouse effect, although necessary to sustain life in the Earth can also cause serious problem in the form of global warming that is threatening environment by melting polar ice caps and raising sea levels around the world and brings about the so-called Climate Change Anthropogenic activities are one of the important sources of three GHGs, namely CO_2, CH_4, and N_2O. Among these GHGs, CO_2 is the most important accounting for 60% of global warming. The concentration of CO_2 has increased from 280 μmol/mol at the beginning of the industrial revolution to the present-day concentration of 417 μmol/mol. This increase is attributed to anthropogenic activities, including fossil fuel burning, deforestation, emission from automobiles, forest fires, etc. The rise in atmospheric CO_2 followed by global warming may be buffered to some extent by increased absorption of C in vegetation, soil, and water. However, there is much uncertainty about the capacity of such sinks to limit rises in CO_2, and the problem is much acute with the soil as a buffer. These GHGs do not only cause global warming/climate change but also cause immense constraints on natural resources, including water resource and groundwater system in particular.

13.3.2 Relationship of climate variability and vulnerability of groundwater quality

As per the estimation there is 1.386 billion cubic kilometers of water in our planet Earth. Out of the total water resource, we have around 97% as saline water found in the world oceans, and the remaining 3% serves as the non-saline water. Thus, groundwater is a fundamental resource on which lives of entire Earth sustains (Sivarajan et al., 2019). However, the non-judicious use and overexploitation of groundwater resource owing to rapid urbanization and population explosion leave it in a very critical condition (Taylor et al., 2013). In addition, the climatic anomalies and disturbances result in severe depletion of groundwater resource. A dozen of scientific studies have been carried out that sets the correlations between climate distresses and groundwater depletion (Asoka et al., 2017). The global warming brings about more severe and recurrent climatic disturbances, intense rainfall, and evaporation that lead to significant variability in surface water resource and ultimately threatens groundwater recharge and rejuvenation (Mustafa et al., 2019). Undeniably, growing demand of groundwater will have significant impact on water conservation (Zaveri et al., 2016).

Climate change upsets the entire groundwater system, obstructs flow and movement of groundwater, constrains on recharge of groundwater and storage (Tillman et al., 2017), poses hindrance in the discharge of groundwater, and above all affects the quality of groundwater. Groundwater recharge is highly dependent on the regional precipitation pattern, and leakage of groundwater (Jayakumar and Lee, 2016) to the subsurface is also dependent on the regional lithology and geological attributes (Tague et al., 2008). The climate change and associated factors like temperature and humidity directly impact on the hydrogeological factors and causes abstraction of the groundwater resource (Whittemore et al., 2016). Land use change including changes in type of vegetation and agriculture activities is significantly affected by climate change, and it alters the evaporation pattern and puts strain on groundwater resource (Alam et al., 2019). Assessment on global scale shows worldwide outline of anticipated trends in groundwater recharge and level of groundwater, but region-based studies on aquifers are important to understand the variation of groundwater level under climate change condition (Meixner et al., 2016).

13.3.3 Metal mobilization in groundwater influenced by climatic factors

Heavy metal and metalloid pollution in topsoil due to deposition from atmosphere and discharge from sanitary landfills, farmlands, and industries is a common issue that is projected to affect terrestrial ecosystems, land surface, and wide regions of the Earth. There is growing apprehension that the existing contaminant stocking in groundwater as well as in surface water bodies could be changed, and hypothetically elevated, by ongoing trend of climate change. The impact would be on large scale on the level of groundwater and recharge. Researchers have investigated the occurrence of metal mobilization through hydrogeological and geochemical simulation model and studied on various metals and metalloids like lead and arsenic (Rodell et al., 2018). The approach of the study was to define initial condition and boundary condition in a particular climate zone, for example, tropical or temperate climate zone (Su et al., 2017). In European continent under temperate climate zone the metal mobilization was found to be significantly increased at the modest depth of 200 cm, and it has become projected that the metal mobilization can be enhanced at the rate of 2–10 in next 20 years or so. The fluctuation in groundwater level in the aquifer could pose more threat to further aggravation of metal mobilization in groundwater and ultimately contaminate multiple groundwater sites (Musolff et al., 2016).

13.3.4 Aquifer contamination risk owing to urban expansion

Risk assessment allows the documentation of regions/locations where aquifers are more vulnerable for chemical contamination owing to their hydrological and geological attributes (Matzeu et al., 2017). A myriad of researchers have projected methodologies to assess the vulnerability to contamination of aquifers. Primarily the vulnerability can be classified as intrinsic vulnerability and specific vulnerability. The intrinsic vulnerability depends on geological, hydrological, and combination of both attributes. The specific vulnerability depends on the nature of pollutants and their chemical characterization (Gogu and Dassargues, 2000). The overlay index methods include "DRASTIC", "DRISTPI", "EPIK", "SINTACS", etc. These models are mainly based on ratings and weightage of input parameters. The process-based mathematical models include "GLEAMS" and "HYDRUS". However, the process design and selection solely depends on the aquifer characteristics and lithology of the region (Langrudi et al., 2016; Busico et al., 2017; Jarray et al., 2017).

13.4 MASS FLOW MODELING OF EMERGING CONTAMINANTS IN GROUNDWATER

Kumar & Singh (2011) worked on the "MODFLOW" model that simulates flow, surface runoff, and leaching of nutrients. Qiu et al. (2015) worked on "GMS" model which was the variant of "MODFLOW" model. A unique model "FEFLOW" simulates flow of groundwater which depends on density-dependent groundwater flow along with the mass and heat transport process of the contaminants (Diersch, 2013). Kumar (2012) worked on "CHEMFLOW" model that describes the mobility of chemical contaminants. Although these models are unique in nature, having some limitations and further modifications and improvement is required in order to simulate the mobility of contaminants in better manner (Portmann et al., 2009; Shiu et al., 2012; Kruse and Eslamian 2017; Dirnböck et al., 2016; Liu et al., 2016).

13.5 CONCLUSION & FUTURE SCOPE

This chapter shows a direction toward the scenario of groundwater contamination under changed climatic condition. This chapter also shows the impact of climate change on groundwater in both direct and indirect pathways. Moreover, we evaluate the utilization of simulation models based on groundwater, climate, and land surface. Each of these components is the integral part of the simulation models that provides information on the fluctuation of groundwater level and mobilization of contaminants. We have also put emphasis to correlate the recharge and storage of groundwater under climate change conditions, but some lacunae exists owing to region-specific variations. In future, our priority is to develop a suitable projection model depending on the nature of aquifer, lithology, microclimate, and other associated factors that foretell the fluctuation of groundwater level and mobilization of contaminants in the groundwater system.

REFERENCES

Adimalla, N., Qian, H., Tiwari, D. M., 2021. Groundwater chemistry, distribution and potential health risk appraisal of nitrate enriched groundwater: A case study from the semi-urban region of South India. *Ecotoxicology and Environmental Safety*, 207, 111277.

Alam, S., Gebremichael, M., Li, R., Dozier, J., Lettenmaier, D. P., 2019. Climate change impacts on groundwater storage in the Central Valley, California. *Climatic Change*, 157(3), 387–406.

Aoshima, K., 2016. Itai-itai disease: Renal tubular osteomalacia induced by environmental exposure to cadmium—historical review and perspectives. *Soil Science and Plant Nutrition*, 62(4), 319–326.

Asoka, A., Gleeson, T., Wada, Y., Mishra, V., 2017. Relative contribution of monsoon precipitation and pumping to changes in groundwater storage in India. *Nature Geoscience*, 10(2), 109–117.

Bigalke, M., Ulrich, A., Rehmus, A., Keller, A., 2017. Accumulation of cadmium and uranium in arable soils in Switzerland. *Environmental Pollution*, 221, 85–93.

Busico, G., Cuoco, E., Sirna, M., Mastrocicco, M., Tedesco, D., 2017. Aquifer vulnerability and potential risk assessment: application to an intensely cultivated and densely populated area in Southern Italy. *Arabian Journal of Geosciences*, 10(10), 222.

Chae, G. T., Yun, S. T., Mayer, B., Kim, K. H., Kim, S. Y., Kwon, J. S., Koh, Y. K., 2007. Fluorine geochemistry in bedrock groundwater of South Korea. *Science of the Total Environment*, 385(1–3), 272–283.

Conant Jr, B., Cherry, J. A., Gillham, R. W., 2004. A PCE groundwater plume discharging to a river: Influence of the streambed and near-river zone on contaminant distributions. *Journal of Contaminant Hydrology*, 73(1–4), 249–279.

Curt, M. D., Aguado, P., Sánchez, G., Bigeriego, M., Fernández, J., 2004. Nitrogen isotope ratios of synthetic and organic sources of nitrate water contamination in Spain. *Water, Air, and Soil Pollution*, 151(1), 135–142.

De Giglio, O., Napoli, C., Apollonio, F., Brigida, S., Marzella, A., Diella, G., Montagna, M. T., 2019. Occurrence of *Legionella* in groundwater used for sprinkler irrigation in Southern Italy. *Environmental Research*, 170, 215–221.

Dezsi, Ş., Mîndrescu, M., Petrea, D., Rai, P. K., Hamann, A., Nistor, M. M., 2018. High-resolution projections of evapotranspiration and water availability for Europe under climate change. *International Journal of Climatology*, 38(10), 3832–3841.

Diersch, H. J. G., 2013. *FEFLOW: Finite element modeling of flow, mass and heat transport in porous and fractured media*. Springer Science & Business Media, Berlin.

Dirnböck, T., Kobler, J., Kraus, D., Grote, R., Kiese, R., 2016. Impacts of management and climate change on nitrate leaching in a forested karst area. *Journal of Environmental Management*, 165, 243–252.

Farooqi, A., Masuda, H., Firdous, N., 2007. Toxic fluoride and arsenic contaminated groundwater in the Lahore and Kasur districts, Punjab, Pakistan and possible contaminant sources. *Environmental Pollution*, 145(3), 839–849.

Fohrer, N., Haverkamp, S., Eckhardt, K., Frede, H. G., 2001. Hydrologic response to land use changes on the catchment scale. *Physics and Chemistry of the Earth, Part B: Hydrology, Oceans and Atmosphere*, 26(7–8), 577–582.

Foley, J. A., DeFries, R., Asner, G. P., Barford, C., Bonan, G., Carpenter, S. R., Snyder, P. K., 2005. Global consequences of land use. *Science*, 309(5734), 570–574.

Freitas, J. G., Rivett, M. O., Roche, R. S., Durrant, M., Walker, C., Tellam, J. H., 2015. Heterogeneous hyporheic zone dechlorination of a TCE groundwater plume discharging to an urban river reach. *Science of the Total Environment*, 505, 236–252.

Garzon-Vidueira, R., Rial-Otero, R., Garcia-Nocelo, M. L., Rivas-Gonzalez, E., Moure-Gonzalez, D., Fompedriña-Roca, D., Simal-Gandara, J., 2020. Identification of nitrates origin in Limia river basin and pollution-determinant factors. *Agriculture, Ecosystems & Environment*, 290, 106775.

Gogu, R. C., Dassargues, A., 2000. Current trends and future challenges in groundwater vulnerability assessment using overlay and index methods. *Environmental Geology*, 39(6), 549–559.

Govindan, K., Noel, M., Mohan, R., 2015. Removal of nitrate ion from water by electrochemical approaches. *Journal of Water Process Engineering*, 6, 58–63.

Green, T. R., Taniguchi, M., Kooi, H., Gurdak, J. J., Allen, D. M., Hiscock, K. M., Aureli, A., 2011. Beneath the surface of global change: Impacts of climate change on groundwater. *Journal of Hydrology*, 405(3–4), 532–560.

Hall, N. D., Stuntz, B. B., Abrams, R. H., 2008. Climate change and freshwater resources. *Natural Resources & Environment*, 22(3), 30–35.

Hamann, A., Wang, T., Spittlehouse, D. L., Murdock, T. Q., 2013. A comprehensive, high-resolution database of historical and projected climate surfaces for western North America. *Bulletin of the American Meteorological Society*, 94(9), 1307–1309.

Hayat, E., Baba, A., 2017. Quality of groundwater resources in Afghanistan. *Environmental Monitoring and Assessment*, 189(7), 1–16.

Jarray, H., Zammouri, M., Ouessar, M., Hamzaoui-Azaza, F., Barbieri, M., Zerrim, A., Yahyaoui, H., 2017. Groundwater vulnerability based on GIS approach: Case study of Zeuss-Koutine aquifer, South-Eastern Tunisia. *Geofísica Internacional*, 56(2), 157–172.

Jayakumar, R., Lee, E., 2016. Climate change and groundwater conditions in the Mekong Region–A review. *Journal of Groundwater Science and Engineering*, 4(2), 14–30.

Khan, M. A., Khan, S., Khan, A., Alam, M., 2017. Soil contamination with cadmium, consequences and remediation using organic amendments. *Science of the Total Environment*, 601, 1591–1605.

Kim, H., Kaown, D., Mayer, B., Lee, J. Y., Hyun, Y., Lee, K. K., 2015. Identifying the sources of nitrate contamination of groundwater in an agricultural area (Haean basin, Korea) using isotope and microbial community analyses. *Science of the Total Environment*, 533, 566–575.

Kløve, B., Ala-Aho, P., Bertrand, G., Gurdak, J. J., Kupfersberger, H., Kværner, J., Muotka, T., Mykrä, H., Preda, E., Rossi, P., Uvo, C. B., Velasco, E., Pulido-Velazquez, M., 2014. Climate change impacts on groundwater and dependent ecosystems. *Journal of Hydrology*, 518(Part B), 250–266.

Kruse, E., Eslamian, S., 2017. Groundwater management in drought conditions. *Handbook of Drought and Water Scarcity, Management of Drought and Water Scarcity*, 3, 275–282.

Kubier, A., Pichler, T., 2019. Cadmium in groundwater– A synopsis based on a large hydrogeochemical data set. *Science of the Total Environment*, 689, 831–842.

Kumar, C. P., 2012. Climate change and its impact on groundwater resources. *International Journal of Engineering and Science*, 1(5), 43–60.

Kumar, C. P., Singh, S., 2011. Impact of climate change on water resources and national water mission of India. In: *Proceedings of National Seminar on Global Warming and its Impact on Water Resources* (Vol. 14, pp. 1–9).

Kumar, M., Das, A., Das, N., Goswami, R., Singh, U. K., 2016. Co-occurrence perspective of arsenic and fluoride in the groundwater of Diphu, Assam, Northeastern India. *Chemosphere*, 150, 227–238.

Langrudi, M. A. O., Siuki, A. K., Javadi, S., Hashemi, S. R., 2016. Evaluation of vulnerability of aquifers by improved fuzzy drastic method: Case study: Aastane Kochesfahan plain in Iran. *Ain Shams Engineering Journal*, 7(1), 11–20.

Liu, M., Xu, X., Wang, D., Sun, A. Y., Wang, K., 2016. Karst catchments exhibited higher degradation stress from climate change than the non-karst catchments in southwest China: An ecohydrological perspective. *Journal of Hydrology*, 535, 173–180.

Lockhart, K. M., King, A. M., Harter, T., 2013. Identifying sources of groundwater nitrate contamination in a large alluvial groundwater basin with highly diversified intensive agricultural production. *Journal of Contaminant Hydrology*, 151, 140–154.

Malki, M., Bouchaou, L., Hirich, A., Brahim, Y. A., Choukr-Allah, R., 2017. Impact of agricultural practices on groundwater quality in intensive irrigated area of Chtouka-Massa, Morocco. *Science of the Total Environment*, 574, 760–770.

Marghade, D., 2020. Detailed geochemical assessment & indexing of shallow groundwater resources in metropolitan city of Nagpur (western Maharashtra, India) with potential health risk assessment of nitrate enriched groundwater for sustainable development. *Geochemistry*, 80(4), 125627.

Matzeu, A., Secci, R., Uras, G., 2017. Methodological approach to assessment of groundwater contamination risk in an agricultural area. *Agricultural Water Management*, 184, 46–58.

Meixner, T., Manning, A. H., Stonestrom, D. A., Allen, D. M., Ajami, H., Blasch, K. W., Walvoord, M. A., 2016. Implications of projected climate change for groundwater recharge in the western United States. *Journal of Hydrology*, 534, 124–138.

Menon, N. G., Mohapatra, S., Padhye, L. P., Tatiparti, S. S. V., Mukherji, S., 2020. Review on Occurrence and Toxicity of Pharmaceutical Contamination in Southeast Asia. In: Kumar, M., Snow, D., Honda, R. (eds) *Emerging issues in the water environment during anthropocene. Springer transactions in civil and environmental engineering*. Springer, Singapore.

Mohapatra, S., Padhye, L. P., Mukherji, S., 2018. Challenges in detection of antibiotics in wastewater matrix. In: Gupta, T., Agarwal, A., Agarwal, R., Labhsetwar, N. (eds) *Environmental contaminants. Energy, environment, and sustainability*. Springer, Singapore.

Mohapatra, S., Menon, N., Padhye, L.P., Tatiparti, S. S. V., Mukherji, S., 2021. Natural attenuation of pharmaceuticals in the aquatic environment and role of phototransformation. In: Kumar, M., Snow, D., Honda, R., Mukherjee, S. (eds) *Contaminants in drinking and wastewater sources. Springer transactions in civil and environmental engineering.* Springer, Singapore. https://doi.org/10.1007/978-981-15-4599-3_3

Moeck, C., Affolter, A., Radny, D., Dressmann, H., Auckenthaler, A., Huggenberger, P.,Schirmer, M., 2018. Improved water resource management for a highly complex environment using three-dimensional groundwater modelling. *Hydrogeology Journal*, 26(1), 133–146.

Msonda, K. W. M., Masamba, W. R. L., Fabiano, E., 2007. A study of fluoride groundwater occurrence in Nathenje, Lilongwe, Malawi. *Physics and Chemistry of the Earth, Parts A/B/C*, 32(15–18), 1178–1184.

Mukherjee, I., Singh, U. K., 2018. Groundwater fluoride contamination, probable release, and containment mechanisms: a review on Indian context. *Environmental Geochemistry and Health*, 40(6), 2259–2301.

Musolff, A., Schmidt, C., Rode, M., Lischeid, G., Weise, S. M., Fleckenstein, J. H., 2016. Groundwater head controls nitrate export from an agricultural lowland catchment. *Advances in Water Resources*, 96, 95–107.

Mustafa, S. M. T., Hasan, M. M., Saha, A. K., Rannu, R. P., Uytven, E. V., Willems, P., Huysmans, M., 2019. Multi-model approach to quantify groundwater-level prediction uncertainty using an ensemble of global climate models and multiple abstraction scenarios. *Hydrology and Earth System Sciences*, 23(5), 2279–2303.

Nies, D. H., 2003. Efflux-mediated heavy metal resistance in prokaryotes. *FEMS Microbiology Reviews*, 27(2–3), 313–339.

Portmann, R. W., Solomon, S., Hegerl, G. C., 2009. Spatial and seasonal patterns in climate change, temperatures, and precipitation across the United States. *Proceedings of the National Academy of Sciences of the United States of America*, 106(18), 7324–7329.

Priyanka, M. S., Mohapatra, S., 2020. Shifts and trends in analysis of contaminants of emerging concern: Sulfonamides. In: Kumar, M., Munoz-Arriola, F., Furumai, H., Chaminda, T. (eds) *Resilience, response, and risk in water systems. Springer transactions in civil and environmental engineering.* Springer, Singapore.

Priyanka, M. S., Mohapatra, S., Menon, N. G., 2021. Hybrid membrane technology: an alternative to industrial wastewater treatment. In: Maulin P. Shah and Susana Rodriguez-Couto (eds) *Membrane-based hybrid processes for wastewater treatment* (pp. 481–501). Elsevier, Netherlands.

Qiu, S., Liang, X., Xiao, C., Huang, H., Fang, Z., Lv, F., 2015. Numerical simulation of groundwater flow in a river valley basin in Jilin urban area, China. *Water*, 7(10), 5768–5787.

Rajeswari, A., Amalraj, A., Pius, A., 2016. Adsorption studies for the removal of nitrate using chitosan/PEG and chitosan/PVA polymer composites. *Journal of Water Process Engineering*, 9, 123–134.

Reddy, A. G. S., Reddy, D. V., Kumar, M. S., Naik, P. K., 2016. Evaluation of fluoride enrichment processes in groundwater of Chimakurthy granitic pluton complex in Prakasam District India. *African Journal of Environmental Science and Technology*, 10(10), 350–379.

Rodell, M., Famiglietti, J. S., Wiese, D. N., Reager, J. T., Beaudoing, H. K., Landerer, F. W., Lo, M. H., 2018. Emerging trends in global freshwater availability. *Nature*, 557(7707), 651–659.

Samantara, M. K., Padhi, R. K., Sowmya, M., Kumaran, P., Satpathy, K. K., 2017. Heavy metal contamination, major ion chemistry and appraisal of the groundwater status in coastal aquifer, Kalpakkam, Tamil Nadu, India. *Groundwater for Sustainable Development*, 5, 49–58.

Sampat, P., 2000. Groundwater shock. *World Watch*, 13(1), 10–22.

Satyanarayana, E., Dhakate, R., Kumar, D. L., Ravindar, P., Muralidhar, M., 2017. Hydrochemical characteristics of groundwater quality with special reference to fluoride concentration in parts of Mulugu-VenkatapurMandals, Warangal district, Telangana. *Journal of the Geological Society of India*, 89(3), 247–258.

Selvakumar, S., Chandrasekar, N., Kumar, G., 2017. Hydrogeochemical characteristics and groundwater contamination in the rapid urban development areas of Coimbatore, India. *Water Resources and Industry*, 17, 26–33.

Shamsuddin, M. K. N., Suratman, S., Ramli, M. F., Sulaiman, W. N. A., Sefie, A., 2016. Hydrochemical assessment of surfacewater and groundwater quality at bank infiltration site. In: *IOP Conference Series: Materials Science and Engineering* (Vol. 136, No. 1, p. 012073). IOP Publishing.

Sharma, P., Sarma, H. P., Mahanta, C., 2012. Evaluation of groundwater quality with emphasis on fluoride concentration in Nalbari district, Assam, Northeast India. *Environmental Earth Sciences*, 65(7), 2147–2159.

Sharma, S., Kumar Yadav, K., Gupta, N., Verma, C., Kumar, V., Arya, S., 2015. Effects of seasonal variation on major ion chemistry of Pahuj Reservoir, Jhansi, Uttar Pradesh, India. *Universal Journal of Environmental Research & Technology*, 5(2), 79–89.

Shiu, C. J., Liu, S. C., Fu, C., Dai, A., Sun, Y., 2012. How much do precipitation extremes change in a warming climate? *Geophysical Research Letters*, 39(17), L17707.

Sivarajan, N. A., Mishra, A. K., Rafiq, M., Nagraju, V., Chandra, S., 2019. Examining climate change impact on the variability of ground water level: A case study of Ahmednagar district, India. *Journal of Earth System Science*, 128(5), 1–7.

Sprynskyy, M., Kowalkowski, T., Tutu, H., Cozmuta, L. M., Cukrowska, E. M., Buszewski, B., 2011. The adsorption properties of agricultural and forest soils towards heavy metal ions (Ni, Cu, Zn, and Cd). *Soil and Sediment Contamination*, 20(1), 12–29.

Srivastava, S. K., 2019. Assessment of groundwater quality for the suitability of irrigation and its impacts on crop yields in the Guna district, India. *Agricultural Water Management*, 216, 224–241.

Su, Y., Langhammer, J., Jarsjö, J., 2017. Geochemical responses of forested catchments to bark beetle infestation: Evidence from high frequency in-stream electrical conductivity monitoring. *Journal of Hydrology*, 550, 635–649.

Suvarna, B., Sunitha, V., Reddy, Y. S., Reddy, N. R., 2020. Data health risk assessment of nitrate contamination in groundwater of rural region in the Yerraguntla Mandal, South India. *Data in Brief*, 30, 105374.

Tague, C., Grant, G., Farrell, M., Choate, J., Jefferson, A., 2008. Deep groundwater mediates streamflow response to climate warming in the Oregon Cascades. *Climatic Change*, 86(1), 189–210.

Taylor, R. G., Scanlon, B., Döll, P., Rodell, M., Van Beek, R., Wada, Y., Treidel, H., 2013. Ground water and climate change. *Nature Climate Change*, 3(4), 322–329.

Thapa, R., Gupta, S., Kaur, H., Baski, R., 2019. Assessment of groundwater quality scenario in respect of fluoride and nitrate contamination in and around Gharbar village, Jharkhand, India. *HydroResearch*, 2, 60–68.

Thivya, C., Chidambaram, S., Rao, M. S., Thilagavathi, R., Prasanna, M. V., Manikandan, S. J. A. W. S., 2017. Assessment of fluoride contaminations in groundwater of hard rock aquifers in Madurai district, Tamil Nadu (India). *Applied Water Science*, 7(2), 1011–1023.

Tillman, F. D., Gangopadhyay, S.,Pruitt, T., 2017. Changes in projected spatial and seasonal groundwater recharge in the upper Colorado River basin. *Groundwater*, 55(4), 506–518.

Tirkey, P., Bhattacharya, T., Chakraborty, S., Baraik, S., 2017. Assessment of groundwater quality and associated health risks: a case study of Ranchi city, Jharkhand, India. *Groundwater for Sustainable Development*, 5, 85–100.

Van Drecht, G., Bouwman, A. F., Knoop, J. M., Meinardi, C., Beusen, A., 2001. Global pollution of surface waters from point and nonpoint sources of nitrogen. *The Scientific World Journal*, 1, 632–641.

Verburg, P. H., Schot, P. P., Dijst, M. J., Veldkamp, A., 2004. Land use change modelling: Current practice and research priorities. *GeoJournal*, 61(4), 309–324.

Whittemore, D. O., Butler Jr, J. J., Wilson, B. B., 2016. Assessing the major drivers of water-level declines: New insights into the future of heavily stressed aquifers. *Hydrological Sciences Journal*, 61(1), 134–145.

World Health Organization, 2011. *Guidelines for drinking-water quality, fourth edition*. Available: http://www.who.int/water_sanitation_health/publications/2011/dwq_chapters/en/.

Yadav, A., Sahu, Y. K., Rajhans, K. P., Sahu, P. K., Chakradhari, S., Sahu, B. L.,Patel, K. S., 2016. Fluoride contamination of groundwater and skeleton fluorosis in Central India. *Journal of Environmental Protection*, 7(06), 784.

Yadav, K. K., Gupta, N., Kumar, A., Reece, L. M., Singh, N., Rezania, S.,Khan, S. A., 2018. Mechanistic understanding and holistic approach of phytoremediation: a review on application and future prospects. *Ecological Engineering*, 120, 274–298.

Zaveri, E., Grogan, D. S., Fisher-Vanden, K., Frolking, S., Lammers, R. B., Wrenn, D. H.,Nicholas, R. E., 2016. Invisible water, visible impact: groundwater use and Indian agriculture under climate change. *Environmental Research Letters*, 11(8), 084005.

Chapter 14

Root causes of groundwater contamination for analyzing resource management and policy framework

Navneet Munoth and Chandra Prabha
Maulana Azad National Institute of Technology

CONTENTS

DOI: 10.1201/9781003247869-17

14.1 INTRODUCTION

Water is a single entity upon which survival, sustenance and preservation of all the species depends. It is the life blood of our planet and its inhabitant. Its mere presence and need for every single organism states and presents a vital clue of its crucialness for everyone. We all require water; thereby, the water crisis will eventually become food crisis for all plants and animals (Robert, 2017). Tragically, on our planet, the fresh water is drying at an alarming rate. India has '4% of the water resources of the world' with 18% of the world population (Srivastava, 2017). Groundwater is portrayed as one of the world's genuine covered-up treasures. In times of drought this groundwater is playing key role in providing secure drinking water and vocational security. When we talk about accessibility of water supply, groundwater is the largest storage of water constituting 94% of all the fresh water supplies excluding ice caps and glaciers. Ground (soil) water is the main authentic available source of portable water which justifies a broad range of economic and environmental setup (Chevalking, 2008). Groundwater ensures the security of livelihood across the world, especially in countries where economy depends on agriculture. Soil water is the nation's principle reserve of drinking water and marks its representation of the benchmark for the future supply of water. It is the major contributor to the flow of streams and rivers, and thus holds strong influence on health and diversity of plants and animals. Groundwater and surface water are interconnected and interdependent in all ecosystems, and the demand of safe drinking water and its requirement to maintain and sustain health of these ecosystems are increasing (Agriculture, 2007). Groundwater consumption in India has drastically increased over the span of past few decades. Groundwater has become the backbone of the Indian Economy in terms of both food security and drinking water security. Groundwater is the major source of approximately 80% of the total drinking water needed. Rapid development of a number of groundwater extraction structures coupled with their expanding deliberation control has made a circumstance where groundwater is overused due to energetic asset accessibility. The widening gap between its availability and demand has often held the governance in precarious and problematic situations, which are the results of the misguided policies, inefficient bureaucracy and unenforced legislations. Water sector has also been affected due to the activities and decisions and developments outside the reach of the water sector leading to the current groundwater situation. Rechargeable annual groundwater potential of India has been overviewed at ground 431 bcm. It is evaluated that around 30% of the groundwater potential has been utilized. Punjab and Haryana have already used around 94% of their groundwater resources (Chatterjee, 2014). Groundwater tables are weak in Gujarat, Rajasthan, all states of the Deccan, and most of western Uttar Pradesh. India faces a progressively critical circumstance where its limited and delicate water assets are draining, whereas its sectorial requirements are developed quickly. So, because of improvements in the Indian economy and infrastructure, the demand of water is increasing rapidly. Consequently, the utilization and management of groundwater is playing a key role for our future survival.

14.2 CHALLENGES AND HURDLES IN GROUNDWATER RESOURCE MANAGEMENT

14.2.1 Unequal spatial and temporal distribution

In groundwater resource management, there is a large spatial variation in the availability of groundwater in India. It accounts to $398 \times 10^9 m^3$, which translates to 120mm for the country. In spite of having stores of freshwater resources as compared to other nations, India is still marked as a 'water-stressed region' as per the current utilizable freshwater guidelines. Evaluation of spatial and transitory classification in quality of groundwater in an alluvial rustic plain in local and fitting groundwater for temperate farming and people's right to water structures and drinking purposes is being done (Tlili-Zrelli et al., 2018). The difference in level is such that it changes from 500mm in the northern alluvial zones to less than 50mm in the nation's western part as shown in Figure 14.1.

As per the information given in Central Ground Water Board 2012, more than 60% of the accessible assets are limited to the northern territory which covers 20% of the topographical zone. The Unequal Spatial distribution and accessibility of assets shifts in wide extend. Decreased groundwater asset is predominately amid the period of (April to June) particularly in the peninsular locale as a result of the nearness of the difficult rocks within the locale. Water, being an important asset to the lives and jobs of all Indian citizens and has essentially contributed to India's wealth-related development since autonomy, the water litters have long pleaded in India's advancement plans. The complete overflow locale grew fourfold from 23 million ha in 1955 to about 90 million ha in 1959, with a large asset commitment to the water infrastructure since the autonomy of 5.9 billion rupees, providing fully comprehensive and stabilizing benefits to the millions of farmers across the country. Apart from these, significant achievements have been realized in the field of drinking water. Around 85% of Indians are in need of a water supply, and 75% of the country's population is depended on open water sources. More than 1,000 million persons in India (about 70% of the total population) lives in water stressed region. In all sections, in any case, there is significant scope for the conservation of water assets. Such preservation can be a reality, when clean assessment of water is done constantly, especially within the agricultural division (Table 14.1).

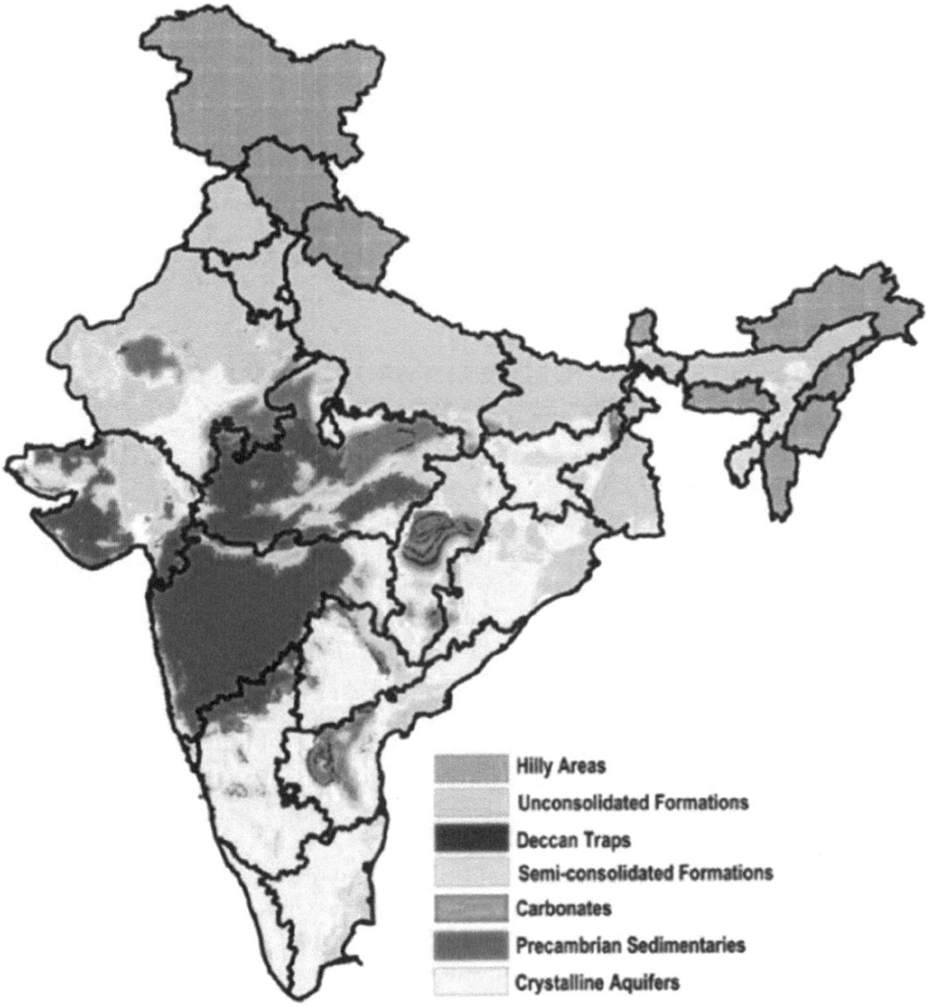

Figure 14.1 Major hydrogeological units in India. (Source: Saha and Ray, 2019.)

14.2.2 Dependency on groundwater

Low accessibility and openness of drinking water is itself a challenge in numerous parts of India, which inevitably applies weight and reliance on groundwater. Subsequently, the dependency on groundwater in last few decades driven to 80% of the available groundwater in the country. India consumes approximately 230 km^3 of groundwater per year, making it the world's biggest consumer of groundwater (Grafton and Chindarkar, 2018). Our focus is on how climate change will affect methods of utilizing water and how change in people's behavior and the physical changes in climate will modify both crop water necessities and accessible water resources (Zaveri et al., 2016).

Table 14.1 Sector-wise projected water demand in India (Press Information Bureau 2013)

Sector	*Water Demand in BCM (billion cubic meter)*					
	2010		*2025*		*2050*	
	High	*Low*	*High*	*Low*	*High*	*Low*
Irrigation	543	557	561	611	628	807
Drinking Water	42	43	55	62	90	111
Industry	37	37	67	67	81	81
Energy	18	19	31	33	63	70
Others	54	54	70	70	111	111
Total	694	710	784	843	973	1,180

Source: (ADRI), A. D. (n.d.).

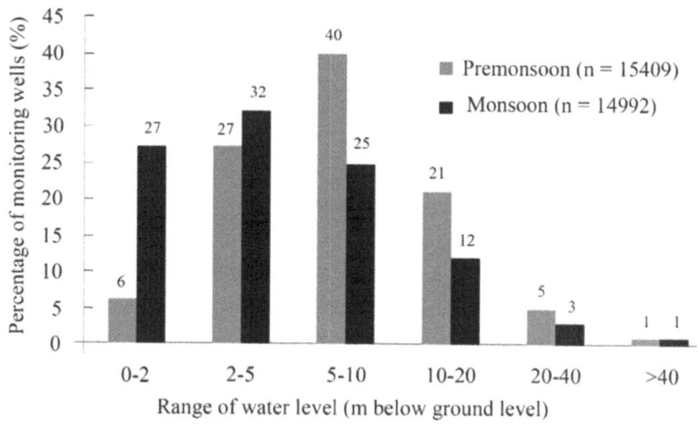

Figure 14.2 Comparison of occasions of water levels in several ranges amid pre-monsoon and rainstorm periods. (Source: Saha and Ray, 2019.)

14.2.3 Over-exploitation

In India, overuse of groundwater resources has led to fall in level of groundwater deficiency in water supply, degradation in groundwater quality and inter-mixing of saline water in coastal regions. These components have genuine suggestions on the environment and financial conditions of the people. Approximately 50% of urban population and 85% of rural population are directly depended on the groundwater for their everyday needs such as drinking water, domestic activities, and irrigation purpose. The effect of over-exploitation of groundwater resources on its quantity and quality can be seen in numerous states; for example, in Maharashtra, the agrarian efficiency has downgraded (Anil and Pophare, 2014). India is the highest in the world in extracting groundwater, and it even surpasses the US and China. Out of 6,607 appraisal reports, in about 2,000 appraisal units, groundwater extraction has as of now surpassed 70% of the yearly accessibility (Figure 14.2).

Around 16% (i.e., 1,071) of the appraisal units have been categorized as over-exploited, and the signs of unsustainable extractions are as of now demonstrative with proofs that water levels are declining, wells are drying up, streams of waterways are diminishing, etc.

14.2.4 Groundwater recharge

In the regions of over-utilization, groundwater level is sub-optimal, and such zones accounts for the central, eastern central and northern eastern region of the nation. Such zones are denied of effective groundwater management system. Thus, State and Central agency have taken proactive steps towards water conservation. The area for artificial recharge has been identified based on post monsoon water level and long-term post monsoon water trend in most of the states and UTs. Large investment in groundwater management to improve the water security is done by government of India through various government schemes. National project on aquifer management (NAQUIM) launched by ministry of water resources, for mapping and managing the entire aquifer system in the country.

14.2.5 Groundwater contamination

Groundwater defilement has become a major challenge for guaranteeing supply of secure and clean drinking water; the need of provincial zones accounts for 85% and the need of urban zones accounts for 50%, which are being met by the groundwater. In such locations, open burrowed wells have been utilized as a major source of drinking water since the 1970s. Anthropogenic and geogenic contaminations are the principal source and causes of groundwater quality degradation. In recent years, aquifers are contaminated from various chemical constituents throughout the country. It gave rise to contamination of other pollutants such as arsenic, fluoride, nitrate, and saltiness, which has been detailed from groundwater of numerous parts of the nation. There are numerous causes of water contamination, and they can be classified into two common categories, namely, direct and indirect contaminant sources (Ask-An-Earth-Scientist, n.d.). Soils and groundwaters contain residues of agricultural activities (fertilizers, pesticides, etc.) and industrial wastes. Air contaminants are moreover determined from human hones (such as vaporous outflows from automobiles, industrial facilities, and indeed bakery shops). Apart from the above-mentioned categories, there are six main sources of water contamination: household effluents and sewage, industrial effluents, agrarian effluents, radioactive wastes, thermal contamination and oil contamination (Kumar, n.d.).

14.2.5.1 Sources of groundwater contamination in the Indian context

Nowadays, water scarcity is one of the major issues facing people of India. Untreated sewage is the biggest source of water scarcity and pollution in India. Around 80% of water bodies in India are highly contaminated. This can be particularly applicable to every aspect of the human settlement in their prominent region. The Ganges and the

Yamuna are the most important rivers in India (Samudranil, 2019). About 80% of the water has been used for household purposes, thus converting it to wastewater. In most cases, contaminated water is not treated properly, which leads to a severe depletion of surface-level potable water (Mohapatra et al., 2014; 2021; Mogili and Mohapatra, 2021). Contaminated water leaks through the surface. Estimated figures say that people with a population of more than one lakh generate about 16,662 million L of wastewater a day. Usually, 70% of the people in the cities use the facilities of sewerage system. About 33% of the wastewater from cities and towns are released into the Ganges.

The following are few important reasons for expanding levels of water contamination in India:

- Industrial garbage
- Improper homes in the farm sector
- Loss in water amount in streams in the plain regions
- Social and religious hones like dumping dead bodies, bathing, throwing wastes in water
- Oil spills from ships
- Acidic rainfall
- Global warming
- Eutrophication
- Inadequate industrial treatment of wastes
- Denitrification.

The concern in the context of India that has a tremendous impact on people's wellbeing is contamination of drinking water due to disposal of untreated sewage into freshwater ecosystem. Further, waterways like the Ganges, Yamuna, and Kaveri, streaming through exceedingly populated regions, are extremely contaminated (Kurunthachalam, 2013). Groundwater is a decentralized supply of drinking water for millions of provincial and urban households that cannot be overstated. Groundwater is safe from contamination when compared with surface water bodies (Shah, 2016). Water contamination is disastrous and life threatening.

This means undesirable changes within the chemical and physical properties of water are not favorable to all living beings who use water (Anju et al., 2010).

There are two fundamental aspects of toxic water:

I. Changing the qualities and quantities of materials which are carried by water.
II. Modifying the physical characteristics of water bodies.

14.3 IMPLEMENTATION AND COMPLIANCE

In India, the entire replenishable groundwater resource is surveyed as 431.89 km^3. Around 15% of this is used for drinking water, 6 km^3 is used for commercial purposes, and the remaining are utilized for agricultural purposes (Rakesh Kumar, 2005). So, the challenges for lawmakers and concerned agencies is to bridge the gap between policy making and effective implementation. Research conducted in the field of groundwater conservation is too limited to provide wealthy reasons and lessons about the implementations, success and failures of the plans implemented. The information available is not enough for execution at the grass-root level. The case studies of of Karnataka, Maharashtra and Gujarat is mentioned below. These are all important in numerous aspects of groundwater management and institutional arrangement where some issues have been highlighted with respect to the execution of plans at different state levels.

14.3.1 Command and control – a case of Bangalore, Karnataka

Karnataka is a drought-prone region in peninsular India, as it falls in the rain shadow region of the Deccan plateau. The zone is the most urbanized region of the states in India with 40% of peoples living in towns and cities. Bangalore is a quickly developing state capital pulling IT technicians from all over the country, but it has witnessed lots of issues pertaining to drought. As a result, the city is largely dependent on groundwater resources for the sufficient supply of water. The introduction of the Karnataka Groundwater (Direction for Security of Source of Drinking water) Act of 1999 alongside the Karnataka Groundwater (Regulation and Control of Development and Management) Act of 2011 paved the way to regulate and control the development and management of groundwater. Quality of potable water systems available is variable from location to location. Groundwater quality was assessed within the Markandeya stream basin for drinking water infrastructure purposes by collecting 47 open borewell tests in the post-monsoon period of 2008 (Ravikumar et al., 2010a). Gulbarga is a rapidly developing city. A GIS-based study of 76 locations was conducted on the basis of water quality issues and their drinking parameters (Balakrishnan et al., 2011). In accordance to regulation of groundwater extraction, there are three main aspects:

- Owners of all existing sorts of wells were required to apply for enrollments.
- Drilling companies were to enlist themselves and their apparatus sometime recently June 2013.
- Anyone needs to drill a modern borewell is required to obtain permission priorly.

Groundwater sources of Karnataka are classified as specific chloride (32.14%) and specific sulfate (100%) waters based on Cl and SO_4 concentrations. The bigger portion of the tests (64.3%) includes a place for C3–S1 water lesson, depressurized water with high saltiness and low sodium (Ravikumar et al., 2010b). Without fruitful campaigns to communicate the more extensive concept behind the sanctioning of the law, person as well as proprietors would find no individual advantage from enrolling their existing groundwater structures. Numerous houses too fizzled to get the need of water gathering and energizing of the groundwater for the more noteworthy community.

14.3.2 Main issues in Maharashtra

The state of Maharashtra is a great extent underlain by the Deccan trap basalts where groundwater abdicate depends on the rate of weathering locally. Groundwater tests appear quality exceedance in terms of chloride, hardness, TDS, and saltiness (Anbazhagan and Nair, 2004). The agriculture and allied activities contribute a few 13% of the state's wage but more than half of the inhabitants are within the provincial ranges. Maharashtra is controlling utilization of groundwater by few parallel laws in Maharashtra Groundwater (Controlling for Drinking Water Region) Act, 1993.

The Central Ground Water Board (CGWB) built 72 borewells down to a center on the profundity of 208 meters in the Nanded region in Maharashtra which is primarily underlain by hard rocks, specifically basalt (Madhnure, 2014). Most substantive arrangements apply within the zones where locale collector pronounced the genuine shortage influenced range based on the perceptions amid certain times of the year. It also focuses on removing of pumps, closing of wells, and disconnection of the power supply to achieve the objective of protecting groundwater for drinking. Fifty types of borewell water tests were conducted and analyzed to select the geochemical and physical quality of the groundwater framework (Wagh et al., 2016). The high-level committee was defined for exploring the operations. The usage disappointment advanced the state to rethink the policy making and the committee suggested to set up the groundwater Administration Specialist. The committee focused on the requirement for free information collection and water utilization debate to be carried out at community level, who shapes a critical portion of town self-governance framework in Maharashtra. Field base analysis depicted that there was a clear absence of social support for the legislation. There was lack of understanding of the law, its arrangements and procedural things, and endeavors from UNICEF to raise mindfulness of people's rights had regularly not been either executed or accomplished. As per the rules surrounded beneath the Maharashtra groundwater (Development and Management) act 2009, the district authority shall be assisted by gram Panchayat in sustainable management of groundwater sources and protection of drinking water sources. The Villages did not have satisfactory specialize understanding to connect the spending shortage to persistence withdrawal of water. By and large, it was appeared that the lawmakers and authorities alike have surrendered their commitments to act to the common long-term interest of all which the CGWA has no implies of pushing them.

14.3.3 Tending to water exhaustion through electricity in Gujarat

The State of Gujarat has expansive precipitation vacillations spatially as well as between a long time and decades, and a history of dry season. Water Demand Management, ensuring the optimum water utilization, water resource arrangements, and groundwater consumption, is the common concern in the megacities, and challenges abound worldwide and within nations to address and mitigate such basic water issues (Arfanuzzaman and Rahman, 2017).

The groundwater exhaustion was considered genuine in mid-1980 within the hard-rock eastern portion – was informed by the CGWA until three more were included to the list in 2012. Found by the western coast, salt water interruption to the aquifers is an increasing problem, and the characteristic levels of fluoride within the bedrock are major concerns. The legislator passed a formal groundwater preservation law in 1973, as the primary of all Indian States. It was nevertheless not marked by the Chief Serve due to fears with respect to its effect on the state's (at that point) 300,000 agriculturists. The 1970s law intents to inquire about elective ways of directing groundwater reflection. One portion of this took up the nexus between water and vitality.

An unmistakable share of the vitality utilization may be connected to agriculturist boring ever more profound wells and introducing more effective pumps. Expanding water pumping coupled with tall power appropriations took its toll on the funds of the state and its Power Board which was inevitably incapable to meet the requirements of the quickly developing request. The high-level political backing of Jyoti gram was 'necessary but not adequate to control the anarchy' on the ground: staff individuals were hesitant to venture into towns for fear of viciousness from angry swarms, they were regularly taken prisoner and kept in servitude (National Groundwater Management Improvement Program, 2016). As portion of the elective approach to governing groundwater, the Gujrat Water Resources Development Enterprise Ltd. was set up in 1975 and has been invested with certain administration obligation. The State Power Board instructing not to energize unused water system tube wells without consent from the groundwater specialist. Nearly half a million unused control associations for little and minimal agriculturists have in any case been allowed amid the past decade. A few lessons from power supply may be learned from Gujrat. It clearly appears that versatile learning and eagerness to continuously overhauling the framework was crucial. The 'lighted Village' approach is highlighted in twelfth Five year plan and within the national water approach, with the proposal that power isolation be rolled out in arrange to put a halt to free or intensely subsidized control supply. No other states in India, have so distant taken after Gujrat's endeavors, to control power supply as well as groundwater pumping.

14.4 METHODOLOGY FOR ANALYZING GWRM (GROUND WATER RESOURCE MANAGEMENT)

14.4.1 National ground water resource management

Understanding the main issues encompassing groundwater administration may be precondition for creating approach suggestions for both the national- and state-level groundwater administration (Theesfeld, 2009). The World Bank proposed NGMIP (National Groundwater Management Improvement Program) for the Government of India to reinforce GWRM within the chosen states to back the existing endeavors of the government to move forward groundwater administration. NGWIP centered on reinforcing organization competencies and trans-sectorial linkages between key groundwater organizations checking the CGWB, state divisions, and neighborhood practices such as PRIs. The Central government will strengthen by giving the qualified specialists in hydrogeology, geochemistry and ground water sciences enabled with made strides state groundwater enactment, rules and controls. Further, the activity was to create information of the status of the reflection and utilize real-time ground water observing and expanded aquifer mapping. Evidence-based and groundwater-based venture plans are created which are able to coordinate the significant segments and advanced community interest.

Thus, efficient interventions will be broadly focused on the protection of groundwater and surface water resources and the main points of organization, which incorporate supply-side and demand-side techniques and address the attributes of climate change (World Bank, 2016).

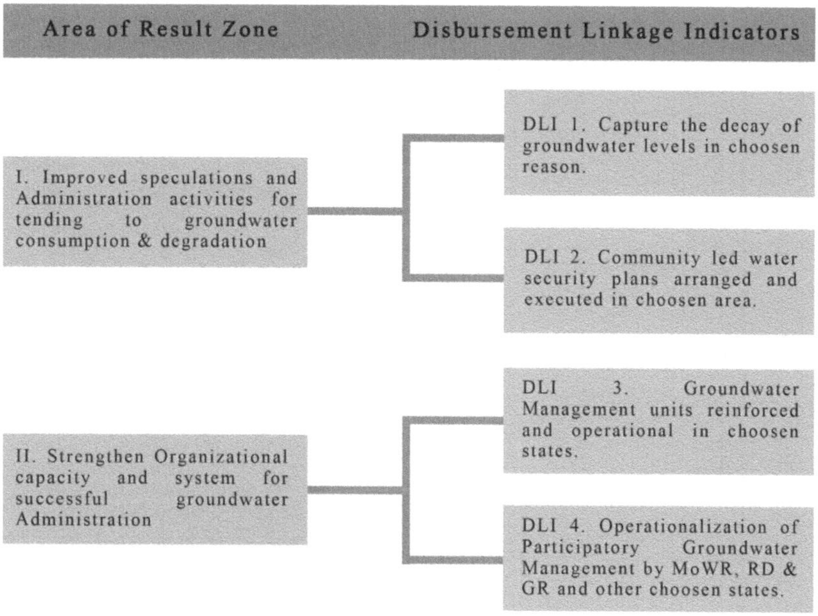

Figure 14.3 Result zones and disbursement linkage indicators (DLI) of National Groundwater Management & Improvement Plan. (Source: World Bank, 2016.)

To further the advanced nature of the program, a program for A – 'Outside Financing' has been proposed, which relies on the key's accomplishment and appears within the markers related to the distribution (Figure 14.3; Table 14.2).

14.4.2 Scope of the current program

Groundwater exploitation circumstance compels to grasp administration practices for the enhancement of financial groundwater resources. Groundwater conditions reflected through distributed writing have been talked about, explaining the reasons for scarcity, aquifer status, and groundwater development in India (Khare and Varade, 2018). The scope of the NGMIP was focused on the states that took interest in its execution, such as Rajasthan, Gujarat, Haryana, Maharashtra, Karnataka, the Bundelkhand region of both Uttar Pradesh and Madhya Pradesh, and the other four regions of western Uttar Pradesh. The specific states were the ones with the most intensely misused groundwater regions in India and stand up to genuine groundwater accessibility and quality issues that are anticipated to induce strongly in the future. CGWB has executed Groundwater Management & Regulation (GWMR), which could be a central segment plot of the services of water resource (Sekhar, 2017). The total of the states speaks to 25% of the entire number of over-exploited, basic and semi- basic squares in India. The program includes a course of activity, with unused states to be included within the execution period. The program was executed inside the states where majorly over-exploited blocks were first taken into consideration. Programs

Table 14.2 NGMIP natural and social risks - chosen organization viewpoint

Elements of present scheme	*Implications on the situation based*	*Link of NGMIP center and result zone areas*	*Risk levels*	*NGMIP recommended diminishing function*
1. Scheme for groundwater management and regulation (GWMR)				
3D Mapping and characterization of aquifers	Active program – low risks from installation of new monitoring stations which is construction related	Increase the no. of monitoring stations, comprehensive assessment and overall	Low	Build up environmental and social guidelines for minor civil works
Technological upgradation of monitoring system				
Develop aquifer management plans (AMP)	Limited progress to date and limited modality for public participation	Strengthen attention to broader and more comprehensive GP water security plan (Result area 1)	Low to moderate	N/A as water security plans directly reduce risk and increase public participation
Implement Amps through regulatory and non-regulatory means	Some progress to date in a few stales on regulatory approaches despite fundamental gaps in national and state law and policy, equity and access questions and some successes on PGWM at pilot scale	Provide significant support for progressive states to take important steps at improving groundwater management and reverse decline in groundwater level (Result areas 1 and 2)	Moderate	Set of recommended studies and activities to improve get to, equity, community participation and decision-making at the appropriate level and build on fruitful PGWM pilots in India
Advance participatory groundwater management				

(*Continued*)

Table 14.2 (Continued) NGMIP natural and social risks - chosen organization viewpoint

Elements of present scheme	*Implications on the situation based*	*Link of NGMIP center and result zone areas*	*Risk levels*	*NGMIP recommended diminishing function*
2. Master plan for artificial recharge to groundwater in India				
Increase groundwater supply via small-scale artificial recharge structures and rainwater harvesting	Active program with national and linked state funding. Environmental and social screening weak though more of a concern at cumulative level given small-scale structure	State-by-state listing in 2013 provides candidate recharge structures for support under the NGMIP (DLI 1) with potential for master plan to be integrated into GP water security plans (Result area 1)	Moderate to Substantia l	Strengthen environmental screening (including studies on downstream river flows and river morphology) and maintenance of structures, improve social management processes through better public participation and accountability, state-specific studies
3. Model bill to regulate and control the development and management for groundwater, 2005				
Registration of existing users. Regulate and control the development and management of groundwater in notified over-exploited blocks (control drilling of new water wells and pumping from existing wells)	Only Maharashtra and Karnataka have developed and enacted these regulations. Enforcing these regulations is challenging due to the presence of millions of wells in every state	Provide support to other participating states (Haryana, Rajasthan, Gujarat and Madhya Pradesh) to develop and enact similar groundwater regulations (Result area 2)	Substantia l	Community-based management for enforcement of the regulation of constraining the scope of control to only few endangered aquifers

Source: World Bank (2016).

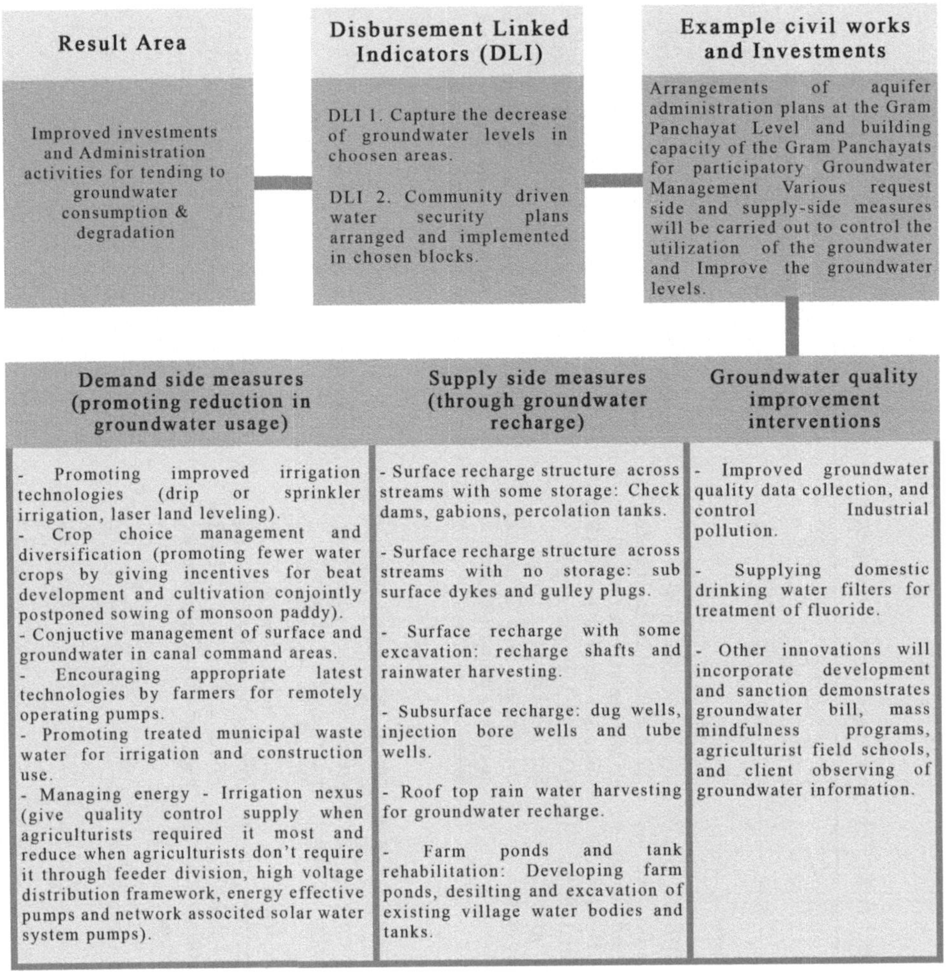

Figure 14.4 NGMIP measures – chosen organization viewpoint. (Source: World Bank, 2016.)

were implemented in the states where majorly over-exploited blocks were first taken into account.

Figure 14.4 shows the typology proposed by the states for civil work and investments under the program.

14.4.3 Groundwater management at the grass-root level

Groundwater management and governance as a lawful tool for the overuse of groundwater remained largely as a periphery issue as the number of users and approach formulation were near to impossible in the first place. The potential opportunity to address this situation to the depth of ground water crisis required the steps needed to intervene right at the grass-root level where decentralization system and the authority is needed to monitor and execute the plan to have any chances of making the program a success. Groundwater is a basic resource for agriculturists, particularly in dry and semi-dry regions where the precipitation could be the most non-reliant phenomenon. In India, 'post green revolution water system' has energized water system farming at first through surface water dams and then through groundwater (Jadeja et al., 2015). Gram Panchayat is one institution that fits in perfectly to the criteria of decentralizing the system and monitoring the process and delegating it right to the point.

In expansion, the decentralization action specified the Gram Panchayat to be capable for village area water supply, environment, agribusiness, and water system in the case at a regional scale, Gram Panchayat can be unified at a minor basin and assurance, it can surrender great outcomes almost on the ground. Specific functions are need to be discussed in groundwater management, the nodal organizations have a significant frame in that scenario. Facilitated endeavors between Gram Panchayat and policy divisions on the preface of guidelines of subsidiarity finished up a perfect way onward. The model of decentralized groundwater acts and regulations addresses the capacity of building on the basis of physiography- Wastes, hydrology, aquifer, and other common frameworks. Out of the many states underlain by hard-rock aquifers, Andhra Pradesh has faced a great amount of groundwater depletion in previous decades. The APFAMGS (Andhra Pradesh Farmer-Managed Groundwater Infrastructure) project has taken a gentle approach related to this issue. The primary consideration of APFAMGS is that when the customer achieves its presence, duration, and compulsion, a conservative framework of groundwater is achievable. Farmers are engaged in these projects to collect information and analyze the data to achieve this target. Hence, an understanding will develop about the status and framework of groundwater. In reality, farmers with limited education capabilities illustrate their potential for collecting and analyzing rainfall data and groundwater data, controlling their annual water use based on accurate cropping patterns, and participation in agricultural water training. These government sponsored awareness programs increase their knowledge of agricultural practices.

14.4.3.1 Recommendation for mitigating the environmental and social risks

- Potential investment categories for betterment of groundwater management could incorporate the construction of dams and new large scale irrigation system & treatment of waste water collection, treatment and recharge systems.
- Procedure for training and capacity building along with periodic impact assessment at state level.
- Induction of staffs with numbers and competencies for the state- and national-level projects.
- Screening of sites for both environmental and social issues and focus on site-specific impacts.
- Adequate budgeting and arrangements for maintenance of recharge structures.
- Formulation of district-level committees for addressing grievances related to program interventions.

- Support local authorities and focus on intensive capacity development on groundwater management.
- Recognizing and working on the institutional reforms and removing the conflicts.

14.5 CAPACITY IDENTIFICATION AND ASSESSMENT OF THE PROGRAM

14.5.1 National-level assessment

14.5.1.1 Adequacy of administrative framework

There is an urgent need for strong laws in India which can provide proper water-sharing platform to avoid interstate water conflicts. Local bodies like Gram Panchayat are assigned with the role to organize and maintain proper budget for water resources in order to maintain clear ground on which interest of different group of peoples can be looked after (Article 243 G). In right to fair compassion and transparency in land acquisition, rehabilitation and resettlement act 2013, procedures and rules for granting compensation, rehabilitation and resettlement are laid down so securing lands is not an issue for the practices coming under CGWB. The above mentioned act provides that wherever multidrop irrigated land is acquired and equivalent area of cultivable wasteland shall be developed by the state for agricultural purposes.

14.5.1.2 Adequacy of regulatory and organizational environment

The union and the state have different levels of responsibilities for the organization and management of groundwater. Importance has been given to different regions for groundwater management, but proper coordination and official supervision are required. Full commitment is not shown by many states for managing their groundwater resources. In many instances, organizations committed to groundwater lacks understanding, capacity requirement, marginalization, and obsolete orders that prioritize further improvement of observation and efficient management of resources.

14.5.1.3 Performance in implementation and support in groundwater program

CGWB (Central Ground Water Board) is responsible for planning aquifer management plans, evaluate assets, and screen quality and amount across India. CGWB is frequently observing groundwater quality on a territorial scale through an organization. The groundwater information and data created are made accessible through the Internet-Enabled Ground Water Data Framework (WEGWIS). CGWB collects data, maps information, and given investigation results to the states for educating them about feasible arrangement and management of GW resources. A few controls of industrial supply wells are input in few states, but compliance and authorization are generally powerless. There has been a paradigm shift from groundwater development to groundwater management over the last decades.

14.5.1.4 *Institutional capacities to address environmental and social issues*

I. **Environmental screening of program interventions:** Preparation of a three-page checklist for screening has been done by 'CGWB' which is considered as its main blueprint for picking up appropriate revival structure. In this screening handle, habitual access immersion, disintegration of groundwater quality, and open support need to be taken into account. CGWB and implementing organizations don't have environmental and social issues in arranging and planning the program interventions.

II. **Social incorporation:** While the system of authorities includes inbuilt security measures to ensure that helpless categories are to be considered within the arrangements as well as execution procedure, there are no components to track or there is no existing declaration framework, no existing management information system, the narrative reports which are under the Point Project Report (PPR), are vulnerable to manipulation.

III. **Participatory planning, mapping and monitoring:** The stake holders include farmers, landless farmworkers and women representatives. Consensus-based decision making is the goal of local water users association (WUA). A national-level independent agency is entrusted for evaluation of the projects. National aquifer mapping and management programme commenced from year 2012. Four major components are the following:
   1. Mapping of aquifers, resource, quality, recharge potential
   2. Groundwater modelling and management plan
   3. Dissemination of data
   4. Participatory management of the resource

IV. **Agriculturists participatory action research program:** However, the free assessment is to be done, but the internal evaluation and assessment of CGWB in these schemes have allowed agriculturists to use water resources and expand their earnings. Community organizations of Panchayati Raj system, Rural officers and other Nobel do not participate successfully in the mapping and tracking of aquifers, and also administrative bodies do not have the capacity and professional who can provide the proper platform to channelize and mobilize different groups of the society. The Rajiv Gandhi National Groundwater Preparation and Investigation Establishment (RGNGWTRI) is a nodal organization for conducting training for state and central government officials and providing training to groundwater professionals and sub-professionals in groundwater investigations, development and management techniques.

14.5.2 State-level assessment

14.5.2.1 *Adequacy of administration framework*

A group of experts on ground water management and ownership in the planning commission have considered environment good enough for the management of the ground water resources:

- Sufficient government activities held by Union and State have got the obligation to ensure groundwater against over-the-top misuse.
- Central groundwater authority has been constituted under section 3 (3) of the environment protection act 1986 to regulate and control development and management of ground water resources.
- Implementation limitations related to administrative measures stated the need for community engagement and management of resources with adequate technical support.
- Advocacy in contention to replenishment of specific groundwater blocks, with respect to command and control as per the arrangements of environment Act announcing it as an Environmental Threat.

14.5.2.2 Adequacy of the legislative system for environmental evaluation

According to the Pollution Control Board notice, 2015, the development of modern projects or existing projects requires first environmental clearance based on the EIA plan. In 2006, the notice under the Environment Protection Act 1986 examines the activities of central and state governments' projects, on the basis of which environmental clearance was given. In NGMIP centers, the requirement of formal EIA report is rare. In any case, surveying the total impacts from any concentration of structures needs to focus on consideration in each state.

14.5.2.3 Adequacy of an administrative system for social safeguards

According to the state water policy, there is a system of ensuring equal spread, assurance of interest of the helpless people, participatory system, observation of the community, progressive governance and transparency, etc. All states give social considerations, decentralized participatory development, and also checks and balances. Community-based organizations coming under the irrigation acts, farming acts, rehabilitation acts and decentralization act are made strong.

14.5.2.4 Adequacy of regulatory and organizational environment

Groundwater management, for the most part, situated at a moderate range within the state's hierarchy, is focusing on one of the most water planning departments for water systems or water supplies. The structure, function and staffing of groundwater organizations fundamentally accommodate long-standing orders to study and build groundwater assets, and customers are not given any consideration and arrangement for financial measurements of groundwater use. The management of resource capabilities still receives exceptionally inadequate human resource and funding compared to the study, observation and advancement of the groundwater resources.

I. **Performance in implementation and support icon groundwater programs:**
   States do not have enough staff in the groundwater divisions to facilitate the subsidy to other important organizations to monitor compliance by groundwater users and to get grants to meet power connections and power supplies.

Excessive water shortages in notified areas make it politically unsafe for direct extraction for residential and water system purposes – this may be one of the reasons behind the high number of licenses granted for modern borewells. In addition, the framework for mapping and enrollment of existing users exists but is not comprehensive for users or for offices offering boring services. Due to the absence of viable participation of neighborhood bodies, general management is powerless for both top-down and bottom-up.

II. **Institutional capacities to address environmental and social issues:**
Key technical issues are abandoned staff positions, lack of a framework to recruit experts and organizations, a limited number of research facilities, adhoc trainings, social and environmental capacity in any of the participating states. While various schemes promote participatory water management, little speculation has been put into creating the capacity of associations to handle the administration and support of these schemes and water bodies. Most organization challenges are created in the absence of a clear governance structure, limited inter-agency coordination, covering capabilities, imparting single education, and managing groundwater areas.

14.5.3 Assessment of the program

National and state governments have a well-developed environmental act, but the project proposed under the policy and acts does not demand any environmental clearance from state-level EIA officials.

Environmental staff in state-level implementation organizations are suggested to increase the capacity of the environmental framework, to deal with the impacts on environmental programs. Environmental programs are community-based approaches toward improvements of groundwater and focused on the water supply and demand.

14.5.3.1 Benefits for potential environment

In the NGMIP program, state participation will take the significant process to mitigate all harmful impacts on the environment and advance resources of groundwater. So, the benefits of disbursement linkage indicators (DLI) for the schemes are as follows:

a. Improvement in sustainable good yields within areas where over-development has drained aquifers.
b. Increasing the groundwater level will reduce the cost of pumping and energy savings, and will also cut greenhouse gas.
c. Better agricultural yields for each unit of water applied, and hence the socio-economic condition of the local people improved.
d. Possibility of better groundwater quality due to dilution.

14.5.3.2 Potential environment impacts and risks

Based on the team's evaluation, and as shown in the table, in relation to environmental and social risks, the environmental risks of the program are typical, 'low-medium', but the type of investment is considered significant. The risks and impacts related to the individual investment of the program may be 'low to medium', including the general effects of small-scale civil work activities as well as the effects related to lack of maintenance, the cumulative effects of an investment in greater watersheds and it just maybe more negotiable.

14.6 GAP IDENTIFICATION IN GROUNDWATER MANAGEMENT

14.6.1 Legal gap in groundwater resource management

Controlling groundwater with the first basic regime has not changed since the 19th century, restoration of basic regulations was needed as it played a limited role, and there has been no change since the 1970s. Governance of control and access to groundwater in India is derived from English Affairs (Soman, 2008). Hence, simple statements hide more complex situations due to the fact that the statutory mention of Indian Statutory Act 1882 is found to be the groundwater right. The limitation of the groundwater allocation rule led to a dramatic increase in groundwater exploitation, eventually resulting in the failure of water tables in many regions of the country. It became a crisis before the Brief Model Bill 1970/2005 made sense of the need to address an alleged issue, and it was welcomed in 1970 but was not addressed as the most important. Problems were identified, and the same have been highlighted since then. The bill talked about the control over groundwater and land ownership, but eventually, it became completely unable to propose a regulatory scheme in favor of regulation at the groundwater level.

14.6.2 Policy-related issues pertaining to groundwater management

Groundwater improvement for water systems has been recognized over the past decade as having the potential to make a major contribution to rural poverty alleviation. There is considerable pressure to meet developmental needs and increase groundwater exploitation as a mechanism for poverty. At present, the groundwater water system usually has half the water areas of India (Moench, 1996). According to the reviewer, the gaps in the policy related to groundwater management are shown in Figure 14.5.

1. A wide range of essential services depend on groundwater including irrigation, municipal and industrial water supplies and environmental maintenance	2. Pressure on groundwater resources are growing and depletion, quality and pollution problems are emerging as a result	3. Basic data on the nature of groundwater resources and on the extent and characteristics of problems are lacking	4. Physical management options are often available the most critical issues lie in the development of institutional arrangements through which responses could be implimented	5. Tensions exist between current efforts to utilize groundwater as a tool for poverty allevitation and sustainable management of the resources

Figure 14.5 NGMIP – gaps in policy – chosen organization viewpoint. (Source: World Bank, 2016.)

14.7 COUNTER- AND RETROSPECTIVE MEASURES

Counter-measures for groundwater management revolve around ideas and strategies which could be planned well and executed for the better outcome and maintenance and focus on groundwater resource management (Tables 14.3 and 14.4).

14.7.1 Promoting recharge

- Safeguarding natural recharge capacity.
- Value of wetlands in strong groundwater levels.

Table 14.3 Regulating ground water use: main policies affecting price incentives

Main policies affecting price incentives	
Trade policies	Regulating of export and import of high and low water consuming crops
Public investment policies	Public investment in water resource infrastructure: – Surface water storage and diversion – Groundwater extraction facilities – Water conservation measures – Recharge and reuse facilities
Crop pricing policies	Pricing control on different crops Government procurement of agricultural crops
Agricultural input policies	Pricing and policies on availability of major agricultural inputs, such as seeds, fertilizers, pesticides and tractors
Energy pricing	– Pricing of electricity and diesel (including control on black market or pilferage) – Effective collection of bills structure of energy pricing (flat rates or consumption-based rates)

Table 14.4 Regulating ground water use: main policies affecting non-price incentives

Main policies affecting non-price incentives	
Rural and peri-urban infrastructure policies	Public investments in roads, drinking water support, agro-processing and link to water management
Groundwater governance and regulation	Implementation of water law and complementary regulation
Policies supporting local regulation	Sharing of information on aquifer characteristics, effectiveness of awareness and education program, policies on land and water rights, particularly as they relate to runoff rights and to groundwater use
Policies affecting water markets	Availability of group or individual credit, rules on right of way, promotion of high-capacity well systems, direct promotion of water sharing
Policies on crop diversification	Attention to crop choices and water conservation in agricultural research, extension and demonstration programs

Source: National Groundwater Management Improvement Program (2016).

- Reactivating the flood plains.
- Focus over wet watershed management.
- Porous pavements.
- Diversion of storm from the sewage system.
- Flood water disposal and recharge.
- Managing aquifer recharge: percolation tanks and leaking ditch irrigation.
- Establishment of groundwater management organization.

14.8 CONCLUSIONS

Considering their vital significance, the management of India's groundwater resources is severely required for improvements. The resource improvement happens generally within the private field and without effective specialist control. At the policy framework level, there is no void of investigation ponders and master reports with uncommon center on the circumstances: the awareness is high that huge parts of the nation are already seriously focused with distressing future under a trade as a regular worldwide. However, there is inadequate activity beyond the reports and arrangements. Groundwater plays a central role in India and has been linked in increasing exploitation of the resource over the past few decades. In the field of groundwater conservation, government laws and policies have very bad reputation. The anticipated resistance from landowners, particularly agriculturists, against confinements of their right to theoretical boundless volumes of water from wells contributes to an unwillingness among lawmakers to indeed propose changes to the prevailing arrangements. There is lack of concern among policy-makers, legislators and chairman charged with obligation of usage. Limited man power, administration, and technical and financial capacity remain to be the major road blocks for every initiative in the run for ground water resource management as this could be eventually the primary source of drinking water for our future generation taking into consideration the use of our primary and current resource of fresh water and its conditional status. Not to mention the basic and primary movement prior to all the advancement is the need and importance of creating an awareness in regard to groundwater resource management without which all the initiatives and plans are incomplete from the get go. Progressed understanding of hydrology is assumed to lead us to organize where the framework is fundamentally modified and the recognition of groundwater as a shared, common property enters activities and decision-making at all levels. Looking for the participation of able administration on constrained sentiments of the community and progressed attempt to alter in demeanor toward control for the common. Behavioral alter ought to be based on a 'leading by example' approach where high-ranking authorities and lawmakers grandstand how they comply with the appropriate arrangements. Only when the requirements for change are internalized and interpreted into modified conduct at the best can law become a practical apparatus for change toward a society that takes its groundwater resources genuinely.

REFERENCES

(ADRI), A. D. (n.d.). *India Water Facts*. Retrieved from Asian Development Research Institute: https://www.adriindia.org/adri/india_water_facts

Agriculture, U. S. (2007). *Techinical Guide to Managing Ground Water Resources.* United States Department of Agriculture, Forest Services, United States of America.

Anbazhagan, S. and Nair, A.M. (2004). Geographic Information System and groundwater quality mapping in Panvel Basin, Maharashtra, India. *Environmental Geology*, 45, 753–761.

Anil M., and Pophare, B. L. (2014). Impact of over-exploitation on groundwater quality: A case study from WR-2 watershed, India. *Journal of Earth System Science*, 123, 1541–1566.

Anju, A., Ravi S. P. and Bechan, S. (2010). Water pollution with special reference to pesticide contamination in India. *Journal of Water Resource and Protection*, 2, 432–448.

Arfanuzzaman, M., and Rahman, A. A. (2017). Sustainable water demand management in the face of rapid urbanization and ground water depletion for social–ecological resilience building. *Global Ecology and Conservation*, 10, 9–22.

Ask-An-Earth-Scientist, G. (n.d.). *University of Manoa.* Retrieved from https://www.gdrc.org/uem/water/water-pollution.html

Balakrishnan, P., Saleem, A., and Mallikarjun, N. D. (2011). Groundwater quality mapping using geographic information system (GIS): A case study of Gulbarga City, Karnataka, India. *African Journal of Environmental Science and Technology*, 5, 1069–1084.

Chatterjee, K. (2014, January 31). *Water resources of India.* Retrieved from Climate Change Centre: http://climatechangecentre.net/pdf/WaterResources.pdf

Chevalking, S. K. (2008). Ideas for groundwater management. *MetaMeta and IUCN*, 1–144.

Grafton, Q. and Chindarkar, N. (2018). India's depleting groundwater: When sciencemeets policy. *Asia & the Pacific Policy Studies* 6(1), 108–124.

Jadeja, Y., Maheshwari, B., Packham, R., Purohit, R., Thaker, B., Goradiya, V., Oza, S., Dave, S., Soni, P., Dashora, Y., and Dashora, R. (2015). Participatory groundwater management at village level in india – Empowering communities with science for effective decision making. Paper presented at the Australian Groundwater Conference 2015, November 3-5, 2015, Canberra, Australia.

Khare, Y.D. and Varade, A.M. (2018). Approach to Groundwater Management towards Sustainable Development in India. *Acque Sotterranee-Italian Journal of Groundwater*, 308, 29–36.

Kumar (n.d.). *Six main sources of water pollution.* Retrieved from Environmental Pollution: Direct sources include effluent outfalls from factories, refineries, waste treatment plants etc.. that emit fluids of varying quality directly into urban water supplies. In the United States and other countries, these practices are regulated, although thi

Kurunthachalam*, S. K. (2013). *Indian Waters: Past and Present.* Retrieved from Hydrology Current Research: https://www.hilarispublisher.com/open-access/indian-waters-past-and-present-2157-7587-S10-001.pdf

Madhnure, P. (2014). Groundwater exploration and drilling problems encountered in basaltic and granitic terrain of Nanded district, Maharashtra. *Journal of the Geological Society of India*, 84(3), 341–351.

Moench, D. M. (1996). *Groundwater Policy: Issues and Alternatives in India.* Colombo: International Irrigation Management Institute.

Mogili, N.V. and Mohapatra, S., 2021. Microbial community analysis of wastewater during blackwater treatment. In *Wastewater Treatment Reactors* (pp. 483–507). Amsterdam: Joe Hayton.

Mohapatra, S., Sargaonkar, A., and Labhasetwar, P.K., 2014. Distribution network assessment using EPANET for intermittent and continuous water supply. *Water Resources Management*, 28(11), 3745–3759.

Mohapatra, S., Sharma, N., Mohapatra, G., Padhye, L.P. and Mukherji, S., 2021. Seasonal variation in fluorescence characteristics of dissolved organic matter in wastewater and identification of proteins through HRLC-MS/MS. *Journal of Hazardous Materials*, 413, 125453.

Rakesh Kumar, R. J. (2005). Availability and management of surface and ground water resources in India. *Current Science*, 89, 794–811.

Ravikumar, P., Somashekar, R. K., and Angami, M. (2010a). Hydrochemistry and evaluation of groundwater suitability for irrigation and drinking purposes in the Markandeya River basin, Belgaum District, Karnataka State, India. *Environmental Monitoring and Assessment*, 173, 459–487.

Ravikumar, P., Venkatesharaju, K., Prakash, K. L., and Somashekar, R. K. (2010b). Geochemistry of groundwater and groundwater prospects evaluation, Anekal Taluk, Bangalore Urban District, Karnataka, India. *Environmental Monitoring and Assessment*, 179, 93–112.

Robert, C. (2017, August 25). *Why is Conensation Important?* Retrieved from Sciencing: https://sciencing.com/why-do-we-need-to-conserve-water-4855101.html

Saha, D., and Ray, R. K. (2019). Groundwater resources of India: Potential, challenges and management. In: *Groundwater Development and Management*, P. K. Sikdar (ed.) (pp. 19–42). Cham: Springer.

Shah, M. D. (2016, February 12). *Groundwater Pollution and Contamination in India*. Retrieved from indiawaterportal: https://www.indiawaterportal.org/sites/indiawaterportal.org/files/ground-pollute4_FULL_.pdf

Samudranil. (2019, February 2). *Water Pollution in India; Cause, Effects and Solutions*. Retrieved from My India: https://www.mapsofindia.com/my-india/education/water-pollution-in-india-causes-effects-solutions

Sekhar, M. (2017). *Evaluation of Scheme of Ground Water*. Central Board: Central Ground Water Board.

Soman, N. S. (2008). Legal regime of underground water resources. *Cochin University Law Review*, 147, 150.

Srivastava, D. S. (2017, May 11). Water: Its Conservation, Management and Governance. 25. Retrieved from National Water Academy: https://nwa.mah.nic.in/resources/downloads/NWA_Compilation_WR.pdf

Tlili-Zrelli, B., Gueddari, M., and Bouhlila, R. (2018). Spatial and temporal variations of water quality of mateur. *Journal of Chemistry*, 2018, 2408632.

Theesfeld, I. (2009). Institutional challenges for national groundwater governance: Policies and issues. *Groundwater*, 48, 131–142.

Wagh, V. M., Panaskar, D. B., Varade, A. M., Mukate, S. V., Gaikwad, S. K., Pawar, R. S., Muley, A. A., and Aamalawar, M. L. (2016). Major ion chemistry and quality assessment of the groundwater resources of Nanded tehsil, a part of southeast Deccan Volcanic Province, Maharashtra, India. *Environmental Earth Sciences*, 75, 1–26.

World Bank, W. (2016). *National Groundwater Management Improvement Program*. India: Government of India.

Zaveri, E., Grogan, D.S., Fisher-Vanden, K., Frolking, S., Lammers, R.B., Wrenn, D.H., Prusevich, A., and Nicholas, R.E., (2016). Invisible water, visible impact: Groundwater use and Indian. *Environmental Rezsearch Letters*, 11, 084005.

Chapter 15

Pesticides and fertilisers contamination of groundwater

Health effects, treatment approaches, and legal aspects

Manish Yadav
Central Mine Planning and design Institute Limited

Nitin Kumar Singh
Marwadi University

Neeraj Kumar Singh
Central Mine Planning and design Institute Limited

Tushali Jagwani
Malviya National Institute of Technology

Suman Yadav
APS University

CONTENTS

DOI: 10.1201/9781003247869-18

15.1 INTRODUCTION

In the recent decades, intensive farming and industrial agriculture along with limitations on expanding the land for crop production necessitated the increased application of agrochemicals (Wolpert et al., 2020). These agrochemicals include fertilisers, pesticides, liming or acidifying agents, insecticide, fungicide, and soil conditioners (Tripathi et al., 2020). The application of such chemicals is reported to be helpful in plants' growth and nutrition, disease control, and higher yield of crops (Ab Rahman et al., 2018). In spite of these benefits, it is also proven in several studies that excessive use of agrochemicals had a harmful effect on human health and environment (Cimino et al., 2017; Sankhla et al., 2018, Singh et al., 2018). The use of agrochemicals simultaneously introduced the several toxic contaminants in the soil, water, or air environment (Jallow et al., 2017). The role of pesticide is to not only kill pests but also to destroy the non-pest population and disturb the microbial flora of soil, supporting the growth of plants and conservation of ecosystem (Bambaradeniya & Amerasinghe, 2004). Moreover, the long-term usage of pesticides not only causes imbalance of the microbial population but also causes extinction of various microbes present in the soil. On the other side, heavy application fertilisers may also have ill effect on environment, especially on the structure of soil (Riley et al., 2008). More specifically, hardpan is formed when fertilisers leach out into sub-soil and react with the various components of soils. Due to formation of hardpan, downward movement of water is hindered and root growth of plants is affected significantly. Another known ill effect of fertiliser's application is soil friability in which chemicals inhibit the formation of soil crumbs, formed from humus and soil organic matter (Barak et al., 1997). Like pesticides, fertilisers also affect the growth of microbes and mycelium which intact the properties of soil required for the plant growth. Other than the soil environment, another major environmental concern is the groundwater pollution by pesticides and fertilisers as they consist of poisonous chemical, namely, nitrate, cyanide, lindane, Malathion, sulphate, etc. Leaching of chemical agents from pesticides and fertilisers into ground and surface water sources causes serious health problems. The release of such chemicals has been a growing concern especially in terms of drinking water contamination (Arias-Estévez et al., 2008). The presence of these chemicals in water environment led to the disturbance in normal metabolic activities of living beings, if entered in biotic system through food chain. If a person gets exposed to these chemicals over certain period of time, then it may result into some acute effect even at very low level of contaminants. Such observations are mainly associated with release of toxins which can alter the hormonal and developmental growth (Biswas et al., 2014). Besides, application of uncontrolled and haphazard way of agrochemicals also reflected in the form of negative effects on various life forms and environment. Most of these compounds are non-biodegradable in nature and can easily enter in our food chain (Mandal et al., 2020). Reactive oxygen species generation and haemoglobin arrest in blood are among the reported issues which can be linked with intake of nitrate contaminated water (Chabot et al 1998). Another reported issue is leaching of agrochemicals into groundwater and water runoff in lakes and ponds which increases the growth of algae which have an antagonistic effect on aquatic animal. Agrochemical causes a serious threat to human health due to its high biological stability and lipophilicity (Rani & Dhania, 2014). Most of these chemicals inhibit the functioning of nervous system by retarding the activity of enzyme acetylcholinesterase. The toxicity of pesticides and its sensitivity towards organisms mainly determine the extent of damage caused to living beings (Yuan et al., 2016). As the toxicity of such chemicals is very high, small traces in water bodies can cause serious problems, and hence water must be treated before its final use. In this view, various physical and chemical water treatment technologies have been proven effective in removing these chemicals from water; however, most of them are quite expensive and require huge investment. Metabolic activities of living organisms such as microbes and earthworms may be helpful in reducing the toxicity of synthetic agrochemicals. When compared to physicochemical technologies, bioremediation approach is reported to have appreciable advantages including being cost-effective and environmentally friendly. These benefits led to the huge demand in organic farming in which agriculture practices are focused on enhanced use of natural manure (Schmid et al., 2011).

Along with treatment approaches, some well-established legal framework may be helpful in this regard. Till now, many developed countries have legal framework for pesticide associated environmental risk assessments, but no such frameworks are available in developing countries (Lewis et al., 2016; Kabaluk et al., 2010). Hence, a strong regulatory body is required for this to regulate the types and usage of pesticides. Developing countries are still using many cheap and environmentally persistent and toxic compounds which are banned in most of developed nation like atrazine and Acephate, banned by the European Union (EU) (Daam et al., 2019). Although banning certain types of pesticides and fertilisers can be a solution to groundwater contamination, the benefits of agrochemicals by increasing the productivity of agricultural sector outweigh the ban. Further, such ban may lead to increased cost of other agrochemicals which will impose the economic burden on consumers via lower crop yield and higher product price. In this situation, there is need to understand the trade-off between productivity increase and groundwater contamination (Babu et al., 1992).

Use of an appropriate technology is always helpful to reduce the leaching potential of agrochemicals through immobilisation such as conservation tillage and no-till method. Also, application of biofertiliser is a better alternative of commercial chemical fertiliser. Therefore, many developed as well as developing countries are promoting the use of biofertilisers and organic farming. One other method is integrated pest management, under which traditional agricultural practices are adopted to reduce the use of chemical fertilisers and pesticides. However, the potential of these methods is still unexplored, and their impact on crop yield and economy of farmers must be analysed. Having these considerations in mind, this chapter is aimed to review and summarise various types of market leading agrochemicals, their usage pattern across the world, effects on groundwater quality, permissible limits of agrochemicals in drinking water, and possible remediation and management approaches.

15.2 TYPES OF POPULAR PESTICIDES AND FERTILISERS

Pesticides and fertilisers are commonly used for agricultural and horticultural purpose. Fertilisers are mainly used to provide nutrient to plants, while pesticides are used to protect crop from pests. In general, agrochemicals refer to pesticides which include insecticides, fungicides, rodenticides, nematicides, herbicides and molluscicides (Tripathi et al., 2020). In addition to this, agrochemicals also include fertilisers and soil conditioners (Baker et al., 1992). Agrochemicals are broadly classified into three categories, as discussed below:

i. Based on function of pesticide function and its target pest.
ii. Based on entry route and/or path.
iii. Based on composition of the pesticide.

15.2.1 Classification based on pesticide function and target pest organism

This classification of pesticides is based on target pest's organism, and names are given to pesticides reflecting their role (Yadav and Devi, 2017). The group name of these pesticides is "cide" which is a Latin word having a meaning "to kill"; cides are used as suffix after corresponding name of pests they kill. Table 15.1 shows the list of pesticides classified based on the organism that they kill.

15.2.2 Classification based on mode of entry

The mode of entry is the way through which pesticides either come in contact with target or enter into the target. It includes systemic, non-systemic, stomach poisons, fumigants and repellents (Alam et al., 2020). Table 15.2 shows the list of pesticides classified based on mode of entry.

15.2.3 Classification based on chemical composition of pesticides

Classifying pesticide based on their chemical composition and nature of active ingredients is one of most common and useful method. In this method, one can know about the physical and chemical properties of the respective pesticides and its efficacy (Kaur et al., 2019). Some of the popular pesticides classified based on their composition are tabulated in Table 15.3.

Table 15.1 Classification of pesticides based on target pest organism

S. No.	*Types of pesticides*	*Target pests*	*Example*	*Structure*
1	Acari+cides	Mites that nourish on animals and plants.	Bifenazate	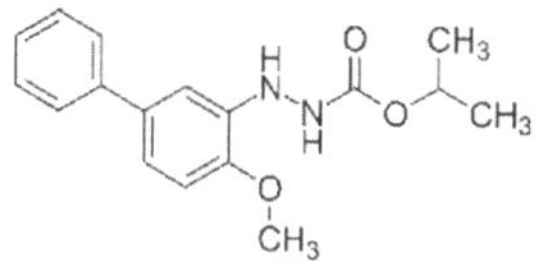
2	Algae+cides	Eliminate or control the growth of algae	Copper sulphate	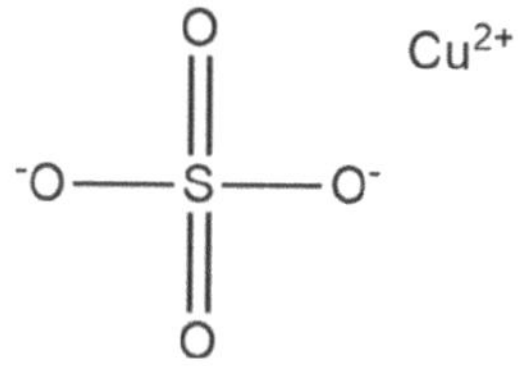
3	Avi+cides	Birds	Avitrol	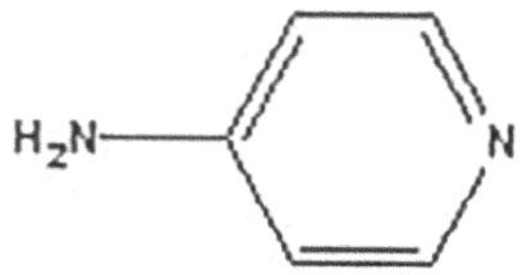
4	Bacteri+cides	Bacteria	Copper complexes	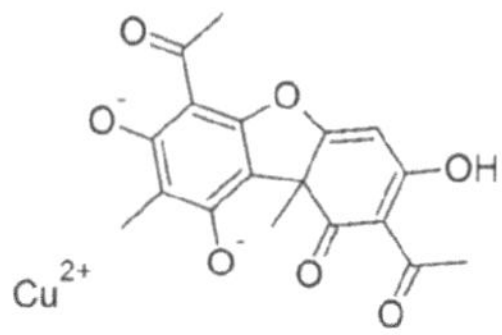
5	Desiccants	Drying plant tissue	Boric acid	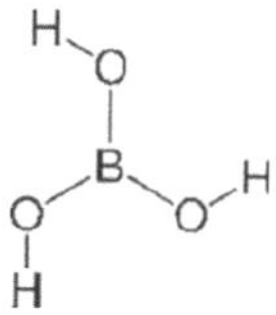
6	Fungi+cides	Fungi	Azoxystrobin	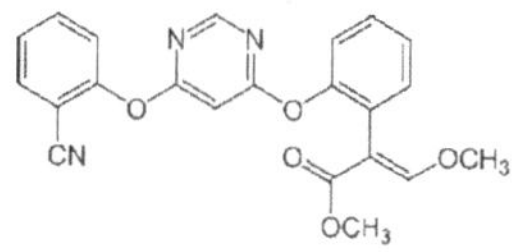
7	Herbi+cides	Weeds and other plants	Atrazine	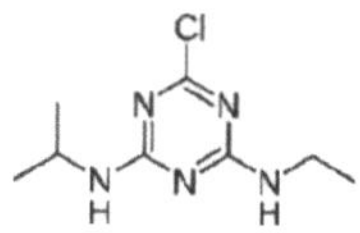
8	Insecti+cides	Insects and other arthropods	Aldicarb	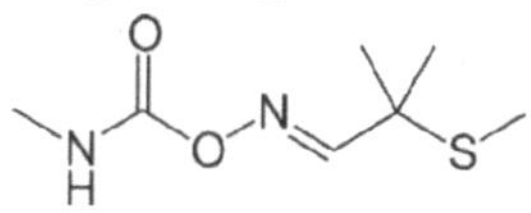
9	Lampri+cides	Larvae of lampreys	Nitrophenol Trifluromethyl	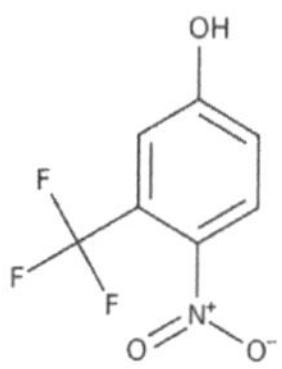
10	Larvi+cides	Larvae	Methoprene	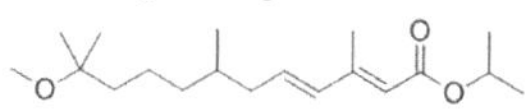

(*Continued*)

Table 15.1 (Continued) Classification of pesticides based on target pest organism

S. No.	*Types of pesticides*	*Target pests*	*Example*	*Structure*
11	Mollusci+cides	Molluscs, i.e. snails	Metaldehyde	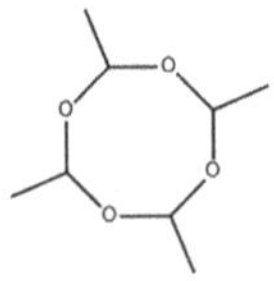
12	Moth balls	Moth larvae or moulds	Dichlorobenzene	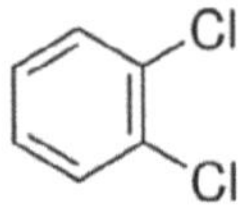
13	Nemati+cides	Nematodes	Aldicarb	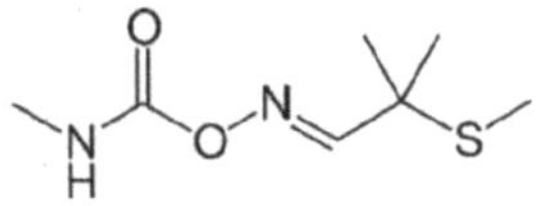
14	Ovi+cides	Eggs of insects and mites	Benzoxazin	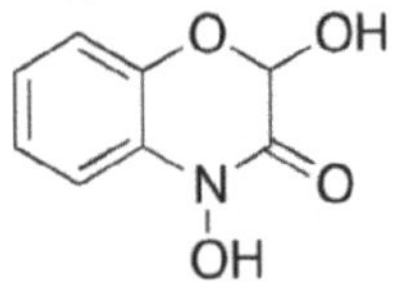
15	Pisci+cides	Fishes	Rotenone	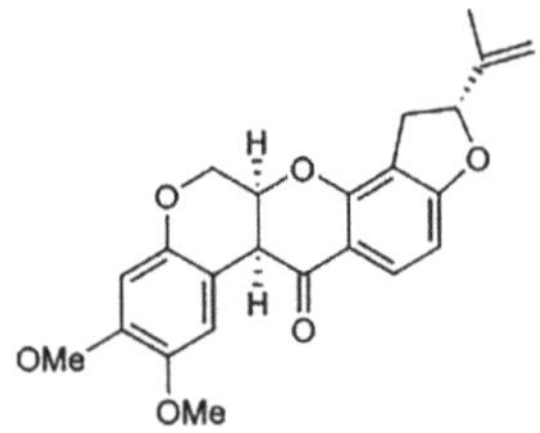
16	Repellents	Pest repellent	Methiocarb	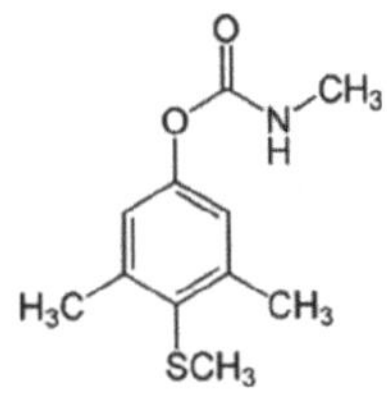
17	Rodenti+cides	Mice and other rodents	Warfarin	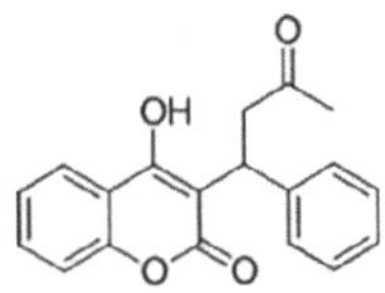
18	Silvi+cides	Woody vegetation	Tebuthiuron	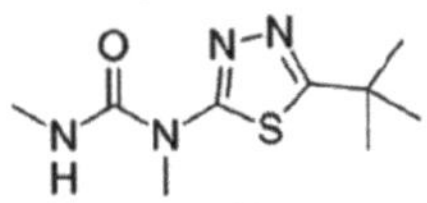
19	Termiti+cides	Termites	Fipronil	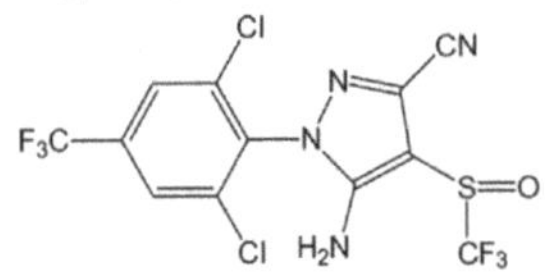
20	Viru+cides	Viruses	Scytovirin	Not available

Table 15.2 Classification of pesticides based on mode of entry

S. No.	*Classification*	*Working*	*Example*	*Structure*
1	Systemic Pesticides	This type of pesticide is absorbed by animals and plants and transferred to untreated tissue.	Glyphosate	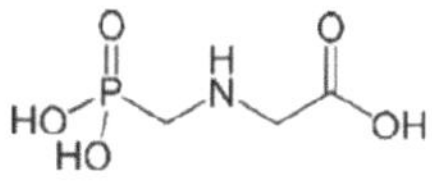
2	Non-Systemic (Contact) Pesticides	It kills pest when they come in contact with pesticide.	Paraquat and Diquat Dibromide	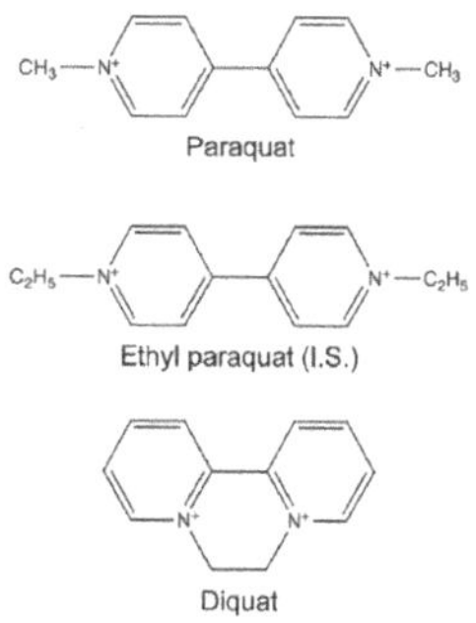
3	Stomach Toxicants	They kill pest through poisoning of stomach.	Malathion	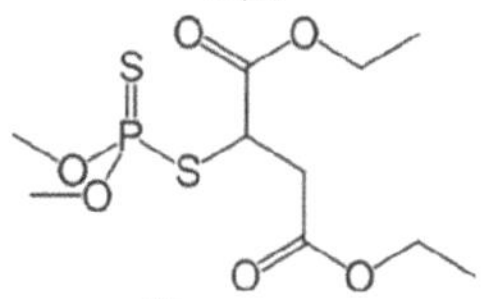
4	Fumigants	Fumigants produce vapour to kill target pests.	Methyl Bromide	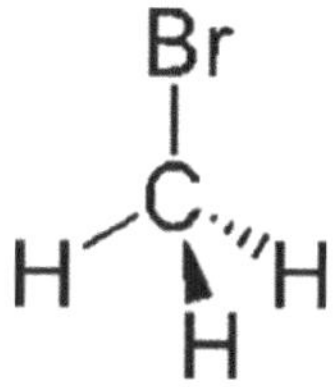
5	Repellents	They do not kill, but their taste and smell are so bad that they keep pests away from treated areas.	N,N-Diethyl-meta-toluamide (DEET)	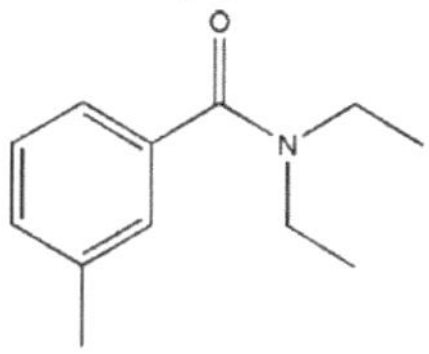

Table 15.3 Classification of pesticides based on its composition

S. No.	*Classification*	*Chemical composition*	*Examples*
1	Organochlorine	Organic compounds attached with five or more chlorine atoms	DDT, lindane, Endosulfan, etc.
2	Organophosphates	Esters of phosphoric acid	Parathion, Malathion
3	Carbamates	Similar to organophosphates, derivatives of phosphoric acid, while carbamates derived from carbamic acid	Carbaryl, Carbofuran
4	Synthetic pyrethroids	Group of organic pesticide, similar structure of natural pyrethrins	Cypermethrin and Permethrin

15.3 USAGE OF PESTICIDES AND FERTILISERS

Groundwater contamination by agrochemicals is the top priority of environmental policy makers of developed as well as developing countries. In this view, banning of different pesticides and reduced use of agrochemicals are general recommendations especially for those places where water table is high. However, the impacts of banning or reduced use on cost economics and crop yield are still not investigated thoroughly. It is need of time to set groundwater standards for different agrochemicals. Further, there is a strong need to consider a system of consumer and producer relationship for crop production and irrigation water usage policies. Use of fertiliser has been increased throughout the world due to the stress on production of crops. The names of some popular nitrogen-based fertilisers are urea, ammonium nitrate, calcium ammonium nitrate, etc. The top 10 nitrogen-based fertiliser consumer countries are given in Table 15.4. Egypt is the highest consumer with usage of 366.87 kg of nitrogen-based fertiliser per hectare of land. Countries like the USA and India are using 72.62 and 100.08 kg/ha, respectively. Usage trend of nitrogen-based fertilisers from 2002 to 2017 is also shown in Figure 15.1. However, based on usage of pesticides, China is ranked one with the usage of 1.77 million t, followed by the USA and India with 0.40 and 0.05 million t, respectively. Table 15.5 and Figure 15.2 elaborate these trends.

15.4 PATHWAY TO GROUNDWATER CONTAMINATION

Broadly two types of sources are responsible for the groundwater contamination by pesticides and fertilisers. The details of such sources are given below:

15.4.1 Point sources

Point sources are associated with specific point of release. Manufacturing unit producing pesticide/fertilisers, waste disposal sites, spillage, water treatment plants, loading and mixing facilities, etc. can be classified under point sources (Bloomfield et al., 2006). Table 15.6 shows the typical concentration of a few pesticides/fertilisers observed at point sources.

Table 15.4 Top ten consumers of nitrogen-based fertilisers

Country	*Start in kg*	*End in kg*	*Absolute change (kg)*	*Relative change (%)*
Egypt	312.51 in 2002	366.87 in 2017	+54.36	+17
Netherlands	307.90 in 2002	242.84 in 2017	−65.06	−21
China	199.21 in 2002	219.56 in 2017	+20.35	+10
Eastern Asia	194.86 in 2002	213.74 in 2017	+18.88	+10
Luxembourg	204.84 in 2002	213.47 in 2017	+8.63	+4
Belgium	177.11 in 2002	198.43 in 2017	+21.32	+12
Colombia	95.46 in 2002	182.50 in 2017	+87.04	+91
Armenia	31.98 in 2002	178.39 in 2017	+146.41	+458
Suriname	75.23 in 2002	174.89 in 2017	+99.66	+132
United Kingdom	193.80 in 2002	168.50 in 2017	−25.30	−0.13

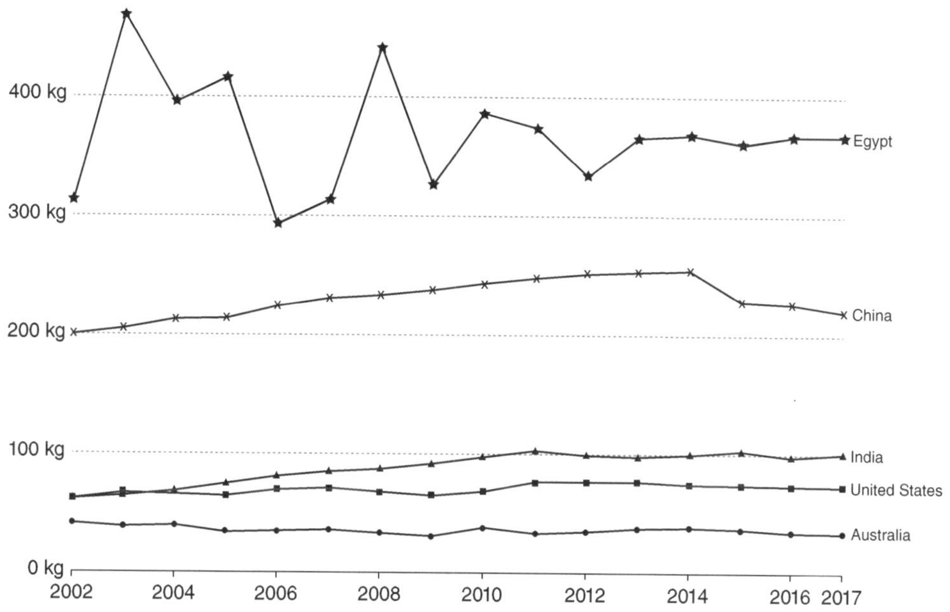

Figure 15.1 Trend (2002–2017) of nitrogen-based fertiliser usages by some major countries. (Source: UN Food and Agricultural Organization.)

Table 15.5 Top ten consumers of pesticides

Country	*Start in t*	*End in t*	*Absolute change (t)*	*Relative change (%)*
China	775,408 in 1990	1.77 million t in 2017	+998,226	+129
United States	400,976 in 1990	407,779 in 2017	+6,803	+2
Brazil	49,695 in 1990	377,176 in 2017	+327,481	+659
Argentina	26,156 in 1990	196,009 in 2017	+169,853	+649
Canada	29,568 in 1990	90,839 in 2017	+61,271	+207
Ukraine	66,772 in 1992	78,201 in 2017	+11,429	+17
France	97,701 in 1990	70,589 in 2017	−27,112	−28
Australia & New Zealand	21,356 in 1990	68,502 in 2017	+47,146	+221
Malaysia	39,406 in 1990	67,288 in 2017	+27,882	+71
Australia	17,866 in 1990	63,416 in 2017	+45,550	+255

15.4.2 Non-point sources

Non-point sources are dispersed or widely spread. Use of fertilisers and pesticides in agricultural land is an example of non-point source. Non-point sources are major contributors of groundwater contamination. Runoff to streams from agricultural and urban land, seepage to groundwater in areas where fertilisers and pesticides are used, and deposition of agrochemicals from the atmosphere are few examples of non-point sources (Arias-Estévez et al., 2008). Table 15.7 shows the pesticides/fertilisers concentration reported for non-point sources.

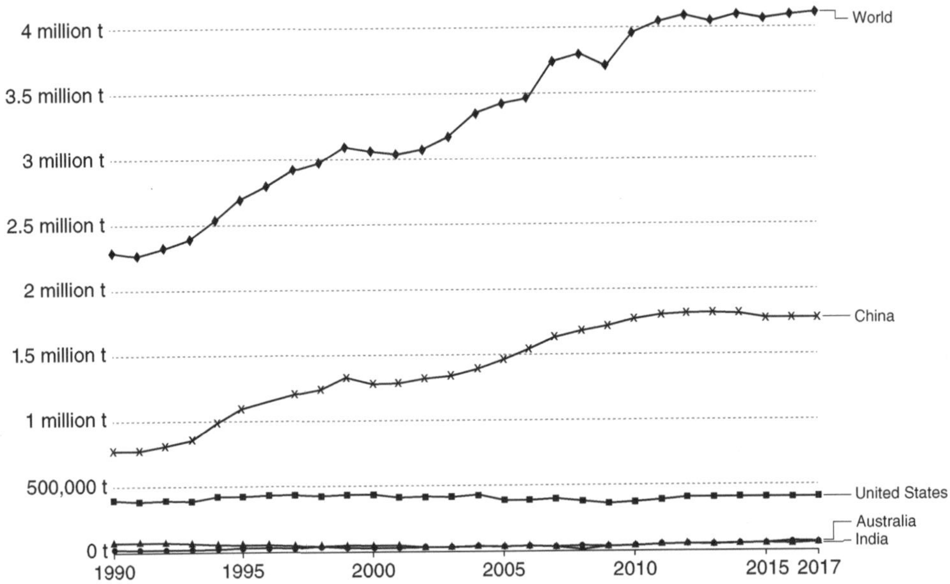

Figure 15.2 Trend (1990–2017) of pesticide usages by some major countries. (Source: UN Food and Agricultural Organization.)

Table 15.6 Typical concentration of investigated pesticides from point sources

Name/type of chemical	*Typical conc. (mg/L)*	*Reference*
Dimethoate	14.71 ± 0.79	Cheng et al. (2015)
Triazophos	5.87 ± 0.89	
Malathion	24.53 ± 4.16	
Simazine	4.675	Mezzanotte et al. (2005)
Prometryn	7.397	
Ametryn	107.525	
Simetryn	5.360	
Tetraconazole	0.716	

Table 15.7 Typical concentrations of pesticides, reported for non-point sources

Name/type of chemical	*Typical conc. (μg/L)*	*Reference*
Atrazine	0.13–1.68	Müller et al. (2002)
Carbetamide	0.14–1.95	
Chlortoluron	0.31–0.37	
Diuron	0.04–5.09	
Triadimenol	1.00–1.60	
Simazine	0.05–0.07	
Chlorpyrifos	0–1.189	Chowdhury et al. (2012)
Carbofuran	0–3.395	
Carbaryl	0–0.163	

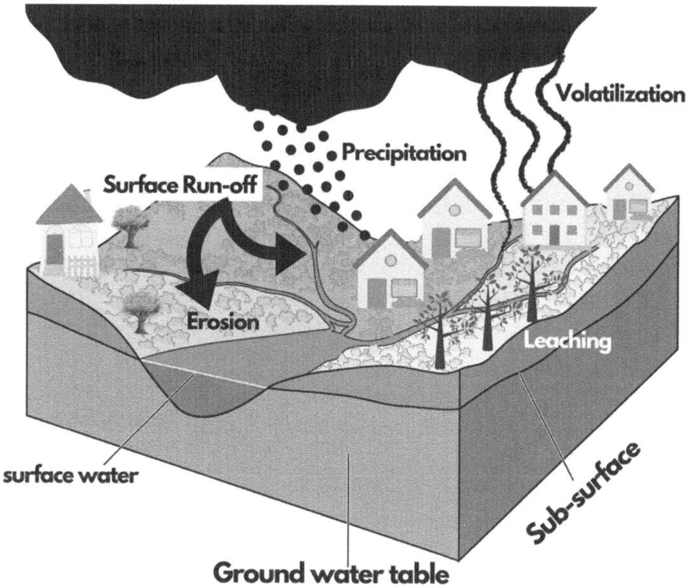

Figure 15.3 Fate of agrochemicals in groundwater and surface water.

Herbicides, mainly triazine and urea compounds, have been the most detected pesticides in the groundwater (Worrall et al., 2002).

The toxic chemicals from agrochemicals take many decades to travel from ground to aquifer and contaminate the whole ground water. Hence, it is important to monitor the lag time between application of agrochemicals to the land and reappearance of agrochemical in groundwater sources as well. It is important to mention here that if permeability of aquifer increases and water table is shallow, then lag time decreases and chances of contamination become high. Contrary, the shallow aquifer takes less time to improve compared to deep aquifer, in case of reduced use of agrochemicals. Figure 15.3 shows the fate of agrochemicals in ground and surface water sources. In case of ground water, leaching effect is reported to play major role in contaminated water quality, while surface runoff is mainly responsible factor for surface water contamination. Detailed information about various pathways of non-point sources contamination is discussed below.

Movement of a pollutant, due to differences in its concentration, is called diffusion. Movement of agrochemical from higher concentrated area to lower concentrated area either in soil or in groundwater is an example of diffusion process. The volatilisation of pesticides is dissipation of pesticide into the atmosphere. Erosion is the process by which the surface of earth is worn away. Loose soil can act as carrier for the movement of agrochemicals through wind erosion or through surface runoff. This could lead to the contamination of surface water like pond, lake and river. Surface runoff is a phenomenon where water is moved over sloping area, and loose soil gets moved with water and enters the surface water body. Through surface runoff heavy amount of agrochemicals are washed from field to the surface water body. Aquatic plants are very helpful in uptake of pesticides from surface water. This approach is known as phytoremediation. The uptake depends on various physicochemical properties of soil and pesticides. Leaching process refers to the downward movement of pesticides through the soil profile and unsaturated zone, and finally to aquifers, and thus they are vulnerable to pollution. Leaching of pesticides is mainly initiated by rainwater or water used in agricultural fields. Heavy precipitation, low temperature, sandy soil, low content of organic matter and persistent compounds collectively promote the leaching of agrochemicals. On the other hand, low persistent compounds disappear quickly from environment and possess less threat to groundwater contamination. Among various factors, leaching is the most responsible factor associated with groundwater contamination. Further, the rate of leaching also depends on plant root uptake, permeability and texture of soil, and volatilisation rate of compound. The pesticides can leach typically by two flow types:

i. **Preferential-flow** – in which water uses larger voids, root channels and cracks to flow.
ii. **Matrix-flow** – in which water movement is slowed down due to the presence of small pores in soil.

Solubility of pesticides in water is an important factor for the leaching of pesticides along with rainwater or irrigation water. Water may enter the soil by infiltration. High molecular weight compounds can easily be carried away by high velocity water and thus have the potential to take these compounds farther.

15.4.3 Factors affecting the fate of pesticides in soil environment

Temperature and precipitation are two main factors that govern the degradation of pesticide in soil environment. It has been reported in literature that pesticides dissipate rapidly in soil, and then move to groundwater more easily in tropical region compared to temperate region. This enhanced degradation may be associated with temperature of the region rather than any other specific condition. In a study, it is estimated that an increase in degradation of pesticides by a factor of 2 and volatilisation rates of 3–4 for a 10°C rise in soil temperature. Another responsible factor for higher contamination in soil and water is intensive irrigation and heavy rainfall (Daam et al., 2019). Properties of soil can also affect the degradation of pesticides. Organic matter content is the most important property of the soil, affecting breakdown of contaminants by microbes. Higher the organic matter content of soil, higher is the adsorption of water and microbial activity which degrades the pesticides and reduces leaching of pesticides into groundwater. The ratio of sand, silt and clay determines the soil permeability which in turn governs the rate of movement of water from soil to groundwater table. Coarse-textured soils have high proportion of sand and thus have high permeability-favouring movement of water more rapidly (Kah et al., 2007). Pesticides carried by water through coarse-textured soil are more likely to reach groundwater, and contamination takes place easily. Soils with high proportion of clay typically have low permeability and hence can hold more water with low seepage, thus increasing the chance of degradation and adsorption of pesticide onto soil particles (Arias-Estévez et al., 2008). Pesticides with permanent anions and weak acids can be weakly adsorbed, and hence may easily reach groundwater. For example, atrazine is one of the most widely used herbicides in the world, and in most of the drinking water, it is present. In the USA alone, seven million people are exposed to atrazine through drinking water. Similarly, organochlorine is widely used pesticides in India. It has high persistence in external environment due to which it is potent to contaminate groundwater through infiltration. Leaching rate of the pesticides is influenced by physical, chemical and biological properties of pesticides. More specifically, the permeability, soil texture and organic matter of soil also mainly influence the leaching rate (Gavrilescu, 2005). However, it can also be influenced by temperature and precipitation pattern as discussed above. Other than this, high water table also increases the chances of contamination in groundwater. Closer the water table to the surface of water, higher are the chances of contamination. Type of bedrock also determines the flow of recharge water and groundwater condition. Limestone bedrock increases the chances of contamination of groundwater as it allows quick passage of recharge water. With reference to slope, steep slopes promote fast surface runoff and reduced chances for water to infiltrate into the ground (Macpherson & Sophocleous, 2004).

15.5 HEALTH EFFECTS

Health impacts of pesticides depend on the route through which it comes into human contact, its toxicity and concentration levels. The main routes of human exposure to pesticides are via food chain, air, water, soil, flora and fauna. Pesticides enter into the human body through skin, eyes, nose and mouth (Wimalawansa & Wimalawansa, 2014). Below expression represents the risk expression for the pesticide exposure.

$$\text{Risk} = \text{Exposure}(\text{amount and / or duration}) \times \text{Toxicity}$$

Pesticides that bear similar structure mostly exhibit similar health effects. Below section discusses the various effects.

15.5.1 Acute effects

Some pesticides show harmful effects on health with single exposure for short duration and termed as acute effects. The dose/concentration required to kill 50% of the test population is termed as acute toxicity. This is denoted as LC50 (lethal concentration 50) or LD50 (lethal dose 50). Acute health problems like abdominal pain, dizziness, headaches, nausea, vomiting, as well as skin and eye problems are mostly seen in people that handle pesticides (Henriques et al., 1997). For example, Aldrin and dieldrin at concentration of 0.03 μg/L in water can cause acute health effects like headache, dizziness and uncontrolled muscle movement. Further, long-term exposure of high concentration can lead to chronic condition like kidney damage. Endosulfan at concentration of 0.4 μg/L shows acute toxicity, and it has also bioaccumulation potential. Long-term exposure can lead to endocrine disruption (Taiwo, 2019).

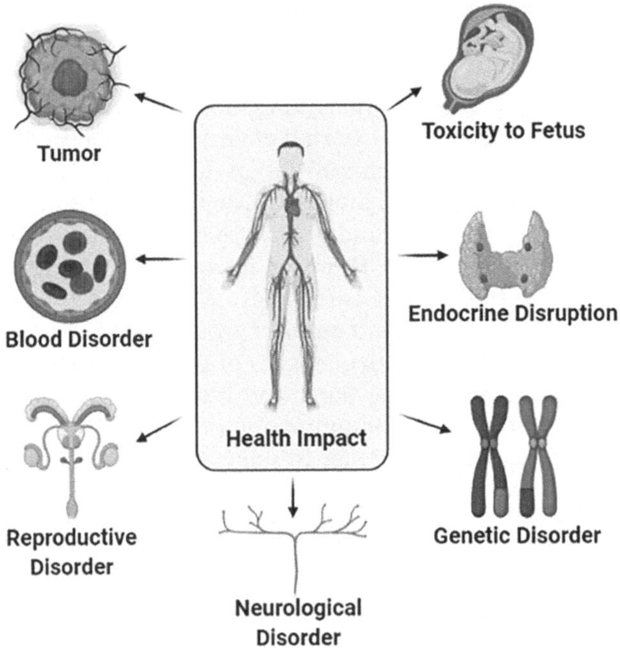

Figure 15.4 List of some major health issues due to exposure of agrochemicals. (Ref: Sekhotha et al., 2016; Kim et al., 2017.)

15.5.2 Chronic effects

Long-term exposure to small pesticide doses is correlated with the chronic effects. The recurrent alleged consequences of exposure to particular pesticides include birth defects, infant toxicity and benign or malignant tumours, genetic alterations, blood abnormalities, nerve disorders, endocrine disorders and reproductive effects. Chronic pesticide toxicity is more difficult to determine through laboratory analysis than acute toxicity (Reinert et al., 2002; Prudente et al., 2018; Kamel & Hoppin, 2004; Magauzi et al., 2011). The health effects related to agrochemicals exposures are presented in Figure 15.4.

DDT at concentration of 1 µg/L shows chronic disorders in both vertebrate and invertebrate species like neurotoxicity and reproductive disorder (Taiwo, 2019). Chronic human exposure at a concentration 2 µg/L to γ-HCH in water has been linked to haematological problems such as anaemia, leukopenia, leukocytosis, granulocytosis, eosinophilia, monocytosis and thrombocytopenia (Taiwo, 2019).

15.5.2.1 Neurological impact

Organophosphorus compounds like diazinon, malathion and parathion are responsible for the development of various neurotoxic effects. Other impacts on nervous system include tremors, memory and sleep disorders, headache, coma, dizziness, ataxia, seizures, anxiety and depression (Florencia and Cora, 2020).

15.5.2.2 Intoxication

The main symptoms of pesticides intoxication can be categorised into two major syndromes, namely nicotine syndrome and muscarinic syndrome. Nicotine syndrome, in which acetylcholine is gathered on the motor nerve endings in the autonomic ganglia, causes spasms, hyperreflexia, hypertonicity, tremors, paralysis or muscle weakness and stimulates the sympathetic autonomic ganglia to promote pallor, hyperglycemia, tachycardia and hypertension. In muscarinic syndrome, acetylcholine strikes on the heart, smooth muscle and exocrine glands increasing the tearing, bronchial secretion and sweating; disrupts the gastrointestinal tone causing vomiting, nausea and diarrhoea; and elicits bronchospasm, urinary incontinence, miosis and bradycardia (Hendges et al., 2019).

15.5.2.3 Other effects

Pesticides like DDT have oestrogenic action which can affect the reproductive system of birds and mammals (Welch et al., 1969). Pesticides from family like organochlorines have been reported to link with diseases like asthma, diabetes, cancer and growth disorders in children. The main symptoms of organochlorine intoxication are altered reflexes, gait disorders, headache, dizziness, nausea, anorexia, vomiting, malaise, diarrhoea, dermatitis, excitement, apprehension, irritability, excessive sweating, muscle weakness, tremors, spasms, mental confusion, coma, anxiety, seizures and death (Wimalawansa & Wimalawansa, 2014). Few studies claim that female exposure to herbicides can cause spontaneous abortion, birth defects and infertility. A detailed list of health impact due to specific compounds is provided in Table 15.8.

15.6 MONITORING METHODOLOGY AND ASSESSMENT

Due care should be taken before analysis of pesticide sample. First is the identification of target analyte and its properties followed by the appropriate detection method such as liquid chromatography or mass spectrophotometry, sampling procedure, sample preparation like solubilisation, extraction and separation, calibration of instrument, and finally calculation and evaluation of results.

15.6.1 Sampling

Samples should be collected in coloured glass bottles to avoid any interference of light. Also, all sample bottles should be rinsed well with water and must be carefully filled to overflow, without trapping air bubbles in sealed bottles. The samples must be transported in a cool box with ice packs. Preparation of the containers includes washing with detergent, rinsing with tap water, ultrapure water (Millipore) and air drying. After transportation to the laboratory, samples must be stored at 4°C and extraction should be done within 48 h. During sample storage over long periods of time, it must be ensured that no degradation or external contamination occurs in samples (Thompson et al., 2020).

Table 15.8 Health impacts of some active ingredients of pesticides

Active ingredient	*Health impacts*
Fungicide	
Azoxystrobin	Skin, eyes, respiratory tract–related issues
Captan	Skin, eyes, respiratory tract–related issues
Chlorothalonil	Skin, eyes, allergic contact dermatitis, respiratory tract–related issues
Thiram	Skin, eyes, respiratory mucous membranes
Ziram	Skin, eyes, neural and visual disturbances
Copper compounds	Skin, eyes, respiratory tract related, metallic taste, nausea, vomiting, intestinal pain
Sulphur	Skin, eyes, respiratory tract issue, diarrhoea, irritant dermatitis
Herbicide	
2,4-Dichlorophenoxyacetic acid	Skin irritation, mucosal membranes. Headache, vomiting, diarrhoea, confusion. Odd or violent behaviour. Muscle weakness in people who are professionally exposed
Acetochlor	Skin, eyes, breathing tract irritation
Atrazine	Skin, eyes, breathing tract irritation. Abdomen, diarrhoea, vomiting, abdominal pain. Eye discomfort, mucous membranes irritation, skin reactions
Dicamba	Irritating to skin, respiratory tract. Loss of appetite (anorexia), vomiting, muscle weakness, slowed heart rate, shortness of breath. Central nervous system effects
Glyphosate	Irritating to skin, eyes, respiratory tract
Propanil	Irritating to skin, eyes, respiratory tract
Pendimethalin	Irritating to skin, eyes, respiratory tract
Insecticide	
Pyrethrins (natural origin)	Irritating to skin and upper respiratory tract. Contact dermatitis and allergic reactions – asthma
Pyrethroids (synthetic pyrethrin)	Abnormal facial sensation, dizziness, salivation, headache, fatigue, vomiting, diarrhoea. Irritability to sounds or touch. Seizures, numbness
Endosulfan	Itching, burning, tingling of skin. Headache, dizziness, nausea, vomiting, lack of coordination, tremor, mental confusion. Seizures, respiratory depression, coma
Malathion	Headache, excessive salivation and tearing, muscle twitching, nausea, diarrhoea. Respiratory depression, seizures, loss of consciousness. Pinpoint pupils
Acephate	Headache, excessive salivation and tearing, muscle twitching, nausea, diarrhoea. Respiratory depression, seizures, loss of consciousness. Pinpoint pupils
Aldicarb (N-methyl carbamate)	Malaise, muscle weakness, dizziness, sweating. Headache, salivation, nausea, vomiting, abdominal pain, diarrhoea. Nervous system depression, pulmonary oedema in serious cases

The analysis of pesticides in water requires a purification step to clean the sample and pre-concentrate the analytes. The extraction process separates the desired compounds for further analysis. Most commonly used methods are liquid–liquid extraction, solid-phase extraction and solid-phase microextraction. The extraction of sample residues requires suitable solvent selection for high extracting efficiency and minimum co-extraction interference. Following extraction, the cleansing phase eliminates the interfering material with minimum analyte loss. The final solution should provide a suitable solvent for analyte determination by the chosen method (Shoda et al., 2018).

15.6.2 Analytical methodology

Analytical methods available for the analysis of pesticides are gas chromatography (GC) with electron capture detection, flame ionisation detection or nitrogen-phosphorus detection and/or liquid chromatography (LC) with ultraviolet, diode array, fluorescence or electrochemical detection (Zanella et al, 2012). Details about these methods are discussed in this section.

15.6.2.1 Physicochemical separation methods

Several studies have demonstrated the simultaneous determination of 245–400 pesticides by GC–EI–MS (Gas Chromatograph-Electron Impact-Mass Spectrometry) with single quadrupole mass filters. Preferentially, GC–MS with chemical ionisation is focused on special substance classes only and is not considered as universal ionisation technique, while mass spectra is reported to have smaller number of fragments, offering less information.

For analyses of volatile or semi-volatile pesticides, GC is highly recommended as it can detect very low quantities of pesticides and offer high resolution. Multiple detectors are used in GC as per the characteristics of the target analyte. There are two phases, mobile and stationary in GC. Mobile phase is a carrier gas (inert gas like helium), and stationary phase is liquid or solid support inside a column. The sample after volatilisation is introduced into the chromatographic column. LC is suitable for thermally unstable and non-volatile pesticides which are not compatible with gas chromatograph. In LC, compound's extract is passed through different adsorbent column which segregates the component of matrix and analyte. Column can be used in combination or separately. The degree of selectivity varies with the kind of adsorbent present in column (silica gel, alumina or florisil), activity strength and mesh size. Capillary electrophoresis (CE) is useful in identification and separation of wide range of agrochemical compounds. It provides large efficiency with high resolution and can even work with small sample size. The solvent consumption is also low in this method. This technique is very cost-effective and gives results in a short period of time (Hercegová et al., 2007).

15.6.2.2 Detection methods

Nowadays, different types of detectors are available in market for the detection of agrochemicals. These detectors are flexible in nature and can be coupled with separation devices. The main techniques used by these detectors are thermal conductivity, electron capture, photometric, diode array detection, atomic absorption and mass evaluation. For detection of pesticides, electron capture and mass spectrometry are the most popular methods of detection. A suitable detector should exhibit the following properties:

- It should have high sensitivity
- It should be stable and reproducible
- It should work in a wide range of temperatures
- It should have low response time and
- It must be easy to handle

Nowadays, micro electron capture detector (μECD) is also available in market which is more reliable compared to electron capture detector (ECD) in terms of stability, robustness and sensitivity. ECD is thousand times sensitive to flame ionisation detection (FID). ECD is helpful in the detection of organochlorine pesticides, poly chlorinated biphenyls (PCB), herbicides and various halogenated hydrocarbons. Mass spectrometry (MS) separates the ions based on mass-to-charge ratio which permits unambiguous identification of molecules. The only limitation of this technique is that it destroys the analyte. Liquid chromatography and mass spectrometry (LC-MS) and gas chromatography and mass spectrometry (GC–MS) are most suitable ways of detection that usually detect and segregate pesticides. Studies conducted by Zanella et al. (2012) and Hercegová et al. (2007) summarise the different pesticides and their extraction techniques, which are shown in Table 15.9.

electrospray ionisation (ESI), diode array detector (DAD), ECD, matrix solid-phase dispersion (MSPD), gel permeation chromatography (GPC), limit of quantification (LOQ), solid-phase extraction (SPE), selected ion monitoring mode (SIM), pressurised liquid extraction (PLE); gas chromatography (GC), liquid chromatography (LC), Electron capture detection (ECD), electrospray ionisation (ESI).

15.7 PERMISSIBLE LIMIT OF GROUNDWATER QUALITY FOR PESTICIDES AND FERTILISER

Concerns are growing to set water quality standards with respect to agrochemicals to monitor groundwater quality. However, in current scenario it is very hard for farmers to comply with recommended permissible limit for groundwater protection. However, monitoring of groundwater and meeting the guidelines may inhibit the usage of agrochemical and due to that there will be significant loss in crop yield. Also, seasonal usage of agrochemicals is different which again creates a barrier in setting up annual limits. Application of fertilisers can lead to higher concentration of nitrates in groundwater. Therefore, regular monitoring of groundwater is necessary with respect to agrochemical. World Health Organization (WHO) and Bureau of Indian Standard (BIS) set the permissible limit for pesticide (BIS, 2012, World Health Organization, 2017), as listed in Table 15.10.

Table 15.9 Different methods applied for the determination of pesticide and others analytes in different samples

Analytes	*Sample*	*Analytical technique*	*Extraction procedure*	*Clean-up*	*LOQ (μg/kg)*	*Recovery (%) (RSD %)*
Benzoylureas	Animal products, cereals, fruits and vegetable	LC-MS/MS	PLE	-	2.0–10.0	58–97 (5–19)
Acetanilide herbicide	Maize, wheat and rice	GC-ECD	PLE	SPE	2.4–5.3	82.3–115.8 (1.1–13.6)
22 Pesticides	White and wild rice	GC-ECD	SFE	SPE	-	74–98
Chlorpyrifos, malathion,4,4_-DDE,4,4_-DDD,4,4_-DDT	Broccoli, carrot, cheese	GC–MS;EI, SIM	PLE	GPC	--	95–110
Acetochlor and propisochlor	Soil and corn	GC-ECD		methanol/ water/SPE	10	73.8–115.5 (<11.1)
Bispyribac sodium	Rice	HPLC-DAD	liquid–liquid partition and anion exchange solid phase		10	83.98–98.51 (0.56–6.36)
Penoxsulan, tricyclazole, propanil, azoxystrobin, molinate, profoxydim, cyhalofop-butyl, deltamethrin and 3,4-dichloroaniline (the main metabolite of propanil)	Rice	HPLC-DAD	MSPD		6.0–600.0	74–127 (<12)
Carbendazim	Wheat grain	HPLC-DAD	MSPD		40	87.3–3.3
22 Organochlorines pesticides	Human milk	GC–MS	Liquid–liquid extraction	SPE	1.00–4.83	(<20)
20 Multiclass pesticides	Peach	GC–MS	Liquid–liquid extraction	GPC	0.5–25	(2.7–8.1)
3*N*-Methylcarbamate	Apple-strawberry	HPLC–MS/ MS; ESI+, ion trap, SRM	Liquid–liquid extraction	GPC	1	120–160 (<18.7)
Bisphenol A, 4-tert-butylbenzoic acid, 4-nonylphenol, 4-tert-butylphenol, 2,4-dichlorophenol, 2,4,5-trichlorophenol and pentachlorophenol	Breakfast cereals	LC–ESI-MS	PLE	-	3.0–43.0	81–104 (4–9)

Table 15.10 Permissible limit of fertilisers and pesticides in water

S. No.	*Chemical*	*WHO permissible limit in µg/L*	*BIS,10500 permissible limit in µg/L*
A	Non-pesticides		
1.	Nitrate	50,000	45,000
2.	Nitrite	3,000	—
B	Pesticides used in agriculture		
1.	Alachlor	20	20
2.	Aldicarb	10	
3.	Aldrin and dieldrin	0.03	0.03
4.	Alpha HCH (Hexachlorocyclohexane)	—	0.01
5.	Beta HCH	—	0.04
6.	Atrazine and its chloro-s-triazine metabolites	100	—
7.	Butachlor	125	—
8.	Carbofuran	7	—
9.	Chlordane	0.2	—
10.	Chlorotoluron	30	30
11.	Chlorpyrifos	30	—
12.	Cyanazine	0.6	—
13.	2,4-Db (2,4-dichlorophenoxy) butyric acid	30	—
14.	2,4-DBc	90	—
15.	1,2-Dibromo-3-chloropropane	1a	—
16.	1,2-Dibromoethane	0.4a (P)	—
17.	1,2-Dichloropropane	40 (P)	—
18.	1,3-Dichloropropene	20a	—
19.	Dichlorprop	100	—
20.	Delta HCH	—	0.04
21.	2,4- Dichlorophenoxyacetic acid	—	30
22.	DDT (*o*, *p* and *p*, *p* – Isomers of DDT, DDE and DDD)	—	1
23.	Dimethoate	6	—
24.	Endrin	0.6	—
25.	Endosulfan (alpha, beta and sulphate)	—	0.4
26.	Ethion	—	3
27.	Fenoprop	9	—
28.	Gamma – HCH (Lindane)	—	2
29.	Hydroxyatrazine	200	—
30.	Isoproturon	9	9
31.	Lindane	2	—
32.	Malathion	—	190
33.	MCPAd	2	—
34.	Mecoprop	10	—
35.	Methoxychlor	20	—
36.	Metolachlor	10	—
37.	Methyl parathion	—	0.3
38.	Monocrotophos	—	1
39.	Phorate	—	2

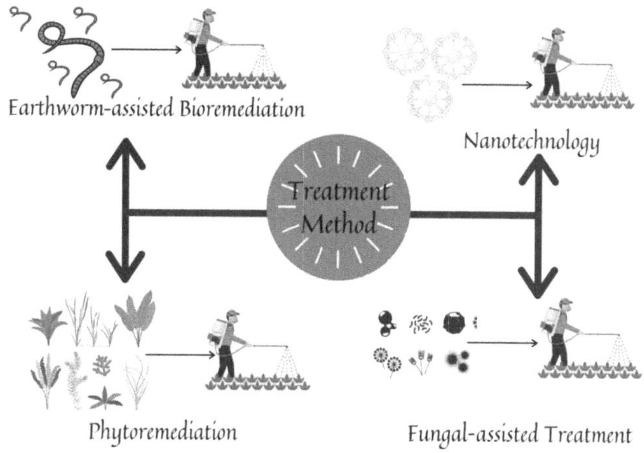

Figure 15.5 Popular treatment technology for agrochemicals.

15.8 AVAILABLE TREATMENT TECHNOLOGY

Various conventional approaches (physical, biological & chemical) exist for the water and wastewater treatment (Eregowda and Mohapatra, 2020; Mohapatra et al., 2021; Mogili & Mohapatra, 2021), but none of them are efficient for the treatment of agrochemical-based water/wastewater. If we use treatment technologies like filtration, adsorption by activated charcoal and reverse osmosis, then this could be fatal. However, in recent past, bioremediation and nanotechnology has proven effective for the treatment of agrochemical. Removal of undesirable contaminants from environment using biological components like plants, algae, microbes, fungi and enzymes is called as bioremediation. Several approaches exist for bioremediation based on which kind of species is utilised for the treatment of specific chemicals. On the other hand, nanotechnology uses nanoparticles for absorption and adsorption, photocatalysis and chemical reaction to clean the environment (Priyanka et al., 2020; 2021). In case of agrochemical treatment, it uses adsorption and absorption phenomena. Small size of nanoparticles enhanced the activity using larger surface area-to-volume ratio. In this section, some of popular treatment methodologies have been discussed and also shown in Figure 15.5.

15.8.1 Earthworm-assisted treatment

Agrochemicals adversely affect the bacterial population in the soil, which finally affects global ecosystem functioning. For the treatment of agrochemicals, earthworm plays a vital role in the treatment of soil. An earthworm belongs to Annelida phylum which has a tube-shape worm with segmented body. They have 7,000 species and one of the most important macrofauna of the soil. They can change physical, biological and chemical properties of soil, which makes them ecosystem engineer. They increase the nutrient content of the soil by degradation and mineralisation of organic matter. One of the major activities of earthworm, i.e. burrowing and casting activity, assists in recycling of nutrients in the soil to make them readily available to plants. Aeration and porosity are increased by the formation of burrows which finally promoted the growth of bacteria. Earthworms have the potential to store heavy metals in their body. This accumulation property of earthworm has taken attention of many researchers which later has proven as good remediation tool for the soil. All breakdown mechanisms take place in guts of earthworm. It has been found that polychlorinated biphenyl (PCB) and atrazine are degraded due to the presence of earthworm which shows that earthworms are capable of removing agrochemical contamination from soil (Singh et al., 2020).

15.8.2 Removal of agrochemicals with nanoparticles

Nano-sorbents is helpful in the treatment of agrochemicals from soil and water. Nano-material uses adsorption or absorption for removal of contaminant which is a surface phenomenon. It helps in the complete removal of pollutants from the environment. Degradation of DDT, lindane, aldrin, atrazine and endosulfan is well documented using nano-sorbent. Even enhanced results can be seen if these nano-sorbents are coated with biological materials. Low-cost adsorbents are available in market for the removal of agrochemicals which are made up of oxide of Mg, Mn, Ti and Zn. These particles have larger surface reaction, shorter intraparticle diffusion and higher specific surface area. These metal oxides not only absorb the toxins of agrochemicals but also convert the toxin into simpler products. One more advantage of nanoparticles is it works in a wide range of temperatures due to their thermostability (Sebastian et al., 2020, Kashyap et al., 2015).

15.8.3 Removal of agrochemicals by aquatic plants

There are huge numbers of aquatic plants which have high potential to remove agrochemical from water. Some of these plants are *Ceratophyllum demersum*, *Typha latifolia*, *Cyperus papyrus*, *Phragmites mauritianus*, etc. The mechanisms that involve in the removal of agrochemical by plants are mainly metabolisation and accumulation. Uptake of toxic chemical depends on physicochemical properties of agrochemicals like molecular weight, hydrophobicity, solubility and polarity. The rate of uptake depends on environmental condition and plant species. If water partition (Kow) of chemical is between 0.5 and 3.0, then it helps in the uptake of that chemical. Similarly, bioaccumulation and translocation of chemical depend on hydrophobicity. Different species use different mechanisms like some species using rhizofiltration, while some use phytoaccumulation for removal of agrochemicals (Dhir, 2020).

15.8.4 Fungal-assisted treatment

Trichoderma sp., *Beauveria bassiana*, and *Paecilomyces lilacinus* are popular fungal species used for bioremediation. Many research works are ongoing for detection of new species for the removal of contaminant of agrochemicals from environment. Lindane can be degraded by using *Fusarium* fungal species. Agave tequilana is a procedure which is most suitable for isolation of fungal strain. It has been seen that *Penicillium brevicompactum*, *Lentinula edodes* and *Fusarium oxysporum* are very good strains for the removal of toxic chemicals like pendimethalin, difenoconazole and terbuthylazine. Methomyl and diazinon are easily degraded by the white rot fungus. Similarly, agrochemicals like chlorpyrifos and DDT that contaminate land could be remediated using strains of different fungi. In a study it was demonstrated that higher rate of degradation of agrochemical can be achieved by fungal consortia at even lower concentrations (Bhadouria et al., 2020).

15.8.5 Enzymatic degradation

Picking up combination of microbes of different strain at pesticide contaminated land for bioremediation is very beneficial as it increases the diversity of microbes, and it is fast acclimatising and provides a multiple metabolic process which is helpful for bioremediation. An important tool for management of agrochemical contamination is the presence of different pools of enzymes released by microbes in soil (Leng et al., 2020). Enzymatic degradation of lindane is well documented which is released through fungal species *Conidiobolus*. Similarly, 30 fugal strains are evaluated for enzymatic degradation of agrochemicals (Bhadouria et al., 2020).

15.9 POSSIBLE MANAGEMENT UNDER LEGAL FRAMEWORK

- Apply the pesticide according to company manual.
- Avoid use of persistent agrochemicals and slow release/encapsulated formulations. Ideally, an effective pesticide is only those which are applied to kill individual and then quickly get degraded in environment.
- Before mixing of pesticides check the pH of water and adjust the pH accordingly.
- Early morning and evening time is good for application of pesticides.
- In pest management programme incorporate organic method like biological and cultural to reduce infestation from outside.
- Irrigated land is good for application of pesticides when it shows no signs of moisture stress.
- Local fire department must be informed before application of smoke formulation agrochemicals.
- Pesticides together with fertiliser should not be used until and unless indicated on the label.
- Previous used sprayer of insecticides should not be used to apply herbicides.
- Reduce the frequency of application whenever possible, particularly nerve toxins. Always evaluate the cost-benefit analysis to justify the application of agrochemicals.
- Rotation of fertiliser and pesticide is good practice for better efficiency.
- Tank mixing should be avoided (mixing two or more insecticides together to control a single pest) until and unless research don't support that practice.
- Until and unless it is labelled on pesticide don't use surfactants.
- Use agrochemical at only small part as much as possible, and whenever possible apply pesticide only on infested plant(s) rather than treating all plants in the field.
- Ban on restricted agrochemical.
- More and more use of compost, green manure, vermicompost and promote organic farming.

15.10 SUGGESTED FRAMEWORK FOR THE MANAGEMENT OF AGROCHEMICALS

While purchasing

- Buy agrochemical from only government registered dealers who are having valid licence.
- Buy only limited amount of agrochemical which is required for single use.
- Always check approved label on the packing.
- Check before buying, date of expiry, batch no., registration no., etc.
- Before buying check the seal of packet that must be sealed.

During storage

- Agrochemical storage should be away from home.
- Agrochemical must be kept in original containers.
- Warning sign should be marked where agrochemicals are stored.
- Reach of children and livestock should be away from agrochemical.
- Direct sunlight and rain must be avoided at storage place.

While handling

- Bulk quantity should be taken to site with proper care.
- During transportation agrochemical should be kept separate.

While preparing spray solution

- Clean water must be used.
- Protective kit like hand glove, mask, cap and full pants must be used.
- Protection must be taken for the nose, ear, hand and eyes from spillage.
- Spillage must be avoided during filling.
- Use recommended dose of agrochemicals.
- Solution must be prepared as per requirement.

Selection of equipment

- Selection of proper equipment is important.
- Proper size nozzle should be used.
- Different sprayer should be used for different pesticides.

While applying spray solutions

- Dilution must be used as per recommendation.
- Cool period must be chosen for application.
- Entry of animals and workers should be prohibited immediately after spray.
- Wind direction must be considered before application.

After spray operation

- Isolated or barren area should be identified for safe disposal of left over.
- Used container should be crushed and must be buried in soil.
- Before eating, washing of hand and body with soap and clean water is must.
- Having any problem of poisoning, immediately contact doctor showing label of used agrochemical.

15.11 CONCLUSIONS

Extensive use of agrochemicals helps to increase the production of crops by many folds. Although, excessive use of fertilisers and pesticides becomes serious threat to environment and human health. Soil, groundwater and surface water are primary environmental component that are at risk of contamination through the use of agrochemicals. Through bioaccumulation from plants and intake of agrochemical contaminated groundwater serious health impacts are caused to the humans. Some of the agrochemicals exhibit carcinogenic properties which can cause cancer and other major health effects. Due to such serious threat regulatory bodies are setting up drinking water standards specifically related to pesticides and fertilisers. Nowadays, many instruments and techniques are available for detection of agrochemicals like LC, GC and mass spectrophotometry. As use of agrochemical causes serious problem to human health and to environment, we need to look for alternative like organic farming, and already contaminated sites can be treated through nanotechnology and bioremediation (earthworm, microbial, fungal-assisted treatment), phytoremediation, etc. Other than treatment technology we need to follow some best pest management practices to overcome the ill effect of agrochemical.

REFERENCES

Ab Rahman, S. F. S., Singh, E., Pieterse, C. M., & Schenk, P. M., 2018. Emerging microbial biocontrol strategies for plant pathogens. *Plant Science, 267*, 102–111.

Alam, M. S., Akhtar, M. Q., Albratty, M., Alhazmi, H. A., & Ahsan, W., 2020. Chemical pesticides: Integrated methods in assessment and monitoring. In: Rouf A.B., Khalid R.H., Najla B.S.A.S. (eds) *Bioremediation and biotechnology*, Vol *3* (pp. 39–63). Springer, Cham.

Arias-Estévez, M., López-Periago, E., Martínez-Carballo, E., Simal-Gándara, J., Mejuto, J. C., & García-Río, L., 2008. The mobility and degradation of pesticides in soils and the pollution of groundwater resources. *Agriculture, Ecosystems & Environment, 123*(4), 247–260.

Babu, S. C., Nivas, B. T., & Rajasekaran, B., 1992. Groundwater pollution from agrochemicals—A dynamic model of externalities and policy options. *Water Resources Management, 6*(1), 1–13.

Baker, E. A., Hayes, A. L., & Butler, R. C., 1992. Physicochemical properties of agrochemicals: Their effects on foliar penetration. *Pesticide Science, 34*(2), 167–182.

Bambaradeniya, C. N., & Amerasinghe, F. P., 2004. *Biodiversity associated with the rice field agroecosystem in Asian countries: A brief review* (Vol. 63). IWMI, Colombo, Srilanka.

Barak, P., Jobe, B. O., Krueger, A. R., Peterson, L. A., & Laird, D. A., 1997. Effects of long-term soil acidification due to nitrogen fertilizer inputs in Wisconsin. *Plant and Soil, 197*(1), 61–69.

Bhadouria, R., Das, S., Kumar, A., Singh, R., & Singh, V. K., 2020. Mycoremediation of agrochemicals. In *Agrochemicals detection, treatment and remediation* (pp. 593–620). Butterworth-Heinemann, Amsterdam, Netherland.

BIS, 2012. *Indian standard drinking water-specification (second revision)*, Bureau of Indian Standards, New Delhi, IS 10500:2012.

Biswas, S. K., Rahman, S., Kobir, S. M. A., Ferdous, T., & Banu, N. A., 2014. A review on impact of agrochemicals on human health and environment: Bangladesh perspective. *Plant Environment Development, 3*(2), 31–35.

Bloomfield, J. P., Williams, R. J., Gooddy, D. C., Cape, J. N., & Guha, P. M., 2006. Impacts of climate change on the fate and behaviour of pesticides in surface and groundwater—A UK perspective. *Science of the Total Environment, 369*(1–3), 163–177.

Chabot, F., Mitchell, J. A., Gutteridge, J. M., & Evans, T. W., 1998. Reactive oxygen species in acute lung injury. *European Respiratory Journal, 11*(3), 745–757.

Cheng, G., Lin, J., Lu, J., Zhao, X., Cai, Z., & Fu, J., 2015. Advanced treatment of pesticide-containing wastewater using Fenton reagent enhanced by microwave electrodeless ultraviolet. *BioMed Research International, 2015*, 205903.

Chowdhury, A. Z., Jahan, S. A., Islam, M. N., Moniruzzaman, M., Alam, M. K., Zaman, M. A., ... & Gan, S. H., 2012. Occurrence of organophosphorus and carbamate pesticide residues in surface water samples from the Rangpur district of Bangladesh. *Bulletin of Environmental Contamination and Toxicology, 89*(1), 202–207.

Cimino, A. M., Boyles, A. L., Thayer, K. A., & Perry, M. J., 2017. Effects of neonicotinoid pesticide exposure on human health: A systematic review. *Environmental Health Perspectives, 125*(2), 155–162.

Daam, M. A., Chelinho, S., Niemeyer, J. C., Owojori, O. J., De Silva, P. M. C., Sousa, J. P., ... & Römbke, J., 2019. Environmental risk assessment of pesticides in tropical terrestrial ecosystems: Test procedures, current status and future perspectives. *Ecotoxicology and Environmental Safety, 181*, 534–547.

Dhir, B., 2020. Green technologies for the removal of agrochemicals by aquatic plants. In: Prasad M.N.V (ed) *Agrochemicals detection, treatment and remediation* (pp. 569–591). Butterworth-Heinemann, Amsterdam.

Eregowda T., Mohapatra S. (2020) Fate of micropollutants in engineered and natural environment. In: Kumar M., Munoz-Arriola F., Furumai H., Chaminda T. (eds) *Resilience, response, and risk in water systems. Springer transactions in civil and environmental engineering.* Springer, Singapore. https://doi.org/10.1007/978-981-15-4668-6_15

Florencia, R. M., & Cora, S., 2020. Agrochemicals and neurogenesis. *Molecular and Cellular Endocrinology, 510*, 110820.

Gavrilescu, M., 2005. Fate of pesticides in the environment and its bioremediation. *Engineering in life sciences, 5*(6), 497–526.

Hendges, C., Schiller, A. D. P., Manfrin, J., Macedo Jr, E. K., Gonçalves Jr, A. C., & Stangarlin, J. R., 2019. Human intoxication by agrochemicals in the region of South Brazil between 1999 and 2014. *Journal of Environmental Science and Health, Part B, 54*(4), 219–225.

Henriques, W., Jeffers, R. D., Lacher Jr, T. E., & Kendall, R. J., 1997. Agrochemical use on banana plantations in Latin America: Perspectives on ecological risk. *Environmental Toxicology and Chemistry: An International Journal, 16*(1), 91–99.

Hercegová, A., Dömötörová, M., & Matisova, E., 2007. Sample preparation methods in the analysis of pesticide residues in baby food with subsequent chromatographic determination. *Journal of Chromatography A, 1153*(1–2), 54–73.

Jallow, M. F., Awadh, D. G., Albaho, M. S., Devi, V. Y., & Thomas, B. M., 2017. Pesticide risk behaviors and factors influencing pesticide use among farmers in Kuwait. *Science of the Total Environment, 574*, 490–498.

Kabaluk, J. T., Svircev, A. M., Goettel, M. S., & Woo, S. G. (Eds.), 2010. *The use and regulation of microbial pesticides in representative jurisdictions worldwide* (p. 99). International Organization for Biological Control of Noxious Animals and Plants (IOBC).

Kah, M., Beulke, S., & Brown, C. D., 2007. Factors influencing degradation of pesticides in soil. *Journal of agricultural and food chemistry, 55*(11), 4487–4492.

Kamel, F., & Hoppin, J. A. 2004. Association of pesticide exposure with neurologic dysfunction and disease. *Environmental Health Perspectives, 112*(9), 950–958.

Kashyap, P. L., Xiang, X., & Heiden, P., 2015. Chitosan nanoparticle-based delivery systems for sustainable agriculture. *International Journal of Biological Macromolecules, 77*, 36–51.

Kaur, R., Mavi, G. K., Raghav, S., & Khan, I., 2019. Pesticides classification and its impact on environment. *International Journal of Current Microbiology and Applied Sciences, 8*, 1889–1897.

Kim, K. H., Kabir, E., & Jahan, S. A., 2017. Exposure to pesticides and the associated human health effects. *Science of the Total Environment, 575*, 525–535.

Leng, Y., Bao, J., Xiao, H., Song, D., Du, J., Mohapatra, S., Werner, D. and Wang, J., 2020. Transformation mechanisms of tetracycline by horseradish peroxidase with/without redox mediator ABTS for variable water chemistry. *Chemosphere, 258*, p. 127306.

Lewis, K. A., Tzilivakis, J., Warner, D. J., & Green, A., 2016. An international database for pesticide risk assessments and management. *Human and Ecological Risk Assessment: An International Journal, 22*(4), 1050–1064.

Macpherson, G. L., & Sophocleous, M., 2004. Fast ground-water mixing and basal recharge in an unconfined, alluvial aquifer, Konza LTER Site, Northeastern Kansas. *Journal of Hydrology, 286*(1–4), 271–299.

Magauzi, R., Mabaera, B., Rusakaniko, S., Chimusoro, A., Ndlovu, N., Tshimanga, M., ... & Gombe, N., 2011. Health effects of agrochemicals among farm workers in commercial farms of Kwekwe district, Zimbabwe. *The Pan African Medical Journal, 9*(1), 26. https://doi.org/10.4314/pamj.v9i1.71201

Mandal, A., Sarkar, B., Mandal, S., Vithanage, M., Patra, A. K., & Manna, M. C., 2020. Impact of agrochemicals on soil health. In: Prasad M.N.V (ed) *Agrochemicals detection, treatment and remediation* (pp. 161–187). Butterworth-Heinemann, Amsterdam.

Mezzanotte, V., Canziani, R., Sardi, E., & Spada, L., 2005. Removal of pesticides by a combined ozonation/attached biomass process sequence. *Ozone: Science and Engineering, 27*(4), 327–331.

Mogili, N.V. and Mohapatra, S., 2021. Microbial community analysis of wastewater during blackwater treatment. In: Maulin Shah, Susana Rodriguez-Couto (eds) *Wastewater treatment reactors* (pp. 483–507). Elsevier, Amsterdam.

Mohapatra, S., Sharma, N., Mohapatra, G., Padhye, L.P. and Mukherji, S., 2021. Seasonal variation in fluorescence characteristics of dissolved organic matter in wastewater and identification of proteins through HRLC-MS/MS. *Journal of Hazardous Materials, 413*, 125453.

Müller, K., Bach, M., Hartmann, H., Spiteller, M., & Frede, H. G., 2002. Point-and nonpoint-source pesticide contamination in the Zwester Ohm catchment, Germany. *Journal of Environmental Quality, 31*(1), 309–318.

Priyanka M.S., Mohapatra S. 2020. Shifts and Trends in Analysis of Contaminants of Emerging Concern: Sulfonamides. In: Kumar M., Munoz-Arriola F., Furumai H., Chaminda T. (eds) *Resilience, response, and risk in water systems* (pp. 65–94). Transactions in Civil and Environmental Engineering. Springer, Singapore. https://doi.org/10.1007/978-981-15-4668-6_14.

Priyanka, M. S., Mohapatra, S., & Menon, N. G. 2021. Hybrid membrane technology: an alternative to industrial wastewater treatment. In Maulin P. Shah and Susana Rodriguez-Couto (eds) *Membrane-based hybrid processes for wastewater treatment* (pp. 481–501), Elsevier, Netherlands.

Prudente, I. R. G., Cruz, C. L., de Carvalho Nascimento, L., Kaiser, C. C., & Guimaraes, A. G., 2018. Evidence of risks of renal function reduction due to occupational exposure to agrochemicals: A systematic review. *Environmental Toxicology and Pharmacology, 63*, 21–28.

Rani, K., & Dhania, G., 2014. Bioremediation and biodegradation of pesticide from contaminated soil and water—A noval approach. *International Journal of Current Microbiology and Applied Sciences, 3*(10), 23–33.

Reinert, K. H., Giddings, J. M., & Judd, L., 2002. Effects analysis of time-varying or repeated exposures in aquatic ecological risk assessment of agrochemicals. *Environmental Toxicology and Chemistry: An International Journal, 21*(9), 1977–1992.

Riley, H., Pommeresche, R., Eltun, R., Hansen, S., & Korsaeth, A., 2008. Soil structure, organic matter and earthworm activity in a comparison of cropping systems with contrasting tillage, rotations, fertilizer levels and manure use. *Agriculture, Ecosystems & Environment, 124*(3–4), 275–284.

Sankhla, M. S., Kumari, M., Sharma, K., Kushwah, R. S., & Kumar, R., 2018. Water contamination through pesticide & their toxic effect on human health. *International Journal for Research in Applied Science & Engineering Technology (IJRASET), 6*(1), 967–970.

Schmid, F., Moser, G., Müller, H., & Berg, G., 2011. Functional and structural microbial diversity in organic and conventional viticulture: organic farming benefits natural biocontrol agents. *Applied and Environmental Microbiology, 77*(6), 2188–2191.

Sebastian, A., Nangia, A., & Prasad, M. N. V., 2020. Advances in agrochemical remediation using nanoparticles. In: Prasad M.N.V. (ed) *Agrochemicals detection, treatment and remediation* (pp. 465–485). Butterworth-Heinemann, Amsterdam.

Sekhotha, M. M., Monyeki, K. D., & Sibuyi, M. E., 2016. Exposure to agrochemicals and cardiovascular disease: a review. *International Journal of Environmental Research and Public Health, 13*(2), 229.

Shoda, M. E., Nowell, L. H., Stone, W. W., Sandstrom, M. W., & Bexfield, L. M., 2018. Data analysis considerations for pesticides determined by National Water Quality Laboratory schedule 2437. Scientific Investigations Report (2018–5007).

Singh, N. S., Sharma, R., Parween, T., & Patanjali, P. K., 2018. Pesticide contamination and human health risk factor. In: Mohammad Oves, Mohammad Zain Khan, Iqbal M.I. Ismail (eds) *Modern age environmental problems and their remediation* (pp. 49–68). Springer, Cham.

Singh, S. I., Singh, S., & Vig, A. P., 2020. Earthworm-assisted bioremediation of agrochemicals. In: Prasad M.N.V (ed) *Agrochemicals detection, treatment and remediation* (pp. 307–327). Butterworth-Heinemann, Amsterdam.

Taiwo, A. M., 2019. A review of environmental and health effects of organochlorine pesticide residues in Africa. *Chemosphere, 220*, 1126–1140.

Thompson, J., Pelc, C. E., & Jordan, T. E., 2020. Water quality sampling methods may bias evaluations of watershed management practices. *Science of the Total Environment, 765*, 142739.

Tripathi, S., Srivastava, P., Devi, R. S., & Bhadouria, R., 2020. Influence of synthetic fertilizers and pesticides on soil health and soil microbiology. In: Prasad M.N.V (ed) *Agrochemicals detection, treatment and remediation* (pp. 25–54). Butterworth-Heinemann, Amsterdam.

Welch, R. M., Levin, W., & Conney, A. H., 1969. Estrogenic action of DDT and its analogs. *Toxicology and Applied Pharmacology, 14*(2), 358–367.

Wimalawansa, S. A., & Wimalawansa, S. J., 2014. Agrochemical-related environmental pollution: Effects on human health. *Global Journal of Biology, Agriculture and Health Sciences, 3*(3), 72–83.

Wolpert, F., Quintas-Soriano, C., & Plieninger, T., 2020. Exploring land-use histories of tree-crop landscapes: A cross-site comparison in the Mediterranean Basin. *Sustainability Science*, 1–17.

World Health Organization, 2017. *Guidelines for Drinking-Water Quality: First Addendum to the Fourth Edition*, Geneva.

Worrall, F., Besien, T., & Kolpin, D. W., 2002. Groundwater vulnerability: Interactions of chemical and site properties. *Science of the Total Environment, 299*(1–3), 131–143.

Yadav, I. C., & Devi, N. L., 2017. Pesticides classification and its impact on human and environment. *Environmental Science and Engineering, 6*, 140–158.

Yuan, L., Li, J., Zha, J., & Wang, Z., 2016. Targeting neurotrophic factors and their receptors, but not cholinesterase or neurotransmitter, in the neurotoxicity of TDCPP in Chinese rare minnow adults (*Gobiocypris rarus*). *Environmental Pollution, 208*, 670–677.

Zanella, R., Prestes, O. D., Friggi, C. D. A., Martins, M. L., & Adaime, M. B., 2012. An overview about recent advances in sample preparation techniques for pesticide residues analysis in cereals and feedstuffs. In: Soundararajan R.P. (ed) *Pesticides: Recent trends in pesticide residue assay* (p. 131). IntechOpen Publisher, London, UK.

Chapter 16

Nanomaterials and devices for the provision of safe drinking water in rural communities

David F. L. Jenkins and Jonathan M. Bloor
University of Plymouth

Bing Li
Imperial College London
UK Dementia Research Institute

Vikram S. Raghaven
Indian Institute of Science

Richard D. Handy
University of Plymouth

Shivaraju Harikaranahalli Puttaiah
JSS AHER

Awadhesh N. Jha
University of Plymouth

Sai S. Gorthi
Indian Institute of Science

CONTENTS

DOI: 10.1201/9781003247869-19

16.1 INTRODUCTION

With growing human population and discharge of a range of contaminants in the environment, there is an increasing demand for clean water to protect human and environmental health. Potable water supplies are, however, too frequently of insufficient quality and negatively impact human health in different parts of the world. This is especially true in rural regions of countries with a low-income economy and where the infrastructure for water purification may be lacking or even missing. The main problems include contamination of potable water with anthropogenic as well as natural (e.g. arsenic) toxic pollutants and pathogens. In addition, increasing salinity of groundwater, systemic shortages through infrastructure neglect and the impact of climate change causing extreme events in the water cycle also contribute to this shortage globally. The unwanted presence of toxic metals, biocides, organics (e.g. polycyclic aromatic hydrocarbons), pharmaceuticals (e.g. antibiotics) and pathogens in ground water supplies is pivotal in the battle to provide simple access to safe drinking water particularly in rural communities including in India. Toxic metals can be attributed to geological factors (arsenic and fluoride, for example), but more concerning are those attributed to illegal industrial activities that pollute rivers and lakes. It is not uncommon for industries to discharge chemical waste into rivers illegally or beyond the permissible levels due to unguarded monitoring processes or law enforcement. Poorly designed land-fill sites are also contributory factors for such contaminations. In a recent study, pollution evaluation indices were carried out to assess the intensity of pollution in the Marar industrial area of Ramgarh in Jharkhand, India (Kumar et al., 2020). Groundwater samples were collected and the physicochemical parameters (e.g. pH and EC, HCO_3^-, Cl^-, SO_4^{2-}, $NO_3{}^-$, F^-, Ca^{2+}, Mg^{2+}, Na^+, K^+) and metals (e.g. Fe, Mn, As, Zn, Cu, Cr, Ni, Se, Al, Ba, Pb and Cd) were analysed. The concentration of metals in the study area in decreasing order was reported as follows: Fe > Zn> Mn > Al> Ba > Ni> Cu > Pb> Cr > Se> As > Cd. The results showed that Fe, Al, Mn and Ni were present in water samples at levels greater than the desirable limit as per BIS,10,500 (2012). Concentration of As, Cu, Cr, Se, Pb, Ba and Cd was within the desirable limit in each sample. However, the presence and concentration levels of metals like As, Cd, Cr, Pb and Co are above BIS levels in many regions of India, and these metals are not essential for human health. In fact they may prove detrimental even at low concentrations (Giri et al., 2019). Setia et al. (2020) analysed the presence of metals in surface water of Sutlej River, India, contaminated by industrial waste, sewage and agricultural runoff. Samples were analysed for ten metals (Zn, Cu, Fe, Mn, Ni, Cd, Pb, Co, Cr and As) using Atomic Absorption Spectroscopy (AAS). Metal contamination was higher along the transboundary followed by the area from the confluence of contaminated drain with river to transboundary along with localised spots in the river flowing in Indian Punjab. The likelihood of cancer risk due to ingestion of metals through water was in the order: Cd > N I > C r>. In another study, Pandiyan et al. (2021) determined metal pollution in different wetlands of Tamil Nadu. They assessed the levels of cadmium (Cd), copper (Cu), chromium (Cr), cobalt (Co), lead (Pb), mercury (Hg), nickel (Ni) and zinc (Zn) in water, sediment and in aquatic organisms. As a result, this can impact human health via food security and the associated food chain. The presence of these metals was found in water, sediment, molluscs, crabs, prawns and fish, and notably higher in sediment. This also threatens the existence of various benthic and food organisms (e.g. prawns, fish) which are the preferred prey for various migratory shorebirds, with long-term implications for environmental sustainability.

Pathogens in ground water are of growing concern. One common pathogen is *Escherichia coli*. The signs and symptoms of *E. coli* infection usually begin 3 or 4 days after exposure to the bacteria, which include diarrhoea, stomach cramps, pain or tenderness, nausea and vomiting, which could be life threatening to young children. The presence of pathogens in water supplies throughout the developing world has resulted in over 1 million deaths due to diarrhoea-related diseases. A large proportion of which are children (Sankar et al., 2013) giving rise to unacceptably high levels of infant mortality. *Escherichia coli*, a thermotolerant member of the coliform group, has emerged as the most widely used indicator of faecal contamination in drinking water (Pantha et al., 2021). According to the Sustainable Development Goals, it is required that a 100 mL water sample contain no culturable *E. coli*. This is a global problem, with an estimated 1.8 billion people worldwide reliant on faecally contaminated water. In one recent study in Rajasthan, northwestern India, where the city of Jaipur receives drinking water exclusively from groundwater (Bivins, 2020), the presence of *E. coli* was investigated. Despite the absence of culturable *E. coli* in 90% of the 100 mL grab samples (n = 20) during the 10-day sampling period, they detected genes associated with protozoan and bacterial pathogens (*Giardia*, *Cryptosporidium* and enterotoxigenic *E. coli*) in three dead-end ultrafiltration samples of groundwater (n = 9; volume 59 to 122.4 L). Of the three groundwater samples positive for waterborne pathogens, two were associated with 100 mL grab samples that were negative for culturable *E. coli*. As part of this study 22 samples were taken from ground water, surface water and where the sources were mixed, and 19 of those 22 dead-end ultrafiltration samples were positive for the *E. coli* associated gene *ybbW*, with no false positives. If chlorine is added in urban water networks, the situation becomes very different (Mohapatra et al., 2014). Disinfection with chlorine might be considered, but the high concentrations of halides and organic contaminants observed in the groundwater increase the likelihood of forming disinfection by-products (Sharma et al., 2021).

Water quality is generally very high in developed countries with good infrastructure, but even these nations can have incidents that may affect public health. For example, lead exposure in the drinking water of the residents of the city of Flint in Michigan, USA, in 2014 was associated with a change in pipeline supply, reiterating the importance of monitoring raw and treated water for toxic metals (Hanna-Attisha et al., 2016). The problem of water quality is, however, global. Water quality issues do vary from nation to nation, for example; across the African continent, high levels of antivirals are present due to the Aids and Ebola epidemics (Wood et al., 2015). In Bangladesh the natural geology contains high concentrations of arsenic that can be released into the water table, and with concerns about arsenic bioaccumulating in crops such as rice (Kundu et al., 2018), and in the UK, 83% of tap water is contaminated by microplastics (Kosuth et al., 2018). Even in countries with good enforcement of water quality standards, the limits can be breached, for example due to intermittent sampling that may 'miss' a water quality event. Chemical water analysis and remediation is a very complex process to manage; each contaminant species may require a multi-stage approach to ensure adherence to the international acceptable limits (Richardson, 2009). A robust water infrastructure in developed economies allows the purification steps, chemical analysis and the monitoring of compliance to be harmonised so that action can be taken swiftly and effectively. For example, according to data from the Drinking Water Inspectorate (DWI) in the UK, compliance is still above 99.96% (DWI, 2018). Realistically, it is difficult to remove all contaminants from water sources without a cohesive infrastructure. Waste management, sewage and water treatment works and a healthy water supply network all require sustained capital investment. In rural locations where the population may be dispersed, the infrastructure costs per head of population can be especially expensive, and so alternative practical solutions are sought for water supply and measuring water quality. One such approach is to improve monitoring capabilities in real-time using 5G Internet of Things (IOT) low-cost sensor technology and novel remediation techniques to overcome failures in infrastructure (Glasgow et al., 2004). Throughout this chapter, the contrasts between a developed water infrastructure and rural supplies (UK–India) are drawn, and our discussion spans sensing and novel filtration technology to help close the gap between what should be a human right for all, potable drinking water.

In this chapter, we focus on the key issues that affect water quality. In most cases, levels found in water exceed Indian and globally accepted standards:

- Metals
- Endocrine disrupting chemicals
- Persistent organic pollutants (POPs)
- Dyes
- Pathogens
- Pharmaceuticals and emerging contaminants

16.2 CHEMICAL OF CONCERN FOR HUMAN HEALTH AND WATER SUPPLY

16.2.1 Common approaches to water purification

Arguably, access to clean drinking water is steadily reducing per capita as the global population (currently 7.8 billion) increases due to the amount available from the hydrological cycle remaining largely fixed. The factors which affect access to safe drinking water include high levels of contamination, which can be compounded by severe droughts followed by extreme rainfall. Water management is severely lacking too in many developing countries. The global picture of this is evident across the Indian subcontinent. For example, over 85% of the fresh water in India is available as groundwater, which is subject to contamination via flooding, the mixing of waste and potable water, industrial discharges, agricultural run-off, urban run-off and direct contamination through improper storage (Khurana and Sen, 2008). Developed municipal infrastructure and methodologies for the treatment of potable water are well recognised and shared from country to country. The India Standards Drinking Water Specification (2012) is based on the Manual of Water Supply Treatment, Third edition (GOV India, 1999), New Delhi. This guideline is in line with the WHO Guidelines for Drinking Water Quality (2008), National Primary Drinking Water Standard (2002) and the EU Directives relating to quality of water for human consumption (CPHEEO, 1999, WHO, 2011, WHO, 2008, Directive, 1980). The common underpinning methods of treatment for potable water include preliminary screening to remove debris, coagulation and flocculation to remove solids/particulates, sand filtration, and then disinfection (e.g. chlorination) to kill microbes.

Management of any natural resources is complicated by geography, geopolitics and unstable climatic conditions. The supply of suitable water will be influenced by factors in climate change such as rising sea levels and the ingress of salt into groundwater, and cycles of drought and flooding that add an extra challenge to the management of water infrastructure (Khurana and Sen, 2008). Potable water infrastructure is expensive to maintain and has to be tailored to the relevant geographic and 'raw' water conditions, and yet India spends only around 0.1% of Gross Domestic Product (GDP) on water (Tortajada, 2016). Urbanisation, especially illegal urbanisation, of the land brings additional challenges with percolation of contaminants into the groundwater (Wakode et al., 2018) and the infrastructure not keeping pace with movements of the populations into the cities. Notwithstanding climate change and the impact of rural-to-urban migration, the implementation of an infrastructure is a challenging task for policy makers and planners.

A vast majority of rural towns and villages in India are without municipal water or wastewater treatment facilities, while cities have infrastructure that is at the theoretical limits of capacity and beyond, in particular drilling of ever deeper boreholes to tap into groundwater supplies (Bhardwaj, 2005). The propagation of raw sewage into rivers, lakes and onto agriculture land is a common practice. It is estimated by the Indian Central Pollution Control Board (CPCB) in 2011 that 269 sewage treatment plants (STPs) around 40 are not operational, resulting in a treatment capacity reduction of 14% for the total urban wastewater treatment (Kamyotra and Bhardwaj, 2011). In 2020 though, according to the CPCB report on Common Effluent Treatment Plant (CETPs) and STPs, there were a total 191 CETPs, out of which 129 CETPs are complying with environmental standards and 62 CETPs are non-complying. There were also a total of 15,730 STPs (including municipal and other than municipal (non-municipal/standalone) STPs), out of which 15,200 STPs are complying with environmental standards and 530 STPs are non-complying (CPCB, 2020).

Boosting crop yields with the use of treated water as the final effluent from the sewage treatment works can produce some positive results, an increased yield by 10%–30%, but due to the lack of infrastructure and cost, the use of untreated sewage wastewater is prevalent. In the UK, however, final effluent is the cleaned and treated water from Water Recycling Centres which is permitted by the Environment Agency for discharge directly into the environment. In some cases, this final effluent is suitable for industrial or business activities and can be an alternative to potable water (Anglian Water, 2020). With such a significant gap in water treatment infrastructure, agricultural systems are therefore exposed to both the chemical pollutants and pathogens from untreated wastewater.

16.2.2 Pathogens in drinking water

The transmission of waterborne pathogenic diseases through the use of untreated wastewater in agriculture is a concern, coming from either direct contamination from the effluent or incidentally via the unwashed agricultural products or soil. Also, the absence of basic sanitation in rural communities compounds this waste problem, with faecal matter inevitably mixing with water than may also be used for drinking water supply. Sewage treatment largely mitigates the risk of diseases such as cholera and typhoid, but more often than not the infrastructure is either not present in rural locations or it is beyond the financial means of the local authorities (Bradford et al., 2003, Blumenthal et al., 2000). The Indian government's launch of the 'Swachh Bharat Mission' in 2014, to emphasise higher levels of sanitation, is due to be completed by October 2019. The Ministry of Drinking Water and Sanitation (MDWS) used this mission to make rural areas open defaecation free (ODF), firstly by attaining no visible signs of faeces in the village environment and secondly by providing every dwelling with sanitation facilities, including public and community spaces (Mission and Aayog, 2017). The MDWS is also in the process of evaluating new innovative technologies to assist the Swachh Bharat Mission and future projects, and the idea is to use the project as a springboard and build momentum to raise the levels of sanitation, access to potable water and prepare for the future requirements. The Compendium of Innovative Technologies on Rural Drinking Water & Sanitation (CITRDWS) provides a platform for state government officials to showcase and innovative technologies that provide robust solutions to water and sanitation problems within their region, the Compendium of Innovative Technologies on Rural Drinking Water & Sanitation (CITRDWS) document lists the future technologies that have already been vetted by the ministry, of which carbon aerogels are near the top of the list (Mission and Aayog, 2017).

Developing countries like India face issues with maintaining drinking water quality, specifically in rural regions. The WHO standards for microbiology in drinking water are somewhat risk based and depend on the pathogen load needed to cause ill health (WHO, 2017), but generally for bacteria a load of ten colony-forming units (cfu) per 100 mL or less would be considered 'low risk'. In the UK, the microbiology standards for drinking water at the consumer's tap are no detection of bacterial counts for coliform species (organisms associated with faecal waste) and the absence of *E. coli*. In rural India, the detection of microbes is far above the low risk category. Suthar et al. (2009) reported 25–41 cfu/100 mL in a survey of drinking water in rural regions of Rajasthan in India. Coliform bacteria groups such as *E. coli* and *Enterobacter aerogenes* were found in the drinking water above the recommended levels. Recently, McGuinness et al. (2020) categorised microbial risks from drinking water in rural India, with the finding that river water was often the main source of drinking water, and with only 5% of the river water samples meeting the WHO standard for *E. coli*. Notably, the same study also showed that well water and point of use water from simple domestic water storage systems were much safer (69% and 31% meeting the WHO standard, respectively). Chauhan et al. (2017) reported the microbiological levels on drinking water sold by vendors in Delhi, India. Of the 36 samples examined, the bacterial counts ranged from 14 to 1600 cfu/100 mL. The prominent coliform was *E. coli* with 61% occurrence in the samples, and none were within the WHO guidelines for microbes in drinking water.

Viruses in drinking water are also a concern for the waterborne transmission of disease. Viruses are too small (~100 nm) to be removed by standard filtration methods used in water treatment, and instead there is reliance on disinfecting the water before it is supplied to the consumer. The disinfectant methods can include the use of UV light, ozone, chloramines and oxidants such as chlorine at the treatment works for water (Gall et al., 2015). There are no specific numerical standards on viral loads in drinking water from the WHO, but there is guidance of the types of viruses of concern and their effects on health (WHO, 2017). In developed countries, the emphasis is on removal or inactivation of viruses by disinfection. In the United States, the Environmental Protection Agency (EPA) requires, for example, the inactivation or removal of 99.99% of enteric viruses. However, while these disinfection methods may have utility in a treatment work, in rural locations in developing countries the options are mainly limited to adding disinfectants such as chlorine tablets directly to the collected water. Outbreaks of hepatitis E virus in the human population in India have been attributed to faecal contamination of the water supply (Kaur et al., 2017), and viruses that cause gastroenteritis are also a major concern (e.g. rotavirus, Joshi et al., 2019). However, such studies often do not confirm cause and effect by measuring the viral loads in the suspect water. Viral loads in drinking water in India remain largely unreported, partly because of the difficulty and expense of the molecular detection methods. There are also some concerns regarding the transmission of the COVID types of viruses in water supplies in India (Bhowmick et al., 2020), but these remain unsubstantiated.

16.2.3 Chemical pollutants and toxic metals

Toxic metals, endocrine disrupting chemicals (EDC), POPs and salinity are also a major concern for large parts of India. Coastal regions that have depleted groundwater levels through excessive bore hole drilling, coupled with porous rock formations as a coastal barrier are susceptible to ingress of seawater to redress the pressure differential, turning the drinking water brackish and no longer potable (Barlow and Reichard, 2010). Dichlorodiphenyltrichloroethane (DDT) was originally used as an insecticide and is now regarded as a POP. In India, despite legally banning the use of POPs, except for DDT, which is used as a control vector for malaria, a 2001 nationwide survey revealed that 75% of food samples had a detectable level of DDT, 10%–15% of which were over the limits set by the WHO (IPEN, 2006). Mean DDT concentrations in ground water across India varies from 6904 ng/L in Nagaon, Assam (Mishra and Sharma, 2011), to as low as 10 ng/L in Greater Kolkata, West Bengal (Ghose et al., 2009) both of which exceed the limit set out by the India Bureau of Standards which is 1 μg/L, 1,000 ng/L (DWS, 2012). Surface water resources also has high levels of DDT, in Jaipur, Rajasthan, the Ramgarh water reservoir reached 133,00 ng/L in 2004 (Bakore et al., 2004) and a drainage source near a Lindane production plant dump reached levels exceeding 17,430,000 ng/L (Jit et al., 2011). Exposure to POPs is endemic due to their ability for long distance transport in a gaseous phase, persistence in the environment, lipophilic nature resulting in bioaccumulation in fat-rich tissue and their continued use in some countries, which still have problems controlling insect transmitted diseases. Despite removal of use in many countries, the International Agency for the Research on Cancer (IARC) has concluded that DDT is not considered toxic to humans, but has nevertheless been demonstrated in mammals, rats and mice (WHO, 2011).

Some metals are well-known for their hazard to human health, including arsenic (Chakraborti et al., 2009), lead (Bellinger, 2008), chromium (George et al., 2016), mercury (Counter and Buchanan, 2004) and cadmium (Al-Khaldi et al., 2015), which can pose a threat to public health even at μg/L concentrations. Inorganic arsenic, which occurs naturally in the soil, is now prevalent in groundwater throughout India and Bangladesh (Chowdhury et al., 2000). For example, of the 19 districts of West Bengal in India, 48%, 24% and 3% had total arsenic concentrations above 10, 50 and 300 μg/L, respectively (Chakraborti et al., 2009). Symptomatic of arsenic exposure are a variety of cancer-based diseases, vascular disease and diabetes (Chakmeng, 2005). In the recent years, pollution by arsenic (As) has become a serious issue of concern in view of its toxicity to humans and it is prevalent across India today (Shaji et al., 2020), and research is underway to map the distribution of inorganic arsenic using Machine Learning Geospatial Modelling (Podgorski et al., 2020).

Lead poisoning has been revisited recently resulting in a lowering of the blood lead levels that are deemed safe, from 25 to 10 μg/L as of December 2013, calling into question, is there any safe lower limit for lead exposure (Hanna-Attisha et al., 2016). Lead can enter the body via the digestive tract airborne and absorbed through lung tissue but the primary route to exposure is through drinking water. The concerns are regarding long-term exposure to the central nervous system, causing intellectual impairment especially during development in children, also there are the long-term health effects on renal function and cancer risks (Canfield et al., 2003). Lead concentrations in drinking water should not exceed 10 μg/L (WHO, 2017). Chromium (Cr) is an abundant element in the earth's crust, occurring in divalent, trivalent and hexavalent forms. Cr (III) is an essential element for the correct functioning of the human body, but is severely toxic in the Cr (VI) form (Pavesi and Moreira, 2020). Generally the measure for total chromium is made due to the difficulties isolating the hexavalent form. Cr (VI) is classified as a Group 1 human carcinogen by the IARC; in contrast, Cr (III) is classed in Group 3, non-carcinogenic to human. The mining of Pb, Cr and other metals and their use in industrial process including leather tanning, metal plating and the textile industry are a main concern (Sharma et al., 2009). Consequently, the exposure via water supplies is often associated with industrialised or urban areas. However, metal pollution can also percolate into the water table, with subsequent contamination of the groundwater. In rural India, the concern is about long-term exposure to low concentrations of bioaccumulative metals in the μg/L range, which sometimes exceed WHO guideline values such as Pb, Cr and Cd (e.g. Kumar et al., 2020; Tiwari et al., 2020).

Mercury is a particular concern for rural environments because of global atmospheric deposition, where even apparently pristine environments can be contaminated (Lyman et al., 2020). Inorganic Hg(II) is hazardous to health via oral exposure, but this form is also methylated by microbes to form methylmercury which is readily absorbed across the gut. The mechanisms of Hg absorption across the gut are well-known (e.g. Hoyle and Handy, 2005) and both mercury species are bioaccumulative. Oral exposure to Hg therefore represents a long-term health risk to humans directly, and via aquatic food webs where biomagnification of the Hg can occur, with risks especially associated with eating fish or shellfish (Counter and Buchanan, 2004). The WHO recommends that total inorganic Hg concentrations should not exceed 6 μg/L in drinking water (WHO, 2017) and in rural locations the Hg concentrations can be measurable, but are often below the WHO guidance in groundwater (e.g. Khandare et al., 2020). Nonetheless, there are concerns about surface water contamination in rural areas, especially those in northern parts of India that will be influenced by atmospheric Hg deposition via geochemical cycling from the Tibetan plateau (Sun et al., 2008).

16.3 SENSING IN RURAL INDIA

Currently, the monitoring of water quality and its chemical contaminants is mainly achieved through onsite sampling and then performing qualitative and quantitative measurements in the centralised laboratories from time to time (Tambekar and Neware, 2012; Mohapatra et al., 2018). This relies heavily on the analytical capability of the laboratory and requires professional skills to conduct the measurement and interpret the results (Suess, 1985). Therefore, a rapid, reliable and cost-effective technique for continuously monitoring water quality and contaminations is urgently needed, especially for those resource-limited rural areas where contaminations levels are often unknown.

A fully integrated electrochemical sensor array could potentially meet all the requirements of this challenge. It offers better analytical performance over the colorimetric sensors, but with convenience and at a cost much lower than using dedicated analytical laboratories (Nag et al., 2021). The most common strategy is to split the sensing system into two components: (1) a portable, reusable and low-cost readout device, which contains all essential electrics and power supplies for performing the electrochemical measurement (CV or EIS) and transducing the chemical recognition into electrical signals. Many groups have miniaturised their readout devices (Cruz et al., 2014; Jung et al.; 2017) and empowered the devices with multiple functions, such as low energy consumption (Shi et al., 2008), multiplexed detection (Wu et al., 2018) and the capability to send the signals to a central data hub or smartphone (Alam et al., 2020), making them portable, practical and sensitive enough to be used in the point-of-need rural environment, and (2) a low-cost sensing unit in the form of chip, strip or microfluidics, consisting of a solid substrate, a functionalised working electrode surface and 2–3 essential electrodes for performing electrochemical measurement (Li et al., 2020). These sensing units are normally produced in large quantity and low price using cost-effective materials and methods. Further work is required for their implementation.

From the perspective of low-cost substrates, synthetic polymers and natural cellulose-based substrates, such as polydimethylsiloxane (PDMS), poly (methyl methacrylate) (PMMA) and cellulose paper, are the most suitable alternatives to conventional microelectromechanical systems (MEMS) substrates for the construction of electrochemical sensors (Dincer et al., 2019). In addition to the low cost, these materials are also chemically stable, offer ease-of-fabrication and possess many suitable properties, such as biodegradability, flexibility, recyclability and stretchability, and they can be easily tuned with additives for constructing a specific sensing unit. To functionalise the working electrodes, a range of cost-effective materials have been discovered and adopted for sensing applications (Maduraiveeran and Jin, 2017). Metallic nanoparticles (e.g. gold, platinum, silver), carbon-based nanomaterials (e.g. carbon nanotube, graphene, quantum dot) and their hybridised composites have been widely used to enhance the electrochemical conductivity and surface-to-volume ratio at the electrode-electrolyte interfaces, due to their unique size, facile synthesis, in ease of functionalisation and stress-free fabrication approach. On the other hand, conductive polymers, especially self-assembled, electro-polymerised and dendrimers-based polymers, can be used alone as a coating or grafting layer or used as hybridised materials with metallic nanoparticles to improve the stability and sensitivity of the electrochemical sensors. As for the fabrication method of low-cost electrochemical sensors, the most promising and well-established technique is screen-printing. In a typical screen-printing setup, different inks are squeezed or sprayed through pre-patterned mask-like screens, layer by layer, to define the function and geometry of the final device. A wide range of inks have been developed to meet the requirement of different sensitivities, stabilities and reliabilities in various sensing environment. The most commonly used ink materials, however, are carbon-based nanomaterials, Ag, Ag/AgCl and Au (Hayat and Marty, 2014; Baptista-Pires et al., 2019). Building on the screen-printing, inkjet printing has been recently developed as an alternative technology to offer maskless fabrication of printed electrochemical sensors. As a more digital technique, inkjet printing can not only deposit various patterns of inks with minimal mask modifications but also reduce the fabrication cost by depositing the ink only where is it required, making it more versatile and cost-effective for large-scale sensor production (Das et al., 2016).

Modern technologies have emerged for diagnosis of pathogenic microorganisms (Rajapaksha et al., 2019 and Leonard et al., 2003). A class of pathogenic viruses, bacteria and fungi exist in the environment, which can cause potential harm to humans and animals (Pachepsky et al., 2011). Early diagnostics of pathogenic microbes is highly essential to ensure proper therapy. Immunological (Schloter et al., 1995) and molecular diagnostics (Toze, 1999) act as promising tools and are widely used to identify the pathogens. Fluorescence spectroscopy was explored in combination with antibodies and DNA-based detection of pathogenic microbes (Cumberland et al., 2012). This technique typically utilises specific antibodies or DNA sequences to identify microorganisms of interest. Diagnostics methods such as mass spectrometry (Sauer and Kliem, 2010), polymerase chain reaction (PCR) and loop-mediated isothermal amplification (LAMP) (Martzy et al., 2017) can replace time-consuming techniques, which also require rigorous sample preparation mechanisms. These techniques can rapidly detect pathogenic microorganisms with high-level accuracy. Integrating automation of sample processing with Fluorescence, PCR, LAMP and mass spectrometry techniques can drastically reduce analysis time and the results can be achieved at high accuracy (Zhang et al., 2018).

Microfluidics technique requires samples in very small volume usually in microliters, and the samples are passed through micron sized channels. This technique consumes a very small amount of samples, with rapid analysis capabilities. Multiple inlets allow us to process samples in different channels simultaneously. Ottesen et al. (2006) demonstrated a model for analysis of single bacterial cells by fabricating microfluidic digital PCR platforms. This model consists of 12 panels with 1176 PCR chambers. This led to identification of unknown ribosomal RNA-based species. Wang et al. explored the quantitative detection of nucleic acids by using fluorescence spectroscopy. They have identified low-abundance nucleic acid using molecular beacons and molecular confinement microfluidic reactors (Wang et al., 2005). LAMP-based detection is gaining increasing popularity for amplification of DNA and to detect certain diseases. Recently, RT-LAMP-based assay is explored for detection of SARS-CoV-2 RNA in clinical samples (Thi et al., 2020). Microfluidic devices can be utilised for detection of pathogenic bacteria. Compared to mass spectrometric techniques, fluorescence microscopy is more convenient to visualise the bacteria in microfluidic channels (Jing et al., 2013). Guo et al. explored the magnetophoretic separation technique for trapping

Salmonella typhimurium using magnetic controlled microfluidic devices. The target pathogens were detected using inverted fluorescence microscopy with limit of detection 5.4×10^3 CFU/mL. A high separation efficiency is achieved at 100 μL/min, and this trend shows that almost 1 mL of sample can be processed in just 10 min (Guo et al., 2015). Qiao Zhang et al. demonstrated the importance of microfluidic immunomagnetic fluorescence assay for detection of avian influenza virus (AIV). This set-up comprises an optical fibre spectrometer integrated with microfluidics devices. The reported limit of detection (LOD) is 3.7×10^4 copies/μL with minimum sample volume of 2 μL. This method is a simple and rapid point-of-care method, which can be utilised for portable experimental studies and can be extended for detection of other viral pathogens (Zhang et al., 2013). Water quality monitoring is highly essential due to the increasing crisis of water in developing countries. Contaminants ingested can cause a serious threat to lives. Considering these aspects, frequent monitoring of drinking water should be undertaken. This process has challenges like depending on sophisticated instruments and increased process time. Advancement in technology like microfluidics can overcome many challenges since microfluidics technique requires a less sample volume and minimal time for the analysis. Integrating the microfluidics with portable sensors can easily detect pathogenic microorganisms. This combination can be highly useful for real-time monitoring of environmental contaminants. Jaywant et al. summarise the importance of microfluidic technique, which can be used as water quality monitoring sensors. The importance of portable devices, and sensitivity and selectivity of sensors for continuous water quality monitoring were specifically discussed. Merits of integrating microfluidics with optical and electrochemical sensors were also highlighted exclusively. A microfluidic-based sensor can be highly useful for monitoring the pathogenic microorganisms. Microfluidics-based detection has several advantages like rapid analysis with reduced wastage of samples, compactness and cost-effectiveness (Jaywant and Arif, 2019).

Optical sensing techniques are simple to process and easily integrated with microfluidics for sample analysis. Optical sensing techniques such as colorimetry, surface plasmon resonance (SPR) and surface-enhanced Raman spectroscopy (SERS) are explored to detect many target molecules using microfluidics devices (Li et al., 2020). Colorimetric analysis is a popular method for detecting metals using paper-based microfluidics platforms. Mentele et al. fabricated microfluidic paper-based analytical device (μPAD) using wax printing. Rapid digestion of particulate metals was performed by adding microliters of acid, and the digested metals were passed through reservoirs containing reagents. The colour change during this reaction is analysed using colorimetric technique. This method is used for detection of Fe, Cu, and Ni with a detection limit of 1.0–1.5 μg of each metal, which is the total mass present on the microfluidic paper-based analytical device (Mentele et al., 2012). This technique is very much cost-effective, and it can be highly useful for analysis of metals in water samples. Wang et al. proposed a method for easy detection of *E. coli* at a highly sensitive level (single cell/mL) using SERS microfluidic biosensor. The samples are passed through a microfluidic device with 1 μL/min flowrate with a dynamic detection range of 1–10 CFU/mL. SERS measurements were performed using a Raman microscope with 780 nm excitation wavelength at 10 mW. Bacterial concentration mechanism Nano-dielectrophoretic enrichment is used along with SERS signal analysis. Gold nanoparticles were used as SERS probes for detection of bacterial samples. This method can be used as an alternative to polymerase chain reaction (PCR), enzyme-linked immunosorbent assay (ELISA), which requires intensive sample preparation mechanism and depends on highly skilled users (Wang et al., 2017). Kraft et al. developed an on-chip microfluidic device to detect *E. coli* DH5α and *Pseudomonas taiwanensis* VLB120 in spiked tap water samples using SERS technique. The surface enhancement is achieved using silver nanoparticles, and SERS spectra were measured using a Raman system with 633 nm excitation source. Typical polydimethylsiloxane (PDMS) was used for fabrication of the microfluidic device, and oxygen plasma treatment is provided to enhance the mechanical stability of the device. This device utilises electrokinetic trapping of bacteria for rapid SERS analysis. The free silver nanoparticles with 60–70 nm size pass through the membrane without any interference and only the nanoparticle clusters and bacteria are trapped and concentrated to exhibit their presence leading to enhancement of SERS signals. SERS spectra recorded at various locations in the membranes show good repeatability of the signals without significant variations (Kraft et al., n.d.). Park et al. developed a smartphone-based technique for detection of *E. coli* from field water samples. A three-channel paper microfluidics set-up was designed with preloaded bovine serum albumin (BSA) conjugated beads. The water samples were subjected to pass through paper chips either by dipping or pipetting process and the antigens from the *E. coli* pass through fibres in the paper by capillary actions. This design also ensures effective filtration of other dust or soil particles. This process initiates immunoagglutination with the presence of *E. coli*, which is further quantified using Mie scattering intensity. This method could be utilised for real-time monitoring of pathogenic *E. coli* in water samples by using existing smartphones with simple modifications (Park and Yoon, 2015).

Water quality measurements require novel sensors to assess the required parameters rapidly. The sensors can be utilised to detect pathogenic microorganisms with high sensitivity and specificity. At present most of the water samples are analysed using spectroscopic and chromatographic techniques. A range of pathogenic microbes such as coliforms and viruses were detected using these techniques. Advancement in nanotechnology materials shows a promising trend for highly effective sensing of pathogenic microbes. Nanomaterials can be a potential alternative to replace existing water quality sensors. Nanomaterials-based sensors can be fabricated to provide high-efficiency with multiplexed detection capacity. These sensors can be easily integrated with portable devices and data logging facilities. These devices can be highly useful for rural regions in developing countries to monitor water quality. The supporting tools should be user-friendly for ease of operation (Vikesland, 2018). Vanegas et al. reported emerging trends on biorecognition and transduction schemes for rapid detection of pathogenic microorganisms. This technique explores the biorecognition-transduction process using simple detectors, which can replace the traditional plate counting technique for the detection of pathogenic microorganisms. The biosensors are tailored to recognise target pathogens using target capture probes such as DNA and antibodies. The interaction between the pathogens and receptor creates transduction; this signal is further examined by acquisition tools. Nucleotide-based detection acts as a promising tool for multiplexed detection with DNA microarray and target probes. These probes are labelled with fluorescent dye to indicate the capture of pathogenic species. Immunological-based detection is another method for the detection of pathogens using lateral flow assay. The samples are allowed to pass through capillaries along an immunochromatographic line, which can detect the change in colour with the presence of pathogenic species. Aptamer-based detection is another technique, which was used as a biorecognition element from the past two decades. Aptamers are inexpensive to produce, and they can be stable even up to 2 years. Aptamers exhibit structural conformations such as pseudo-knot, G-quartet, hairpin and stem or loop bulge categories. The structural changes indicate the binding of pathogenic elements. Aptasensors are immobilised on the surface of devices through covalent bonding for the detection of target elements. SPR and fluorescence are common detection techniques, which can confirm the biorecognition activity of target pathogens (Vanegas et al., 2017). Figure 16.1 shows how such sensors can be implemented and integrated into a digital platform. Mobile devices are used to process and transmit sensed data to a server. This is being developed by University of Plymouth (UK) and Indian Institute of Science (IISc) Bangalore (India).

Fan et al. reported a simple and effective detection of waterborne bacteria such as *E. coli* and *S. epidermidis* using SERS technique coupled with silver nano substrates. TEM analysis shows the accumulation of silver nanoparticles within the cells of bacteria and deposition closer to the cell membranes. SERS plots showed an intense enhancement of signal due to silver nanoparticles and the peaks attributed to functional groups present in bacteria such as proteins or amino acids (1003, 1239 and 1590 cm), nucleic acids (713 and 1322 cm) and lipids (1239 and 1450 cm). The enhancement factor due to the influence of the silver nanoparticles was 1×10^4. This method is suitable for rapid and early detection of bacteria in food and water samples. The data was consistent and reproducible with the analysis of bacterial species. This technique proves

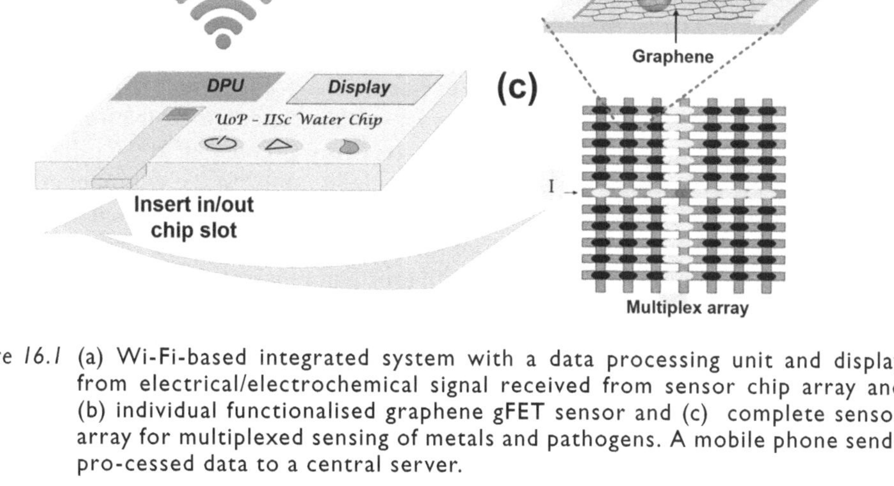

Figure 16.1 (a) Wi-Fi-based integrated system with a data processing unit and display from electrical/electrochemical signal received from sensor chip array and (b) individual functionalised graphene gFET sensor and (c) complete sensor array for multiplexed sensing of metals and pathogens. A mobile phone sends pro-cessed data to a central server.

the efficiency of detecting bacterial samples with high sensitivity and specificity (Fan et al., 2011). Chen et al. reported one-step detection of pathogenic Newcastle disease virus (NDV) using magnetic nanoparticles of small size (30 nm) by magnetic separation (MS) and magnetic relaxation switching (MRS) technique. The target pathogens attached to magnetic beads were separated using a very small magnetic field (0.01 T). The antibody-conjugated magnetic nanoparticles can selectively capture targets because of the antibody-antigen reaction. This convenient technique has detection sensitivity of 102 cfu/mL compared to laborious ELISA technique with sensitivity 103 cfu/mL. This technique can be deployed for detection of NDV to prevent contamination through food and water sources (Chen et al., 2015).

16.4 RECENT ADVANCES IN FILTRATION WITH NOVEL NANOMATERIALS

The dominant physio-chemical water remediation mechanisms are forms of sorption (adsorption, absorption and ion exchange), and they are easy to implement and cost effective on a large scale (Worch, 2012; Juang et al., 2002). Adsorption can be physical and chemical. Chemical adsorption is largely attributed to covalent bonding, where the adsorbate is assimilated beyond the surface of the absorbent (often referred to as absorption), and physical adsorption is largely a charge-based reversible process that is limited to the surface of the adsorbent. Physical adsorption is typically governed by electrostatic interaction (cations and anions in ionic bonding), van der Waals forces (an uneven net distribution of charge across a molecule due to quantum fluctuations) or hydrogen bonding (caused by polar molecules attracting neighbouring hydrogen ions that are already in a permanent dipole moment). Chemical adsorption can also assume a catalytic property and is out of the scope of this section (photocatalytic section). Furthermore, the key factor for the success of physical adsorption is the specific surface area (SSA) to mass ratio of the adsorbent available per unit weight and if large enough the mechanism for 'filtration' is assumed entirely an adsorption process. Nanomaterials due to their inherent surface area to volume ratio is the primary reason why the 'nano effect' has such a dramatic impact on the properties and functionality of the material, and when put into a filtration context the results can be extraordinary. The emerging field of genetic and protein engineering is paving the way for protein-based nano-biomaterials for the specific remediation of toxic metals. The metal-binding polymers are based on elastin-like polypeptides (ELP) which have similarities with mammalian protein elastin, and they have reversible phase transition properties ideal for meta l–polymer complex recovery (MacEwan and Chilkoti, 2010). The phase transition characteristics are tuneable by altering the chain length of the peptide sequence. This simple modification increases the binding capacity of the protein. Also reports argued that ELPs containing a terminal hexahistidine domain have shown a binding affinity of 1:1 for Cd^{+2} with excellent reusability (Scanlon et al., 2007). The regeneration of the binding site for ELP was achieved by incubating in an acetate buffer of pH 4. However, it is noted that the omission of a bulk material control leads to the question, is the remediation due to a 'nano effect' just a property of the bulk material itself? Aerogel filtration structures that incorporate both the 'nanomaterial' and a structured network with extremely high SSA are now a cost-effective alternative to typical activated carbon. The weight ratio to volume of a typical aerogel is approximately 0.1–0.5 g/cm^3, and these can be made from a variety of materials, including, silica, metal oxides and graphene/polymer-based materials (Zuo et al., 2015). Aerogels can also be made of metal alkoxide precursors, produce metal oxide aerogels, for example, alumina (Al_2O_3), magnesium oxide (MgO), titania (TiO_2) and zirconia (ZrO_2). Potable and wastewater treatment is of particular interest for aerogel research as the material and the efficacy of this principle have already been demonstrated in a deconstructed graphene foam for wastewater treatment (Tabish et al., 2018). Graphene oxide (GO) and cellulose nanofibril hybrid aerogel demonstrated the adsorptive removal of 21 antibiotics simultaneously (Yao et al., 2017). Figure 16.2 shows lead adsorbing by charge transfer, to a layer of a graphene oxide aerogel structure, developed by the authors at the University of Plymouth. These biopolymer aerogels can remediate highly concentrated Pb pollution (100 mg/L), with a capacity of over 500 mg/g (Bloor et al., 2021).

Further studies reveal the ability to produce robust aerogels with a high fatigue resistance of over a million cycles, whilst under elongation of 200% is reported (Guo et al., 2018). The technology proposed is scalable and based upon organic biomaterials, such as graphene oxide and biopolymers, which are processed to form the lightest material known to man – aerogels. These structures are 3d in form yet based around a labyrinth of 2d materials, and the basis of filtration is not 'sieving' but adsorbing the contaminants within the host matrix whilst maintaining a high flux rate.

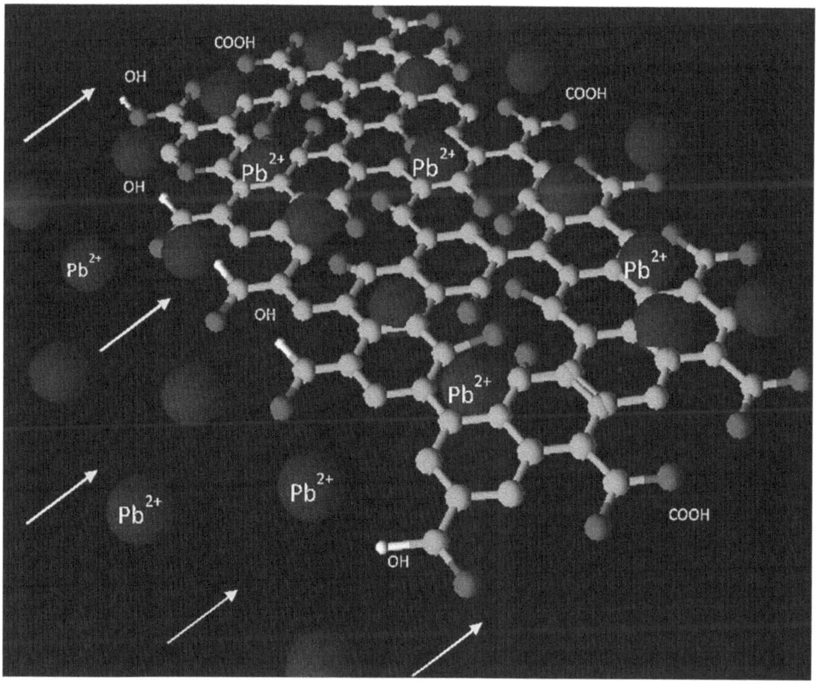

Figure 16.2 Schematic of lead adsorbing by charge transfer to a graphene oxide structure.

16.5 REMOVAL BY PHOTOCATALYSIS AND OTHER DEGRADATION METHODS USING NANOMATERIALS

Water is unevenly distributed in most parts of the world, not only in quantity but also quality, which is also a big factor in India, with varied water quality across the country, especially in non-urban/rural communities. Hence, the major challenge is to supply safe and clean water to the population (Harikaranahalli Puttaiah et al. 2020). Nowadays the scientific community is not only contributing towards the removal of conventional pollutants but the remediation of complex, refectory and persistent chemicals (Mohamad and Mohamed 2019). After a critical review of every treatment techniques in place to remediate water, advanced oxidation processes (AOPs) hold the top position in terms of sustainability and efficiency (Betancourt-Buitrago et al. 2019). AOP-based treatment has attracted every scientist because of its easy adaption and selectivity. Photocatalysis is one such AOP that is initiated by a functional photocatalyst to generate reactive oxygen species, resulting in mineralization of the pollutant (Shivaraju et al. 2020). To date, photocatalysis has been employed to remediate pollutants such as aromatic hydrocarbons and dyes, which arise from illegal industrial discharge into rivers and lakes. Nevertheless, in recent years, synthesis of engineered nano-photocatalysts have made it possible to decontaminate more complex contaminants, like pharmaceutical agents, antibiotics, musk compounds, endocrine disrupting compounds, pesticides, pathogens, etc. (Ebrahimi et al. 2020; Ravindra et al. 2020). Nanotechnology has enabled nano-sized catalysts that are far efficient over the bulk materials to be produced. Their properties, such as availability of active surface area, rapid electron mobility and compatibility of nano-photocatalysts, are significant (Nagaraju et al. 2020). For some time, TiO_2 and ZnO have been the most widely applied photocatalyst, as they possess a wide band gap in the range of 3.0–3.50 eV. Typically, these are UV active semiconductors with no toxicity, cost-effective, ease of handling and high resistance to photo-recombination (Nagaraju et al. 2017). Figure 16.3 represents the traditional pollutant removal mechanism involved in photocatalysis (Ebrahimi et al. 2019).

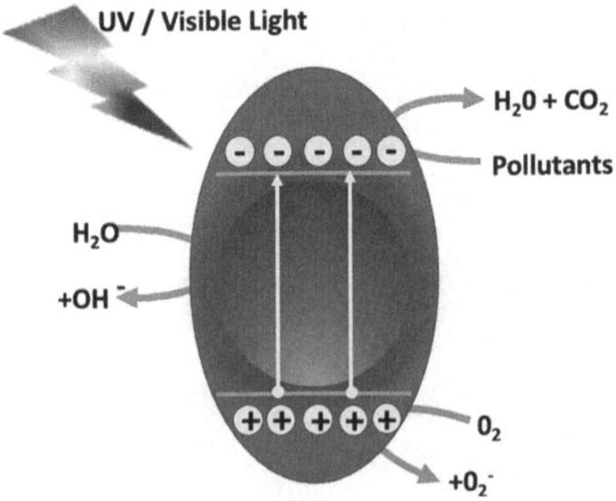

Figure 16.3 Typical photodegradation mechanism.

Further, with this casual understanding of photocatalysis, the synthesis and structure of a photocatalyst is very influential in the treatment process. Recent methods used to construct/synthesise photocatalysts are by the preparation of nanocomposites, using electro/photo-deposition, solution-combustion, electrospinning, sonochemical methods, etc. However, conventional approaches like sol-gel, co-precipitation and hydrothermal methods are also used (Wang and Zhuan 2020). Materials like metal-oxides, carbonaceous, organometallic, polymers, clay, etc. are used to prepare these catalysts. More recently, newer techniques have been developed to obtain large surface-active sites, oxygen-facets and a significant coordination number of surface atoms, which is compatible and essential for the fabrication of hybrid catalysts used for degradation with suitable photo/light source (Afzal et al. 2019). In summary, an overview of recent photocatalysts, preparation methods, targeted contaminant, light source and the characteristic results is provided in Table 16.1. To appreciate the efficiency of the system (photocatalysis), our previous reports are described here. Catalysts, tin-oxide (SnO_2) (Yashas et al. 2020b) and manganese oxide coupled with graphitic carbon nitride (Yashas et al. 2020a) were applied to mineralise the antibiotic compound, tetracycline hydrochloride in water. Here, an eco-friendly and energy-efficient light source, light-emitting diodes (LEDs) light, was used to irradiated the catalysts. Interestingly, SnO_2 was applied for sonophotocatalytic mode of treatment. In both the cases, tetracycline was degraded considerably, up to 93% within 135 min. Similarly, titanium dioxide was coupled with the conducting polymer, polypyrrole, to remediate Reactive Red 45 dye in water (Kratofil Krehula et al. 2019). Here, the composite was prepared by chemical polymerisation with titania nanoparticles. Uniquely, the work used two sources of photo-irradiation: natural sunlight and UV-A. The dye was degraded up to

Table 16.1 Reports of recent photocatalytic systems

Photocatalyst	*Synthesis Method*	*Pollutant*	*Efficiency*	*Reference*
$CuBi_2O_4/MoS_2$	One-pot hydrothermal method	Tetracycline	76% in 120 min	Guo et al. (2019)
Fe/CdS	Co-precipitation	Methylene blue	70% in 25 min	Junaid et al. (2019)
$Gd{:}MgZrO_3$	Solvothermal	Rhodamine B	75% in 45 min	Akshatha et al. (2020)
$g\text{-}C_3N_4/Ti_3C_2$	Pyrolysis and sonochemical combustion	Ciprofloxacin	95% in 150 min	Liu et al. (2019)
$SnSe/SnO_2$	Co-precipitation	Methylene blue and Rhodamine B	>90% in 90 min	Kaur et al. (2019)
Zn-Co-LDH@ biochar nanocomposite	Hydrothermal process	Gemifloxacin	92.7% in 130 min	Gholami et al. (2020)
MoS_2@ZnAl LDH	Coprecipitation and hydrothermal method	Methylene blue	93% in 150 min	Chen et al. (2020)

~99% in both the cases within 60 min. All these aforementioned efficiencies are traditionally attributed to the enhanced absorption band ability, prolonged photo-induced charge separation and electronic compatibility of the catalysts.

Apart from conventional photocatalysis, photo-Fenton, photoelectrocatalysis, solar photocatalysis and solar photo-Fenton techniques are being used of late to address complex pollutants. All these methods by and large work on the same principle except for few technical modifications. It is very interesting to note that interaction of solar irradiation with iron ions (Fe^{3+}) in photo-Fenton and solar-photo-Fenton improves the rate of pollutant degradation as a result of rapid photochemical reactions (Pignatello et al. 2006). Typically, a Fenton reaction is between hydrogen peroxide and an iron radical (Fe^{2+}) in acidic solution to generate hydroxyl radicals (OH), which is an oxidant responsible for contaminant degradation. Here, natural-light/UV/visible irradiation endorses photochemical reactions in presence of intermediate species like $Fe(OH)^{2+}$. The light enhancement is mainly by photolysis of hydroxide complexes of Fe^{3+} and photochemical interaction of complexes formed between Fe^{3+}/Fe^{2+} (Gutierrez-Mata et al. 2017). Hence, modified photocatalysis process is advantageous to abet a vast number of drinking water pollutants. Another important system is photoelectrocatalysis of pollutant degradation. The combined effect of electrocatalytic and photocatalytic functionalities of the catalyst is employed to mineralise the pollutants. Photoelectrocatalytic oxidation is an ecologically benign methodology that utilises both light and electric energy to generate strong oxidants (hydroxyl, superoxide, electron-hole pairs) that target the organic matter. Even persistent organics, such as pharmaceuticals or hazardous chemicals, are broken down to smaller non-toxic molecules and over time reduces to carbon dioxide and water (Orimolade and Arotiba 2020). Hence, coupling other promising oxidation techniques with photocatalysis could also be investigated in order to develop advanced treatments. These processes can be made efficient and feasible for drinking water treatment with a simple pre-treatment. The bacterial loads and organic loads could be simultaneously addressed with improved design of photocatalysts.

16.6 CONCLUSIONS

In UK, for example, consumers expect that potable water is safe and that quality is monitored to ensure that the relevant standards are met. Good infrastructure also ensures that water is continuously available to consumers. This is not the case in many developing countries, and yet the guidance on safe drinking water from organisations such as the WHO should be applied globally. For India, the local climate includes periods of drought and then heavy rain (monsoon), and these weather events are now exacerbated by climate change. The current infrastructure in India cannot harvest surface water by any simple means, and in any case, in the rainy season much of the flood water will flow into the sea. In large cities where municipal water systems are in place, there are still significant issues for supplying water. For example, private sector companies are competing with the municipal organisations for the infrastructure and supply. Consequently, more agreement on the governance of the water resources is needed. In Chennai, it is not uncommon to see consumers waiting with plastic vessels to obtain safe water from a municipal tanker. Such observations highlight that even in urban areas the infrastructure is not adequate for a continuous supply at the consumer's tap. This also presents a problem for enforcing water quality, as the drinking water regulations are intended to be applied at the point of consumption – usually the consumer's tap. In rural communities, where municipal infrastructure is lacking, the problems with safe supply and ensuring quality standards are much harder to resolve.

In rural communities where water is accessed from wells, rivers and lakes, water quality is not often acceptable (i.e., does not meet WHO standards), and non-compliance is a factor in the subsequent health risks. In this chapter we have focused on the key issues that affect water quality. In most cases, concentrations of hazardous substances found in water exceed Indian and globally accepted standards for metals, endocrine disrupters, POPs and industrial dyes. There are concerns for pathogens in potable water.

Two of the many challenges in rural area are the detection and remediation of such contaminants. Sensors have to be sensitive, selective and specific and fit within an easy to use system. This is not about analysing samples in laboratories, but 'in the field' in village communities. Current research by the team here is exploring low cost, multiplexed sensor systems that can work with mobile platforms to form a sophisticated point of diagnostics sensing system. A small water sample, dropped onto the sensor chip via a microfluidic platform, will be processed to provide a clear indication of water contaminant levels and relay this data to a central server for further analysis. The mobile platform will inform the user as to how 'safe' the water is for human consumption. The cost of the sensor chip should be a 'few rupees' and would be single use only.

Until water meets the required standards, mitigation will be required. In most cases considered, the water may appear clean, with good clarity and no odour as required by the regulations. However, soluble hazardous substances may be present, such as those outlined above. One approach to improving water quality is the removal of individual substances of concern. For example, removal of lead from water using low cost, environmentally friendly materials, formed into aerogels that can be incorporated into water delivery systems, such as pump stations or water bottles. Water throughput is high due to the relatively large pore size of the aerogels. These aerogels can be adaptively designed according to the requirement. This requirement maybe to remove lead (metals), oil and water separation or dye removal. Dye removal has been demonstrated for methylene blue, but there are other methods which may be more effective for industrial waste. The chapter shows how photocatalysis can remediate a range of pollutants such as aromatic hydrocarbons and dyes, and that engineered nano-photocatalysts can extend this remediation to pharmaceuticals endocrine disrupting compounds, pesticides and pathogens.

REFERENCES

Afzal S, Quan X, Lu S (2019) Catalytic performance and an insight into the mechanism of CeO_2 nanocrystals with different exposed facets in catalytic ozonation of p-nitrophenol. *Appl Catal B Environ* 248:526–537 doi: 10.1016/j.apcatb.2019.02.010

Akshatha S, Sreenivasa S, Parashuram L, et al. (2020) Solvothermal synthesis of nanoscale disc-like gadolinium doped magnesium zirconate for highly efficient photocatalytic degradation of rhodamine B in water. *SN Appl Sci* 2:876. doi: 10.1007/s42452-020-2686-3

Alam AU, Clyne D, Jin H, Hu N-X, Deen MJ (2020) Fully integrated, simple, and low-cost electrochemical sensor array for in situ water quality monitoring, *ACS Sens* 5(2):412–422.

Al-Khaldi FA, Abusharkh B, Khaled M, et al. (2015) Adsorptive removal of cadmium (II) ions from liquid phase using acid modified carbon-based adsorbents. *J Mol Liq* 204:255–263.

Anglian Water (2020). https://www.anglianwater.co.uk/business/services/final-effluent/

Bakore N, John P, Bhatnagar P (2004) Organochlorine pesticide residues in wheat and drinking water samples from Jaipur, Rajasthan, India. *Environ Monit Assess* 98:381–389.

Baptista-Pires L, de la Escosura-Muñiz A, Balsells M, Zuaznabar-Gardona JC, Merkoçi A. (2019) Production and printing of graphene oxide foam ink for electrocatalytic applications, *Electrochem Commun* 98:6–9.

Barlow PM, Reichard, EG (2010) Saltwater intrusion in coastal regions of North America. *Hydrogeol J* 18:247–260.

Bellinger DC (2008) Very low lead exposures and children's neurodevelopment. *Curr Opin Pediatr* 20:172–177.

Betancourt-Buitrago LA, Hernandez-Ramirez A, Colina-Marquez JA, et al. (2019) Recent developments in the photocatalytic treatment of cyanide wastewater: an approach to remediation and recovery of metals. *Processes* 7:225. doi: 10.3390/pr7040225

Bhardwaj RM (2005) Status of wastewater generation and treatment in India. Intersecretariat Working Group on Environment Statistics (Iwg–Env) Joint Work Session on Water Statistics, Vienna, 20–22.

Bivins AW., et al. (2017) Estimating infection risks and the global burden of diarrheal disease attributable to intermittent water supply using QMRA. *Environ Sci Techno* 51: 7542–7551.

Bloor JM, Handy RD, Awan SA, Jenkins DF (2021) Graphene oxide biopolymer aerogels for the removal of lead from drinking water using a novel nano-enhanced ion exchange cascade. *Ecotoxicol Environ Saf*, 208:111422.

Blumenthal UJ, Mara DD, Peasey A, Ruiz-Palacios G, Stott R (2000). Guidelines for the microbiological quality of treated wastewater used in agriculture: recommendations for revising who guidelines. *Bull World Health Organ* 78:1104–1116.

Bradford A, Brook R, Hunshal C (2003) Wastewater irrigation in Hubli–Dharwad, India: implications for health and livelihoods. *Environ Urban* 15:157–170.

Canfield RL, Henderson Jr CR, Cory-Slechta DA, Cox C, Jusko TA, Lanphear BP (2003) Intellectual impairment in children with blood lead concentrations below 10 μg per deciliter. *N Engl J Med* 348:1517–1526.

Chakmeng (2005) Environmental contamination of arsenic and its toxicological impact on humans. *Environ Chem* 2:146–160.

Chakraborti D, Das B, Rahman MM, et al. (2009) Status of groundwater arsenic contamination in the state of West Bengal, India: a 20-year study report. *Mol Nutr Food Res* 53:542–551.

Chauhan A, Goyal P, Varma A, Jindal T (2017) Microbiological evaluation of drinking water sold by roadside vendors of Delhi, India. *Appl Water Sci* 7:1635–1644. doi: 10.1007/S13201-015-0315-X

Chen S, Yang F, Cao Z, et al. (2020) Enhanced photocatalytic activity of molybdenum disulfide by compositing ZnAl–LDH. *Colloids Surfaces A Physicochem Eng Asp* 586:124140. doi: 10.1016/j.colsurfa.2019.124140

Chen Y, Xianyu, Y, Wang, Y et al. (2015) One-step detection of pathogens and viruses: combining magnetic relaxation switching and magnetic separation. *Acs Nano* 9:3184–3191. doi: 10.1021/Acsnano.5b00240

Chowdhury, TR, Basu, GK, Mandal, BK, Biswas, BK, Samanta, G, Chowdhury, UK, Chanda, CR, Lodh, D, Roy, SL, Saha, KC, Roy, S (1999) Arsenic poisoning in the Ganges delta. *Nature* 401 (6753): 545–546.

Counter SA, Buchanan LH (2004) Mercury exposure in children: a review. *Toxicol Appl Pharmacol*, 198:209–230.

CPCB (2020) http://www.indiaenvironmentportal.org.in/content/468537/cpcb-report-on-cetps-and-stps-16092020/

Cpheeo G (1999) *Manual on Water Supply and Treatment*. New Delhi: Goi.

Cruz AF, Norena N, Kaushik A, Bhansali S (2014) A low-cost miniaturized Cruz potentiostat for point-of-care diagnosis, *Biosens Bioelectron* 62:249–254.

Cumberland S, Bridgeman J, Baker A, Sterling M, Ward D (2012) Fluorescence spectroscopy as a tool for determining microbial quality in potable water applications. *Environ Technol* 33(6): 687–693. doi: 10.1080/09593330.2011.588401

Das SR, Nian Q, Cargill AA, et al. (2016) 3D nanostructured inkjet printed graphene via UV-pulsed laser irradiation enables paper-based electronics and electrochemical devices. *Nanoscale* 8(35) (2016) 15870–15879.

Dincer C, Bruch R, Costa-Rama E, et al. (2019) Disposable sensors in diagnostics, food, and environmental monitoring. *Adv Mater* 31(30):1806739.

Directive C (1980) 80/778/Eec Of 15 July 1980 Relating to the Quality Of Water Intended For Human Consumption. Oj L, 229.

DWI (2018) Drinking water 2017 summary of the chief inspector's report for drinking water in England.

DWS, I. 2012. Bureau of Indian Standards drinking water specifications. *BIS* 10500:2012.

Ebrahimi R, Hossienzadeh K, Maleki A, et al. (2019) Effects of doping zinc oxide nanoparticles with transition metals (Ag, Cu, Mn) on photocatalytic degradation of Direct Blue 15 dye under UV and visible light irradiation. *J Environ Heal Sci Eng* 17:479–492. doi: 10.1007/s40201-019-00366–x

Ebrahimi R, Mohammadi M, Maleki A, et al. (2020) Photocatalytic degradation of 2,4-dichlorophenoxyacetic acid in aqueous solution using Mn-doped ZnO/graphene nanocomposite under LED radiation. *J Inorg Organomet Polym Mater* 30:923–934 doi: 10.1007/s10904-019-01280–3

Fan C, Hu Z, Mustapha A, Lin, M (2011) Rapid detection of food- and waterborne bacteria using surface-enhanced Raman spectroscopy coupled with silver nanosubstrates. *Appl Microbiol Biotechnol* 92:1053–1061. doi: 10.1007/S00253–011–3634–3

Gall AM, Mariñas BJ, Lu Y, Shisler JL (2015) Waterborne viruses: a barrier to safe drinking water. *PLoS Pathog* 11(6):e1004867.

George KS, Revathi KB, Deepa N, et al. (2016). A Study on the potential of moringa leaf and bark extract in bioremediation of heavy metals from water collected from various lakes in Bangalore. *Procedia Environ Sci* 35:869–880.

Gholami P, Khataee A, Soltani RDC, et al. (2020) Photocatalytic degradation of gemifloxacin antibiotic using Zn-Co-LDH@biochar nanocomposite. *J Hazard Mater* 382:121070. doi: 10.1016/j.jhazmat.2019.121070

Ghose NC, Dipankar S, Gupta A (2009) Synthetic detergents (surfactants) and organochlorine pesticide signatures in surface water and groundwater of greater Kolkata, India. *J Water Resour Prot* 1:290.

Giri S, Singh AK, Mahato MK (2019) Monte Carlo simulation-based probabilistic health risk assessment of metals in groundwater via ingestion pathway in the mining areas of Singhbhum copper belt, India. *Int J Environ Health Res* 30:447–460. doi:10.1080/09603123.2019.

Glasgow HB, Burkholder JM, Reed RE, Lewitus AJ, Kleinman, JE (2004) Real-time remote monitoring of water quality: a review of current applications, and advancements in sensor, telemetry, and computing technologies. *J Exp Mar Biol Ecol* 300:409–448.

Gov India (1999) http://cpheeo.gov.in/cms/manual-on-water-supply-and-treatment.php

Guo F, Li M, Ren H, et al. (2019) Fabrication of p-n $CuBi_2O_4/MoS_2$ heterojunction with nanosheets-on-microrods structure for enhanced photocatalytic activity towards tetracycline degradation. *Appl Surf Sci* 491:88–94 doi: 10.1016/j.apsusc.2019.06.158

Guo F, Jiang Y, Xu Z, et al. (2018) Highly stretchable carbon aerogels. *Nat Commun* 9:881.

Gutierrez-Mata AG, Velazquez-Martínez S, Álvarez-Gallegos A, et al. (2017) Recent overview of solar photocatalysis and solar photo-Fenton processes for wastewater treatment. *Int J Photoenergy* 2017:8528063. doi: 10.1155/2017/8528063

Hanna-Attisha, M, Lachance, J, Sadler, RC, Champney Schnepp, A (2016) Elevated blood lead levels in children associated with the flint drinking water crisis: a spatial analysis of risk and public health response. *Am J Public Health* 106:283–290.

Harikaranahalli Puttaiah S, Menon S, Shivamurthy Ravindra Y, et al. (2020) Preparation of supporting photocatalysts for water treatment using natural sunlight as an alternative driving energy. *Mat Today: Proc*45 (2021): 3936–3944. doi: 10.1016/j.matpr.2020.08.020

Hayat A, Marty JL (2014) Disposable screen printed electrochemical sensors: tools for environmental monitoring. *Sensors* 14(6):10432–10453.

Hoyle I, Handy RD (2005) Dose-dependent inorganic mercury absorption by isolated perfused intestine of rainbow trout, *Oncorhynchus mykiss*, involves both amiloride-sensitive and energy-dependent pathways. *Aquat Toxicol* 72(1–2):147–159.

IPEN (2006) Country Situation on Persistent Organic Pollutants (Pops) In India.

Jaywant S, Arif K, (2019) A comprehensive review of microfluidic water quality monitoring sensors. *Sensors* 19(21):478.

Jing W et al. (2013) Microfluidic device for efficient airborne bacteria capture and enrichment. *Anal Chem* 85(10):5255–5262.

Jit S, Dadhwal M, Kumari, H, et al. (2011) Evaluation of hexachlorocyclohexane contamination from the last lindane production plant operating in India. *Environ Sci Poll Res* 18:586–597.

Joshi MS, Lole KS, Barve, US, et al. (2019) Investigation of a large waterborne acute gastroenteritis outbreak caused by group B rotavirus in Maharashtra state, India. *J Med Virol* 91(10):1877–1881.

Juang R-S, Wu F-C, Tseng R-L (2002) Characterization and use of activated carbons prepared from bagasses for liquid-phase adsorption. *Colloids Surf A Physicochem Eng Asp* 201:191–199.

Junaid M, Imran M, Ikram M, et al. (2019) The study of Fe-doped CdS nanoparticle-assisted photocatalytic degradation of organic dye in wastewater. *Appl Nanosci* 9:1593–1602 doi: 10.1007/s13204-018-0933-3

Jung J, Lee J, Shin S, Kim YT (2017) Development of a telemetric, miniaturized electrochemical amperometric analyser *Sensors* 17(10):2416.

Kamyotra J, Bhardwaj R (2011). Municipal Wastewater Management in India. India Infrastructure Report, 299.

Kaur D, Bagga V, Behera N, et al. (2019) SnSe/SnO_2 nanocomposites: novel material for photocatalytic degradation of industrial waste dyes. *Adv Compos Hybrid Mater* 2:763–776. doi: 10.1007/s42114-019-00130-7

Kaur M, Sidhu SK, Singh K, Devi P, Kaur M, Singh NJ (2017) Hepatitis E virus: a leading cause of waterborne viral hepatitis in northwest districts of Punjab, India. *J Lab Phys* 9(2):121–124.

Khandare AL, Validandi V, Rajendran A, et al. (2020) Health risk assessment of heavy metals and strontium in groundwater used for drinking and cooking in 58 villages of Prakasam district, Andhra Pradesh, India. *Environ Geochem Health* 42(11):3675–3701. doi: 10.1007/s10653-020-00596-1

Khurana I, Sen R (2008) Drinking water quality in rural India: issues and approaches. *WaterAid India* 89:1–23.

Kosuth M, Mason SA, Wattenberg EV (2018) Anthropogenic contamination of tap water, beer, and sea salt. *PLoS One* 13:E0194970.

Kratofil Krehula L, Stjepanović J, Perlog M, et al. (2019) Conducting polymer polypyrrole and titanium dioxide nanocomposites for photocatalysis of RR45 dye under visible light. *Polym Bull* 76:1697–1715. doi: 10.1007/s00289-018-2463-2

Kundu, DK, Gupta A, Mol AP, Rahman MM, Van Halem D (2018) Experimenting with a novel technology for provision of safe drinking water in rural Bangladesh: the case of sub-surface arsenic removal (SAR). *Technol Soc* 53, 161–172.

Kumar R, Mittal S, Sahoo PK, Sahoo SK (2020) Source apportionment, chemometric pattern recognition and health risk assessment of groundwater from southwestern Punjab, India. *Environ Geochem Health* 43(2), 733–755.

Leonard P, Hearty S, Brennan J, et al. (2003) Advances in biosensors for detection of pathogens in food and water. *Enzyme Microb Technol* 32(1):3–13.

Li B, Tan H, Jenkins D, et al. (2020) Clinical detection of neurodegenerative blood biomarkers using graphene immunosensor. *Carbon* 168:144–162.

Liu N, Lu N, Su Y, et al. (2019) Fabrication of g-C_3N_4/Ti_3C_2 composite and its visible-light photocatalytic capability for ciprofloxacin degradation. *Sep Purif Technol* 211:782–789. doi: 10.1016/j.seppur.2018.10.027

Lyman SN, Cheng I, Gratz LE, Weiss-Penzias P, Zhang L (2020) An updated review of atmospheric mercury. *Sci Total Environ* 707:135575. doi: 10.1016/j.scitotenv.2019.135575

Macewan SR, Chilkoti A (2010) Elastin-like polypeptides: biomedical applications of tunable biopolymers. *Biopolymers* 94:60–77.

Maduraiveeran G, Jin W (2017) Nanomaterials based electrochemical sensor and biosensor platforms for environmental applications. *Trends Environ Anal Chem* 13:10–23.

McGuinness SL, O'Toole J, Barker SF, et al. (2020) Household water storage management, hygiene practices, and associated drinking water quality in rural India. *Environ Sci Technol* 54(8):4963–4973.

Mentele MM et al. (2012) Microfluidic paper-based analytical device for particulate metals. *Anal Chem* 84(10):4474–4480.

Mishra K, Sharma RC (2011) Assessment of organochlorine pesticides in human milk and risk exposure to infants from North-East India. *Sci Total Environ* 409:4939–4949.

Mohamad P, Mohamed ST (2019) Fate and transport of ppcps in the environment: a review on occurrences, sources, and cases. *Mater Sci Forum* 967:179–188. doi: 10.4028/www.scientific.net/MSF.967.179

Mohapatra S, Padhye LP, Mukherji S (2018) Challenges in detection of antibiotics in wastewater matrix. In: Gupta T, Agarwal A, Agarwal R, Labhsetwar N (eds) *Environmental Contaminants. Energy, Environment, and Sustainability*. Singapore: Springer. doi: 10.1007/978-981-10-7332-8_1

Martzy R, Kolm C, Brunner K, et al. (2017) A loop-mediated isothermal amplification (LAMP) assay for the rapid detection of *Enterococcus* spp. in water. *Water Res* 122:62–69

Mohapatra S, Sargaonkar A, Labhasetwar PK (2014) Distribution network assessment using EPANET for intermittent and continuous water supply. *Water Resour Manag* 28(11):3745–3759.

Mission SC, Aayog N (2017) https://agricoop.nic.in. Krishi AR 2017-18-1

Nag P, Sadani K, Mohapatra S, Mukherji S, Mukherji S (2021). Evanescent wave optical fiber sensors using enzymatic hydrolysis on nanostructured polyaniline for detection of β-lactam antibiotics in food and environment. *Anal Chem* 93(4):2299–2308.

Nagaraju P, Khunphonoi R, Puttaiah SH, et al. (2017) Photocatalytic paraquat degradation over TiO2 modified by hydrothermal technique in alkaline solution. *J Adv Oxid Technol* 20:10–11. doi: 10.1515/jaots–2017–0004

Nagaraju P, Puttaiah SH, Wantala K, Shahmoradi B (2020) Preparation of modified ZnO nanoparticles for photocatalytic degradation of chlorobenzene. *Appl Water Sci* 10:1–15. doi: 10.1007/s13201-020-01228–w

Orimolade BO, Arotiba OA (2020) Towards visible light driven photoelectrocatalysis for water treatment: application of a FTO/$BiVO_4$/Ag_2S heterojunction anode for the removal of emerging pharmaceutical pollutants. *Sci Rep* 10:1–13. doi: 10.1038/s41598-020-62425–w

Ottesen EA, Hong JW, Quake SR, Leadbetter JR (2006) Microfluidic digital PCR enables multigene analysis of individual environmental bacteria. *Science* 314(5804):1464–1467. doi: 10.1126/science.1131370

Pachepsky Y, Shelton DR, McLain JE, Patel J, Mandrell RE (2011) Chapter two - irrigation waters as a source of pathogenic microorganisms in produce: a review. *Adv Agron* 113:75–141.

Pandiyan J, Mahboob S, Govindarajan M, et al. (2021) An assessment of level of heavy metals pollution in the water, sediment and aquatic organisms: a perspective of tackling environmental threats for food security. *Saudi J Biol Sci* 28(2):1218–1225. doi: 10.1016/j.sjbs.2020.11.072
Pantha K, Acharya K, Mohapatra, S et al. (2021) Faecal pollution source tracking in the holy Bagmati River by portable 16S rRNA gene sequencing. *npj Clean Water* 4(1):1–10.
Park TS, Yoon, J-Y (2015) Smartphone detection of *Escherichia coli* from field water samples on paper microfluidics. *IEEE Sens J* 15(3):1902–1907. doi: 10.1109/JSEN.2014.2367039
Pavesi T, Moreira JC (2020) Mechanisms and individuality in chromium toxicity in humans. *J Appl Toxicol* 40(9):1183–1197. doi: 10.1002/jat.3965
Pignatello JJ, Oliveros E, MacKay A (2006) Advanced oxidation processes for organic contaminant destruction based on the fenton reaction and related chemistry. *Crit Rev Environ Sci Technol* 36:1–84. doi: 10.1080/10643380500326564
Podgorski J, Wu R, Chakravorty B, Polya DA (2020) Groundwater arsenic distribution in India by machine learning geospatial modeling. *Int J Environ Res Public Health* 17: 7119. doi: 10.3390/ijerph17197119
Rajapaksha P, Elbourne A, Gangadoo S, Brown R, Cozzolino D, Chapman J (2019) A review of methods for the detection of pathogenic microorganisms. *Analyst* 144(2):396–411. doi: 10.1039/C8AN01488D
Ravindra YS, Puttaiah SH, Yadav S, Prabagar JS (2020) Evaluation of polymeric g-C_3N_4 contained non-hierarchical ZnV_2O_6 composite for energy-efficient LED assisted photocatalytic mineralization of organic pollutant. *J Mater Sci Mater Electron* 31:16806–16818. doi: 10.1007/s10854-020-04235–4
Richardson SD (2009) Water analysis: emerging contaminants and current issues. *Anal Chem* 81:4645–4677.
Sankar MU, Aigal S, Maliyekkal SM, et al. (2013) Biopolymer-reinforced synthetic granular nanocomposites for affordable point-of-use water purification. *Proc Natl Acad Sci U S A* 110:8459–64.
Sauer S, Kliem M (2010) Mass spectrometry tools for the classification and identification of bacteria. *Nat Rev Microbiol* 8:74–82.
Scanlon S, Aggeli A, Boden N, et al. (2007) Peptide aerogels comprising self-assembling nanofibrils. *Micro Nano Lett* 2:24–29.
Schloter M, Aßmus B, Hartmann A (1995) The use of immunological methods to detect and identify bacteria in the environment. *Biotechnol Adv* 13(1):75–90.
Setia R, Singh Dhaliwal S, Kumar V, Singh R, Kukal SS, Pateriya B (2020) Impact assessment of metal contamination in surface water of Sutlej River (India) on human health risks. *Environ Poll* 265:114907. doi: 10.1016/j.envpol.2020.114907
Shaji E, Santosh M, Sarath KV, Prakash P, Deepch V, Divya BV (2020) Arsenic contamination of groundwater: a global synopsis with focus on the Indian Peninsula. *Geosci Front* 12(3):101079. doi: 10.1016/j.gsf.2020.08.015
Sharma N, Mohapatra S, Padhye LP, Mukherji S (2021) Role of precursors in the formation of trihalomethanes during chlorination of drinking water and wastewater effluents from a metropolitan region in western India. *J Water Process Eng* 40:101928.
Sharma Y, Srivastava V, Singh V, Kaul S, Weng C (2009) Nano-adsorbents for the removal of metallic pollutants from water and wastewater. *Environm Technol* 30:583–609.
Shi Y, Dou H, Zhou A, Chen Y (2008) Design and fabrication of a miniaturized electrochemical instrument and its preliminary evaluation. *Sens Actuators B Chem* 131(2):516–524.
Shivaraju HP, Yashas SR, Harini R (2020) Application of Mg-doped TiO_2 coated buoyant clay hollow-spheres for photodegradation of organic pollutants in wastewater. *Mater Today Proc* 27:1369–1374. doi: 10.1016/j.matpr.2020.02.754
Suess M (1985) Examination of water for pollution control: A reference handbook. Sampling, data analysis and laboratory equipment. United States.

Sun R, Sun G, Kwon SY (2008). Mercury biogeochemistry over the Tibetan Plateau: an overview. *Crit Rev Environ Sci Technol* 51(6), 577–602.

Suthar S, Chhimpa V, Singh S (2009) Bacterial contamination in drinking water: a case study in rural areas of northern Rajasthan, India. *Environ Monit Assess* 159:43. doi: 10.1007/S10661–008–0611–0

Tabish TA, Memon FA, Gomez DE, Horsell DW, Zhang S (2018) A facile synthesis of porous graphene for efficient water and wastewater treatment. *Sci Rep* 8:1817.

Tambekar D, Neware B (2012) Water quality index and multivariate analysis for groundwater quality assessment of villages of rural India. *Sci Res Rep* 2(3):229–235.

Thi D, et al. (2020) A colorimetric RT-LAMP assay and LAMP-sequencing for detecting SARS-CoV-2 RNA in clinical samples. *Sci Transl Med*, 12(556). doi: 10.1126/scitranslmed.abc7075

Tiwari R, Aulakh RS, Bedi JS, Gill, JPS, Kumar A (2020) Occurrence and spatial distribution of metals and arsenic in groundwater sources of Punjab (India), and their health risk assessment. *Toxicol Environ Chem* 102(9):425–438.

Tortajada C (2016) Policy dimensions of development and financing of water infrastructure: the cases of China and India. *Environ Sci Policy* 64:177–187.

Toze S (1999) PCR and the detection of microbial pathogens in water and wastewater. *Water Res* 33(17):3545–3556

Vanegas DC, Gomes CL, Cavallaro ND, Giraldo-Escobar D, Mclamore ES (2017) Emerging biorecognition and transduction schemes for rapid detection of pathogenic bacteria in food. *Compr Rev Food Sci Food Saf* 16:1188–1205. doi: 10.1111/1541–4337.12294

Vikesland PJ (2018) Nanosensors for water quality monitoring. *Nat Nanotechnol* 13:651–660. doi: 10.1038/S41565–018–0209–9

Wakode HB, Baier K, Jha R, Azzam, R (2018) Impact of urbanization on groundwater recharge and urban water balance for the city of Hyderabad, India. *Int Soil Water Conserv Res* 6(1):51–62.

Wang TH, Peng Y, Zhang C, Wong PK, Ho CM (2005) Single-molecule tracing on a fluidic microchip for quantitative detection of low-abundance nucleic acids. *J Am Chem Soc* 127(15): 5354–5535.

Wang et al. (2015) Development of ELISA for detection of Rh1 and Rg2 and potential method of immunoaffinity chromatography for separation of epimers. *J Chromatogr B* 985:197–205. doi: 10.1016/j.jchromb.2015.01.037.

Wang J, Zhuan R (2020) Degradation of antibiotics by advanced oxidation processes: an overview. *Sci Total Environ* 701:135023. doi: 10.1016/j.scitotenv.2019.135023

WHO (2008) https://www.who.int/water_sanitation_health/dwq/fulltext.pdf

WHO (2011) Guidelines for drinking-water quality. *Who Chronicle* 38:104–108.

WHO (2017) Guidelines for drinking-water quality fourth edition incorporating the first addendum. Available at: https://www.who.int/publications/i/item/9789241549950

Wood TP, Duvenage CS, Rohwer E (2015) The occurrence of anti-retroviral compounds used for HIV treatment in South African surface water. *Environ Pollut* 199:235–243.

Worch E (2012) *Adsorption Technology in Water Treatment: Fundamentals, Processes, and Modeling, Walter De Gruyter.* doi: 10.1515/9783110240238

Wu D, Rios-Aguirre D, Chounlakone M, Camacho-Leon S, Voldman J (2018) Sequentially multiplexed amperometry for electrochemical biosensors, *Biosens Bioelectron* 117:522–529.

Yao Q, Fan B, Xiong Y, Jin C, Sun Q, Sheng C (2017) 3d assembly based on 2d structure of cellulose nanofibril/graphene oxide hybrid aerogel for adsorptive removal of antibiotics in water. *Sci Rep* 7:45914.

Yashas SR, Shivaraju HP, Pema G, et al. (2020a) Sonochemical synthesis of graphitic carbon nitride-manganese oxide interfaces for enhanced photocatalytic degradation of tetracycline hydrochloride. *Environ Sci Pollut Res* 28:4778-4789. doi: 10.1007/s11356-020-10813–0

Yashas SR, Shivaraju HP, Thinley T, et al. (2020b) Facile synthesis of SnO2 2D nanoflakes for ultrasound-assisted photodegradation of tetracycline hydrochloride. *Int J Environ Sci Technol* 17:2593–2604. doi: 10.1007/s13762-020-02636–w

Zhang RQ, et al. (2013) A Simple Point-of-Care Microfluidic Immunomagnetic Fluorescence Assay for Pathogens. *Anal Chem* 85: 2645−2651.

Zhang F, Ye Y, Song D, Guo N, Peng Q, Li A, Zhou X, Chen Y, Zhang M, Huang D, Tang Y (2017) A simple and rapid identification method for newly emerged porcine Deltacoronavirus with loop-mediated isothermal amplification. *Biol Res* 50:30.

Zuo L, Zhang Y, Zhang L, Miao Y-E, Fan W, Liu, T (2015) Polymer/carbon-based hybrid aerogels: preparation, properties and applications. *Materials* 8:6806–6848.

Chapter 17

An emerging treatment technology

Exploring deep learning and computer vision approach in revealing biosynthesized nanoparticle size for optimization studies

SR. Abhishek, N. Sneha, and PBH. Karthik
NMAM Institute of Technology

CONTENTS

ABBREVIATIONS

ADAM	Adaptive Moment Estimation Optimizer
AgNPs	Silver Nanoparticle
ANN	Artificial Neural Network
AuNPs	Gold Nanoparticles
CLAHE	Contrast Limited Adaptive Histogram Equalization

DOI: 10.1201/9781003247869-20

CNN	Convolution Neural Network
DIBaS	Digital Image of Bacterial Species
EDT	Euclidean Distance Transform
MSE	Mean Square Error
MTBI	Myers–Briggs Type Indicator
OpenCV	Open Computer Vision
pH	Potential of Hydrogen
PSNR	Peak Signal-to-Noise Ratio
ReLu	Rectified Linear Unit
ResNet	Residual Network
RF	Random Forest
RMSE	Root-Mean-Square Error
RSM	Response Surface Methodology
SEM	Scanning Electron Microscopy
SPR	Surface Plasmon Resonance
STM	Scanning Tunneling Microscopy
SVM	Support Vector Machine
TEM	Transmission Electron Microscopy
UV-Vis	Ultraviolet-Visible spectrophotometry

17.1 INTRODUCTION

In recent years, the rapid synthesis of nanoparticles by physical, chemical, or biological approaches has been broadly explored for its wide range of applications. Nanoparticles synthesized using the green approach have been explored for remediation of emerging pollutants like dyes (Pandey et al. 2020) and pesticides (Khan and Pathak, 2020) through adsorption/degradation mechanisms. Green nanoparticles have also been studied for use in the production of low-cost colorimetric sensors for the detection of toxic heavy metals (Choudhary et al. 2020; Annadhasan et al. 2014). It is a well-known fact that nanoparticle size has a direct effect on its remediation potential as the surface area available for interaction with pollutants increases with smaller size nanoparticles. Nanoparticle size, its surface plasmon resonance (SPR), and color of the formed solution are also related (Nayak et al. 2020). Smaller green synthesized silver nanoparticles (AgNPs) were 70 nm in size with dark brown color and SPR around 420 nm. However, upon agglomeration due to unoptimized synthesis conditions like faster synthesis and partial capping, the color of the solution became dark gray, and nanoparticles were of size 150 nm and SPR shifted to the red side (480 nm). Therefore, understanding nanoparticle size is of utmost importance for exploring its potential for environmental remediation. Characterization by scanning electron microscopy (SEM) imaging is the most important criterion to prove that the synthesized material is in the nanoscale.

There has been tremendous growth in the application of machine learning for aiding in data interpretation which includes novel drug discovery (Nayak et al. 2020), solid-state material science (Borlido et al. 2020), prediction of protein corona formation of nanoparticles (Ban et al. 2020), in silico studies related to the toxicity of nanoparticles (Furxhi et al. 2020), particle sizing and counting (Seyedin et al. 2019), bacterial colony classification (Zieliński et al. 2017), early detection and classification of live pathogenic bacteria (Wang et al. 2020), prediction of disease (Rao et al. 2021), and many more. Using this approach, the present study has tried to explore the use of a machine learning technique for biosynthesized nanoparticle characterization by SEM, as there is a need for a computer-aided tool. This could be applicable for characterizing the nanoparticles for bridging the delays in sample preparation and characterization, thereby giving an approximate size and SPR of the nanoparticles. The unique property of metallic nanoparticles is the SPR, which is due to electronic interactions at a nanoscale and an indicator of a resonance effect as a result of interaction between the conduction electrons of the nanoparticles with the incident photons (light source). Slight changes in this interaction due to the size variation of nanoparticles can lead to color shifts which are exploited in the present study. A colloidal solution of metallic nanoparticles when formed gives a characteristic color to the solution, which is a confirmation for nanoparticle formation (Lima et al. 2016). The colloidal solution of AgNPs is brown and that containing gold nanoparticles (AuNPs) is purple, which is directly dependent on its SPR, a unique property of materials in nanoscale. Based on the increase in the size of nanoparticles, there is a shift in SPR to the red side which indicates agglomeration of nanoparticles. Instability caused in colloidal nanoparticle solution may be due to an increase in nanoparticle size, shape, or variation in synthesis parameters (pH, temperature, metal ion concentration, reducing agent concentration) (Lima et al. 2016).

Previous studies carried out by the research team used seed cake extract of *Pongamia pinnata* and leaf extract of *Cyclea peltata* for the biogenic synthesis of metallic AgNPs and AuNPs, respectively, through the green synthesis approach. Gold chloride and silver nitrate were used as precursor salts, and plant phytochemicals helped in nanoparticle synthesis and capped around the nanoparticles to form stable nanoparticles. The colloidal AgNPs formed were brown when the size of the particles is in the range of 82–91 nm, which changes to dark brown at 47–66 nm due to increased yield, and gray at 80–100 nm which is a confirmation for agglomeration (Lima et al. 2016; Nayak et al. 2019; Nayak et al. 2018a). Similarly, AuNPs are reddish at around 18–20 nm, purple at 20–22 nm, and bluish at 60–100 nm (Nayak et al. 2018b). This property of the material at the nanoscale was seen during optimization studies by response surface methodology. This was chosen as the basis for nanoparticle characterization in the present study by approximating the size of nanoparticles in the given sample.

Machine learning is a set of techniques to make computers better at doing things that humans can do better than machines. It is concerned with the design and development of algorithms and techniques that allow computers to learn. Machine learning applications are widely used in image recognition, object detection, and many more real-time scenarios. In the present study, as we are focused on identifying the nanoparticle from the given SEM image and determining its size, we have applied the machine learning algorithms to learn the features from the given image and to classify the SEM image based on the colloidal nanoparticle solution. In this study, we have used a robust machine learning approach that would help in differentiating AuNPs and AgNPs biosynthesized using plant extracts. Based on the SPR and the color of the solution, a new SEM image will be auto-generated by augmentation technique using the previously loaded SEM images as the basis. Feature extraction methods through machine learning techniques are usually based on artificial neural networks (Rao et al. 2021). Convolution Neural Network (CNN) is a subclass of deep learning technique that is usually applied to analyze the input image by allocating the weights and biases to the image.

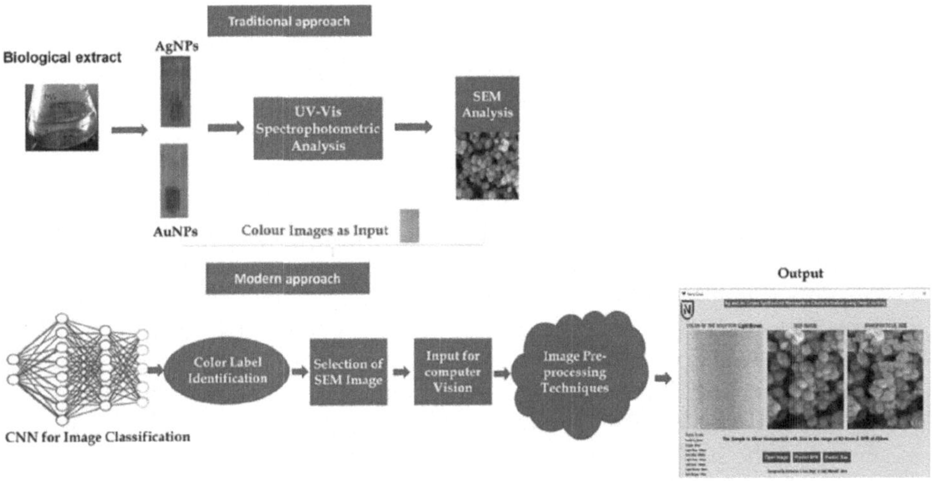

Figure 17.1 Graphical abstract of the proposed model.

Fundamental features of CNN that minimize the number of features in a neural network are sparse interactions, parameter sharing, and equivariant representation. CNN helps in classifying the image effectively through its various layers (Emmert-Streib et al. 2020; Rao et al. 2021). The classified image will be processed further to detect the occurrence of edges of the objects present in the given sample using various image preprocessing techniques (Rao et al. 2017; Sapna and Renuka, 2017; Sapna and Renuka, 2020; Sandhya et al. 2021). These images will be highly useful in finding out the approximate size of nanoparticles in the sample which can be explored during the optimization studies for nanoparticle formation as it requires a lot of SEM analysis, which is highly expensive. Figure 17.1 gives an overview of the proposed model in a graphical format.

Figure 17.1 depicts the traditional approach of nanoparticle characterization by UV-Vis spectroscopic analysis and morphology analysis by SEM which is a time-consuming, expensive, and cumbersome task that could be easily supported by the present approach which used deep learning concepts and computer vision for the identification of SEM image along with its nanoparticle size information, based on the colloidal nanoparticle picture given as input. Nanoparticle optimization by varying different parameters like pH, agitation speed, and temperature has been explored for increasing the yield of nanoparticles. This requires a lot of SEM analysis to check if varying a parameter had a positive or negative impact on synthesis (Lima et al. 2016). Therefore, this approach would be very useful in the case of optimization studies to nullify the nanoparticle characterization required during each case.

17.2 LITERATURE REVIEW

The use of machine learning schemes has been explored in the present era for a wide range of real-time applications in the field of nanoscience and nanotechnology. The use of deep learning for the recognition of metal nanoparticles deposited on the surface of pyrolytic graphite has been explored and compared with STM images. The cascade mask-RCNN neural network was used, and the model gave a precision of 93% (Okunev et al. 2020). Integration of TEM imaging to machine learning models was experimented with using U NET – neural network to simulate the real-time images. The diversity of nanoparticle properties was mapped with the first-order chaining assembly. This approach helped in understanding the properties at the nanoscale through the simulation approach (Yao et al. 2020). The use of AgNPs for energy applications by simulating the electronic structures and improving the transfer of electrons by the machine learning approach has been explored (Sun et al. 2017). Table 17.1 highlights the usage of several machine learning approaches used by researchers for various problems.

17.3 MATERIALS AND METHODS

In this section, we will discuss the methods applied to generate an SEM image based on the colloidal nanoparticle color solution and the procedure involved in identifying the size of the nanoparticles in each SEM image.

17.3.1 Convolution neural network model

In the present study, CNN was used to recognize the color image and extract suitable color features from a colloidal nanoparticle solution and apply appropriate image classification techniques. The following steps will include convolution, pooling, flattening, and the fully connected layer which are then applied for classifying an image which is shown in Figure 17.2. Color coding was used in neural networks to distinguish between different fully connected layers; however, the color codes used had no significance on the results.

17.3.1.1 Convolution operation

In this layer, the given image is convolved through multiple layers by using kernels. These filters are applied across the data through the sliding window. The convolution output of a 3D filter color image is a 2D matrix. Considering the 1D convolution, if input '*f*' and kernel '*g*' are both functions of 1D time-series data, then the convolution is given by the mathematical equation as shown in Equation (17.1).

$$(f * g)(t) \underline{\underline{\text{def}}} \int_{-\infty}^{\infty} f(\tau) g(t-\tau) \mathrm{d}\tau \tag{17.1}$$

Equation (17.1) signifies the percentage (%) area of the filter '*g*' that overlays the input '*f*' at a time 'τ' over the time '*t*'. In practice, we may have multidimensional input tensors that require multidimensional kernel tensors. In this case, we can set an image input '*I*' in a kernel '*h*' whose convolution is defined and shown in Equations (17.2 and 17.3).

$$(I * h)(x, y) = \int_0^x \int_0^y I(i, j) \cdot h(x-i, y-i)\, di\, dj \tag{17.2}$$

Table 17.1 Overview of the use of machine learning techniques by various researchers

Objectives	*Dataset used*	*Methods/algorithms used*	*Results*	*References*
To predict the structure of the protein corona and identification of nanoparticles.	RF-and-PC GitHub repository.	Random Forest algorithm was explored for learning the complex relations between the properties of nanoparticles and their corona structure.	The functional protein in complex coronas was accurately predicted which showed an $R^2 = 0.80$.	Ban et al. (2020)
Review on machine learning application in nanotoxicology.	Through existing literature, databases, and new data experimentation.	Investigation of various machine learning methods applied in nanotoxicology and steps involved in implementing a model.	Identification of numerous models that deliver a prediction to several nanotoxicological outcomes.	Furxhi et al. (2020)
Developing a suitable computer program for particle counting in an image.	Benchmark dataset.	Image processing steps.	The tool was developed to obtain information on particles such as shape and count.	Seyedin et al. (2019)
To classify the types and species of bacteria.	DIBaS dataset (Digital Image of Bacterial Species).	Deep CNN, SVM, RF algorithms.	The accuracy of the recognition was $97.24\% \pm 1.07\%$.	Zieliński et al. (2017)
Detection system for capturing microscopic images of lived bacterial growth.	LIVE images.	Deep learning algorithms.	More than 90% of bacterial colonies were identified within 12 h with an accuracy of 99.2%.	Wang et al. (2020)
To build a predictive model for career guidance.	A web-based questionnaire based on the MBTI category.	Artificial Neural Network (ANN).	Evaluation matrices like sensitivity, specificity, precision, and accuracy for all the MBTI categorizations were evaluated by the model which showed values above 92% in all the cases.	Rao et al. (2020a)
To effectively design a handwritten character recognition system for the Kannada language.	Chars74K dataset.	Image Preprocessing, Convolutional Neural Network (CNN).	The model gave an accuracy of 95.11% during training and 86% during testing.	Rao et al. (2020b)

(*Continued*)

Table 17.1 (*Continued*) Overview of the use of machine learning techniques by various researchers

Objectives	*Dataset used*	*Methods/algorithms used*	*Results*	*References*
To develop an application for a better vision of online shopping using computer vision.	Web images.	Image Processing using OpenCV.	A tool was developed for tracking real-time face detection.	Rao et al. (2017)
To identify objects in real-time video.	CIFAR10, CIFAR100, and MNIST.	The analysis was made to determine the performance using three popular networks such as AlexNet, GoogleNet, and ResNet 5.0.	GoogleNet and ResNet 5.0 gave better precision in recognizing the objects.	Sharma et al. (2018)
To detect edges from X-Ray images of human being.	Standard images from imageprocessingplace.com	Image Preprocessing.	The proposed method has shown better results compared to MSE, RMSE, PSNR in edge detection from X-Ray images of a human being.	Bharodiya and Gonsai (2019)
Digital image edge extraction using image processing techniques.	Benchmark dataset.	Experimented with numerous algorithms for edge detection.	Binary morphology operators obtained better edge features from an image compared to others.	Kalpana et al. (2012)
Performance evaluation of different edge detection algorithms.	Standard datasets on X-Ray images.	Image Preprocessing for noise removal and edge detection.	Results proved the canny algorithm works well for edge detection and noise removal.	Anas et al. (2019)
To optimize the features of the image for better classification.	Standard dataset of dogs and cats.	CNN	Model accuracy of approximately 88.3% was achieved.	Panigrahi et al. (2018)
Nanoparticle drug delivery system design.	chEMBL and public sources.	Perturbation theory, PTML algorithm, and R studio.	Multilabel PTML model that is useful for designing a new drug delivery system.	Santana et al. (2020)
Prediction of nanoparticle in vitro toxicity.	Safe and sustainable technology (S2NANO).	Meta classifier which combines models (neural network and random forest).	Copeland index which ranks the meta classifier and predicts nanoparticle toxicity.	Furxhi et al. (2019)

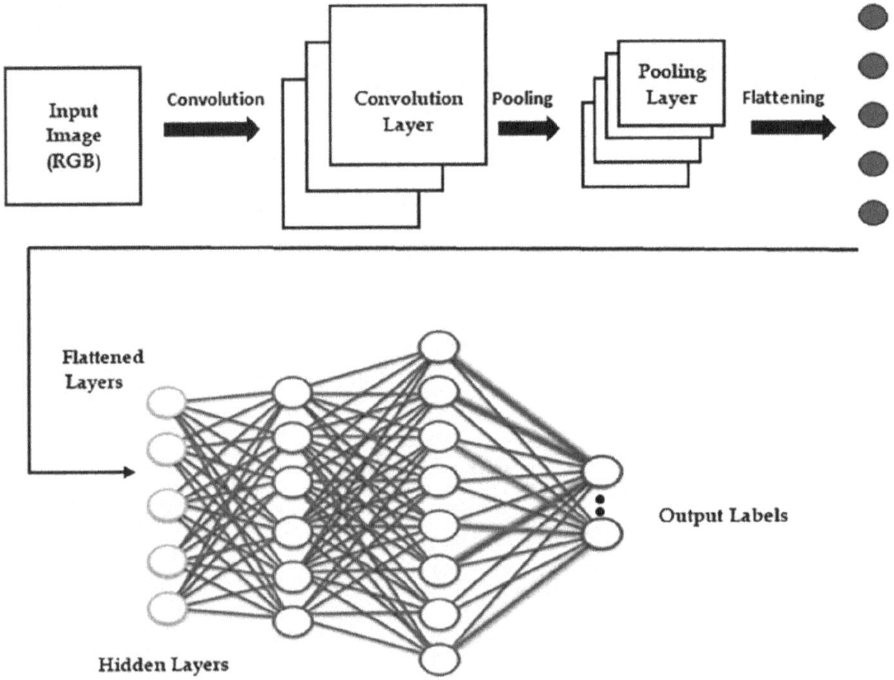

Figure 17.2 CNN architecture.

$$= \int_0^x \int_0^y I(x-i, y-j) \cdot h(i,j)\, di\, dj \tag{17.3}$$

Equation (17.2) performs convolutions by sliding the image over the kernel, and Equation (17.3) slides the kernel over the image. Since there are usually fewer possible values for '*x*' and '*y*' in the kernel than in an image, we use the second form for convolution for which the result will be a scalar value. We then repeat this process for every point '*x*', '*y*' for which the convolution exists on the image. These values are stored in a convolved matrix represented by $(I * h)$. This output constitutes a part of a feature map.

17.3.1.2 Activation layer

Images are usually nonlinear because of the different elements present in them. When we apply the mathematical operation to an image and run the feature detection to create a feature map, we may end up with linearity in an image.

17.3.1.3 Pooling layer

Pooling involves the downsampling of a feature to reduce the learning features during the training stage. The pooling layer consists of two hyperparameters, such as the dimension of spatial extent and stride. Among the pooling filters, the maximum value of the features in the regions is returned by the max filter. The depth of the feature map after applying pooling to each layer will always remain the same. Performing pooling would reduce the chances of overfitting due to fewer parameters involved.

17.3.1.4 Flattening layer

The output from the convolution layers is flattened by its structure to generate a 1D feature vector which can then be utilized by the dense layer for the final classification.

17.3.1.5 Fully connected layer

A vital component of CNN has demonstrated its usefulness in identifying and categorizing images for applications related to computer vision. The output from the convolution layers typically denotes high-level features in data.

17.3.2 Model implementation

Deep learning is the subset of machine learning used for image classification using CNN. To build and train the model, we have used the Keras library in python. In general, image classification intends to categorize all the pixels in a digital image into one or more themes. The first step in this process is to obtain the dataset. In this study, we have collected silver and gold colloidal nanoparticle solution images from the Department of Biotechnology, NMAM Institute of technology, and used them as input for the proposed model. Colloidal AgNPs have different colors like light brown, dark brown, and gray, whereas AuNPs are reddish, purple, and bluish based on their size and SPR as shown in Figure 17.3. The nanoscale values and SPR considered for each colloidal solution colors of AuNPs and AgNPs are as given:

Reddish (18–20nm and SPR of 530nm), Purple (20–22nm and SPR of 550nm), Light Blue (60–100nm and SPR of 570nm), Dark Blue (100–200nm and SPR of 590nm), Light Gray (80–100nm and SPR of 460nm), Dark Gray (100–150nm and SPR of 480nm), Light Brown (82–91nm and SPR of 430nm), Dark Brown (47–66nm and SPR of 420nm). In the present study, we have calculated the height and width of a nanoparticle using the formulas given below.

$$Distance\ in\ pixel = Euclidean\ Distance\left(bottom_{left}, top_{right(BoundingBoxRectangle)}\right)$$

$$Distance\ in\ nanometer = Nanometer\ scale * 3.9370078740157e^{-8}$$

$$Pixel/nanometer = \frac{Distance\ in\ pixel}{Distance\ in\ nanometer}$$

$$Width\ of\ nanoparticle = \frac{Euclidean\ Distance\left(bottom_{left}, bottom_{right(BoundingBoxRectangle)}\right)}{Pixel\ /\ nanometer}$$

Gold Nanoparticles			
Light Blue	**Dark Blue**	**Purple**	**Reddish**

Nanometer-Scale3.937e-6 inches	Nanometer-Scale 7.874e-6 inches	Nanometer-Scale 7.874e-7 inches	Nanometer-Scale1.5748e-6 inches
Silver Nanoparticles			
Light Brown	**DarkBrown**	**Light Grey**	**Dark Grey**

Nanometer-Scale3.5433e-6 inches	Nanometer-Scale2.7559e-6 inches	Nanometer-Scale3.937e-6 inches	Nanometer-Scale5.9055e-6 inches

Figure 17.3 Colloidal nanoparticle solution.

$$Height\ of\ nanoparticle = \frac{Euclidean\ Distance\left(bottom_{right}, top_{right(BoundingBoxRectangle)}\right)}{Pixel\ /\ nanometer}$$

As it is a known fact that the machine learning model requires large datasets for getting better accuracy during prediction. The augmentation technique was used for nanoparticle solution images and SEM images as the samples were limited in numbers. These augmented images were varied based on various parameters like shades, intensity, zoom range, rotation range, and sheer range which helped in broadening the arena of the data samples for enhanced training of the model. The data samples used to train and test the model are 25,772 belonging to eight classes of color solution.

17.3.2.1 Model construction

The color solution image is plotted from the dataset and checked its size using 'plot' and 'shape' functions in python as shown in Figure 17.4.

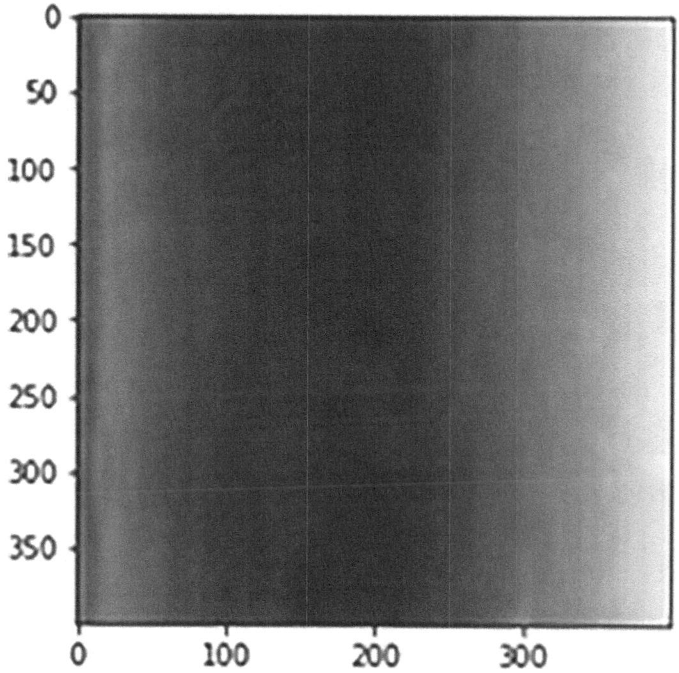

Figure 17.4 Color solution image.

To build our model, we have used the Keras library in python and the type of model used was sequential. Since our model experiments with only eight classification labels, we have imported the VGG16 transfer learning model to construct our model as it comes with predefined neural networks. The model was primarily initialized and later flattened into a 1D feature vector and given as an input to the fully connected layer for further processing. As we know the *Dense* layer is a standard layer used in neural networks for many cases, we have used two dense layers among which the first dense layer contains 100 units and the second dense layer contains 50 units. Since the model was based on labeled categorical classification, we have used a *softmax* activation function for the final dense layer.

17.3.2.2 Compiling the model

The model was compiled using an adaptive moment estimation optimizer (ADAM) optimization algorithm along with categorical cross-entropy loss functions and evaluated the performance of the model with the support of various performance metrics. ADAM is a method that computes an adaptive learning rate for every parameter and can handle any noisy problems. This optimizer helped us in gaining better performance in terms of precision and speed. To minimize the error rate during testing, we have utilized the categorical cross-entropy loss function as we have more than two output class labels for classification. Accuracy and loss of the model are associated with the number of epochs (seven epochs) and the model gave an accuracy of 91.46% as shown below in Figure 17.5. Figure 17.6 depicts a summary of the model.

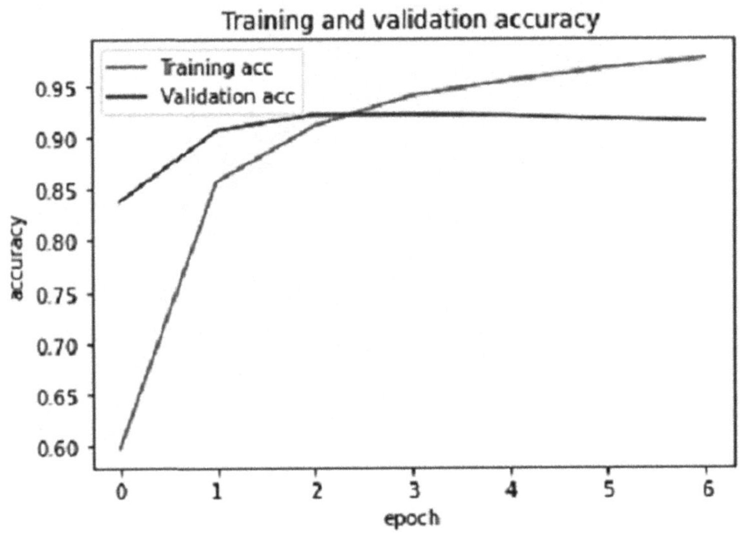

(a)

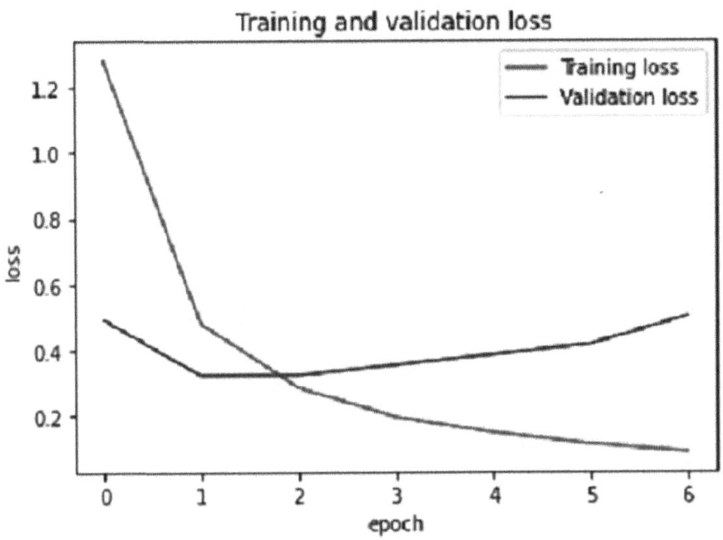

Figure 17.5 Line plots of (a) training and validation accuracy data and (b) training and validation loss concerning the number of epochs.

```
Layer (type)                 Output Shape              Param #
=================================================================
flatten_7 (Flatten)          (None, 2048)              0
_________________________________________________________________
dense_19 (Dense)             (None, 100)               204900
_________________________________________________________________
dropout_13 (Dropout)         (None, 100)               0
_________________________________________________________________
dense_20 (Dense)             (None, 50)                5050
_________________________________________________________________
dropout_14 (Dropout)         (None, 50)                0
_________________________________________________________________
dense_21 (Dense)             (None, 8)                 408
=================================================================
Total params: 210,358
Trainable params: 210,358
Non-trainable params: 0
_________________________________________________________________
```

Figure 17.6 Model summary.

```
[INFO] loading and preprocessing image…
ID: 0, Label: Dark Blue 9.77%
ID: 1, Label: Dark Brown 0.03%
ID: 2, Label: Dark Grey 0.06%
ID: 3, Label: Light Blue 0.51%
ID: 4, Label: Light  Brown 0.09%
ID: 5, Label: Light Grey 0.64%
Final Decision:
.
..
...
ID: 6, Label: Purple
```

Figure 17.7 Prediction report.

17.3.2.3 Model prediction

To predict the unlabeled image, we have used two different functions. The first function will make the model load the image, vary the size, and convert it into an array. The second function will utilize the transfer learning prediction model and iterative function for better prediction of an image. Figure 17.7 shows the summary report of prediction for the purple color solution.

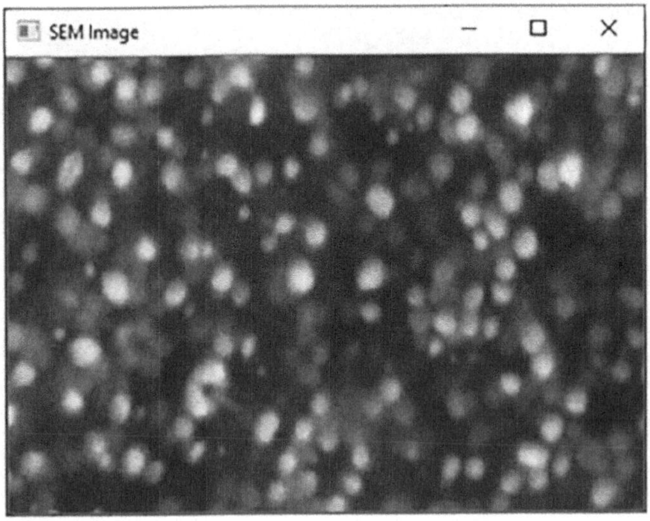

Figure 17.8 SEM image.

Based on the predicted color solution, the SEM image was randomly selected from the augmented SEM images dataset. Figure 17.8 shows the random selection of an SEM image for the purple color solution.

The SEM image was later fed to computer vision for identifying the size (width and height) of nanoparticles in the given image by using various image processing techniques as described below.

17.3.2.1.1 IMAGE PROCESSING

Image processing plays a significant role in refining the quality of an image. The Pre-processing steps include grayscale conversion, image filtering mechanisms, denoising, thresholding, and morphological operations. The image preprocessing steps followed in the proposed work are as shown in Figure 17.9.

17.3.2.1.2 PYRAMID MEAN SHIFT FILTERING ALGORITHM

At primary, the generated SEM image will be loaded and applied with a filtering process to help the accuracy of the thresholding. The filtering approach used in our study is the pyramid mean shift filtering algorithm. This algorithm seeks modes or local maxima of density in the feature space. This algorithm will preserve important edges of an image and help in the segmentation process. The set of neighboring pixels for each pixel of an image is calculated and a new spatial center along with the color mean value is determined for these set of neighboring pixels. The process will iterate until the spatial and color means stop shifting. The final mean color determined will be assigned to the iteration's starting place as shown in Figure 17.10.

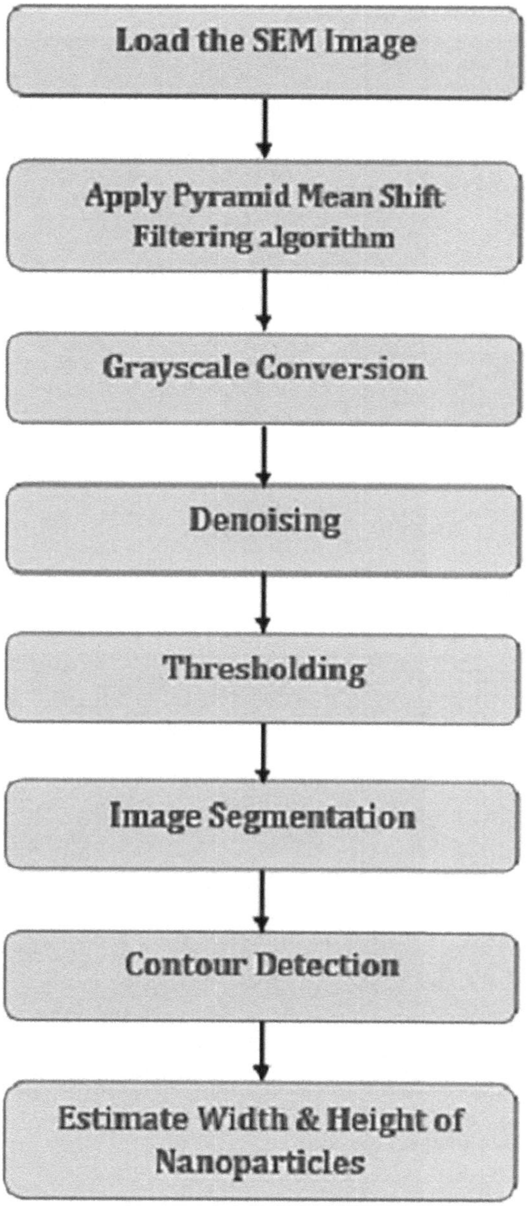

Figure 17.9 Image processing steps.

17.3.2.1.3 GRAYSCALE CONVERSION

In image preprocessing, the grayscale conversion is applied for a colored image to get a grayscale image. This process will help in reducing the complications of the image by projecting the three-dimensional pixel value of an image to a one-dimensional image.

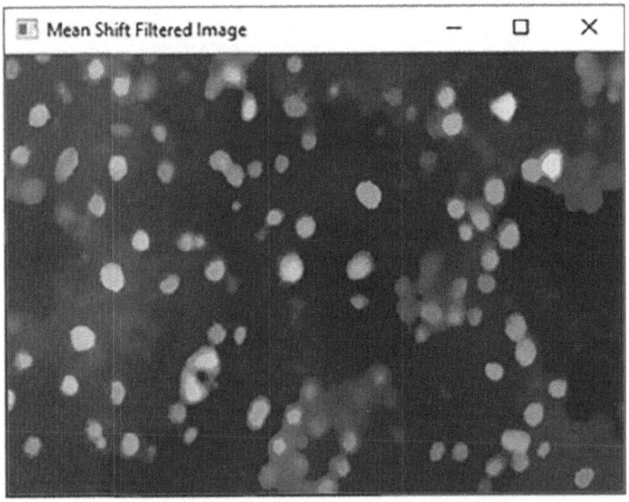

Figure 17.10 Pyramid mean shift filtered image.

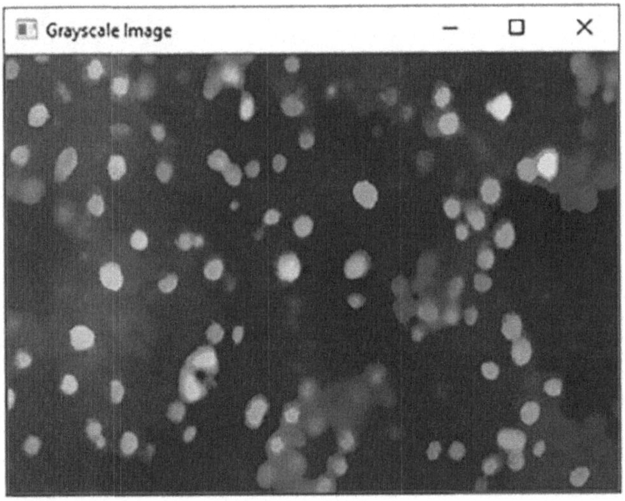

Figure 17.11 Grayscale image.

This approach will help in reducing the pixel information for further computations. In the present study, we have converted the filtered SEM image to grayscale as shown in Figure 17.11.

17.3.2.1.4 DENOISING

Noise is an unfortunate result of data acquisition which comes in many forms and sources. Therefore, image denoising is one of the primary preprocessing operations

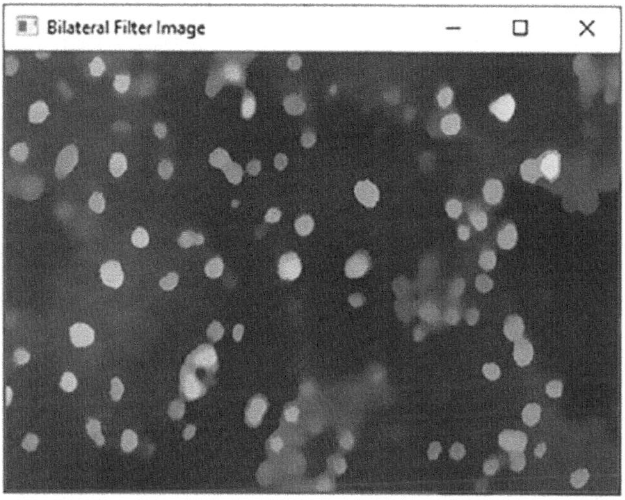

Figure 17.12 Bilateral filter image.

that is performed before the information extraction of images. In the present study, the grayscale image is applied with a bilateral filter for denoising. This filter will replace each pixel value with a weighted average of nearby pixel values. Thus, the noise-removed image shall enable further processing to be smooth and flawless as shown in Figure 17.12.

17.3.2.1.5 THRESHOLDING

Image thresholding is a basic image segmentation technique to convert a grayscale image to a binary image. This is usually done to help image processing by separating the objects or foreground pixels from background pixels. In this study, we have applied Otsu's method for thresholding the image. This algorithm will automatically find the optimal threshold intensities. It works by searching for the threshold intensity that optimally separates the image into foreground and background classes, Figure 17.13 shows the threshold image after applying Otsu's method.

17.3.2.1.6 IMAGE SEGMENTATION USING WATERSHED ALGORITHM

Image segmentation is a sub-field of artificial intelligence and computer vision. This method is highly useful in locating the objects and boundaries within an image. It is mostly used for partitioning an image into its region based on some criteria where the regions are meaningful and disjoint like finding an object in an image and isolating it. Segmentation will help in identifying the significant segments in an image, which could be used for image processing. This approach generates a pixel-wise mask for every single object in the image, thereby helping in attaining a better granular understanding of the objects. Even after applying the thresholding mechanism for a given

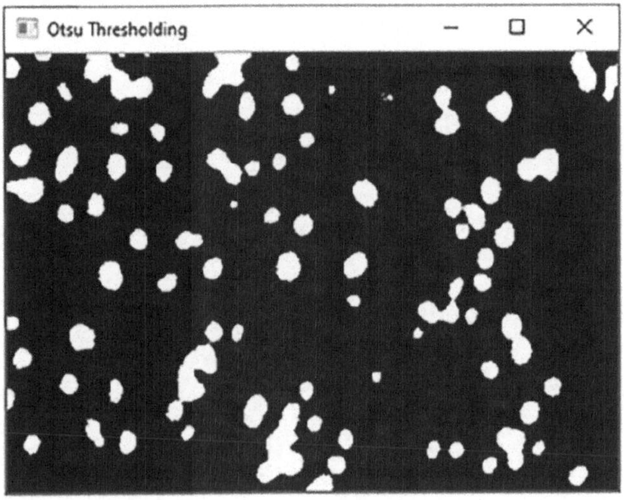

Figure 17.13 Otsu's threshold image.

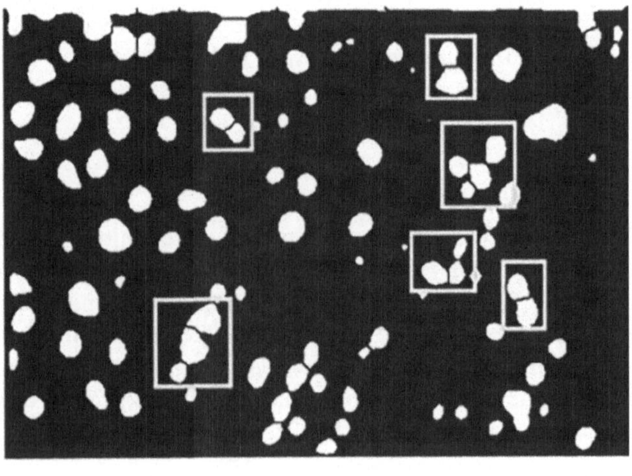

Figure 17.14 Image segmentation using watershed algorithm.

image, the results are pretty much awful that it is not able to depart few overlapped objects. To overcome this issue, we have applied the watershed algorithm for segmentation. This method will compute the euclidean distance transform method to compute the euclidean distance to the closest zero for each of the foreground pixels as shown in Figure 17.14.

It is very clear from the image that the watershed algorithm has successfully helped in separating the overlapping and touching nanoparticles to a major extent. It is also clear from the image that a few of the nanoparticles were unable to segregate as

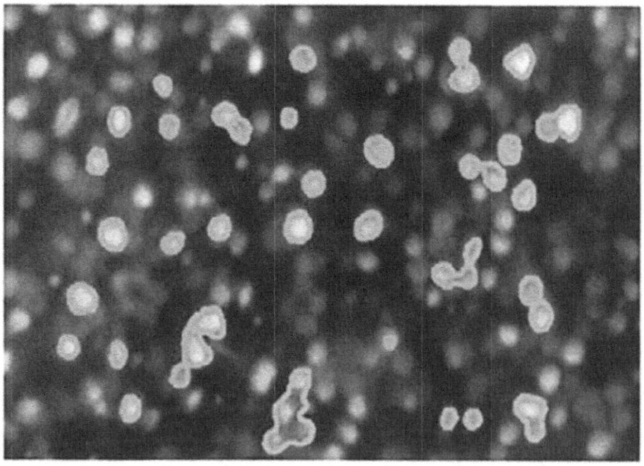

Figure 17.15 Contour detection image.

separate particles due to slight agglomeration seen in the image. Hence, the use of a quality image would nullify this issue and reduce this error drastically.

17.3.2.1.7 CONTOUR DETECTION

For recognizing the shape of an object in an image, the contours are one of the useful tools which are extensively explored. In the present study, the OpenCV contour function was utilized to draw the contours. The contours are later sorted from left to right as the leftmost contour is the reference object. After sorting the contours, we have filtered the incomplete contours and then computed the *bounding boxes* of each contour to determine the object dimensions. Euclidean distance was measured between the two end points of bounding box to compute the width and height of an object in an image. Figure 17.15 shows the contour detection image. The diameter of the nanoparticle in an image was found to be in the range of 10.0–25.0 nm.

17.4 RESULTS

This section will produce the results of model evaluation, confusion matrix, classification metrics, and the output of the model.

17.4.1 Model evaluation

The model was trained and validated with 25,772 samples for seven epochs as shown in Figure 17.16. Accuracy is not always the best way to evaluate the classifiers as it may give random results based on the images tested. Henceforth, the model was evaluated to verify the prediction of the model against unseen labeled data as shown in Figure 17.17. To verify how well our model can predict or classify the images, we have used

```
Train on 25772 samples, validate on 25772 samples
Epoch 1/7
25772/25772 [==============================] - 9s 348us/step - loss: 1.2780 -
acc: 0.5990 - val_loss: 0.4906 - val_acc: 0.8377
Epoch 2/7
25772/25772 [==============================] - 9s 350us/step - loss: 0.4791 -
acc: 0.8568 - val_loss: 0.3216 - val_acc: 0.9064
Epoch 3/7
25772/25772 [==============================] - 8s 309us/step - loss: 0.2842 -
acc: 0.9105 - val_loss: 0.3220 - val_acc: 0.9214
Epoch 4/7
25772/25772 [==============================] - 8s 321us/step - loss: 0.1944 -
acc: 0.9400 - val_loss: 0.3522 - val_acc: 0.9214
Epoch 5/7
25772/25772 [==============================] - 8s 318us/step - loss: 0.1457 -
acc: 0.9546 - val_loss: 0.3836 - val_acc: 0.9201
Epoch 6/7
25772/25772 [==============================] - 9s 332us/step - loss: 0.1109 -
acc: 0.9664 - val_loss: 0.4146 - val_acc: 0.9169
Epoch 7/7
25772/25772 [==============================] - 8s 302us/step - loss: 0.0869 -
acc: 0.9758 - val_loss: 0.5004 - val_acc: 0.9146
25772/25772 [==============================] - 2s 62us/step
[INFO] accuracy: 91.46%
[INFO] Loss: 0.5004300424106393
Time:  0:01:02.038954
```

Figure 17.16 Model accuracy associated with some epochs.

```
25772/25772 [==============================] - 2s 70us/step
Out[26]: [0.5004300424106393, 0.9145584106445312]
```

Figure 17.17 Model evaluation.

two methods such as confusion matrix and classification metrics. A confusion matrix will provide a summary of a prediction result based on a classification problem. As we are using more than two class labels, the model accuracy alone may lead to ambiguity due to the unequal number of observations. Hence, the model can be visually represented by computing the confusion matrix as shown in Figure 17.18. When we build the model to predict a certain category, we need a way to measure the accuracy of the predictions made by the classifier. To achieve this, we have evaluated our model for its precision, recall scores, and f1-score performance metrics. Precision, recall, and f1-score will consider the positives and the negatives for measuring the model's accuracy and making precise predictions. Precision metrics will indicate what percentage of positive predictions made were correct. Recall metrics will indicate what percent of actual positive values were correctly classified by the classifier and the F-1 score will give the harmonic mean of precision and recall. Figure 17.19 shows the classification metrics of the proposed model.

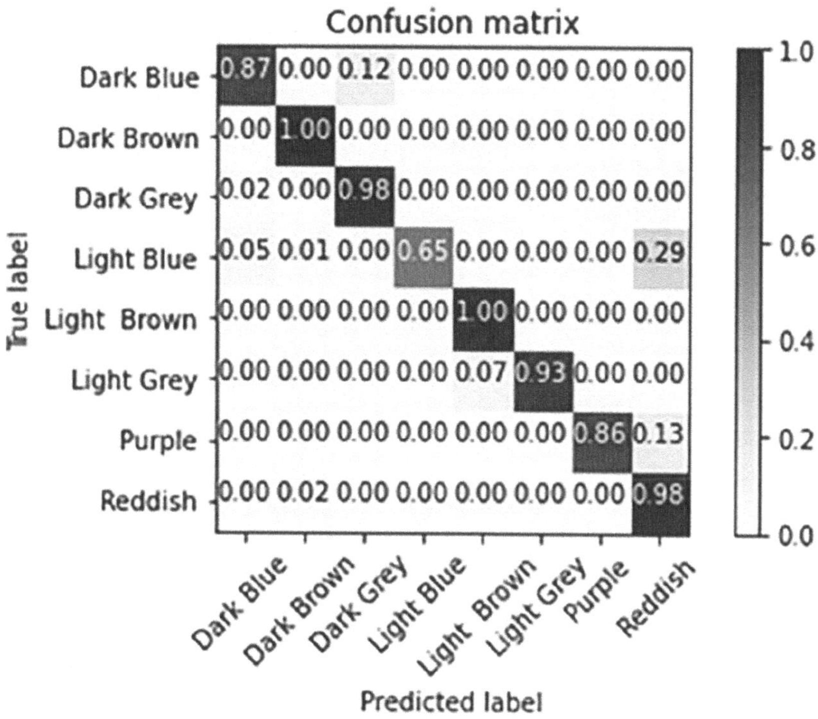

Figure 17.18 Confusion matrix.

	precision	recall	f1-score	support
Dark Blue	0.96	0.86	0.91	3823
Dark Brown	0.97	1.00	0.98	3545
Dark Grey	0.85	0.98	0.91	2598
Light Blue	1.00	0.65	0.78	2573
Light Brown	0.95	1.00	0.97	3442
Light Grey	1.00	0.93	0.96	2597
Purple	1.00	0.86	0.93	3564
Reddish	0.74	0.98	0.85	3630
micro avg	0.92	0.91	0.92	25772
macro avg	0.93	0.91	0.91	25772
weighted avg	0.93	0.91	0.91	25772
samples avg	0.91	0.91	0.91	25772

Figure 17.19 Classification metrics.

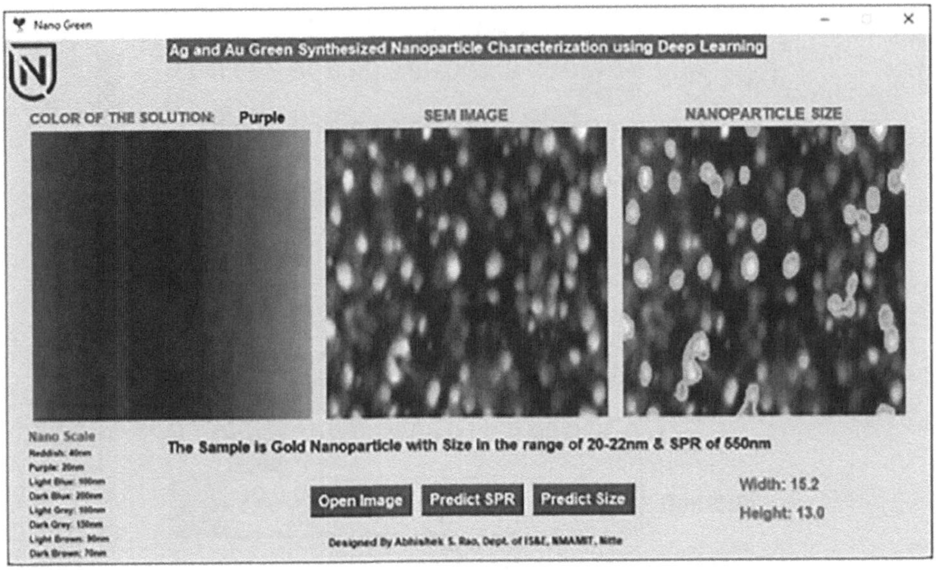

Figure 17.20 SEM image of green synthesized AuNPs with size details.

It is evident from Figure 17.19 that the model has achieved better results for precision, recall, and f1-score which in turn measures the accuracy of the model prediction.

Figure 17.20 shows the outcome of the model which has effectively identified the nanoparticle size based on the color solution and SEM image for AuNPs synthesized using plant extract.

17.5 DISCUSSION

A comparative analysis of nanoparticle diameter and count in each SEM image of biosynthesized nanoparticles was carried out in parallel by ImageJ and the model proposed by Lind (2012). It is evident from Figure 17.21 that both the applications were quite accurate in analyzing the nanoparticle diameter in nm and the count corresponding to that nm size particle. Nanoparticle diameter in the range of 10–25 nm was quite comparable by both approaches. The additional advantage of using the proposed model would be its capability to use colloidal nanoparticle color image as input from which a suitable SEM image would be selected by the model itself and the analysis would be carried out.

17.6 CONCLUSION

Although there is limited literature on deep learning methods for automatic particle recognition using electron microscopy images as data, there is a growing body of knowledge in this area. The current research focuses on using baseline knowledge of the results, such as colloidal nanoparticle color production based on size, localized

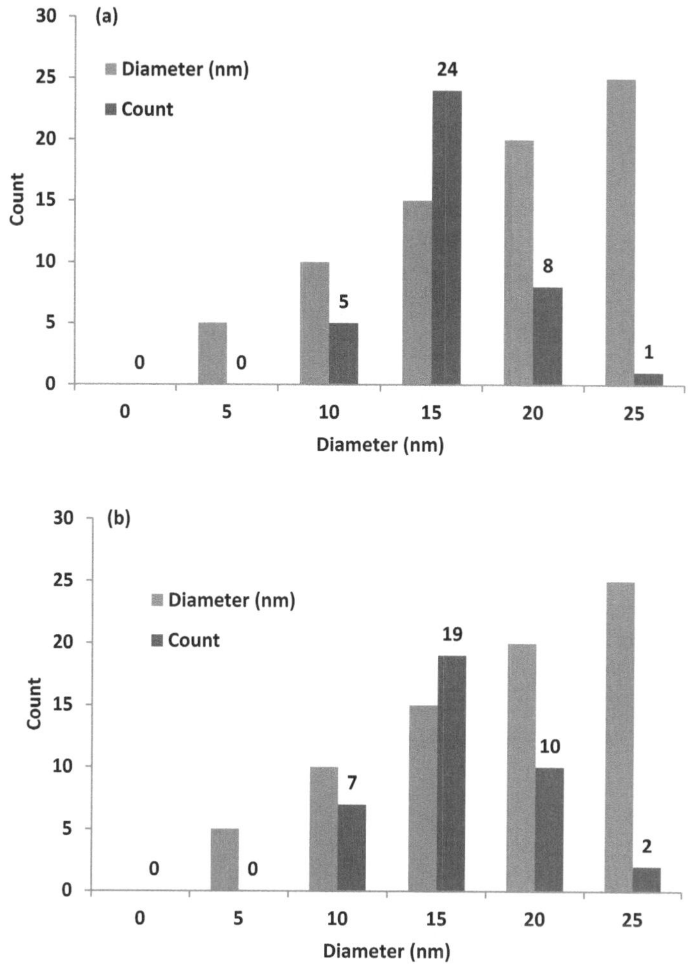

Figure 17.21 Comparative analysis of biosynthesized nanoparticle SEM image analysis by (a) the proposed model and (b) ImageJ software.

SPR information, and baseline SEM images during initial screening, rather than just particle recognition. The main goal of this study is to take these experimental results as inputs and try to create an AI-based automated model that can aid in understanding sample quality based on color achieved during optimization and quality (monodispersed or agglomerated) in order to reduce the need for repetitive SEM. Nanoparticle type and size characterization by the proposed model was quite comparable with the traditional approach results which could be effectively useful in finding out the nanoparticle-related information from a colloidal solution. This can be explored during the optimization studies for nanoparticle formation, which would only require an approximate idea of the size changes occurring in the nanoparticle formed because of a changing parameter like pH or temperature which is very crucial for nanoparticle synthesis. The nanoparticle size obtained using this method may also be explored during bioremediation studies in which green synthesized nanoparticles are investigated for adsorption/degradation mechanisms. The above-said studies would require a lot of SEM analysis, which is expensive and requires a lot of sample preprocessing. Hence the worldwide acceptable nanoparticle characterization, worth in crores of rupees, could be easily substituted to some extent by the proposed model. The application of a deep learning approach for colloidal nanoparticle image classification and the concepts of computer vision techniques to determine the approximate size of the nanoparticles in each SEM image were accomplished by this high-throughput and high-efficiency analysis technique. Previously fed SEM image was taken as a basis which was achieved for a typical color of nanoparticle solution. The error rate during image classification was minimized, by utilizing the categorical cross-entropy loss function. The model accuracy and loss in association with seven epochs gave an accuracy of 91.46%. A comparative analysis of nanoparticle diameter and count in each SEM image of biosynthesized nanoparticles with ImageJ software was quite comparable with an added advantage of using colloidal nanoparticle color image as input.

ACKNOWLEDGMENTS

The authors acknowledge the technical support received from the Department of Information Science and Engineering, NMAM Institute of Technology, Nitte for carrying out this research work. The corresponding author would also like to thank the Department of Biotechnology Engineering, NMAM Institute of Technology, Nitte for providing necessary inputs like colloidal nanoparticle solution photographs and SEM images which helped in carrying out this research project.

REFERENCES

Anas, R., Elhadi, H.A. and Ali, E.S., 2019. Impact of edge detection algorithms in medical image processing. *World Scientific News*, 118, 129–143.

Annadhasan, M., Muthukumarasamyvel, T., Sankar Babu, V.R. and Rajendiran, N., 2014. Green synthesized silver and gold nanoparticles for colorimetric detection of Hg^{2+}, Pb^{2+}, and Mn^{2+} in aqueous medium. *ACS Sustainable Chemistry & Engineering*, 2, 887–896. https://doi.org/10.1021/sc400500z

Ban, Z., Yuan, P., Yu, F., Peng, T., Zhou, Q. and Hu, X., 2020. Machine learning predicts the functional composition of the protein corona and the cellular recognition of nanoparticles. *Proceedings of the National Academy of Sciences of the United States of America*, 117, 10492–10499. https://doi.org/10.1073/pnas.1919755117

Bharodiya, A.K. and Gonsai, A.M., 2019. An improved edge detection algorithm for X-Ray images based on the statistical range. *Heliyon*, 5, e02743. https://doi.org/10.1016/j.heliyon.2019.e02743

Borlido, P., Schmidt, J., Huran, A.W., Tran, F., Marques, M.A. and Botti, S., 2020. Exchange-correlation functionals for band gaps of solids: benchmark, reparametrization and machine learning. *npj Computational Materials*, 6, 1–17. https://doi.org/10.1038/s41524-020-00360–0

Choudhary, M.K., Garg, S., Kaur, A., Kataria, J. and Sharma, S., 2020. Green biomimetic silver nanoparticles as invigorated colorimetric probe for Hg^{2+} ions: A cleaner approach towards recognition of heavy metal ions in aqueous media. *Materials Chemistry and Physics*, 240, 122164. https://doi.org/10.1016/j.matchemphys.2019.122164

Emmert-Streib, F., Yang, Z., Feng, H., Tripathi, S. and Dehmer, M., 2020. An introductory review of deep learning for prediction models with big data. *Frontiers in Artificial Intelligence*, 3, 4. https://doi.org/10.3389/frai.2020.00004

Furxhi, I., Murphy, F., Mullins, M., Arvanitis, A. and Poland, C.A., 2020. Practices and trends of machine learning application in nanotoxicology. *Nanomaterials*, 10, 116. https://doi.org/10.3390/nano10010116

Furxhi, I., Murphy, F., Mullins, M. and Poland, C.A., 2019. Machine learning prediction of nanoparticle in vitro toxicity: A comparative study of classifiers and ensemble-classifiers using the Copeland Index. *Toxicology Letters*, 312, 157–166. https://doi.org/10.1016/j.toxlet.2019.05.016

Kalpana, M., Kishorebabu, G. and Sujatha, K., 2012. Extraction of edge detection using digital image processing techniques. *International Journal of Computational Engineering Research*, 2, 1562–1566.

Khan, S.H. and Pathak, B., 2020. Zinc oxide based photocatalytic degradation of persistent pesticides: A comprehensive review. *Environmental Nanotechnology, Monitoring & Management*, 13, 100290. https://doi.org/10.1016/j.enmm.2020.100290

Lima, A.N., Philot, E.A., Trossini, G.H.G., Scott, L.P.B., Maltarollo, V.G. and Honorio, K.M., 2016. Use of machine learning approaches for novel drug discovery. *Expert Opinion on Drug Discovery*, 11, 225–239. https://doi.org/10.1517/17460441.2016.1146250

Lind, R., 2012. Open source software for image processing and analysis: picture this with ImageJ. In Lee H. Mark F. (eds) *Open source software in life science research* (131–149). Woodhead Publishing. https://doi.org/10.1533/9781908818249.131

Nayak, S., Sajankila, S.P., Goveas, L.C., Rao, V.C., Mutalik, S. and Shreya, B.A., 2020. Two fold increase in synthesis of gold nanoparticles assisted by proteins and phenolic compounds in Pongamia seed cake extract: Response surface methodology approach. *SN Applied Sciences*, 2, 1–12. https://doi.org/10.1007/s42452-020-2348-5

Nayak, S., Sajankila, S.P. and Rao, C.V., 2018a. Green synthesis of gold nanoparticles from banana pith extract and its evaluation of antibacterial activity and catalytic reduction of malachite green dye. *Journal of Microbiology, Biotechnology and Food Sciences*, 7, 641–645. https://doi.org/10.15414/jmbfs.2018.7.6.641–645

Nayak, S., Sajankila, S.P., Rao, C.V., Hegde, A.R. and Mutalik, S., 2019. Biogenic synthesis of silver nanoparticles using Jatropha curcas seed cake extract and characterization: Evaluation of its antibacterial activity. *Energy Sources, Part A: Recovery, Utilization, and Environmental Effects*, 43, 3415–3423. https://doi.org/10.1080/15567036.2019.1632394

Nayak, S., Shet, V.B., Rao, C.V. and Joshi, K., 2018b. Performance evaluation and emission characteristics of a 4 stroke diesel engine using green synthesized silver nanoparticles blended biodiesel. *Materials Today: Proceedings*, 5, 7889–7897. https://doi.org/10.1016/j.matpr.2017.11.470

Okunev, A.G., Mashukov, M.Y., Nartova, A.V. and Matveev, A.V., 2020. Nanoparticle recognition on scanning probe microscopy images using computer vision and deep learning. *Nanomaterials*, 10, 1285. https://doi.org/10.3390/nano10071285

Pandey, A., Shukla, P. and Srivastava, P.K., 2020. Remediation of dyes in water using green synthesized nanoparticles (NPs). *International Journal of Plant and Environment*, 6, 68–84. https://doi.org/10.18811/ijpen.v6i01.08

Panigrahi, S., Nanda, A. and Swarnkar, T., 2018, Deep learning approach for image classification. In *2018 2nd International Conference on Data Science and Business Analytics (ICDSBA)* (511–516). https://doi.org/10.1109/ICDSBA.2018.00101

Rao, A.S., Bhat, S. and Chandavarkar, N.D., 2017. Development of an application for better online shopping using computer vision. In *International Conference on Energy, Communication, Data Analytics and Soft Computing, 2017* (1282–1285). https://doi.org/10.1109/ICECDS.2017.8389649

Rao, A.S., D'Mello, D.A., Anand, R. and Nayak, S., 2021. Clinical significance of measles and its prediction using data mining techniques: A systematic review. In Chiplunkar N., Fukao T. (eds) *Advances in artificial intelligence and data engineering* (737–759). Springer, Singapore

Rao, A.S., Kamath, B.S., Ramya, R., Chowdhury, S., Shreya, A. and Pattan, R.K.K., 2020a. Use of artificial neural network in developing a personality prediction model for career guidance: A boon for career counselors. *International Journal of Control and Automation*, 13, 391–400.

Rao, A.S., Sandhya, S., Anusha, K., Arpitha, C.N. and Meghana, S.N., 2020b. Exploring deep learning techniques for kannada handwritten character recognition: A boon for digitization, *International Journal of Advanced Science and Technology*, 29, 11078–11093.

Sandhya, S., Fernandes, R., Sapna, S. and Rodrigues, A.P., 2021. Segmentation of lip print images using clustering and thresholding techniques. In Chiplunkar N., Fukao T. (eds) *Advances in artificial intelligence and data engineering* (1023–1034). Springer, Singapore. https://doi.org/10.1007/978-981-15-3514-7_76

Santana, R., Zuluaga, R., Gañán, P., Arrasate, S., Onieva, E. and González-Díaz, H., 2020. Predicting coated-nanoparticle drug release systems with perturbation-theory machine learning (PTML) models. *Nanoscale*, 12, 13471–13483. https://doi.org/10.1039/d0nr01849j

Sapna, S. and Renuka, A., 2017. Techniques for segmentation and classification of leukocytes in blood smear images - A Review," In *2017 IEEE International Conference on Computational Intelligence and Computing Research (ICCIC)* (1–5). https://doi.org/10.1109/ICCIC.2017.8524465

Sapna, S. and Renuka, A., 2020. Computer-aided system for Leukocyte nucleus segmentation and Leukocyte classification based on nucleus characteristics. *International Journal of Computers and Applications, 42*, 622–633. https://doi.org/10.1080/1206212X.2020.1726013

Seyedin, S.V., Seyedin, S.H. and Seyedin, A.S., 2019. Designing and programming an efficient software for sizing and counting various particles using image processing technique. *BRAIN. Broad Research in Artificial Intelligence and Neuroscience*, 10(2), 103–118.

Sharma, N., Jain, V. and Mishra, A., 2018. An analysis of convolutional neural networks for image classification. *Procedia Computer Science*, 132, 377–384. https://doi.org/10.1016/j.procs.2018.05.198

Sun, B., Fernandez, M. and Barnard, A.S., 2017. Machine learning for silver nanoparticle electron transfer property prediction. *Journal of Chemical Information and Modeling*, 57, 2413–2423. https://doi.org/10.1021/acs.jcim.7b00272

Wang, H., Koydemir, H.C., Qiu, Y., Bai, B., Zhang, Y., Jin, Y., Tok, S., Yilmaz, E.C., Gumustekin, E., Rivenson, Y. and Ozcan, A., 2020. Early detection and classification of live bacteria using time-lapse coherent imaging and deep learning. *Light: Science & Applications*, 9, 1–17. https://doi.org/10.1038/s41377-020-00358–9

Yao, L., Ou, Z., Luo, B., Xu, C. and Chen, Q., 2020. Machine learning to reveal nanoparticle dynamics from liquid-phase tem videos. *ACS Central Science*, 6, 1421–1430. https://doi.org/10.1021/acscentsci.0c00430

Zieliński, B., Plichta, A., Misztal, K., Spurek, P., Brzychczy-Włoch, M. and Ochońska, D., 2017. Deep learning approach to bacterial colony classification. *PLoS One*, 12, 1–14 https://doi.org/10.1371/journal.pone.0184554

Index

Note: **Bold** page numbers refer to tables and *italic* page numbers refer to figures.